FASTER
THAN LIGHT

THE LIFE CYCLE OF OUR UNIVERSE

BY

C J HARVEY

AuthorHouse™ UK
1663 Liberty Drive
Bloomington, IN 47403 USA
www.authorhouse.co.uk
UK TFN: 0800 0148641 (Toll Free inside the UK)
UK Local: 02036 956322 (+44 20 3695 6322 from outside the UK)

This book is printed on acid-free paper.

ISBN: 978-1-6655-8130-1 (sc)
ISBN: 978-1-6655-8131-8 (e)

Print information available on the last page.

Published by AuthorHouse 11/17/2020

authorHOUSE

CONTENTS

Author's Note 1

Preface 3

Introduction The Early Ethers 9

Chapter 1 The Energy Medium Concept 20

Chapter 2 The Seven Circle Theory 42

Chapter 3 Propagation of EIL Oscillations 70

Chapter 4 Gravitation 143

Chapter 5 Proton Time 189

Chapter 6 The Primary Atom 210

Chapter 7 Forcephoidal Effects 260

Chapter 8 Velocity Effects 337

Chapter 9 Practical Implications 363

Glossary of Terms 389

Appendix Calculation of E-Flow Velocity 397

Epilogue 400

Bibliography 401

About the Author 403

AUTHOR'S NOTE

It is now thirty two years since I completed writing **'Faster Than Light'** as a text in which I presented my Seven Circles Theory of the Life Cycle of our Universe. I had a picture in my head of its reality as each cyclic phase progressed. I was not able to present this in any mathematical format and apart from some revised Michelson-Morley type experiments I had not needed to design any serious practical experiments to prove my concepts. Everything I proposed was simply what we could clearly observe all around us.

I had hopes that as a result of my theoretical presentations a branch of science theory could evolve that was separate from the Big Bang and Steady State philosophies. I sincerely believed that somewhere within all of this correlation of science fact a fundamental truth existed.

Sadly I received no encouragement then from any quarter and so after a period of time my written work was secreted within a computer file and temporarily forgotten. I could not force my conclusions upon an otherwise occupied science community.

Recently I received an enquiry that gave me hope that **'Faster Than Light'** could be resurrected and studied with a renewed degree of interest. Self publication and the internet gave me hope that a reading public could be found.

Science history shows many theories to have been presented and initially accepted but had later been found wanting or replaced by a more enlightening theory. The earth as the centre of the universe with the sun and other planets revolving around it was widely accepted during the second century A.D. Since then science has progressed in leaps and bounds until today we talk of parallel universes, super-strings and wormholes. We have the ten dimension theory of the universe which makes possible the merger between the geometry of Einstein's theory and that of the quantum field theory. Enormously powerful theorems in mathematics now take on physical significance. Physics and mathematics are so intricately interwoven that mathematics leads us in directions we would not normally take if we followed up physical ideas by themselves. Calculus was born from a need by Newton to solve the equations for gravity.

Physics, I believe, is ultimately based on a small set of physical principles. These principles, called first principles, can usually be expressed in plain English without reference to mathematics. From the Copernician theory to Newton's laws of motion, and even Einstein's relativity, the basic physical first principles require just a few sentences that are largely independent of any mathematics. And remarkably, only a handful of first principles are sufficient to summarise most of modern physics. Nevertheless, mathematical equation is still the best way to prove a point.

Cosmologist and mathematician Stephen Hawking has written eloquently about the need to explain to the widest possible audience the physical picture underlying all of physics:

"If we do discover a complete theory, it should in time be understandable by everyone, not just a few scientists. Then we shall all, philosophers, scientist and just ordinary people, be able to take part in the discussion of the question of why it is that we and the universe exist. If we find the answers to that, it would be the ultimate triumph of human reason; for then we would know the mind of God."

My Seven Circle Theory has a correlation with the creation account stated in the first chapter of Genesis in the Bible. The activity that we see in nature all around us was all progressively created in six cyclic periods out of this dark matter or energy medium. Today we live in that sixth cyclic phase. Here all physical matter i.e. protons/atoms are in a progressive decay status which gives us a duration measurement factor called 'Time'. Time only commenced with the start of this sixth phase. The next phase is the seventh cycle phase and total atom decay will have been completed with physical matter non-existent. The seventh cycle phase will be one of total inactivity. The universe will be dormant and devoid of all matter; virtually a period of rest.

Perhaps the 'Ancients' really knew how our universe was created, and that with the passing of time their records seem distorted. The overall seven cycles principle however, has perpetuated.

Today, Big Bang theorists believe that a mysterious fluid existed at 10^{-12} seconds from Big Bang and that in the following moments as expansion occurred and temperatures dropped, a sudden phase change occurred in this mysterious fluid reminiscent of water freezing to ice. Suddenly (they believe) all the familiar particles, protons, electrons, neutrinos, photons, quarks, etc. came into existence!

'**Faster than Light**' is a science theory of invisible Dark Matter and has its relationship with the E in Einstein's formula $E = mC^2$. As such throughout the text I have referred to Dark Matter as that subtle Energy-Medium from which all mass is created. It is a 'Grand Unified Theory' but without rigorous mathematical treatment.

But why have I chosen **Faster Than Light** as a book title?

Light-waves in the text are shown to propagate as a wave front in 'quantum steps', a mechanical type forward stepping action that has three distinct components. An analogy would be that of a child playing on a pogo stick. Each of the springing jumps takes the child forward a set distance. This is then followed by a moment that is stationary but with the spring being compressed. Then along comes the spring release and another jump forward. The overall progress forward is thus a combination of all the above actions. The forward jump is obviously the fastest part and much greater than the overall velocity of progress. Similarly, a component in the light-wave propagation has a corresponding jump or quantum step that far exceeds its average velocity of 186,000 miles per second.

I have attempted to calculate the velocity of this component in the Appendix at the back of the book but here too I have had to make assumptions in key areas.

Birmingham 2020   C J Harvey

PREFACE

The text of this book is an attempt at presenting a Unified Theory to explain the basic structure and Life Cycle of our Universe. What was its origin? When and how was the Atom structure assembled, and what are its basic constituents? What is Time, and does it really vary from place to place? What are Light Waves, and does an Ether-type medium actually exist? What exactly are the Forces that result in gravitation of masses towards one another? And what are the other Forces that cause Electromagnetic effects? How can Electrons behave as both Particles and Waves? Is there a single unified principle that applies to all of Nature, or must we accept different Rules for different phenomena? Questions like these are frequently asked, but so far the answers have remained obscure. Admittedly scientific thought has advanced considerably with the development of new theories, but often this has simply complicated matters even further.

In order to attempt an answer to some of the above questions, we must sift through the 'Pot' of Scientific Facts and correlate the effects of one with those of another. The reader may compare the situation with the problem of fitting together the pieces of a very large Jigsaw Puzzle. Each piece holds a set of clues, and this enables us to fit a few pieces together. As more pieces are assembled a partial picture emerges. Soon we are able to observe a definitive portion of the jigsaw picture and to recognise pictorial features within it. It is through this local recognition of features that we are able to imagine the surroundings and so speculate upon the make-up and overall character of the complete picture. Unfortunately, our Universe Puzzle comprises an infinite number of pieces and therefore we can never hope for a completed picture.

This apparent knitting together of scientific clues has been an activity in the minds of philosophers and scientists over a very long period of time and can be dated back to around 400 BC. However in more recent years a science explosion has resulted in more real clues than ever before, but without getting any nearer to an overall theory. It seems that we must therefore add an element of 'Human Imagination' to project our theories forward into the realms of the unobservable.

In 'Cosmology Today', Sir Herman Bondi wrote that "*the essential thing in science is for the scientist to think up a theory. There is no way of mechanising this process; there is no way of breaking it down into a Science Factory, it always requires human imagination, and indeed in science we pay the highest respect to creativity, to originality,----- we do not honour scientists for being right; it is never given to anybody to be always right. We honour scientists for being original, for being stimulating, for having started a whole line of work*".

The theory presented in this text must be considered on terms similar to an hypothesis that has been built up around the clues of factual science and imaginative logic commencing with the first principles. Every scientific theory must evolve from a central theme, and I consider mine to have its origin in the notions that produced the early ether theories. I have presented this in the Introduction section, where a concise historical account of some of the early ether theorising is given. It would have been quite easy to provide considerably more detail here but since much already exists in other texts, it was considered prudent not to bore the reader with extensive repetition.

In Chapter One the Energy medium concept is presented against a background of mass formation from concentrated forms of an Energy Medium Entity. This is based upon the principle that if Mass is wholly convertible into another identity called Energy, then this must be viewed as a Medium that also has property and form. The essence of this chapter then is to present a theory in which mass formation occurs according to a definitive set of rules or laws, and which must bring new meaning to the relationship represented by $E = MC^2$. So far I have not been able to assign a specific parameter or quantitative value to this energy medium entity or to its varying levels of intensity. However, a scale of conceptual energy medium concentration levels has been presented on a graphical basis that permits direct comparisons to be continuously made with reference to a fixed datum. In order to build the required mental images for visualisation of a particular aspect, I have resorted to comparative examples. Whenever necessary, concepts have been portrayed using graphically illustrated reasoning along with a brief mathematical workout. The basic properties and character of the energy medium have been assumed on the basis of logic, and is the First Principle and Essence of our Theory. It is not possible to prove or to demonstrate the existence of these first principles or essences, but reasoning makes their existence self evident as we progress to other phenomena and find no contradictions in correlated behaviour. The Big Bang and Steady State Theories of the Universe were both discarded as these could not be connected to any such first principle or essence. Instead of the Big Bang I suggest a 'Widespread Crackle'! Big Bang theorists believe that a mysterious fluid existed at 10^{-12} seconds from Big Bang, and that in the following moments as temperatures continued to drop, a sudden phase change occurred reminiscent of water freezing to ice. Suddenly (they believe) all the familiar particles became identifiable. My theory, which I have named as the 'Seven Circle Theory', has as its central theme a comparatively similar phase change from 'energy medium' to 'mass medium', but which occurs at pre-defined levels of energy medium concentrations.

In Chapter Two I have attempted to fit the Energy Medium concept into a predictable set of Rules that could be extended in application. This provided us with The Seven Circle Theory, which represents the behaviour of the energy medium on a theoretical energy medium curve. This Curve is a graphical illustration of the Life-Cycle of our Universe. I have traced each stage of the energy medium curve highlighting its relevance, especially where present-day activity is concerned. It is essential that the reader possess a reasonable understanding of the general principles that make up current physics and cosmology. It would not be possible for me to trace the background to every proposition that I make. Therefore I expect a silent though parallel running between this text and the readers' knowledge of current scientific thought. It is hoped this will enable the reader to anticipate and extend some of the aspects presented.

I believe that an obstacle to the widespread understanding of new theories is usually the fact that they are written for mathematicians, and are therefore loaded with more mathematics than is necessary for obtaining a mental conception of their make-up. Faraday himself had little grasp of technical mathematics, a circumstance which to a lesser man in so mathematical a field, would have proved an insurmountable obstacle. With Faraday though, it was to be an asset, for it forced him to plow a lone furrow and invent a private pictorial system for explaining his experimental results to himself. His 'tubes of force' had an extreme simplicity and peculiarly non-mathematical appearance, and was ridiculed by the professional mathematicians of the time. Yet it was to prove in some ways, superior to their own systems. The two points of view are nicely contrasted in the simple case of a magnet attracting a lump of iron. The mathematicians felt that the essential things here were the magnet, the iron and the number of inches between them. For Faraday on the other hand, the magnet was no ordinary lump of matter but a metal-bellied super-octopus stretching multitudinous invisible tentacles in all directions to the utmost ends of the world.

It was by means of such tentacles, which Faraday called 'magnetic tubes of force', that the magnet was able to pull the iron to itself. The tentacles were the important 'thing' for Faraday; they, and not the incidental bits of metal, were the reality. Many years later Maxwell became deeply interested in Faraday's ideas and translated his seemingly mystical ideas into the more familiar language of mathematics. This in itself was no small task, but when it was accomplished it revealed the idea of Faraday as of the very quintessence of mathematical thought. From this was born an important new physical concept, the 'field', which was later to form the basis of Einstein's general theory of relativity. The electromagnetic field is more or less the refined mathematical form of Faraday's tubes of force.

As such only the barest basics in mathematical formulations has been resorted to as was considered essential in explaining a particular physical relationship. Far more important is the readers' inward eye that mentally sees the flow of Energy medium into and out of the theoretical basic mass particle. That can view the energy medium intensity variations and the quantum step action of light-waves. That can imagine the energy medium to mass, and mass to energy medium inter-conversions that take place so many billions of times per second inside the proton structure. Just as a mechanic is able to mentally see the internal mechanisms of a working machine, so it is hoped will the reader be able to visualise in the minutest detail the past, present and future structural modes of our universe by the end of this text.

The occurrence of all events relating to the seven circle theory are treated on an absolute event basis and not on what an observer is likely to have witnessed. This does not mean that the principles of Relativity have been discarded altogether. On the contrary, the principles annunciated by Einstein in both his special and general theories have been an essential influence in the development of this energy-medium theory. His concepts of Time and Mass variability were accepted by me and resulted in the more defined concept of Proton Time.

Chapter Three explains the mechanism in the propagation phenomenon of the Light-wave. We have shown this to consist of several distinctly separate sequences which then led us to conclude that the overall velocity of wave-front propagation could vary considerably under differing energy medium conditions. From this it became a matter of simple interpretation to explain the other well known light-wave phenomena of reflection and refraction. An explanation of the Red Shift phenomenon is also attempted based upon the above principles, and not solely on the Doppler Effect alone. We consider the Red-Shifting of frequencies to indicate the energy-medium intensity levels that existed so long ago at the location of their origin. At the higher energy medium intensity levels wavefront propagation is also slower. There are reasons to believe that the physical makeup of atoms and molecules of an earlier universe had an energy medium content that was proportionately greater and as indicated in the theoretical lengthening of the wavelengths of the lightwaves of the background radiation that have travelled through the vastness of time and space.

Chapter Four explains the basis for the phenomenon of Gravitation. In the present day phase of the seven circle theory all mass i.e. protons, are considered to be in a continuous state of decay. This means that each mass entity is continually giving up some of its mass as a 'Flow of Energy Medium'. We have shown that this energy medium flow has a concentration or intensity level that obeys the Inverse Square Law. Since the Universe has abundant mass, then so must we assume there are also considerable amounts of these energy medium flows. This then constitutes the make-up of the Universe and is termed an Energy Medium Grid. This Grid is comparable to a 3-Dimension contour map giving high and low energy intensity values as well as gradients. Here we recognise a similarity with Einstein's popular Spatial Grid. We consider that he had a strong notion of the reality of an energy medium grid, but lacked the necessary terminology and theoretical background (of an overall theory) to effectively phrase or connect his ideas. He ultimately presented his concept of gravitation in conventional grammatical terms by indicating that space was curved. Mass objects were then assumed to move "downhill" into spatial "dips" created by this curvature. However, he did not indicate the means whereby masses received impulses towards these "dips" because he considered Gravity as an ever-present and invisible action (somewhat like the early mechanical philosophers). In our theory however, we consider that the impulses already exist in a balanced form inside the structure of the proton. These can become unbalanced if external conditions of energy medium flows and/or intensity levels are not uniform. Protons that find themselves in grid gradient zones experience differential decay conditions on their opposite surfaces, which causes a Resultant impulse in the direction of the higher grid levels. Details of the mechanics of this process is built up from the first principles. Newton's Law of Gravitation is shown as not being based purely upon the notion of attractive impulses, but rather as a resultant of the continuing attractive and repulsive impulses that act both within and upon the proton. These notions are projected forward into the reality of neutron stars and ultimately the proverbial 'black hole'. Black holes have been so named because apparently no light-waves emanate from them. This is not because of the mistaken notion that their gravity is so great that nothing can escape their pull, but rather because the energy medium intensity levels within the 'Event Horizon' of the 'black object' are so high that the mechanics of light-wave propagation fail through loss of their wave-front synchronism. I assume that energy medium intensity levels within the event horizon exceed a critical level and hence conditions of Circle 4 phase would prevail. As such, new matter would be structured and expelled outwards and away from the black object. Black hole objects are considered to be nothing more than proton giants of unimaginable size. The formation of galaxies is shown to be based around such super-massive objects.

In Chapter Five we discuss the concept of Time. Since this is a non-entity function only, we have linked it to the decay process of the proton and so re-defined it as Proton-Time. Time is without property or form, but in our definition we have given it a basic unit of duration. This 'Unit' is the duration of the repetitive cyclic sequence of the most basic mass particle. Once again, variations in the rate of Time are linked to the variations in the cyclic rate of the basic mass particle, which in turn is dependant upon the prevailing grid energy medium intensity levels. Einstein's predictions on Time variation are basically true and we have been able to show exactly how this takes place (from the first principles).

The concept of elapsing event moments or 'Time' has never before been the subject of a serious consideration and it is the intention here to remedy that omission by developing interrelated physical rules that characterise and define its measurement. Incidentally, in our seven circle theory, the 'Time' function is only active below the defined critical energy medium intensity levels and as such becomes effective only in the Circle-6 phase of the universe/realm life cycle. The phrase 'the beginning of Time' has a new meaning in this theory.

Our concept of duration is therefore wholly related to the functional 'beat' or energy medium decay of the protons that make up our material selves and surroundings. Although protons in grouped assembly within an atom nucleus may vary in individual energy medium quantity and thus in 'functional beat' like a pendulum, it is the overall averages within the structural locality which determines for us a pace for Proton-Time. An interesting and very exciting development.

In Chapter Six we develop the Structure of the Atom in a conventional sense. It was considered prudent to review the events that culminated in the formation of the proton and finally the Primary (Hydrogen) Atom. The reader will find our structural combination of protons within the more complex nucleus of considerable new interest. The Atom is not considered to be as depicted by the Bohr Model, in which combinations of Electrons are assumed to be spinning around a central nucleus. We consider the central nucleus (by virtue of its proton decay emissions) to set up a highly repetitive basic mass particle 'Shell' at a defined distance position. The text explains all this in the finest detail possible. The layout of each of these shell sections corresponds to the exact layout of certain protons in the nucleus, which imparts to each atom its unique shape, texture and properties. This also permits the nucleus and the atom to be orientated in preferred directions. We give a new meaning to the term electron, which we have shown to emanate from the nucleus zone of the atom. As such, an understanding of the nucleus configuration was considered essential for the further development of our Theory, and before we could discuss the topic of Electric Phenomena.

Chapter Seven applies the principles of our evolved theory in the explanation of Magnetic and Electric Phenomena. Starting from basic principles we show the constituents of a magnetic and electric field, and how these affect a mass body structure to induce a resultant impulse therein. Also, from basic principles we show the origin and mechanics behind the flow of electric currents in a Conductor Loop. We also explain the difference in the internal action between conductors and non-conductors. It was not possible to offer an explanation of all known phenomena, since this would have meant a considerable extension to our discussions. Instead we have dealt with the main themes only and expect that the principles laid down for these would assist the reader in analysing all the other phenomena.

Chapter Eight defines the term Absolute Velocity, and considers the effects of a high velocity traverse through the energy medium grid by mass bodies such as the proton and Primary Atom. Einstein's predictions of Mass gain and Time variability are proved yet again. Newton's theory of gravitation however, does not hold true for masses that approach or recede from one another at high velocities. We predict these variances and suggest practical experiments in support of our Theory. In this chapter we consider that the Michelson-Morley interferometer experiments produced a wholly negative result because of a totally erroneous basic supposition that the flow of Energy Medium (ether) was horizontal to the surface of the Earth at some stage. We have proposed a series of six new experiments under preset conditions and for which the results are predicted. It may be worth a mention here that the experiments conducted in Japan in the late 1980s on the action of gyroscopes could have benefited from this chapter. The experiments were to compare the weights of a gyroscope at rest and when spinning and which showed a small but measurable decrease of the rotor's weight on some occasions only. Because the exact 'cause of the weight loss' could not be explained the consensus in the scientific community was that there was some flaw in the original experiments.

Chapter Nine indicates the practical uses to which our Theory may be applied, and our expectations from the twenty first century and beyond. We talk about an inexhaustible energy supply, at least until the "end of Time" as defined in our Theory. We discuss the practical possibility of relocating the earth in a more advantageous position (near the fringe of our Galaxy) as an aid to inter-galactic travel. We discuss the spread of mankind across the Universe, and how the problem of communication can be overcome. We conclude on a very futuristic note. Having harnessed the Energy-Medium as a source of near infinite power, man will not only be able to "play with" and "control" the pace of Time, but also travel at phenomenal velocities across the Universe. We have also attempted to establish our current status in the overall Life-Cycle of the Universe, and in so doing brought to light some very interesting theoretical possibilities.

During the write-up of this Theory, which has spanned many years (1980 – 1988), I discovered many new aspects emerging which I had not mentioned earlier. I have managed to induct some of these into later chapters, e.g. the review at the start of Chapter Six. As such the presentations in the early chapters must not be considered on their own, but judged only when the entire text has been read and overall view obtained.

In a highly theoretical presentation of this kind it was difficult to restrict a discussion when the sphere of interacting influence was so extensive. It would have been easy to just continue with the detail of the topic in hand, but this would have been at the expense of the progress of the other topics awaiting discussion. I therefore decided to limit each discussion to a "reasonable" extent, assuming that the reader would extend the relevant examination as a personal project. It would also have been impossible for me to link up the Theory with all known phenomena, as this would again have required volumes of written matter. As such there is plenty of scope for another more specialised write-up at a later date.

Early in the text I considered it appropriate and more relevant to refer the reader to diagram references rather than page numbers. This has been continued throughout the text, and the reader is advised to review the entire section whenever such reference-back is made. Recently, a prominent scientist was heard to remark that our present day physics did not allow sufficient latitude for us to expand beyond the current laws of thermodynamics. That if mankind were to achieve that ultimate dream of colonising the universe, then we would require a 'new physics' to do so. That physics would also need to have its own laws and be in a technical language quite removed from the physics of today. The text above partially meets this criteria, but whether this is the 'new physics' is for you to decide. In a theory of this nature there are bound to be scientific gaps and omissions. It requires the joint effort of the science community for it to evolve fully into an exact and practical science. The reader should consider this effort as a very basic and rough proof of a theory that has yet to undergo considerable refinement. For now though, it leads us in a completely new direction and can be the prompt that starts a whole new line of work.

The theory presented here has required the usage of a completely different terminology from that of current day physics and therefore a 'Glossary of Terms' is included at the end of the text, defining these in some detail.

My thanks go out to my wife Angela and daughters Karen and Nicole, who supported and encouraged me while I was researching and writing all this down. And to Yvonne Smith my mother-in-law for patiently converting my scribbles into a sensible type-script.

Birmingham 1988   C J Harvey.

The Early Ethers

There has never been any reason to doubt that a basic material-building medium has always existed in the minds of science theorists. However, because of the complex character of our world, attempts to explain observable phenomena have led to vague and contradictory formulations of such a 'behind the scenes' medium. Initially the theories were built on a correlated science logic of Plato and Aristotle but which gradually became applicable to the increasing number of unveiled scientific phenomena. As more 'clues' became available, so the structure of this medium altered until eventually today we hold that $E = MC^2$. Yet confusion remains as to the exact interpretation and meaning of this exchange apart from the fact that we believe energy and mass are mutually interchangeable. The exact mechanism of this exchange remains unconnected and it is the object of the following chapters to show how energy relates to mass and to all other phenomena within the infinite cosmos. Although modern theories are happy with their mathematical formulations, they cannot account for the apparently perpetual machine of gravitational force exerted by masses upon each other. Do we look for complicated solutions to the clue of the galactic 'red shifts' when the relation between this and the behaviour of the gravitational constant could be simple and obvious?

At this stage however we cannot make conclusive statements but rather must lay the foundation for the acceptance of a basically simple and all embracing theory. A brief historical account of past ether type subtle medium theorising will indicate the background against which modern theories were constructed and will show an evolutionary trend that has brought us to the present theory. The roots of most existing theories lie in the past and show clearly that man has always theorised upon aspects that were beyond visual observation.

The first documented theory of an ether, as such, was to be made by Aristotle (384 -322 BC) who in essence introduced it as a fifth element to earth, air, fire and water. Aristotle's ether was used to explain effects of the other four and caused itself to link with 'pneuma' which was breath and spirit. I mention this in brief simply to indicate how far back man has thought about an inter-elemental medium in his attempts to explain observed phenomena.

Later, the Stoics integrated ether and pneuma more fully, eventually identifying them explicitly with one another (around 300 BC). Ether was considered to be the embodiment of Nature or God which was everywhere actively mixing with matter, penetrating and shaping it so as to constitute bodies that could be further acted upon.

The neo-platonists (205 BC - 70 AD) avoided any special ether as a fifth element and instead when elaborating accounts of the earth, air, fire and water placed all activation to the World Soul, a theological solution. The thoughts behind these early ethers were simply a series of attempts to explain theories of being and substance and the integration of the heavens and of spirits. Many other vague and analogical chacterisations were made to solve problems that arose as a direct result of explanatory attempts in other areas. This line of thought was used extensively in the medieval centuries whenever doctrines of spirit or of heaven were being given a physical interpretation.

Rene Descartes (1596 - 1650) put forward the hypothesis that matter had been differentiated from the beginning of creation into three elements. He called these simply the first, second and third elements and identified them as fire, air and earth respectively. He insisted that they were not to be equated with what had customarily gone under those names before. He began with the formation of the second element, air, which he considered arose everywhere as minute spheres, formed like grains of sand rubbed around as they rolled about in running water. The other two elements arose as products and residues from this process. The first element, fire, comprised simply the portions rubbed off and the third element, earth, comprised any particles left unrounded because of being too large or packed too firmly in irregular lumps and also formed by conglomeration. The portions of this first matter were, in their very formation, filling the changing and indefinitely small interstitial spaces and so had no fixed shape. In any large swirling vortex the excess first element matter (not needed to fill the interstitial spaces) will have moved towards the centre to form there a liquid of perfect fluidity. The sun it was assumed therefore was formed as a vast turning globe of this first element and as it turned it pressed on and moved the surrounding layers of minuscule second element spheres. Light as a physical action was the transmission of this pressure through the second element, whereas movements of the first element constituted heat i.e. fire without light.

Descartes' subtle matter comprised the second element which as always was permeated by the first. This Cartesian ether was the basis of his science in that all observable changes in the gross physical bodies of the third element were traced to actions upon them by the second and first. The material world according to Descartes was, like space, unlimited in its extent and within it there was no empty space, no vacuum void of bodily substance. And since this substance was everywhere, the heavens and earth were all made of the same continuous material. A body was rarefied not by expanding its own matter, but by distension with invading matter, as cotton wool is swollen by absorbing water. As such Descartes' metaphysical theory of material substance led to a world of bodies moved in swirling vortices of continuous material.

Along with his definitions for substance, extension and motion and the influence of God in all of this, Descartes' ethereal hypotheses were highly speculative and his main influence was in convincing people of the coherence of a mechanical explanation in general. One has only to look at Huygens on gravity, Newton on colours, Leibniz on planetary motion to see that whether later scientists accepted or rejected, modified or replaced Descartes' proposals they often took them very seriously in their own efforts to solve those problems.

Robert Boyle (1627 - 91) in his essay on mechanical Philosophers includes Descartes but denied that extension was the one property essential to matter, distinguishes solid matter from empty space, and admits that much of space is void. However he accepted the Cartesian ether and stated that in principle it was detectable through its mechanical effects although his experiments to examine its motions and sensibility gave only negative results.

During the latter half of the seventeenth century the neo-platonic position was supported more extensively by Henry More, Ralph Cudworth and their Cambridge Associates. Initially an admiring correspondent of Descartes, More subsequently opposed Cartesian physics as being inconsistent with true theology. This was the background of ether theorising during Newton's early years at Cambridge, though Newton's ethers were also fraught with difficulties. He constructed several different, even incompatible theories, yet worked out none of them in any detail.

For Newton (1642 - 1727), ethers were the cause of a wide variety of phenomena, and in almost all cases according to him ether's role was principally as an active agent initiating in all bodies new motions that they would not otherwise have acquired. He was thus led to reject Descartes' identification of extension as the essence of matter and space alike, and hence to reject also the Cartesian material plenum and restriction of action to contact action. Newton followed the mechanical philosophers in supposing ordinary material bodies to be composed of hard impenetrable particles. In conformity with nature be conceived that aeriform fluids, light rays and ethereal fluids were also formed of such particles, the ether particles being the smallest. Newton often claimed that ether was much the same constitution as air, but far rarer, subtler and more strongly elastic He expressed the possibility of an explanation of gravity in terms of the differential densities and sizes of ether particles. Newton apparently envisaged ether acting by a differential density arising from the repulsive forces exerted by the minute particles of ether. He considered that its tendency to expand itself enabled it to press upon ordinary material bodies and so caused planets to approach or to recede. Finally Newton made a profound suggestion that nature may be nothing but ether condensed by a fermental principle. To quote, he supposed that nature; *"may be nothing but various contextures of certaine aethereall spirits or vapours condens'd as it were by precipitation, much after the manner that vapours are condensed into water or exhalations into grosser substances and after condensation wrought into various formes, at first by the immediate hand of the Creator, and ever since by the power of Nature.............. Thus perhaps may all things be originated from aether".*

The Dutch physician and Chemist Hermann Boerhaave (1668 - 1738) considered fire as a subtle imponderable fluid. Like Descartes' first element fire, it was considered a ubiquitous, imponderable, penetrating, and active material. According to him the particles were naturally indivisible, not transmutable into particles of other elements, were solid, hard, smooth and rounded. Fire was supposedly formed as such at the beginning of things and had not been mechanically producible since from other kinds of bodies. His particles of fire had no one direction of motion natural to them. Fire was thus without gravity or levity. Left to itself, a pure sample of elementary fire expanded and dispersed itself in all directions. However, whether this dispersal arose from forces of repulsion acting between the particles or from mere collisions in their jostling motions was left unclear. The rarefaction of the gross bodies it acted upon was to him a universal and reliable sign of fire as an agent. And this rarefaction was only possible because fire was repelled by the corpuscles of the invaded rarefied body. In being reflected as light from solid surfaces, streams of fire particles were not merely bouncing back upon contact, but were, according to Beorhaave, actively repelled by the particles of the surface. Unlike any other material, fire was equally distributed throughout the world, except where concentrated or dispersed. The convergent motions of ordinary bodies that were attracting one another were the main cause of its concentrations, but exactly how this worked was left obscure. All of this had a remarkably pervasive influence upon other ether theories of that time.

The ethers of the mid-eighteenth century were thus highly diverse in their constitutions and operations. A brief glance at three active sites of ether theorising in this period can provide a general indication of that diversity.

1. It was agreed that electrified bodies could, like magnets, sometimes attract and sometimes repel other bodies. Theories to explain these attractions and repulsions abounded. One set proposed a streaming effluvia whose parts acted on one another and on particles of gross bodies solely by contact. The action of an electrified body was propagated by this intermediary, the effluvium, which was itself in motion as a whole. Benjamin Franklin however considered that a stationary medium surrounding the electrified body was the cause of the repulsive or attractive actions. He considered the medium to consist of particles repelling each other with forces acting across the short distances between them.

2. The caloric theory of the three states of matter was a crucial overlap between physics and chemistry. Caloric itself, the matter of heat, was supposed to consist of small particles that once again repelled one another across the distances between them. The caloric introduced into a body was therefore considered to distribute itself so as to surround each of the particles of that body. If few caloric particles were present then the original mutual attraction between the ordinary matter particles would predominate and the body would remain solid. Add more caloric, more heat, and the body would become fluid when there was a net repulsive tendency but one still small enough to be counteracted by the pressure of the atmosphere. Add still more and there would eventually be a net repulsion great enough to overcome that pressure and the fluid would become gaseous. Although one may compare caloric as heat to Boerhaave's fire, we can go no further because each attempted to explain a different facet of physical nature. Lavoiser's chemistry had the distinction of centralising this conception of bodily states to elements, mixtures and compounds and which had no equivalent in earlier chemistry. He stated two ways in which two elements could leave a solid or fluid and enter the surrounding atmosphere. They could leave separately; that is, all particles could each have their own coating of caloric particles, in which case the gas produced was a mixture of the two elements. Or they could leave together. That is, particles of the elements could be associated in clusters with each cluster coated with caloric particles, in which case the gas formed was a compound of the two elements.

3. In physiology and theology the main problems often involved the transmission of action not between bodies but between body and soul. Relationships between God and the physical universe and between mind (or spirit) and matter brought forth the analogy between the infinite divine spirit or Holy Ghost and the Universal ether or elemental Fire. Philosophers made the ethers responsible for the extraordinary phenomena of matter and the Holy Ghost the cause of all spiritual conduct in consonance with the divine law. Both ether and the Holy Ghost were universally present, yet both were considered unequally diffused. Thus while some material objects supposedly contained an excess of fire, some men by their virtuous acts were considered to have become more endowed with divine spirit. The Dublin clergyman Richard Barton in 1750 attempted to show the essential coherence between natural philosophy and divine revelation. He claimed that although our knowledge of ether and its functions was far from complete, science had shown that the whole economy of nature depended upon ether. Likewise, he conceived a parallel dependence of the moral world on the divine spirit. There were however a significant cluster of theological writers who considered God as the sole source of all activity in nature, so that no role was allowed for unobservable subtle media. While it was difficult for them to identify the theological function of an ether, ether theory could however be conceived as a resource that might help solve some recurrent problems in natural theology and the theology of nature.

Although I have sited only three modes of ether theorising it must be abundantly clear to the reader that in order to rationalize certain unknown factors within a problem or observed phenomena, the physicists, chemists and theologians of the day found consolation in the presence of an unquantified, vaguely characterised and often obscure medium upon whose action the particular behaviour was wholly related. Comparisons between the varying mediums could not be performed easily because each applied itself to a different problem. However there was common ground between them all and that lay upon the very character of the speculation that an as yet obscure but rather basic substance only could account for the character of our universe. Although I have so far shown the uses to which ethers were put in general, there was also a body of opinion who rejected ethers from science.

Physiological animists attributed activity to living matter and thus rejected the option of employing subtle fluids as the source of activity. Another was a form of inductivism which opposed the wave theory of light and hence the presence of a vibrating ether. Of great importance however were those opponents of ethers who held reductionist philosophies of nature founded upon the notion of Force, or later Energy. In the eighteenth century many attempts were made to explain all the activity of matter, and often of mind also, in terms of forces rather than the preceding ether theories. Joseph Priestley, Roger Boscovich, John Michell, John Robison, Henry Cavendish and William Herschel provide well known examples of late eighteenth century force theorists who accounted for activity in the physical world by forces but without evoking ethereal fluids. They argued that perfectly adequate and simple explanations were obtained by referring observable motions to the forces that produced them. It must be stated however that they shared no consensus over whether those forces were centred on points or on hard material particles of finite volume or whether the forces alone were sufficient. Just as forces were used to explain all physical activity in this era so a century later Energy came to fulfil a similar role. By then the principle of the conservation of energy had become a pillar of physics and although used by ether theorists as a necessary postulate governing the propagation of light or electrical disturbances in ether there were a number of writers, particularly in Germany, who sought in 'Energetics' a general metaphysical principle. They considered that science should be re-modelled so as to avoid the untenable hypothesis of material atoms and instead the mathematical formalism of thermodynamics should be given a physical interpretation solely in terms of the ontology of energy.

Faraday seems to have favoured a field theory. The field had an existence in space that was independent of its sources. It carried in itself the power to effect action, (that is, quantity of force or energy), and propagated that power in time from point to point. Fields in the nineteenth century were basically of two kinds. Force fields and ether fields, depending on whether force itself was taken to be a power distributed in space (as in modern electromagnetic theory) or whether the power was carried in the state of a medium, ether. Faraday's mature theory of lines of force provided a classic example of a pure force field. He characterised electrical and magnetic phenomena by the geometrical patterns in space of the vectorial forces that would be exerted on electrical or magnetic test bodies. These spatial patterns were graphically delineated by representative 'lines of force' whose directions represented the directions of the forces, while their spacing indicated the magnitude of the forces. The closer the spacing the stronger the force. Faraday managed to work out a rudimentary calculus of these lines of force which enabled him to deal with known electrical and magnetic phenomena effectively, and which proved immensely fruitful in suggesting new experiments that led to the discovery of novel phenomena.

Faraday's discovery of the magnetic action on light in 1845 led him to adopt a more positive attitude towards an ether medium, conceding that magnetic force "*may be a function of the ether*". It was Thompson and Maxwell who identified Faraday's lines of force with mechanical conditions in a material medium thus providing a background for a unified ether theory of electromagnetic and optical phenomena. The trend was towards an ultimate theory which would display the medium ether as a concrete mechanical system. Many investigators considered the existence of a unifying medium throughout space as a necessity for the rational comprehension of science as it then stood in the nineteenth century. So far no one had been able to integrate the various effects of gravity, light, electricity and magnetism, though in Germany 'energy' had become the symbol of unity in nature.

In 1887 Michelson and Morley conducted their well known experiments to determine the earth's relative motion within the ether. Although the results of the experiments were negative with regards to determining the velocity of the ether medium flow past the earth, the remarkable experimental fact emerged that there was never any change detected in the interference pattern of the two light rays at right angles to each other, even though the experiment was repeated hours or months later. The same was true if two such experiments were carried out simultaneously at different stations on the earth, which being in relative movement, could not both be at absolute rest at the same time. The ultimate laws of the Universe must therefore be such as to resolve this paradox, namely that even when two observers are in relative linear motion, each of them shall have no option but to interpret his own observations as showing that he is at rest. In order that theory may be made to conform to the results of the experiment, it was necessary that the length of any object in the direction of motion away from or towards an observer be treated as if it had contracted. In the 1890s Fitzgerald and Lorentz suggested this and computed that moving bodies contract in the direction of their motion through the ether by a factor of $\sqrt{(1 - V^2/C^2)}$, usually called the contraction coefficient. This formula became central to Einstein's Special Theory of Relativity and Einstein readily acknowledged his debt to Lorentz.

The advent of quantum mechanics postulated that the energy of an oscillating system (like a pendulum) is quantified into discrete energy values separated by equal steps. The magnitude of those energy steps is found by multiplying the frequency or number of oscillations per sec with a certain constant of nature $h = 6.63 \times 10^{-34}$ joule-seconds, usually called Planck's constant. It is only in the atomic domain that quantum steps are important. For the overall movement of larger bodies, even if they are microscopic in size, quantum mechanics is quite irrelevant. If you watch a bacterium under a microscope the slightest quiver you can observe amounts to millions of quantum energy steps.

In 1900 Max Planck suggested that light was emitted and absorbed not in a continuous manner but as energy quanta which he related to the frequency f of the light ($e = hf$ joules). This energy quantum hypothesis accounted accurately for the colour of the light emitted by glowing objects. In 1908 Albert Einstein and Peter Debye accurately accounted for the change in specific heats of solids with temperature and in 1913 Niels Bohr was able to propose a model of the atom by quantifying the angular momentum (i.e. product of radius, mass and velocity) of the orbiting electrons. Energy as a calculated quantity was central to an understanding of behaviour in the atomic regions. In his special theory Einstein modified Newton's second law (of 'Force equals mass times acceleration') for bodies under continued acceleration. In relativity physics the above mass term refers to a mass at rest relative to an observer. For continued acceleration in a straight line the rest mass divided by the contraction coefficient $\sqrt{(1 - V^2/C^2)}$ becomes the new mass being accelerated. For bodies approaching the velocity of light (3×10^8 metres per sec), the ratio V/C approaches unity and Einstein's energy equation of the force required for further acceleration approaches infinity. An accelerating body hence takes on mass and as it approaches the velocity of light its mass becomes nearly infinite. Does it really take on extra mass and if so how does this buildup occur? The occurrence of the expression mass times the square of light velocity (MC^2) as the first term of the expression for energy put into a moving body led Einstein to suggest that energy and mass were mutually convertible. The energy contained in or corresponding to a mass of value M amounts to the above expression MC^2. Einstein followed his 'special theory' of 1905 by the 'general theory of relativity' a decade later. Essentially this is a theory of gravitation and Einstein interpreted this remarkable property as permanent, i.e. gravity cannot be destroyed or neutralised. Under earth's gravity if a lift is falling freely, a man inside it will feel weightless. He may not feel it but the force of gravity continues to accelerate him and the lift downwards. So it is with an astronaut in orbit. The attractive force of the earth is accelerating him in the earth's direction but the ground, because of the earth's curvature, keeps falling away: This is in sharp contrast to electric or magnetic fields which can be destroyed in a permanent way. For example the electric field in a region can be made zero by screening with earthed conducting surfaces.

Now Einstein argued that the non-destructibility of gravity implies that it is in a sense permanent and all pervading and he related it to something else that has the same characteristics, namely space and time. The way in which he related gravitation to space and time was through geometry, by arriving at a new synthesis of space, time and matter. The mathematicians of the nineteenth century had arrived at the conclusion that Euclid's geometry need not be the only possible geometry. By altering the basic axioms of Euclid's geometry, new geometries can be constructed which are entirely self-consistent. To give an example take Euclid's parallel postulate. This states that for a given straight line L and a point P outside it, there is a unique straight line through P that is parallel to L. This is an assumption as it cannot be proved on the basis of the rest of Euclid's postulates or axioms. The nineteenth century mathematicians investigated whether something wrong shows up if this postulate is modified by saying either that no line through P can be drawn parallel to L or that more than one such line can be drawn. In either case they discovered no self contradiction and so the subject of non-Euclidian geometry was born. Theorems in non-Euclidian geometry are different from those of Euclid. For example the three angles of a triangle in such a geometry need not add up to 180^0. The triangle formed on the earth's surface by lines joining the north pole to the equator by the Greenwich line and a 90^0 meridian has each of its internal angles equal to 90^0, a sum total of 270^0.

Einstein suggested that the effect of gravitation on the space-time of any given region was to modify its geometry from Euclidian to non-Euclidian. Since gravitational influence is supposed to be exerted by all forms of matter and energy then they form the sources of the non-Euclidian geometry. Euclid's geometry and the special theory of relativity (1905) therefore hold only in an empty Universe. Hence the necessity for a general theory of relativity. Einstein gave a set of equations to describe quantitatively how the non-Euclidian geometry arises through the presence of matter and energy. Light travels in a straight line but is known to bend towards a large mass when passing close by it. Thus the space around that mass is curved although to an observer the light ray seems to travel through it in a straight line. Also, according to Einstein the effect of gravitation is felt not only in space but also in time, which introduced a fourth dimensional variable to the quality of space. Atomic clocks have been observed to run at discrepant rates in different regions once again proving Einstein correct.

About three centuries ago, Isaac Newton discovered the simple, yet profound, law of gravitation. The law states that two objects with masses m_1 and m_2 situated a distance r apart, attract each other with a force F given by

$$F = (G\, m_1\, m_2)/r^2$$ where G is a constant known as the gravitational constant.

Both Newton's and Einstein's theories of gravitation assume the gravitational constant G to be constant in time. However there are some observational and theoretical reasons for believing that this may not be so. Cosmological models in which G changes with time have been constructed by various theorists, for example, by Dirac as long ago as 1937 and later by Jordan, Brans and Dicke, and Hoyle and Narlikar. In all cases the change of G as a fraction of G is of the order of Hubble's constant. Hubble found a simple linear relationship between the red shift Z in light from distant galaxies and their distance D from earth and expressed as:

$$Z = {}^{DH}/_C$$

where H is Hubble's constant and C the velocity of light. Hubble estimated the constant as approximately 1.5×10^{-17} per second but later work revised this to ten times smaller at 1.5×10^{-18} per second. Thus the gravitational constant G will change by only a few parts in 1000 million in a human lifetime. The measurement of the time variation of G can also be carried out by a method of observation of the moon's orbit around the Earth and any changes over the centuries. The variation of G poses many interesting questions regarding the nature of the solar system when G was greater. However of chief importance is the mechanism within the atom which is causing a gradual decay of the gravitational force of attraction. Could it be that the mass of the atom is actually dissipating and is releasing its mass in a form of gravitational radiation or energy emission?

There was a widely held view that a stationary medium in space was incompatible with Einstein's two theories of relativity and that ether theorising would cease after 1905. Such did not prove to be the case, as along with the quantum theory many physicists urged the necessity of some form of ether type theory. Einstein, in an address delivered at the University of Leyden in 1920, stated that,

"More careful reflection teaches us, however, that the special theory of relativity does not compel us to deny ether. We may assume the existence of an ether; only we must give up ascribing a definite state of motion to it, i.e. we must by abstraction take from it the last mechanical characteristic which Lorentz had still left it........There is a weighty argument to be adduced in favour of the ether hypothesis. To deny ether is ultimately to assume that empty space has no physical qualities whatever. The fundamental facts of mechanics do not harmonise with this view........ According to the general theory of relativity space without ether is unthinkable; for in such space there would not only be no propagation of light, but also no possibility of existence for standards of space and time (measuring-rods and clocks), nor therefore any space time intervals in the physical sense".

Astronomy, like other branches of science, has faced difficult problems from time to time, and out of their solutions have emerged important discoveries. The motion of planets posed a problem for several centuries until it was understood in terms of Newton's law of gravitation. Investigations of faint nebulae that proved to be other distant galaxies led to the concept of an expanding universe. Over the last few decades astronomy has presented the theoreticians with phenomena on a vastly grander scale than could ever be achieved in a terrestrial laboratory. Astronomy is initiating the application of physical laws to unknown territories and it is not surprising therefore, if stumbling blocks are being encountered and will continue to be encountered. J.B.S. Haldane remarked that the universe is *"not only queerer than we suppose, but queerer than we can suppose"*. As a result there are now many theories and cosmological models of the universe. The universe has been defined as the largest possible object in existence and its study is not confined to what it looks like 'now', but also includes its properties in the past and in the future.

At this stage I must digress to comment upon scientific thought regarding 'fundamental forces'. In his book 'Superforce' Paul Davies writes that *"in the beginning the universe was a featureless ferment of quantum energy, a state of exceptionally high symmetry........ It was only as the Universe rapidly expanded and cooled that the familiar structures in the world froze out of this primeval furnace. One by one the four fundamental forces separated out from the Superforce. Step by step the particles which go to build all matter in the world acquired their present identities"*. Subsequently, in order to explain how the atom was held together, physicists came up with their forces theories which can be likened to the ancient gods of the Greeks and Romans. Each has its domain of power and its action is supposedly without question.

The *'Strong force'* binds the atom together. The existence of a strong force was considered essential, as something had to hold the protons together against the repulsion caused by their similar electric charge. No trace of this strong force could be discerned outside the confines of the nucleus. No mathematical description seemed completely satisfactory until the quark theory of nuclear matter was proposed in the early 1960s. An inter-quark force had to be present and it was concluded that the strong force must be its residue on the surface of the proton. No attempt was made to explain how the impulse or linear momentum is generated that holds the particles together.

The existence of a *'weak force'* was used to explain the emanation of alpha and beta particles from the nucleus. The story began in 1896 when Henri Becquerel accidentally discovered radioactivity while investigating the mysterious fogging of photographic plate that had been left in a drawer next to some uranium crystals. A systematic study of radioactive emissions was undertaken by Ernest Rutherford. He demonstrated that two distinct types of particles were emanating from the radioactive atoms. The alphas were heavy positively charged particles, which turned out to be fast moving helium nuclei. The betas were shown to be high speed electrons. Mathematical calculations indicated that the law of conservation of energy was being violated as some energy was missing. Wolfgang Pauli rescued the law by suggesting that another intensely penetrating particle was coming out with the electron. Enrico Fermi called it the nutrino, and it was not spotted until the 1950s. Yet a mystery remained. It was known that, left to themselves, neutrons disintegrated after about 15 minutes leaving behind a proton, an electron and a nutrino. One particle disappears and three particles appear. It was considered that no known forces could break up a neutron in this way and so some other force had to be driving beta decay. So arose the need for a 'weak force'. This is given a character very different from the others. It does not exert a push or pull in the engineering sense, except in cosmic events like supernova explosions. Its task is to drive transmutation in the identity of particles, often propelling them to high speeds. Its action is restricted to an extremely limited region of space and is inoperative beyond about 10^{-16} cm from its source. It is therefore confined to individual sub-atomic particles.

The effects of electromagnetism are clearly visible in everyday life. The force between magnetic poles obeys an inverse square law and can be discerned over large distances. The magnetic field of the earth extends far out into space. The sun also has a magnetic field, which permeates the whole of the solar system. There is even a galactic magnetic field. A strong *'electromagnetic force'* has been designated to operate within this field that causes attraction and repulsion effects upon other magnetic and electric particles. Again no explanation is given of how an impulse is generated. Historically, gravity was the first of the forces to be treated scientifically. Although we have always been aware of gravity, the true role of gravity as a force was not recognised until Newton published his theory of gravitation in 1687. An important feature of gravity is its universality. Nothing in the universe escapes its grip. The force of gravity between particles is always attractive. The most surprising thing is its feebleness. The force of gravity between the components of a hydrogen atom is 10^{-39} of the electric force. Thus this is a weak force. Yet it is the dominant force in the universe. Since every particle of matter gravitates, gravity accumulates as more and more matter is aggregated. Newton provided the laws while Einstein attempted to explain the reason behind gravity in his 1915 'general theory of relativity'. According to him bodies are not forced together but simply follow the straightest downhill path through curved space-time. A simplistic notion that once again does not explain how impulse is generated in each atom to produce this directional acceleration. A notion still held by scientists today.

Then along comes the idea of messenger particles. How did the moon manage to reach out across 230,000 miles of space and grasp hold of the ocean to cause tides? The answer was of course that the moon produces a gravitational field that reaches out to the ocean and forces it to move. Yet scientists felt that there ought to be some sort of signal sent by the moon to the ocean to tell the ocean it was there. Also, the signal ought to contain a message telling the water how to move. This is the modern approach to the action of the forces in which virtual messenger particles travel back and forth relaying information. Each force has its own brand of messenger particle; photons for the electromagnetic force, gluons for the strong force, W and Z particles for the weak beta decay force and in the case of gravity a messenger particle called the 'graviton' was invented. Gravitons travelling at the speed of light traverse back and forth between the earth and sun to keep us in correct orbit. A network of gravitons binds you and me firmly to the earth. So now we not only have another Greek god but also messengers to the gods to do their bidding!

Recently a weak fifth force has come under investigation to account for anomalies in the gravitation effect in bodies when they approach each other. This weak fifth force has a negative gravity effect and is only apparent at short distances of a few metres. (This is dealt with in chapter 4 on gravitation).

In 1911 Rutherford proposed that a positive electric charge was concentrated in the minute heavy nucleus of the atom and that electrons were orbiting around it at relatively enormous distances carrying negative electric charges. The combined electric charges just balancing the positive charge of the nucleus, so that the whole structure exhibited a non-charged neutral condition. Rutherford did not propose his model of the atom till he had mathematically proved the experimental evidence so compelling that acceptance of his conclusion was without question. The unasked question however is, what is a positive or negative charge? And how do all these gods 'work' to produce the impulse actions that we constantly observe all around us?

In our Seven Circle Theory there is no need to ascribe any activity to unknown factors. This theory is based upon a single principle of nature and all activity follows on from it. The intention is not to criticize conventional physics but rather to present new ideas to be considered in parallel with what is known today with the hope that future research is also directed in a new direction.

In 1928, Edwin Hubble observed a remarkable property of light from distant galaxies. This property known as the red shift represented a systematic increase in the wave length of light from its source. Since the lines in the spectrum correspond to absorption processes taking place in the atoms and molecules of those galaxies and there is no reason to suppose their physical make up as differing from those in our own galaxy, then (after making the necessary corrections for red shifts due to gravitation and spin of our own galaxy) we can assume that the observable red shift of the spectral lines results from the Doppler effect caused by a receding galaxy. So far we have observed every other galaxy as receding from us. The further away the galaxy the greater the recession velocity as assumed by the red shift. This led to the concept of an expanding universe as exemplified by the case of a simple balloon getting bigger and bigger. Every point on the balloon skin moves away from its neighbour as the balloon enlarges. Hubble's constant is the ratio of the apparent velocity of the recession of a distant galaxy to its distance from earth and at present is estimated to be about 53 kilometres per second for every mega parsec distance (a parsec being the distance light travels in 3.26 years). Two leading theories are the Big Bang theory and the Steady State theory. The former suggests that about 15 to 20 Billion years ago all the matter in our known universe was located at a single point and consequently underwent a massive explosion. We are assumed to still be on the outward bound stage of that explosion and have evolved into the currently observed galactic systems. The mechanics of such a theory suggests that an eventual slowing down of the outward motion will occur and we may then stop and retrace a path back to the central point of explosive origin (which future generations may confirm if a blue shift in the spectrum lines is observed). We can imagine ourselves in a sort of pulsating universe with a "Big Bang" occurring with regular frequency.

The Steady State theory was put forward as an alternative to the Big Bang model of the universe and also accepts that the universe is expanding. However it was also stated by its authors that in addition to the cosmological principle that the universe be homogeneous and isotropic at a given cosmic time it must also look the same at all times. As the universe expands, its existing galaxies must move apart from one another. This would tend towards a thinning out of the material in the universe. In order for the overall material density of the universe to remain the same it requires the steady or continuous creation of matter to make up for the depletion produced by the expansion. Dependant upon the estimated mean density of matter in the universe and the rate of expansion of the universe, Bondi and Gold estimated the required rate of creation of new matter to be 4.5×10^{-45} kilograms per cubic metre every second. That is, for one kilogram of matter to condense inside a cubical box of one metre side it would take 7 million, million, million, million, million, million (6 times) years. Although this rate of creation is extremely small its explanation in the framework of physics was not easy. Since energy and matter are interchangeable the new matter must be created out of some energy reservoir. As we create matter, the reservoir will be steadily depleted and eventually nothing will be left. To get around this difficulty Hoyle imagined a reservoir of negative energy which he called the C-field. To explain; suppose the C-field has -10 units of energy per unit volume. If we create a unit of matter in that volume we are left with -11 units of energy per unit volume. The expansion of the universe then reduces the magnitude of the C-field to -10 units per unit volume again. Technically, the C-field has no mass, no charge and no spin. It comes into effect only at the time when particles are created. Whenever a particle with a certain energy is created, a C-field of equal but negative energy is also radiated and the overall energy is thus conserved. Hoyle's formulation provided a theoretical description of matter creation but went back to ether type theorising where a particular problem initiated a particular theory. However it is an important step in the field concept in physics for we have not yet formulated a pure field upon which to base all our known laws.

Although the big Bang and the steady-state models have occupied the main cosmological stage there are many other models that have been put forward and can be found in most texts on theories of the universe.

A few decades ago cosmology was a subject with too many theories and too few observations. This is no longer the case as the many observational checks have shown that cosmology is now fulfilling the main requirement of a scientific discipline, that it should make testable predictions. In the past man has attempted explanations of terrestrial behaviour through the action of a subtle medium. Surely then the origins and action of that medium is bound to be found not in the laboratory but in the far reaches of the cosmic 'smelter's pot'. Atoms and molecules are known to have had their basic beginning in the single proton hydrogen nucleus. Yet at the present time we have no theoretical origin for that proton, a basic building block of the universe. Then again we believe that mass is condensed energy; and so we must develop a concept for that medium called energy and the manner in which it is physically interchangeable with the mass of the proton. Since one is a tangible substance then the other must also be quantifiable as having property and form. Could it be possible that in some far distant past the conditions in the cosmos were such that only the energy medium existed in an expansively deflated and dormant state separated from other similar energy medium 'clouds' by the infinite blankness of the cosmic void? The answers are more likely to be found in outer space through the application of Einstein's relativity theories than in the atomic world ruled by the quantum theory of matter and radiation. The two dominions are enormously far apart and attempts at unifying them have so far not achieved much success. It is like having one law for the rich and another law for the poor. Perhaps both theories, although each is highly competent in its own field, are merely approximations of an all embracing theory (yet to come).

Of one thing I am sure physics is not going to become easier. It has never done so in the past. Although theories have made our understanding of nature much easier the details have become far more intricate and have created more unanswered questions. Perhaps in the future the notion of the dual nature of matter as particles and waves will be taught in primary schools as will concepts of space-time curves and naked singularities. In 1930 there was widespread belief that physics was nearly complete. Can we ever assume that the nature of matter is well understood? Will the recent torrent of new knowledge, new concepts and new vistas continue forever? I have no doubt that it will and that in a few hundred years from now men will look back and wonder at some of our outlandish theories. For now, however, let us turn a page and step into a theory of cosmological behaviour that extends beyond time and gravity (and yet is only basic to the construction of a material universe).

Conclusions

Aristotle's fifth element ether, Descarte's subtle matter and vortices, Newton's aeriform fluids and cyclical precipitation thereof, Caloric as a medium and Boerhaave's fire, all led towards metaphysical theories of force and energy to account for activity in the physical world.

Faraday's field theories, Max Planck's quantum energy steps and Albert Einstein's energy equations finally brought us to realise that the ether substance earlier theorised was nothing more than a virtually unseen medium known as energy, whose concentrations into mass obeyed the law $E = MC^2$.

So far there has been very little research directed at this invisible entity but we must nevertheless conclude that it is there, and that it has both property and form. The determination of that 'energy-medium' status is the subject of the following chapters, and it will be shown that there are laws that this medium and its concentrations must obey. And that the entire universe so constructed follows a predictable life cycle of mass reverting back to the energy medium status along a defined mode of behaviour.

CHAPTER 1.

The Energy Medium Concept

The Energy medium.

In the previous chapter an account was given of the theoretical progress made over the years towards a subtle medium that would explain the activity in nature. It was concluded by the author that the search for such an all-pervading ethereal medium should have ended when the law of interchangeability between mass and energy was laid down. Here we had the basic constituent of mass upon which Newton and others had theorised for so long. Mass is made up from a concentration of energy medium and hence the energy medium must be looked upon as a medium having property and form.

The energy medium is to be considered as a perfectly homogeneous entity that has no structure or individual particles to make up the whole. It is indivisible in its volume and yet an infinitesimal part of it can be absorbed into a mass particle. The medium itself has no inertia and is unaffected by any attractive or repulsive forces. It has no constituent other than itself but does have varying energy levels throughout the whole. The energy medium in motion will travel in a perfectly straight line until part of it gets absorbed into a particle of mass. The energy medium is thus comparable to a cloud of the medium floating through a clear blue sky until by some means mass particles condense out of it. A further example to assist in the visualisation of this medium would be to conduct a thought experiment far out in the cosmic void. A cosmic void may be defined as a region in space totally devoid of all matter or medium of any sort. In this void there is no propagation of light or effects of gravity. There is a total absence of magnetic influence and of time itself. In our thought experiment let us by some means cause a particle of matter to become present within this void region. Nothing else, just one particle such as a single proton. Now let this proton be totally converted into its energy medium counterpart as given by the formula $E = MC^2$, where E is the quantity of the energy medium in the proton mass and C the theoretical velocity of light. We now have a situation in the void region where the only existence is of a medium of energy that has been dissipated out from a single proton. A single proton occupies a space as small as a sphere of 10^{-20} metres in diameter. Its energy medium counterpart must occupy a volume very much larger. Let us continue with our thought experiment and assume that the dissipation of energy medium from the former proton is held inside a spherical area of diameter 1 metre.

We now have an energy medium quantity that is still $E = MC^2$ but occupies a much larger space. If the energy medium level is uniform throughout the 1 metre sphere then we can assume a certain level of energy medium intensity inside that sphere. Let us now consider allowing the energy medium from the proton to spread further into a spherical volume that is 2 metres in diameter. The total energy medium is still the same but it is obvious that the Energy-medium Intensity Level (abbreviated EIL) inside the 2 metre sphere is much less than that in a 1 metre sphere. We now see that the energy medium from the mass particle possesses a property of varying EIL which may be compared to but not confused with the term density as we know it.

Now that we have developed a rough concept of energy as a medium occupying space and also possessing a variable level of intensity, let us proceed towards a mathematical relationship between these factors. In our thought experiment the proton is assumed the optimum concentration of energy medium. Assume that this proton suddenly changes from the mass to the energy medium state and that this energy medium expands spherically outwards at a set velocity yet maintaining a uniform EIL throughout. It is obvious that as the spherical volume increases, the uniform EIL will get less and less. If we take the energy medium in the proton as MC^2 and divide it by its spherical volume at any instant, then we shall obtain the intensity level of that energy medium at any point inside the expanding sphere.

$$EIL = \frac{\text{Total Energy}}{\text{Spherical Volume}}$$

$$EIL = MC^2 / \left(\frac{4}{3}\right) \partial\, r^3 \text{ where r is the spherical radius at a set moment.}$$

Therefore $\qquad EIL = \text{constant} \times \frac{1}{r^3}$

Thus the EIL is inversely proportional to the cube of the radius of uniform spherical expansion of a set quantity of energy medium.

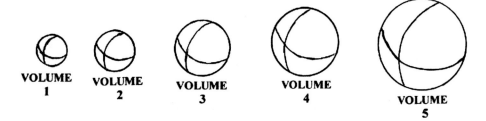

VOLUME 1   VOLUME 2   VOLUME 3   VOLUME 4   VOLUME 5

Fig 1.1

In Fig 1.1 the energy medium ball is shown to expand in successive stages. However in reality, energy medium that is expelled from a central point over a period of time is not uniformly distributed over the expanding volume. In Fig 1.2 the energy medium emitted from the centre Proton P spreads outwards at a progressively diminishing EIL.

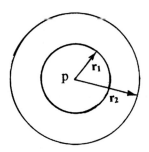

Fig 1.2

Let us assume as part of our thought experiment that the energy medium flowing out from the proton diameter does so at a uniform rate and at a constant velocity. The EIL at any distance from the proton centre will depend upon the spherical surface area that a defined quantity of energy medium must cover. At the radial distance r_1,

$$EIL_1 = {}^{\text{Energy quantity}}/4\,\partial\,r_1^2$$

or $EIL_1 = $ Constant x $^1/r_1^2$

at a radial distance r_2 this will become,

$EIL_2 = $ Constant x $^1/r_2^2$

Thus the Energy Intensity level (EIL) for energy medium expanding radially outwards from a central point, obeys the inverse square law just as it does for light intensity and gravitational forces. This is represented graphically in Fig 1.3.

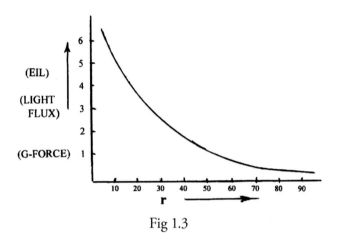

Fig 1.3

The Cosmic Void.

The cosmos may be defined as that infinite volume of space which has no beginning and no end. It extends beyond comprehension and is even larger than imagination allows. If a particle were to speed across the cosmos in a straight line at near infinite velocity, then at the end of an infinite period of time its position may correctly be considered to be still at the approximate cosmic centre. By definition, a void is an area of space that contains 'nothing'. It is devoid of all property or medium. Light cannot propagate through a void, and space and time have no meaning in such an area. Gravitational attraction would be non-existent due to the absence of matter and it is estimated that approximately 99.9% of the cosmos is void.

In contrast a universe is an entity within the cosmos that possesses mass and property. A universe like our own is very much larger than observation allows and is probably speeding through the cosmic void at some immeasurable velocity. For all we know, there may be other universes separated from us by the infinite distances of the cosmic void and tracing paths that are towards or away from us and at or near infinite velocities. The demarcating boundary of a universe would be where its energy medium ends and the cosmic void begins. However it is not that simple as will be shown later on but for now the concept will suffice.

The Energy Medium Realm.

A universe has been described above as consisting of mass and having physical property. In order to establish an origin one must assume that at some stage there was no mass in our universe. However one must also assume that during that stage the present universe was a vast expanse of the basic mass building medium called energy. Let us now omit the word 'universe' from future terminology in this text as it could lead to ambiguity of meaning in later descriptions.

Consider expansive volumes of the energy medium as separate entities moving at set velocities through the cosmic void. Each energy medium entity travels in an absolute straight line and is separated from other such entities by the vast distances of the void. Each has a defined volume, defined relative velocity and has defined levels of energy medium intensity within its volumetric boundary. Each energy medium expanse is termed an 'Energy medium Realm' or simply a 'Realm'. There are no forces acting across the void between them and so each Realm remains an independent entity oblivious of the existence of any other. Let us assume that the EILs inside a realm are in a static mode, i.e. the energy medium is not expanding, at least not appreciably. One may ask where these realms originate from and as to their relative volumes. We shall discuss the former in a later section on the life cycle of the realm. As to their size one may only speculate regarding our own realm. At the present time our observational limit is considered about 15 billion light years, a limit caused by the sky brightness, diffraction of light, and absorption of several essential wavelengths by the earth's atmosphere. It is considered that all this could be compared to less than a pin prick in something the size of our current observational limit. The object here is not an attempt to be accurate but rather to convey to the reader a conceptual image of the possible volume of a realm. Obviously some will be much larger than others but further speculation at this point is futile.

Now consider this wholly energy medium Realm to be moving along a defined path in the cosmic void as visualised in Fig 1.4. Each Realm is like a parcel of the energy-medium entirely on its own inside the vastness of the cosmic void. Max Planck was the first to consider energy to be confined to discrete packets of energy when he initiated quantum mechanics in 1900 for atomic behaviour. We are now dealing with something infinitely larger but there is no reason why the conceptual principles should alter.

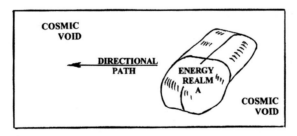

Fig 1.4.

Let us now look at a much larger area of the cosmic void in order to visualise the relativity of Energy medium Realm A with other neighbouring realms.

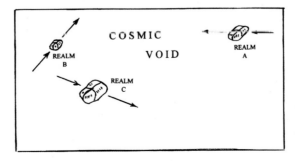

Fig 1.5

In Fig 1.5 there are three realms represented visually by the parcels A, B and C. Each is moving along a defined path in the cosmic void and at a particular instant are shown to be in the relative positions indicated on the diagram. Realm A is moving towards the path of Realm B but is unlikely to intercept B. However, Realm B and C can be seen to be moving away from each other and it appears that their paths may have intersected at a previous point. This does not necessarily mean that the two realms B & C were at that intersecting point at the same instant, but it does raise an interesting aspect to consider had this been so.

Let us go back to our earlier thought experiment in which we had placed a single proton in the cosmic void and allowed it to revert to pure energy medium in a set duration. We saw that the EIL at a particular distance from the original proton centre obeyed the inverse square law. Now into this experiment position a second proton two metres distant from the first and then re-start the thought experiment. Each proton commences to revert to total energy medium at the same instant. When the energy medium has spread to a sphere of one metre radius for each proton then there will be a point between them where the two energy medium spheres meet. This is shown in Fig 1.6(a) and it is clear that at this meeting point O the EIL will be due to that from proton P_1 and also from proton P_2.

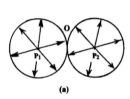 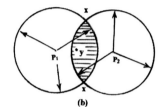

Fig 1.6

$$\Sigma EIL_0 = EIL_{P1} + EIL_{P2}$$

Now energy medium is a perfectly homogeneous medium and we ascribe to it the property that it is able to share the same space with another energy medium without affecting or being affected by the other. Also, if each is in relative motion then they will offer no resistance to the other even though each occupies the same space at the same instant. This characteristic is essential to our theory.

In Fig 1.6(b) of our thought experiment the energy medium from protons P_1 and P_2 have spread to spheres of about 1.5 metres radius and intersect each other at a circle of diameter 'xx'. The energy medium from P_1 continues unaffected into the area of energy medium from P_2 and vice versa. The quantity of the energy medium at point 'y' inside the shaded intersecting area is represented by the EIL at that point. The quantity of energy medium available at point 'y' is therefore the total energy medium that is reaching that point from both energy medium sources P_1 and P_2. Once again the EIL at point 'y' will be;

$$EILy = constant \cdot ((1/(p_1y)^2 + 1/(p_2y)^2)$$

or $\quad EILy = EILy(p_1) + EILy(p_2)$

The phenomenon in which two separate energy medium entities pass through the same space at the same instant is defined as 'energy medium phasing'. Each entity continues in its EIL phase without reaction to the other and when each leaves the shared space it simply continues as it did prior to the phasing. Logic dictates the assumption of this property and this is the first principle of our theory. When referring to very large sections of energy-medium such as a 'realm' then the above phenomena is termed as Realm Phasing.

In Fig 1.5 we mentioned a possible phasing of realms B and C at an instant previous to that shown on the diagram. We assumed earlier that the EILs within a realm were in a static mode but varied in intensity across its volume. Now consider what happens when the boundaries of the two realms B and C intersect. From our thought experiment we know that the two realms will phase through each other and that each will continue along its original path as though the other did not exist. This is exactly what happens except that when the realm phasing is total i.e. the approximate realm centres also intersect, the sum of the EILs of both may attain a rather high level. The EIL that is normal to a particular section of the realm is termed as the Realm EIL Datum and later we will show that such a datum exists within our own structured realm and that there is a gradual and continuous lowering of its level of intensity with elapsing time. A visual conception of the phasing phenomena can be obtained from the analogy of a number of search light beams arcing across the night sky. Every now and again two or more beams cut across each other but the emerging beams remain unaltered. The search light beams appear to pass (phase) through each other without resistance or interference of any kind at the other end. It is noticed that the light intensity at the point of intersection is the sum of the individual light beam intensities. Although we know the mechanics and construction of a beam of light we are not all that far away from a pure energy medium condition. It is essential that the reader should at this stage have a visual conception of energy as a medium that constitutes the vastness of a realm. That energy medium has the property to phase with other energy medium and establish a higher EIL temporarily and that the concentration of energy medium is reflected in its intensity levels and also that the optimum concentration is represented by mass.

Critical Energy Medium Levels.

Realm velocities undergo no restrictions whatsoever in their path through the cosmic void. However it must be considered that even a realm has a cycle of origin and therefore must have been imparted an initial velocity. Since energy medium can only have resulted from a former mass then the energy medium realm as a consequence is similarly limited to a finite velocity. Although Einstein indicated that the velocity of light was the limiting velocity for mass bodies in space this did not apply to the unknown energy medium. We have stated that the energy medium is neither accelerated nor decelerated even when phasing with another energy medium or realm. However a change may be brought about in its velocity by converting the energy medium into mass, accelerating the mass and then reverting the mass back to energy medium. An interesting thought experiment for the moment only but we shall see later how this occurs in reality. Let us assume for now that a mass body cannot travel through an energy medium faster than light can propagate through that same medium. Let us also for the moment assume that the energy medium dissipates from the mass body when converting to energy medium at light propagation velocity (in all directions). The energy medium dissipating in the direction of the original velocity will therefore possess a velocity in excess of C. Although this is not the final relative velocity it will suffice for the current conception of the probable velocity of a realm in the cosmic void. The intention here is to guide the reader towards an idea of the relative duration of a realm phasing sequence. A possible volume of our realm was indicated earlier and it would clearly require countless billions of our years to phase through another such realm.

The possibility now arises of a multiple phasing between three or more energy medium realms. Due to the infinite vastness of the cosmic void this would be a rare and unlikely event. Certain conditions during the early part of the energy medium cycle of the realm however do present a more favourable opportunity for such an occurrence as will be explained in a much later section Fig 2.3. For the moment the reader may assume the probability of multiple realm phasing as being low but nevertheless possible as an event.

Consider the EILs that develop when there is multiple phasing of several energy medium realms at the same event moment. The duration of such a phasing event will be extremely lengthy although 'Time' will not be the same as we know it today. Phasing events are therefore to be measured by the sequence at which events occur and not by non-existent parameters of Time in a faceless energy medium realm. In chapter 5 we shall deal more fully with the Time parameter, but until mass is created and a uniform Time factor is instituted we must relate events by the sequence of their occurrence. Let us for the moment ignore the realm phasing duration and concentrate upon the EIL build up. The EIL at any given intersection point will be the sum of all the individual phasing realm EILs.

$$\Sigma EIL = EIL_1 + EIL_2 + EIL_3 + \ldots\ldots + EIL_N$$

From this we may theorise that the total EIL will reach a very high intensity for some areas. From the equation of mass and energy, $E = MC^2$, we know that an optimum EIL is the mass M and that the energy medium state will prevail for EILs below this. As such we can conclude that the sum of the phasing EILs will ultimately approach a critical intensity that balances the equation $E = MC^2$ on a very fine edge. This critical energy medium intensity is the concentration level at which the energy medium will begin to transform into a mass medium and is denoted by C_1 for Critical Energy medium level one. At or above C_1 EIL the energy medium concentration is sufficient for the phasing medium to undergo a transformation resulting in a fine distribution of unit mass entities. Detail of the exchange from energy medium into mass medium is covered later in the text. The unit mass entities are termed as 'Mass-Points' and may be defined as the primary construction of mass from C_1^+ EILs. A mass-point, denoted Mp, by virtue of its formation into mass is no more an integral part of the realm energy medium. A proportional quantity of energy medium is therefore lost from each of the phasing realms but accounted for in the energy medium structure of the Mp. The law of conservation of energy therefore still holds good. The sequence of events in which a Mp is formed from realm energy medium must be carefully looked at in order that the reader may visualise this phenomenon and be able to relate to it quickly and easily in later descriptions. There are a series of events that occur when a Mass Point is formed. Firstly the EIL in an area must attain a certain intensity or concentration level in order for the energy-medium to convert or sublimate into a mass-medium. This mass-medium then implodes down to a myriad of mass-points (See Fig 4.2 and Fig 8.6). This proposal is not original since physicists have long struggled to come to terms with the phenomenon of the wave-particle duality in the nature of light. In their research into the nature of the photon (conventional usage), scientists at the Desy laboratory just outside Hamburg carried out high energy experiments on the photon in the 1990s. The experiments showed that the photons did indeed behave as if they were complex and made up of an interior jumble of short-lived particles. The essential feature of the discovery is the idea that light or wave energy itself can change into matter and back into light again in a time frame measured in terms of the Planck time of 10^{-43} of a second. In our theory we would stipulate that the life of the Mp be reckoned on a similar time frame. There is much to be explained with regard to the whole process of mass and energy medium inter-changes and this will be dealt with as we progress through the text. To continue, the Mp was constructed out of the combined energy medium intensities of several phasing energy medium realms and as a consequence acquires a linear momentum which is the resultant of all the energy medium velocities that it has absorbed. In Fig 1.7 the phasing of realms A, B, C and D is represented by the corresponding vectors. Let us assume that the realms each have velocities V_A, V_B, V_C and V_D and that these do not necessarily act in the same plane. Also that the respective EIL of each realm approaching the central point M is in the proportion indicated as a percentage of C_1. When a Mp forms at M we can determine its resultant linear momentum by assuming the unit mass to contain four separate components. Let us assume the component due to Realm A has 20% of the unit mass with a velocity V_A as indicated. The component due to realm B has 15% of the unit mass with a velocity V_B. The component due to realm C has 55% of the unit mass and the velocity vector V_C. Finally the component due to realm D has the remaining 10% of the unit mass and the velocity vector V_D. Each component has a linear momentum which is the product of its mass and velocity. Since each is also acting at the same point then a resultant linear momentum many be calculated for the unit mass or mass-point. Linear momentum being a vector quantity it would be a simple matter to draw a vector diagram to scale in the xyz coordinates of Fig 1.7 to obtain a resultant vector.

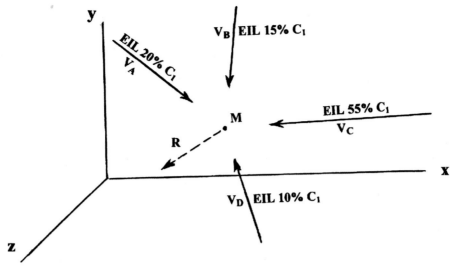

Fig 1.7

Now linear impulse is the change in linear momentum with time and thus the components of the mass-point will achieve the resultant linear momentum of the whole in an event sequence. We now have a mass-point which is reverting to its energy medium state and which has acquired its own linear momentum.

Let us now visualise what happens to the dip or hole in the combined phasing EILs at the location where the Mp was formed. Fig 1.8 assumes the realms A,B,C and D to be represented by plain sheets of 2-dimensional paper of varying thickness and in sliding contact with one another. They have all been placed on parallel planes to simplify the example for easier conception. Let sheet C be sliding upon a flat table-like surface which is not shown on the diagram and therefore not to be considered further. Sheet A slides upon sheet C in the direction shown. Sheet D slides on sheet A and finally sheet B is atop sheet D. For the purpose of this analogy let the sliding velocities be presented in slow motion. At a selected instant imagine a laser beam to burn a hole of diameter d at the spot M. Assume the power of the laser to be strong enough to burn through all four sheets A, B, C and D. The laser acts for an instant only and its action is to create a hole instantly through all the

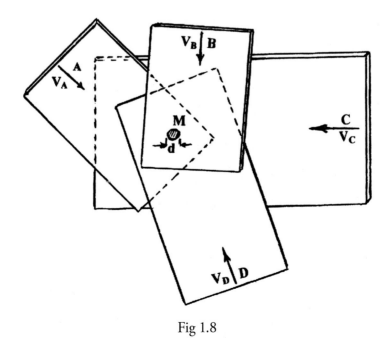

Fig 1.8

27

moving sheets of paper at the overlapping spot M and without causing any restraint to the sliding action. For an instant one is able to see straight through all four sheets at the hole M. However at the next instant with each sheet continuing at its original velocity, the hole is overlapped by a solid part of another sheet.

In the next diagram Fig 1.9 the same sheets of paper are shown at a later sequence and the holes in each are represented at M_A, M_B, M_C, and M_D. The initial spot where the laser burned through is marked at M. The hole in sheet B is visible at M_B, but underlying this are the combined thicknesses of sheets D, A and C. The hole in sheet D is indicated underneath sheet B at the position M_D.

Similarly the hole in Sheet A is indicated at M_A and that in sheet C at M_C. Each of the holes has thus continued in the direction of motion of the sheet of paper it is in and thus gets covered by the paper thickness of the other sheets.

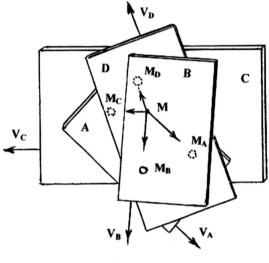

Fig 1.9

Similarly in the phasing of several realms as depicted by Fig 1.7 we may theoretically observe that when a Mp forms, the EIL drop occurring at that point is again quickly raised through the continuing realm phasing, and that the EIL dips in each realm will not coincide again. Therefore when a Mp forms and causes the phasing EIL to dip, it is bound to find itself very quickly in a relatively high realm EIL even if it had not acquired a linear momentum of its own. However in this example it can be clearly seen that the Mp acquires a linear momentum of its own and follows a path that is different in direction to that taken by any of the EIL dips that resulted from its instantaneous formation. This Mp is unable to maintain itself as a mass entity having a life down at the level where a split second is measured in terms of the Planck time, 10^{-43} of a second, and therefore rapidly commences a return to the energy medium state (Fig 4.2). This reversion from mass to energy medium occurs as a continuous flow of the energy medium in an outward direction from the Mp to the surrounding realm. The flow of energy medium ceases when the Mp has totally reverted to the energy medium status. This energy medium flow has an EIL that obeys the inverse square law with distance. As such we can now consider this as an additional energy medium quantity that phases with the realms A, B, C and D in Fig 1.7. Thus an additional EIL will be added to the existing sum total of the phasing realm EILs.

Now if the phasing realms of Fig 1.8 initially resulted in a C_1 EIL at M then the dissipation of an energy medium flow from a Mp at M (an instant later) will result in an EIL that must exceed the C_1 level. As more and more Mps form and revert to the energy medium so will the number of phasing energy medium flows increase. This means that the effective phasing EIL will rise even further. At this stage we consider that Mps have an energy medium quantity that is directly proportional to the total EIL of the phasing energy medium from which they formed. We also consider as an added assumption that Mps with greater energy medium quantities have a greater cyclic duration than those with lesser energy medium. In a zone of C_1^+ EIL we must consider that this creates a situation for a countless number of Mps to form and reform with extreme rapidity. As the realm phasing presents a zone such as at M in Fig 1.8 with a continuous arrival of energy medium, then the energy medium flows from the profusion of Mps there will continuously add small amounts of energy medium to that zone with a resultant continuously increasing EIL. Thus we note that the EIL of the zone in question will rise higher and higher above C_1. At the greater C_1^+ EILs we consider that the profusion of recurring Mps is such that at any instant there is a very large quantity of energy medium locked in Mps. If we may consider the energy medium and Mps as wholly different entities then we may state that as a result of the existence of a large population of Mps within the realm, that its energy medium quantity has subsequently diminished. This aspect brings us to define a second EIL above which the energy medium cannot exceed. We refer to this EIL as C_2 and consider that phasing realms can never exceed this level simply because of the profusion of Mps (and Mp-congregations) that result there on a continuous energy medium replacement basis. Mps that result at EILs approaching C_2 are considered to contain an optimum quantity of energy medium which can never be exceeded. Although one may argue as to the reason for an upper limit to the possible concentrations of the energy medium, we find it logical that the very character of the energy medium (to form Mps) makes this compulsory. To summarise, Mps that form from EILs near C_1 contain a proportionately lesser quantity of energy medium than Mps that form at near C_2 EILs. Also an Mp with greater mass will have a correspondingly longer duration for its overall cyclic sequence than one with lesser mass. As a final conclusion we must state that at the approach to C_2 realm EILs all energy medium will sublimate into a mass medium with the subsequent profusion formation of Mps therefrom.

Let us for a moment review the series of events that have occurred since the realm EILs achieved C_1 and C_1^+ intensities. The Mps that formed in these high EIL zones reverted rapidly to the energy medium status. However an equally rapid re-formation of mass points continued in the locality which made it seem as though Mps were oscillating in a continuous manner between the energy medium and mass point status. Thus an energy medium to mass medium and subsequent mass-point to energy medium exchange takes place as an apparently continuous exchanging process. The energy medium to mass-point exchange is abbreviated E-M/X and the mass-point to energy medium exchange is abbreviated to M-E/X. This then is the basic functional relationship between a mass-entity and its convertible energy medium content as given by the formula $E = MC^2$.

Now as the EIL within a realm zone approaches C_2 levels, Mps form alongside one another in great profusion. A grouping of Mps results with a very high population. The outward energy medium flow from any particular Mp in the group therefore tends to get absorbed into the other Mps that form alongside. Thus when some Mps are in an E-M/X sequence others are in a M-E/X sequence. The cyclic rate of occurrence is usually evenly matched since they all operate within the same realm zone which is at a particular EIL. As a result we may infer that at any instant a large quantity of Mps will be in existence which means that a quantity of energy medium is more or less permanently in the mass state and so lost to the phasing realms. Our realm thus cannot be considered as consisting of energy medium only - it now contains a proportional quantity of a mass entity even if this is of a continuous highly transitional nature.

Let us momentarily return to the basic phenomenon of the energy medium at C_1^+ EIL becoming a mass point. Visualise a sphere of radius r at C_1^+ EIL as in Fig 1.10a. The volume of the energy medium in the sphere is given by the expression $\odot \pi r^3$ and is equivalent to all the energy medium in a Mp. The sequence of events that lead to a Mp forming is a transition of the C_1^+ energy medium in the volume at Fig 1.10(a) to a mass type medium at the same volume. The structure of this mass type medium must at this stage be visualised as wholly homogeneous without any defined particle separations. The mass concentration level is considered extremely low and the entire mass medium in the volume ☺ πr^3 behaves as though it were under some kind of stretch action like an elastic band pulled apart to its limits. As shown in Fig 1.10(b) the entire mass medium consequently implodes inwards to the spherical central point, and there it achieves an optimum mass concentration that we earlier classified as a Mass Point (Mp).

If we assume that the total energy medium in a Mp is very nearly constant then we can infer that the sphere in Fig 1.10(a) would be smaller for higher C_1^+ EILs. Thus the radius r of a sphere containing energy medium at approx. C_2 EIL would be a minimum for a Mp and is termed as the Minimum Energy Radius. More on this in a later section. The essential result is that at near C_2 EILs we have Mps forming at extremely short distances from one another. As such we get Mp groups as stated earlier.

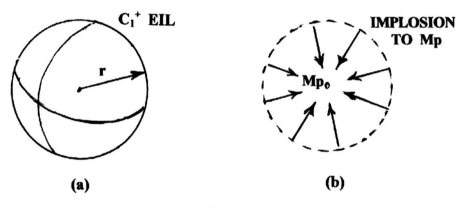

Fig 1.10

The situation has now become rather complex. We have a multiple number of realms phasing and creating EIL high spots. These in turn result in Mps that move with a resultant linear momentum. Mps are short lived (10^{-43} secs) and revert to energy medium. This energy medium adds to the EIL high spots resulting in even higher EILs. A continual exchange is occurring between the energy-medium and Mps. The realm phasing continues and EILs approaching C_2 are attained. Mps form close to one another and an interaction with their energy medium and linear momenta result in a Mp grouping. The Mp group's resultant linear momentum carries it into further areas of near C_2 realm EILs and subsequently more and more Mps form alongside with an overall balance in E~M/Xs being maintained. So long as the realm phasing EIL datum persists at near C_2 levels, then Mps will continue to form and congregate together. Although a single Mp is an extremely unstable mass entity, a congregated assembly of Mps becomes a relatively permanent mass entity through the mutual E~M/X and M~E/X exchanges that are continuously taking place. Hence the energy medium of the phasing process is being depleted with the net result of a subsequent lowering of the realm EIL datum. The EIL datum being the general EIL within a local zone. In a C_1^+ EIL area there is an optimum mass level for Mp-congregations at which the energy medium exchanges acquire a perfect balance. This mass-point congregation entity is termed a proton and can be considerably varied in size depending on the realm EIL at which it was formed. We shall show in a later chapter how this formative mass varies between a size as large in volume as our solar system at one end of the scale down to a fraction of the size of our current proton. Initially the proton exists in a C_1^+ EIL area. However with the realm phasing progression a gradual lowering of the EIL datum will occur as more Mp-congregations/protons are formed and also as the phasing sequence advances into its later stages. The realm EIL datum will thus also gradually diminish to a level below C_1 and at which Mps will not normally form under these conditions.

At this stage it would be appropriate to discuss the flow characteristic of the energy medium. The movement of energy medium from one position to another within the cosmos may be termed as an energy medium flow. The realm is thus flowing through the cosmos. We are however more interested in the relativities of this flowing energy medium within the realm confines and in relation to mass bodies in general. We explained how a Mp resulted from the implosion (inward flow) of a spherical volume of mass medium at C_1^+ EIL as part of the E-M/X (Fig 1.10). The inward flow implosion reaches an impacted limit at the spherical centre when the mass-medium becomes a mass-point particle. Let us now look at the phenomenon whereby mass reverts to energy medium in the M-E/X process. Earlier in the realm phasing sequence we explained how a Mp attained a certain linear momentum and that when it reverted to energy again it caused an addition of EIL to the existing phasing levels. The M-E/X process will now be explained (see also Fig 4.2).

A Mp exists only for an instant, 10^{-43} of a second approximately. By the very nature of its formation the EIL around the Mp for that instant is very low. The Mp was formed by the implosive impact of an energy medium converted to mass medium. The implosive velocities are neutralised at the focal centre, but the impacting effect lasts only for an instant. That instant, 10^{-43} of a second approximately, is the total life event of the Mp. Once the central mass impact effect is released, the high energy medium concentration of the Mp explodes outwards as there are no external forces to prevent this happening. This does not however mean that the entire energy equivalent just explodes out of the Mp. The reverting of the Mp to energy medium follows an event sequence. Just as water evaporates at the surface only so the Mp outer surface inhibits the central part in the M-E/X process. Thus there is no instantaneous conversion to energy medium but an event sequenced outward flow of energy medium (abbreviated as 'E-Flow'). This is similar to the example shown earlier in Fig 1.2 where energy medium is emitted from a central point. The EIL at any point in the E-Flow obeys the inverse square law as depicted in Fig 1.3. The EIL of the E-Flow at the surface of the Mp would be just under C_2 EIL and thereafter would diminish accordingly. There is no relationship between the E-Flow EIL and the earlier Minimum Energy Radius although these are shown together in Fig 1.11. So far we have observed in theory only, the upper limit C_2 as the practical extreme in high realm EIL. The EIL of the E-Flow from a Mp diminishes with distance. The rate of decrease of EIL gets less and less the further one goes from the point of origin and the theory predicts that in the limit the EIL will approach zero. This leads us to conclude that in the limit the energy medium would cease to exist. Since this cannot be true then the former statement must also be modified. As such we can state that the EIL of the E-Flow from a mass particle diminishes with distance from that particle according to the inverse square law and that in the limit it approaches a minimum critical level which we shall denote by C_0. The distance at which this occurs is called the Maximum energy Radius and is independent of other phasing E-Flows. One may speculate regarding what happens to the E-Flow beyond the Maximum Energy Radius. One possible theory maintains that the energy medium continues as an E-Flow which has split into separate and independent radiant lines of energy medium at C_0 EILs as shown in Fig 1.12. These may be referred to as Energy Medium Lines and the path taken by such a line would be an absolute straight line through the cosmos. At this stage we must not confuse the propagation of light with an Energy Medium Line. The former is simply an EIL oscillation of the energy medium whereas the latter is an E-Flow as will be shown in Chapter 3.

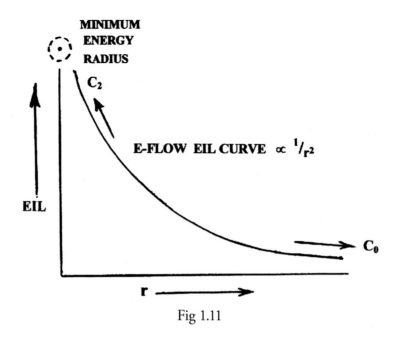

Fig 1.11

We have now concluded that the energy medium of which all mass is constituted is subject to three critical levels of intensity as stated below.

1. The EIL cannot exceed a maximum practical limit defined at level C_2.
2. The EIL does not decrease below a minimum existence intensity defined at level C_0.
3. The energy medium converts to mass medium particles only at or above a critical EIL defined at level C_1.

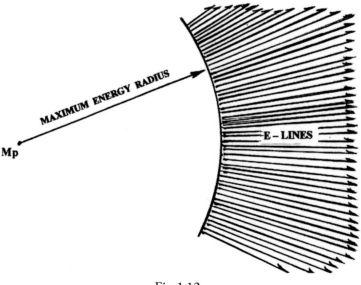

Fig 1.12

Now why cannot the EIL exceed C_2? What is the constraint upon realisation of EILs above C_2? The answer to both of these questions is quite simple in principle, yet becomes exceedingly complex in operation. During an E~M/X we saw how the energy medium did not exchange directly into the mass state but first, by the very nature of its concentration, became a mass-medium occupying the same volume which sectionalised and then contracted into a myriad of Mps. The characteristics of this mass medium is very different from those of the energy medium in the realm. The mass-medium becomes a tangible entity within the realm and cannot phase with other adjacent mass-medium formations. Although the mass medium is only a transitional feature within the E~M/X, nevertheless it controls the behaviour of a majority of the phenomena that we observe around us. We shall deal with some of these features later on. Now when an energy-medium converts into a mass-medium then we may correctly assume that the energy-medium section has ceased to exist. As such there will be no EIL to measure or to phase with adjacent energy medium. As mentioned earlier, an EIL gap has occurred which follows the pattern indicated in Fig 1.8 and Fig 1.9. The net effect upon the realm energy medium is two-fold. Firstly the formation of mass in the realm considerably depletes the phasing energy medium quantities. Secondly the M-E/Xs set up a dissipating E-Flow within the realm which diffuses the energy-medium such that eventual expansion of the realm volume is inevitable. This results in a lower overall average EIL in the phasing realms.

Now these two factors play a critical role when phasing EILs approach C_2. The energy medium absorption and dissipation at these levels is phenomenal and the rate at which the realm EIL tends to build up is evenly matched. Although it is extremely unlikely that C_2 will ever actually be attained under these circumstances let us assume that a section of the phasing realms does achieve a C_2 EIL. When this occurs then the energy medium is unable to exist at this level of concentration and becomes wholly a mass-medium. This resolves itself into a myriad of Mps with a total absorption of all the phased energy medium. Now since the energy medium event rate is higher for an E-M/X than it is for a M-E/X then the resultant EILs from the dissipating Mp E-Flows will never approach the levels prior to mass medium formation. Thus we see that C_2 is an absolute physical upper limit for energy medium concentration in the realm and at which the transition into a mass medium occurs. At this time we are not able to ascribe a parameter to the EIL status and so will continue with its relativity only.

Let us now consider the concept of event sequence or 'time'. In the above sections no mention was made of an independent time factor. Hours, minutes, seconds and micro divisions of these do not apply in realm energy medium phasing or in the E-M/Xs that follow. Realms moving through the cosmic void do so without reference to any independent space factors. Time as we know it is purely relative. The speed of a motor vehicle is measured by the distance it covers in a set time. That time is measured by a set number of quartz or cesium crystal oscillations. Those oscillations alter at greater distances from the earth's gravitational influence and also when under fractional light propagation velocities. Time itself has become a variable under Einstein's principles and as such we are left without an independent duration measuring device.

An alternative to the above is indicated by the approach to relative event changes. One event can usually be related to the event that preceded it and a chain of such events is referred to as an Event-Sequence. An event is defined as the smallest possible change in status that occurs between two relative points. In the example of the motor vehicle an event would be the smallest possible change in position of the vehicle relative to the ground or it could refer to any part of the vehicle relative to some other datum. An event could therefore be the smallest change in the E-M/X during the formation of a mass point or vice versa. A chain of such events becomes an event sequence that follows a step by step forward progression. (In quantum mechanics the dimensions given for the shortest distance or Planck length is 10^{-35} of a metre, and that for the shortest duration interval or Planck time as 10^{-43} of a second). Events follow events and a progressional sequence is established that is non-reversible. This does not mean that events cannot repeat themselves but it does mean that the event sequence does not work back on itself.

Let us for the moment define Time (as we know it) as the rate at which events sequence. During realm phasing there is not an even rate of event sequencing and hence there can be no regular concept of time. However in a Mp-congregation or proton, events do take place in a decided sequence. In Chapter 5 we establish an absolute parameter upon which to base all our Time concepts although Time itself is variable. For the present it will be sufficient to recognise that Time can be measured only in relation to the sequential rate of other occurring events. Since the realm section that we are in consists of large quantities of mass, then it is the event changes that occur within these masses that determine the time rate. Although the unit of mass is the Mp we are unable to use it for determining event sequence since it is a strictly one-off occurrence. However a Mp-congregation or proton, which may be referred to as a basic mass body, does have a non-repetitive event sequence as a whole and so may be used in our Time relationships. The curious reader may refer to Chapter 5 for an insight into the development of a Time rate which is based upon the energy medium changes in the proton body as a whole. In the next chapter we shall discuss the overall energy medium cycle of our realm and also the proton. This infers that the proton mass is not only built up as seen earlier, but also that it must decay back into the energy medium to complete a full energy medium cycle.

Of considerable importance is the phenomenon of proton decay. The phasing of realms produced mass congregations when the EIL exceeded C_1 levels. With the large absorptions of the energy medium into those mass quantities there is a resultant lowering of the overall realm EIL datum. Although realm phasing may continue, the resultant EIL will continue to attain lower and lower values. At some stage the EIL will drop to below C_1 and subsequently Mps will cease to form on their own. However mass formations will continue their existence at these below C_1 EILs and it is this area of existence in the life of the proton which is of great importance to the physical phenomena of today.

Let us first attempt to visualise the event sequence inside the proton body. There are a countless number of M-E/X and E-M/Xs occurring within the proton in the Mp formation mechanism that can be compared visually to the speckled snow storm image on a television screen at a non-channel frequency. The dark dots appear to interchange continuously with the lighter ones and vice versa, in a never ending repetitive sequence. The whole appears as a pulsating effect and this represents only a small section of Mps well inside the proton body. Imagine the proton to be the size of the earth, then this TV image analogy portrayed above could be just about on the right size scale.

Now let us look at a specific Mp well inside the proton. Although the local realm EIL datum is below C_1, the energy medium exchanges within the proton create EILs well above this. As the M-E/X occurs, energy medium flows into the surrounding area. This E-Flow phases with similar energy medium releases from other Mps with resultant EILs in excess of C_1. When a Mp reforms inside this proton complex it not only absorbs the C_1^+ EIL energy medium but also any E-Flows that enter the Minimum Energy Radius spherical volume during the mechanism of Mp formation. As such all the E-Flow from a M-E/X gets to be re-absorbed into other E-M/Xs well before it can reach the surface of the proton. It is estimated that our everyday proton contains some 100 trillion, trillion Mps in constant transition. Thus inside the proton body the M-E/Xs and E-M/Xs remain closely balanced. What about the linear momentum of a re-formed Mp? Within the body of the proton each E-M/X is surrounded by a uniformity of E-Flow from M-E/Xs such that the resultant linear momentum remains zero relative to the proton body as a whole.

Let us now look at a Mp at the surface of the proton. By definition a Mp is short-lived and exchanges to energy medium in a sequence of events resulting in an E-Flow spherically outwards from its centre.

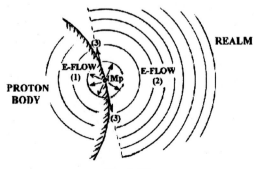

Fig 1.13

In Fig 1.13 a M-E/X is depicted at a point on the surface of the proton. As the exchange occurs a flow of energy medium radiates in all directions. Imagine the diagram to be 3-dimensional so that the E-Flow is visualised coming out of the foreground as well as travelling into the background. This E-Flow can be categorised into 3 parts as follows;

(1) That part of the spherical E-Flow that enters the main body of the proton and which accounts for nearly half the total E-Flow.

(2) That part which flows outside and away from the proton surface and into the surrounding realm. This also accounts for nearly half of the total E-Flow.

(3) That portion of the E-Flow that is in a plane tangential to the proton surface. This is a very small proportion of the total E-Flow and would normally be indistinguishable from the previous E-Flow(2). However it does perform a specific function with regard to the proton congregation and so deserves a section of its own.

Let us now deal with each of these sections in turn and in a bit more detail.

The E-Flow(1) that phases into the proton body simply gets re-absorbed into the countless E-M/Xs occurring within the proton. However the absorbed energy medium flow in the diagram is on the

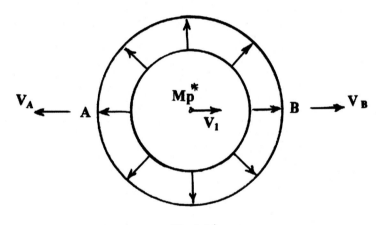

Fig 1.14

proton right hand side only and so must impart a net linear momentum in the Mp-congregation that points toward the centre of the proton. The net effect is therefore a resultant linear momentum that pushes that section of the proton in a direction diametrically opposite to the point on the surface where the M-E/X occurred. There are other Mp surface exchanges all over the proton that also produce a similar resultant linear momentum from their E-Flows. These tend to balance out against one another resulting in a zero linear momentum for the whole proton. There is one major effect however and that is that the linear momenta of the Mps near the proton surface exert an inward push upon the rest of the internal Mp-congregation to compress and maintain the proton as a tightly enclosed mass entity. We shall see later on in chapter 4 on gravitation how this balance can be upset resulting in a net directional linear momentum for the proton as a whole. We must digress a bit farther into resultant effects of linear momentum transfer in order to understand how a directional force can arise inside the proton body. The E-Flow is known as occurring uniformly outwards from a Mp centre. Also the velocity of the E-Flow is uniform in all directions relative to the originating Mp centre. This remains true whether or not the Mp had an original velocity of its own.

Consider the Mp* in Fig 1.14 to have undergone a M~E/X. The E-Flow is in the form of a spherical band of defined width spreading outwards from the centre of origin denoted by Mp*. The symbol * is used to indicate the point of an earlier Mp event sequence. Let us for the moment assume that the E-Flow velocity is C_V relative to Mp*, where C_V is about the velocity of light wave propagation. Explanation for this assumption will be set out in a later chapter. Now consider that prior to the M~E/X the mass-point at Mp* had a linear momentum in the direction A to B represented by a velocity V_1. The subsequent absolute velocity of the E-Flow at A and at B will be given by,

E-Flow $V_A = C_V - V_1$
E-Flow $V_B = C_V + V_1$

Components of the E-Flow absolute velocity in other directions can also be worked out. When a section of the E-Flow is absorbed into a Mp it will impart linear momentum proportional to its absolute velocity and quantity of energy medium. Thus the linear momentum of a previous mass-point is maintained in the energy medium flow from that Mp and re-appears in all the other Mps into which it is absorbed. So far as the Mp is concerned there appears to be a law in existence of the conservation of linear momentum. This is the basic mechanism of force that tends to push the

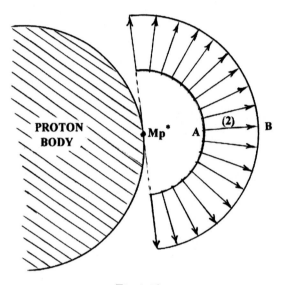

Fig 1.15

mass-medium along a defined direction. Only mass is subject to a 'force' as such which is really the effect of change in the linear momentum of the Mps within that mass body. It is hoped that the reader is now able to visualize the effect of the E-Flow(l) upon the interior of the proton body as a simplified concept.

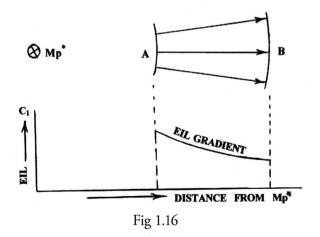

Fig 1.16

The E-Flow(2) that proceeds outward from the Mp* on the proton surface is in the form of a semi spherical E-Flow band of depth that equates to the sequence duration of the M-E/X. In Fig 1.15 the E-Flow is shown at a particular event sequence where,

(i)   the initial E-Flow from the Mp* now occupies the outer spherical position B.

(ii)   the last of the energy medium from the M-E/X occupies the inner spherical position A.

(iii)   the sequence duration of the M-E/X is represented by the E-Flow band width AB.

The velocity of E-Flow at the start and end of the M-E/X is constant. It therefore follows that the E-Flow band will remain the same width up to the maximum energy radius. Also the E-Flow will remain in a semi-spherical band with an EIL that obeys the inverse square law with regard to distance from Mp*. Thus at any instant the EIL on the inside of the semi-spherical band will be greater than that on the outer side. We may subsequently conclude that an EIL gradient exists within the E-Flow band and that the gradient is on a diminishing scale in the E-Flow direction.

This EIL gradient is visually represented in Fig 1.16 for a section AB of the E-Flow band. Numerical quantities are not yet indicated. The EIL scale is in the region below C_1 and so we are not concerned with the complexities of very high energy medium intensities. It will be noticed that the EIL gradient curve obeys the inverse square law for distance as represented in the earlier Fig 1.3. and Fig 1.11. The EIL gradient depicted is really a simplification of the M-E/X sequence in order that the reader may form a visual conception of the events that occur when a Mp reverts to energy medium. The beginning and end of the E-Flow would in reality not be as sharply defined as in the diagram. For the moment however, the above representation will suffice and it will be seen

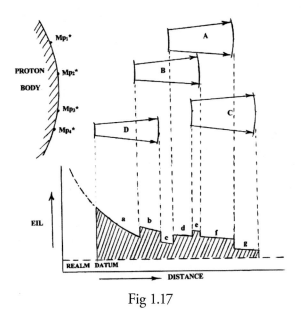

Fig 1.17

in chapter 4 how the EIL gradient affects the linear momenta of another proton body to produce the effect commonly termed as Gravitation. The above EIL gradient will of course be superimposed upon the realm EIL datum of that area.

Let us now look at the case when random M-E/Xs occur on the proton surface to present multiple

E-Flows across the same realm area but at different event sequence moments. As such the E-Flow from each Mp* may partially overlap as represented in Fig 1.17.

Let A, B, C and D sections represent the semi-spherical E-Flows from Mp_1^*, Mp_2^*, Mp_3^* and Mp_4^* on the proton surface. The E-Flows are away from the proton body at the same C_v velocity and obey the inverse square law for intensity. Underneath these is a graphical representation of the overlapping (phasing) E-Flow intensities derived as follows;

Section 'a' represents the intensity of the semi-spherical E-Flow D on its own as there is no over- lapping with other E-Flows.
Section 'b' represents the addition of intensities from the outer part of D and the inner part of B.
Section 'c' represents the EIL of the middle part of B
Section 'd' represents the overlapping intensities of B and A.
Section 'e' is the added EIL peak due to the overlapping of E-Flows A, B and C.
Section 'f' is the EIL due to E-Flow A and C.
Section 'g' is the EIL for the outer part of E-Flow C on its own.

We shall see in a moment that the third portion of the E-Flow from a M~E/X which is tangential to the proton surface has a considerable influence upon the configurative distribution of Mp decay. The E-Flows from Mps on the proton surface do not therefore occur in a random manner but rather in a set sequence from defined positions. This results in an overlapping E-Flow pattern that is represented in Fig 1.18

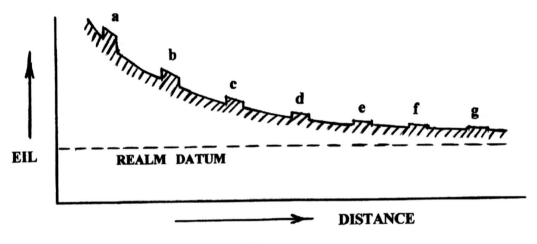

Fig 1.18

Each of the EIL peaks a, b, c, d, e, f, g, etc, are of diminishing prominence (on the graph) and are depicted for concept purposes only and refer to a single proton's action. At this stage it may be concluded that a series of such peaked energy medium spheres exists around the proton with peak intensities that diminish with distance from the proton centre. Also that these spherical peaks of EIL take on a semi-permanent repetitive position around the proton so long as it is in decay. These will be termed as 'forcephoid peaks' or simply 'forcephoids', and we shall be discussing their characteristic effects more fully in chapter 6 in the development of the Primary atom and also in chapter 7 on electric and magnetic effects.

Let us now look at the third part of the E-Flow from a M~E/X on the proton surface. This is the E-Flow(3) that is in a plane tangential to the proton surface and flowing away from the Mp as shown in Fig 1.19. The E-Flows(1) and (2) are not represented here although they are also present. The main consequence of this E-Flow to the proton surface is that a region of high EIL is set up when phasing with similar E-Flows from neighbouring Mps. The intensities combine and it becomes possible for an E~M/X to occur at the proton surface. This means that Mps on the proton surface do not therefore irreversibly disappear but rather re-form repeatedly in a zone of high EIL. The combination of all the E-Flow(3) on the proton surface results in a closed energy medium grid around the proton and extremely close to the surface. This energy medium grid surrounds the proton and is similar (in principle) to the atmosphere that envelopes the earth. The very existence of this energy medium grid is dependant upon the continued mass to energy medium decay at the

Fig 1.19

proton surface and the EIL thus set up within that grid would be at C_1^+ levels with the obvious tendency to induce E-M/Xs and more Mps. There is thus a regulating effect to control the net M-E/Xs that occur at the proton surface. When the M-E/Xs occur profusely on the surface this results in a higher energy medium grid intensity which subsequently induces a higher proportion of E-M/Xs. The net mass to energy medium decay is reduced in this higher EIL and the energy medium for the E-M/X is absorbed from this energy medium grid. The EIL of the grid will then reduce somewhat. When the energy medium grid EIL is relatively low, E-M/Xs do not occur as profusely and the net decay increases. An eventual balance is achieved between the energy medium grid EIL and the net Mp decay rate which suits a particular local realm EIL environment. Other effects of the energy medium grid around the proton will be discussed later in chapter 6 in context with the primary atom state.

We have now shown that there is an energy medium flow from the proton to the surrounding realm. This is achieved by a net M-E/X that is greater than the corresponding E-M/Xs. The imbalance occurs mainly upon the outer surface of the proton, the proton itself being located in a local realm EIL datum that is less than C_1.

We have also seen that the event sequence of a Mp is repetitive within the proton structure and therefore cannot help us to judge overall event duration. However the Mp-congregation as a whole does undergo an event change that is not repetitive but progresses in the direction of a net mass to energy medium interchange or an overall mass decrease. This is referred to as the decay of the proton and the rate of this decay is considered to be the most basic event change rate. We may therefore use this as a parameter to measure the relative duration of all other events that occur at below C_1 realm datum. This parameter is defined as Proton-Time and is directly proportional to the proton decay rate. Each proton has a normal decay rate that is grossly influenced by a number of other factors such as realm EIL, changes due to the proximity of other protons, relative velocity within the realm or simply a differing realm datum. We shall deal with this aspect more fully in chapter 5 but for now it will be sufficient to state that Proton-Time or simply 'Time' as a measure of the duration of all events within the realm is a variable that is related to the decay of the proton mass. We may also state that for an EIL environment at or above C_1 the proton does not decay as such and in fact is more than likely to gain mass. When there is no net gain or loss of proton mass it is to be concluded that 'Proton Time' ceases to exist as a measure for event changes. As mentioned earlier, event changes are progressional in a forward futuristic direction only and as such Proton Time always progresses into the future. Events cannot be undone once they have occurred and therefore Time has become the parallel of Proton mass change. The reader may find some of these concepts difficult to accept but they have been mentioned here simply as an indication of later developments within the text.

Realm Changes.

We have now hypothetically observed certain aspects in the behaviour of phasing realms. From an initial non-expanding energy medium quantity in the realm we now have a mass quantity and a depleted energy medium quantity. This energy medium quantity is also partly endowed with a linear momentum that imparts an expanding character to the realm. Although the original realm phasing will continue around the formed masses, the EIL is never likely to rise to those high initial levels. The proton decay results in a complex network of E-Flows within all of this and as events sequence the realm EILs will gradually and continually reduce. The combination of all the EILs from realm and proton decay will thus get lower and lower till eventually the original realms or what is left of them may complete phasing to continue their respective path through the cosmic void. The masses will then exist in the E-Flows of one another. The resultant EIL datum of this E-Flow will vary from place to place and will prevail to near critical levels at certain mass central locations. That is to say, the EIL would approach critical levels due to the focal effect of the E-Flows from multiple sources (chapter 4). The overall realm EIL will however continually decrease and will tend towards C_0.

In brief we may summarise that the realm status along with its mass and energy medium flows is simply a machine-like entity that operates between higher and lower energy medium levels. The entire structure of our universe (that we observe today) may be attributed to this activity in one way or another, and we shall see how true this really is when we analyse each physical phenomenon in turn. It is hoped that on this basis we may also make predictions of phenomena that we may possibly never observe or even require to observe.

We find ourselves in an energy medium cycle and as a result we are able to represent our existence within the cosmos at a point on a graphical representation of that cyclic exchange. The binding parameters are the EIL values of C_0, C_1, and C_2, for the energy-medium and the mass-point unit for the mass-medium. Let us now enunciate the three most important physical aspects of the realm that control all phenomena.

1. Mass is a higher state in a progression of the energy medium intensity phases.
2. Energy medium intensity levels tend to revert from a higher state to the more uniform lower energy medium state unless additional influence is exerted upon it.
3. The intensity of an energy medium is limited to a maximum EIL of C_2 and a minimum EIL of C_0 in every instance.

In the next chapter we shall indicate the above concepts on an energy medium cyclic graphical representation in order that we may ascertain our relative position in the overall life cycle of our universe.

. . . .

Chapter Summary.

1. Energy medium is a perfectly homogeneous medium without physical structure of any kind.
2. A realm is a marginally expanding energy medium entity that traverses the cosmic void.
3. Two or more realms may phase through one another without a colliding effect. The energy medium intensity levels will however be the arithmetical sum of the individual realm EILs.
4. Realm phasing results in some very high EILs. C_1 denotes the EIL at which mass points commence to form. C_2 EIL denotes the maximum EIL at which the energy medium can theoretically exist. In reality at C_2 EIL the energy medium wholly converts to the mass medium status.
5. A proton is a congregation of phase balanced mass points that exist as a single mass body (a Mp-congregation) and which formed at near C_2 EILs. These can be spectacularly large.

6. At a realm EIL datum below C_1 the surface Mps of a proton result in a net energy medium flow into the surrounding realm and which is termed as Proton decay.

7. The cyclical nature of inter-realm phasing is such that after the initial high EIL effects, there is continued lowering of the realm EIL datum due to energy medium flow from M-E/Xs.

8. The EIL in an E-Flow obeys the inverse square law with distance. The EIL in that flow however, cannot decrease below a minimum EIL denoted by C_0.

9. Mass is an unstable energy medium condition and is in a continued state of exchange back to energy medium. This is proved by the M-E/X status of the Mp. As such mass may be termed as a high energy medium condition that will naturally revert to its former lower energy medium status.

10. A series of concentric energy medium peak spheres exists around a decaying proton body. An energy medium grid also exists very close to the proton surface analogous to the atmosphere surrounding the earth.

11. Proton decay is the most basic measure of event changes and is automatically used as the measure of event duration. This measure is termed as Proton Time and can only apply to events in a realm zone at EIL below C_1.

CHAPTER 2.

The Seven Circle Theory

In the previous chapter we presented a hypothetical concept of energy-medium behaviour within the realm. A test of that concept would be to fit such behaviour into a predictable set of rules. Energy-medium intensities must eventually revert from the higher to the lower levels and so obey a cyclic order of behaviour. It is this cyclic pattern that we are about to establish and present as a support for the realm energy medium concept.

In accordance with previous description we shall present the characteristic behaviour of the energy medium along a defined curve on a graphical representation of realm EILs. Since there is no EIL parameter as yet, this graphical representation must be in comparative terms only.

The problem of showing up cyclic changes of the realm EIL on a graph is not a difficult one if one may assume event duration to be temporarily a non-measurable quantity. This means that a graphical curve may take on any shape without actually going back upon itself. In Fig 2.1 a change is represented in the energy-medium from an EIL at A to an EIL at B. The vertical axis represents the realm EIL to an increasing logarithmic dimension scale. The choice of the logarithmic EIL scale was necessitated for concept purposes. An ordinary linear scale would have indicated a comparison between relative positions on the graph. For example, on an ordinary scale the EIL at B may have been considered to be approximately twice that at A. This would be a false interpretation as the EIL parameter is entirely unknown at this point in the theory. With the logarithmic scale such a dimensional comparison is not immediately possible and so we are able to concentrate upon the concept aspect alone.

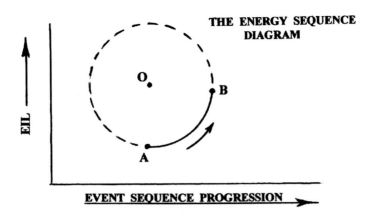

Fig 2.1

The horizontal axis on the other hand does not have a dimensional quantity portrayed but simply indicates the direction in which events occur progressively. In the previous chapter we defined event changes as always occurring in the futuristic direction and that the basic measure of these changes could be carried out by comparison with the proton energy medium decay. Energy medium by itself is a massless entity and therefore Proton-Time does not apply directly to it. Indirectly however changes in realm EIL may be compared to event changes and an event sequence is brought into play. This sequence of realm EIL changes will occur as a succession of EIL events, and as mentioned earlier such events cannot be undone although they can be repetitive. The horizontal axis therefore indicates event sequence progression in the futuristic direction and we are thus able to interpret the curve AB as a conceptual EIL occurrence. Unfortunately we are unable to indicate the rapidity of the EIL change from A to B, but at this stage we are simply concerned with change and not the rate at which it occurs. Later when we further develop the concept of Proton-Time we shall apply this to realm changes to obtain an event sequence rate.

In Fig 2.1 we have shown the curve AB to be part of a circle with centre at O. Since we are concerned only with the curve AB the presence of the fully indicated circle simply pronounces the graphical nature of that curve. The graphical representation of realm EIL change is only to be taken on the basis of curve AB (full line). The dotted line that completes the circle is to be discounted from any EIL representation but must remain on the graph to indicate comparative status with other curves that also show EIL change. We now have a graphical system to portray a conceptual trend in the cyclic nature of the realm energy medium as it occurs between the critical levels indicated by C_0, C_1 and C_2 when energy medium phasing takes place. On this system we should then be able to relate to and achieve explanations for all phenomena within the realm structure. An attempt at some of these explanations is offered in the later chapters which is in addition to an understanding of the character of the energy-medium and of the mass-medium.

The Seven Circle Energy Medium Curve.

The general pattern for the cyclic behaviour of phasing of the realm energy medium can be graphically traced on a set of curves drawn from a link-up of seven equal sized circles in the Energy Medium Sequence diagram shown in Fig 2.2. Although the complete circles are represented, the energy medium event sequence traces a representative path only as shown by the bold portion of each circular quadrant.

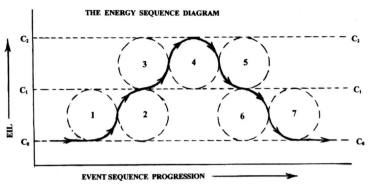

Fig 2.2.

The critical EILs C_0, C_1 and C_2 are indicated on a logarithmic scale on the vertical axis and it is worth noting how the curve flattens out at each level. In theory, C_2 is an EIL that may be approached but in practical terms is not quite attainable. As such the energy medium curve at circle 4 may be assumed as peaking just short of C_2 before dipping down again towards the circle 5 configuration. Circle 6 is of special importance as it is considered that our realm section is currently somewhere along this portion of the energy medium curve. An estimate of an approximate position will be established in a later section when an attempt is made to link the red shift phenomenon with the elapsed Proton Time since the realm was at the preceding C_1 EIL.

As stated earlier the energy medium event sequence rate was an unmeasured quantity and as such the progression of EILs along circles 1, 2 and 3 are not necessarily the same as for circles 5, 6 and 7. In fact it would be expected that the former occurs at a much more rapid sequence rate due to realm phasing being an enforced condition. The latter however is an indication of independent EIL decay which is by its very nature a more prolonged process. Let us now trace the energy medium curve and its impact on the character and behaviour of the energy medium realm from circle 1 through to circle 7. Although the energy medium curve is only a part of each circle it would be simpler to denote each portion of the energy medium curve by the circle it comprises. The definitions and energy medium phenomena discussed in chapter one form an essential part of the Seven Circle Theory and reference will be made to them as necessary. As we progress, some readers would be well advised to refer back to sections in the previous chapter as a matter for conceptual revision.

Energy Medium Sequence Circle 1.

The energy medium curves denoted by the seven circle representation refer to a specific section within the realm structure and not to the realm as a whole. The realm is a relatively vast entity and the phasing EILs would vary across it. Thus when one area is at the circle 1 portion of the energy medium curve, neighbouring sections may be at EILs represented by different parts of that curve. Although every part of the phasing realms will obey the same rules, each will be governed by the local EIL condition that determines its relative position on the energy medium curve.

The circle 1 energy medium curve is represented as an EIL that increases from C_0. This may not necessarily be the case for every part of the phasing realms as some realm datums could have existed at an EIL well above C_0. However the energy medium curve does represent the event sequence for ordinary realm phasing and as such some sections may commence at a higher EIL than others. The initial consequence of realm phasing is a general rise in the EIL. It will be noticed that the energy medium curve approaches the vertical towards the end of circle 1. This reflects the increase in EIL that results from the arrival of E-Flows from multiple Mps* that formed in another section of the realm. Although the horizontal axis does not have a measured sequence duration parameter, nevertheless any approach of the energy medium curve to the vertical definitely indicates a higher relative rate of change.

Let us for a moment extend our concept of a realm. If realms travel through the cosmos at near light propagation C_V speeds, which may also be the approximate E-Flow velocity from Mps* (Fig 1.14), then two or more realms may have their energy mediums blending through one another at twice C_V rate. Assuming that a realm is comparable to a cloud-like entity of the energy-medium, then it is highly probable that the EILs near its more central regions would be above the basic C_0 level. Earlier we assumed the realm to be a non-expanding entity. At this stage we must alter that view in order to surmise upon the origins of each realm. We have mentally observed that a realm traversing the cosmic void (Fig 1.5) is an exceedingly small entity in comparison to the void. The chances of two or more realms phasing is therefore remote. We can however considerably improve upon those chances by assuming the realm to be a breakaway packet of the energy-medium from a much larger energy medium entity. For want of a better term let us refer to this grosser entity as a 'Mother Realm'. Further, let us assume that a large number of the relatively small energy medium packets are continuously being expelled away from the Mother Realm in every direction as shown in Fig 2.3. Each energy medium packet that is expelled away from the Mother Realm may be considered to be a section of E-Flow. In the diagram, countless numbers of these are represented as issuing from the Mother Realm entity and also crossing each other's paths of traverse. It is therefore suggested at this point that realm phasing which results in the energy medium curve of circle 1 takes place relatively near the Mother Realm entity. Also that these emitted realms are in a state of expansion lateral to the direction of traverse. To theorise further upon aspects of the Mother Realm at this stage would not serve any purpose other than to confuse the reader. We can only state that the phenomena within the cosmos simply defies all imagination and until more theoretical data is available it would be useless to speculate upon it.

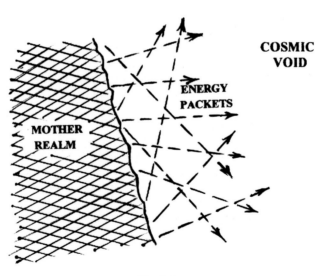

Fig 2.3.

There is a second aspect to the energy medium curve and that arises from the E-Flows emanating from the proton. In relatively mass dense areas within the realm there would be some central regions where the E-Flows from all these masses would tend to focus. The EIL in such locations would be very high and may even exceed C_1. Such an area would subsequently be represented on an energy medium curve that is superimposed upon the seven circle representation. Such a curve is termed a secondary energy medium curve and is purely the result of proton decay.

Hence we have a multiple cause for the EIL to behave according to the energy medium curve of circle 1. The rate of EIL increase has changed from near zero to a very high rate just before the curve adopts the circle 2 profile.

Energy Medium Sequence Circle 2.

During this part of the energy medium curve the EIL is represented as rising towards the C_1 level but at a progressively diminishing rate of increase. The initial high rate at the start of this sequence is due to the same factors as stated for circle 1. However with continued realm phasing there will now also come into play the EIL dips from Mps that had formed earlier at another realm location (see Fig 1.9). Although these EIL dips are from another realm quadrant, they will nevertheless restrain the rate of EIL increases in other realm quadrants through which they must phase. Subsequently as the phasing EILs approach C_1 there will be the added influence of EIL dips from the more immediate surroundings. As is represented by the curve at the top end of circle 2 the EIL rate of change drops to near zero. This is also due to the prolific formation of mass points within the realm section being observed. When mass points form (see Fig 1.10), a quantity of the phasing energy medium is removed by its conversion to mass and the instantaneous EIL at that spot is reduced to near zero. Such EIL reductions have a temporary effect only, since the M-E/Xs that ensue (in addition to the continued realm phasing) ensure the prevalence of the EILs that have already been attained.

The approach to C_1 EIL is not restricted to the exact parameter shown on the graph. C_1 is defined as the critical EIL at which Mp formation commences. This is not quite correct since Mp formation may be stimulated marginally before or after the C_1 EIL line. The 'critical one' EIL is therefore in existence on the graphical representation as a line of defined EIL band width. At this stage however we may assume the C_1 EIL to be at the middle of this zone, the energy medium curve being drawn accordingly. The energy medium sequence for circle 2 therefore enters this C_1 zone at a very low rate of EIL increase. The initial E-M/Xs remove further energy medium and the EIL increase is considerably slowed. With the first M-E/Xs taking place the energy medium curve moves into the circle 3 graphical region shown in Fig 2.4.

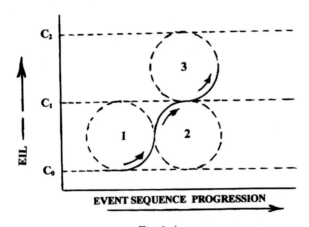

Fig 2.4

Energy Medium Sequence Circle 3.

The trend of the energy medium sequence circle 3 curve commenced when the EIL entered the C_1 zone and just after the initial formation of Mps. The trend is for the energy medium curve to obey an increasing rate of EIL rise. This is caused by an increased rate of Mp formation which subsequently results in an increased E-Flow from their M-E/Xs. This in turn results in the E-Flows phasing with the existing EILs to increase them even further. The situation is one of an accelerating trend between the three phenomena of Mps, E-Flows and EILs, where an increase in any one results in a corresponding increase in the others. We thus have an energy medium curve with a sharply rising characteristic similar to that of circle 1. On this occasion however the curve is due to an accelerating rate of Mp formation and exchange back to energy medium in the C_1^+ zones.

During the circle 3 stage of the energy medium curve we must visualise the nature of Mp formation. These occur as individual Mps and the subsequent M-E/Xs result in E-Flows that add to the realm phasing characteristic. We may assume that at this stage there is no permanent reduction of the energy medium from that particular realm section. Although a continuous exchange between mass and energy is now taking place, the mass cannot be considered a semi-permanent entity. A contrast with this situation is the existence of later Mp-congregations where mass is deemed to exist as a relatively permanent entity at the expense of a reduction in the total energy medium quantity. Since Mp-congregations do not begin to exist at the EILs represented by the circle-3 energy medium curve, we therefore continue to observe a sharp rise in the EIL rate of increase. At the zone where the energy medium curve circle 3 meets circle 4, the graphical representation of the rate of EIL increase approaches the near vertical. Once again as at circle 1 this only indicates a very high rate of change (event sequence duration parameters remain unmeasured).

Energy Medium Sequence Circle 4.

The energy medium curve represented by circle 4 is of special importance to us because of the emergence from it of the relatively permanent mass entities that we term as the proton and including the proton giant. In the circle 3 portion of the graphical representation we mentally observed the phenomenal rate of EIL increase linked to the increasing levels of E-M/Xs and M-E/Xs. The circle 4 curve now continues that rate of EIL rise but having entered into the very much higher energy medium intensity zone results in the formation of Mp groups. For the first time since realm phasing commenced do we have a set quantity of energy medium that remains linked with a Mp even if it is for only a relatively few energy medium exchange cycles. The effect upon the phasing EILs is to reduce the runaway trend to one of an actually diminishing rate of EIL increase as is apparent in Fig 2.5. As the EIL approaches C_2 the energy medium absorbed into the mass formations increases rapidly with the net result that the phasing realms are not able to sustain the increasing EILs. As such the energy medium curve levels off just below C_2 and then commences a decline. This becomes more pronounced towards the end of the circle 4 curve when Mp-congregations are building up at a phenomenal rate and size.

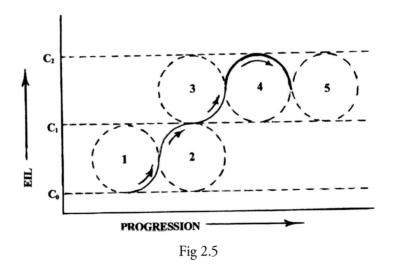

Fig 2.5

Let us look at this sequence of events in a bit more detail. The mechanism whereby Mps synchronise with one another to become Mp groups and on to larger more complex Mp- congregations is based upon the effects from the linear momentum of the energies absorbed. A Mp group becomes complex when the energy medium momenta absorbed by its constituents is considered isotropic. When energy medium flows enter the same absorption area, a neutralisation of the linear momenta occurs in each Mp and the energy medium exchanges of one set of Mps relate in an opposite sequence to those of another set of Mps to result in a balanced and even interchange. The mutuality of the E~M/X and M~E/X phases depend upon the principle in which greater absorption of the energy medium quantity into a Mp results in a marginally longer sequence duration. This sychronisation within a Mp grouping can be brought about by a mutual exchange of energy medium quantities absorbed. The process may be continued indefinitely to produce the more complex Mp-congregations (see also start of Chapter 6). Small Mp groups are not absolute stable entities and are usually in existence for a limited number of energy medium exchange cycles only. However, the greater the number of synchronised Mps in a congregation the greater will be the stability of the whole through their mutual exchanges. Also as described for E-Flow (1) in the previous chapter (Fig 1.13) the larger more complex congregations such as the proton possess a compacting force resulting from the linear momentum of Mps near the surface. This characteristic impulse is greater when a congregation is large and so maintains it as a permanent entity.

Another factor in the stability of the proton entity is the variable energy medium quantities that are absorbed into a Mp. While the Mps are in a C_1^+ EIL zone they are able to construct themselves from a quantity of the energy medium that is dependant upon the phasing realm EIL and a certain spherical volume of that energy medium. The minimum energy radius concept has already been described in the previous chapter (see Fig 1.10). As the EIL rises the energy medium sphere from which the Mp forms gets smaller. At approach to C_2 the EIL is a maximum and the energy medium sphere a minimum. However the quantity of energy medium in a Mp increases with EIL in spite of the diminishing sphere of action. In the limit for EIL at C_2 the mass in the Mp is a maximum. The Mp at this condition is referred to as being at optimum mass and its sphere of absorption is determined by the minimum energy radius. Now in the more central regions of a Mp-congregation the energy medium exchanges there result in EILs that are very near the C_2 level. Subsequently the more central the Mp within a congregation the more likely is it to be at near optimum mass. The Mps near the congregation outer regions however will be at the local realm EIL corresponding mass level. We therefore mentally observe a drop from optimum mass at the centre of the proton to a lower mass level at the surface. Since Mps can only form at or above C_1 EILs then the minimum Mp energy medium level must obviously occur for a realm EIL at C_1. The central Mps are shielded from realm EIL influences rather effectively by the surrounding Mps. A hypothetical curve indicating the nature of this relationship between inner and outer-most Mps is shown in Fig 2.6 for a realm EIL at C_1.

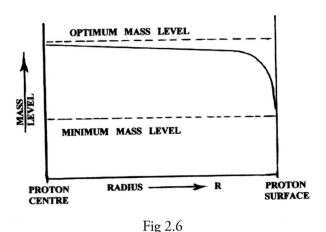

Fig 2.6

As the Mp approaches the minimum mass level it becomes more unstable in its energy medium exchanges until it ceases to become mass when the EIL drops below C_1. In a Mp-congregation however the existence of near optimum mass Mps ensures the energy medium exchanges for those that are not at the surface. The proton therefore as a Mp-congregation retains its stability into the lower EIL levels through its near optimum Mp exchanges within its structure.

Thus far we have conceived that Mps congregate into complex synchronised groups which develop into relatively permanent mass entities. This occurs at the expense of the realm energy medium quantity. The energy medium locked into the mass congregations must therefore be subtracted from the total phasing energies. This results in a reduction in the EIL rate of increase as indicated by the initial part of the circle 4 energy medium curve. The runaway trend between EIL, Mps and E-Flow as described for the latter part of circle 3 has now been arrested. As such the EIL rate of increase diminishes as the energy medium curve progresses towards C_2. Finally, as the energy medium curve approaches C_2 there is no further increase in the realm EIL. At this stage all Mps will be at the optimum mass level thus subtracting very substantial quantities of the phasing energy medium from the phasing realms.

The inducement for the formation of Mp-congregations is considerably accelerated when the EIL approaches C_2. We may define C_2 as the realm EIL beyond which energy medium cannot exist in its current form. It simply must convert into mass-medium. The graphical line representing C_2 is really the central line with a C_2 zone of action similar to that described for C_1. As such the realm EIL may enter this zone without setting off the mass-medium phenomenon. However as an alternative action it does set off an accelerated rate in the formation of Mp-congregations with resulting proton type formations. Once stimulated the development of these proton entities continues profusely. In fact the proton size is not limited to what we see today but goes on building to a size comparable to our current solar system volume and beyond. This accelerated mass inclusion is wide-spread in that high EIL realm zone contributing considerably to the depletion of free energy medium. These proton giants are what will become the central core of future galaxies. Some of the smaller proton giants will remain as what is today termed as a massive dark object or black hole. This phenomenon of proton giant formation only occurs under special high EIL conditions and there is a limit to its size. Once optimum size is attained further enlargement is restricted for two reasons. Firstly, a very strong repulsive action is set up towards all other proton entities and secondly, the realm EIL is depleting rapidly. The repulsive action is set up by the very intense high EIL outward E-Flows from each proton giant. Subsequently, although Mp-congregations continue to form, these can never again attain the proton giant sizes due to the relatively depleted realm EIL conditions as conditions approach the circle 5 curve. It will be here that our current standard protons form with their size controlled by the prevailing realm EIL conditions. They will of course exist as separate entities to eventually form the primary atoms of today. The circle 4 energy medium curve now results in a diminishing EIL as more energy medium is locked into these proton type masses. This rate of reduction of the EIL of the phasing realms is represented on the energy medium curve as approaching a steep downward trend as it nears circle 5. This implies that a very high proportion of the energies arriving at this realm section are being absorbed into mass formations. This additional absorption of the energy-medium into existing mass congregations is based upon the arrival of the various realm energies into the sphere of absorption of an E~M/X already underway as shown in Fig 2.7.

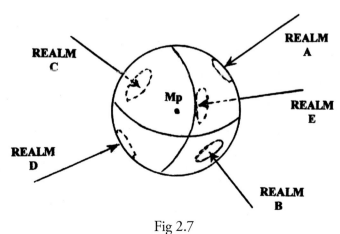

Fig 2.7

Since the realms phase at defined relative velocities it is logical to reason that when a Mp formation or E~M/X is triggered, that a set quantity of energy medium will phase into the sphere of action and be absorbed into the Mp. The quantity of energy medium from each realm that flows into this Mp sphere of action will depend upon the phasing velocities and the individual realm EIL. The E-M/X has an extremely short event sequence duration and therefore the phasing velocity of each realm must be proportionately high to have any effect whatsoever. Individual realm EILs are below C_1 and may therefore be considered as a secondary factor in the above process. An alternative is for the Mp-congregations to acquire linear momentum through the absorption of Mp E-Flows resulting in a sweeping-up action of energy medium absorption while they speed through the realm. This will be dealt with on its own in chapter 8 when we discuss 'velocity effects'.

Although we have mentally observed the EIL rate of change for the energy medium curve of circle 4, of greater importance is the net effect that it has had upon the structure of the realm's contents. For the first time vast quantities of the realm energy medium is permanently encased in proton and giant proton type structures which are to become the basis for all matter within the realm. Although these are similar in basic structure, their effects have a vastly differing impact within the realm.

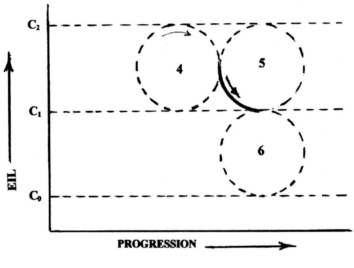

Fig 2.8

Energy Medium Sequence Circle 5

The energy medium curve leaves the circle 4 graphical area at an EIL that is being reduced at a phenomenal rate. However once the energy medium curve approaches the lower EILs indicated for circle 5 the energy absorption into mass bodies diminishes. This is because the spontaneity of the E~M/X is reduced and the rate at which Mp-congregations form is decreased. Below are listed 4 of the factors that effect the realm EIL at this stage:-

1. The energy medium absorption rate into newly created Mps is marginally diminished as the realm EIL is reduced. This results in a slowing down in the rate of EIL drop.
2. The arrival of E-Flows from mass exchanges will continue but also at a reduced quantity in line with Mp formation at the lesser realm EILs.
3. The phasing realms will by now have a progressively reduced energy medium content. Also the smaller energy medium realms may have completed their traverse through the sector being represented and so will cease to contribute to the net EIL. This will have the effect of a gradual but continual EIL decrease.

4. As indicated by Fig 2.6 the mass level of Mps in the proton varies between a maximum at its centre to a minimum at the surface. As the realm EIL drops, the centre is relatively unaffected whereas the surface Mps' mass level diminishes further. The overall mass level of the proton averages out at something less than the optimum level. As such it is clear that a proton in a higher EIL will have a greater mass content than one in a reduced realm EIL. The difference is the energy medium that is returned to the realm as the net EIL drops. The proton does not lose energy medium from decayed Mps but rather through the process of an existing surface Mp adjusting its E~M/X quantity to suit the prevailing realm condition. This mechanism permits the proton to retain a fairly constant mass relative to the EIL situation. As the energy medium curve enters the C_1 zone, these surface Mps will be at minimum mass levels and will find it increasingly difficult to maintain a synchronism with other exchanges. The first indications of a Mp decay on the proton surface are now apparent.

Each of these factors has a changing effect upon the realm EIL. As the energy medium curve circle 5 progresses, each factor will have a greater corresponding effect upon the EIL and its subsequent rate of decrease. As a result it can be seen that the energy medium curve circle 5 represents a slowing down in the rate of EIL drop. In the C_1 zone this EIL drop is practically negligible as the curve is represented by its approach to the horizontal. There is nevertheless a continual decrease in the realm EIL which brings us into the circle 6 portion of the realm energy medium curve.

Let us for a moment look back at a further aspect of conditions in a realm EIL at above C_1. The E-Flows that are received by a proton or other mass congregation get absorbed into the mass structure. There is a balance between the energy medium absorbed and that exuded by the surface Mps. The E-Flow quantities that are absorbed however impart a linear momentum to the Mps formed. The net linear momentum imparted depends upon the net E-Flow quantities absorbed and will be in the net E-Flow direction. As such the Mp-congregation will increase in mass to correspond to the equivalent higher realm EIL and will also have a velocity imparted by the additional mass. When the E-Flow ceases the proton will be in a lowered realm EIL and will adjust its mass level accordingly. This occurs as an E-Flow of surplus energy medium that is not re-absorbed into the surface Mps and results in a further addition to the linear velocity received from the earlier E-Flow. The net result of the whole process has been for the proton to reverse the direction of initial E-Flow and so absorb twice the linear momentum from it. This is visually represented in Fig 2.9.

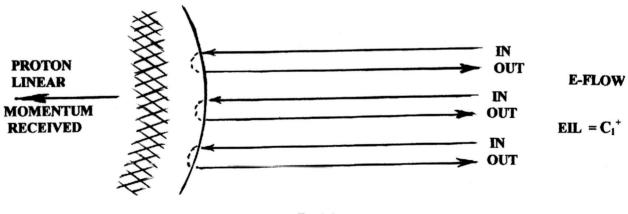

Fig 2.9

The above is based upon the E-Flow (1) absorption mechanism described in the previous chapter (Fig 1.13) and we may thus conclude that a proton in a C_1^+ EIL appears to be repelled away from another E-Flow producing entity. For protons in an EIL environment below C_1 we shall observe a completely different characteristic to this phenomenon as discussed in chapter 4 on gravitation.

Energy Medium Sequence Circle 6.

The energy medium curve depicted by this phase of the seven circle representation is specially important because of its relevance to the phenomena in the realm as observed today. The previous phases of the energy medium curve represented an impressed condition within the realm with the

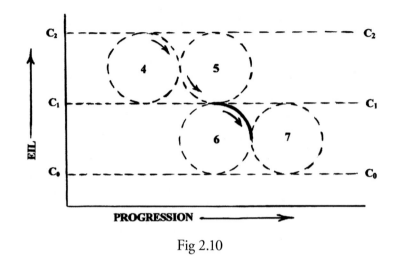

Fig 2.10

resultant high EILs and mass formations. So far all the proton structures had existed at C_1^+ EILs only and behaved as totally isolated entities with an overall event sequence that was nil. One could say that all protons in a realm EIL above C_1 would appear to be in a state of timelessness corresponding to a continuity of status gained at those levels.

The circle 6 energy medium curve however represents a lowering of the realm EIL to below C_1 which subsequently results in a continual decay of the proton mass back into the energy medium (Fig 1.13). The process of proton decay is the essence of all activity in the realm and in itself is the base on which all event duration is measured. Before venturing into a discussion on some of these effects let us for a moment observe the profile of the circle 6 energy medium curve as shown in Fig 2.10. It can be clearly seen in the diagram that the energy medium curve gradually drops below the C_1 zone and into the sub-critical area of realm EIL. Realm phasing is at an advanced stage with a

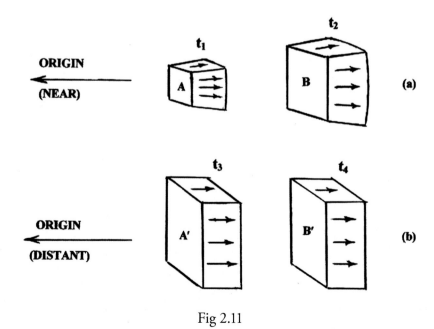

Fig 2.11

52

considerable proportion of the total energy medium being locked inside the mass formations produced earlier. Realm EILs although still additive cannot achieve the previous high levels. This is due to the several factors pointed out earlier although one additional factor now comes into prominence. This factor, which is the expanding effect of each realm as it proceeds away from the Mother Realm, was of little importance when considering the extremely high EILs at peak phasing. Earlier we hypothesised that an energy medium realm is a packet of energy-medium that had separated from a larger entity termed the Mother Realm. As the packet of energy medium (our realm entity) traverses the cosmic void in a direction generally away from the Mother Realm (Fig 2.3) it has the tendency to expand. This expansion is comparable in principle to that in a section of E-Flow from a Mp. The rate of expansion decreases as the packet gets further away from its point of origin as highlighted by the example in Fig 2.11.

At (a) is represented a section of an E-Flow that is relatively close to its Mp source. It is quite apparent that section A at event t_1 will traverse to and occupy the position shown by section B at event t_2. The rate of lateral expansion is high as indicated by the visual volumetric difference between sections A and B. At (b) however is represented another section of the same E-Flow but at a much greater traversed distance from the Mp source. At event t_3 the section A' is only marginally different from the section B' at event t_4 as is conceptually observed. The rate of lateral expansion is considerably smaller. This fact may be summarised as a law in which the rate of lateral expansion of an E-Flow is inversely proportional to the square of the distance from its point of origin. This may be represented by the curve in Fig 1.11 which also applies to the EIL of an E-Flow.

This rule may be applied to our packet of energy medium as it proceeds away from the Mother Realm source. The distances involved are exceedingly great and as such the rate of volumetric expansion of the realm energy medium packet will be small as represented by Fig 2.11(b). You will remember that we assumed earlier in chapter one (for the sake of simplicity) to regard the realms as relatively non-expanding entities. However we now observe that a marginal rate of expansion does occur and until now its effects could be ignored with impunity. Such is no longer the case as the trend of the energy medium curve of circle 6 is influenced by this effect. As the realm expands it follows that the EILs within it must decrease (other factors remaining constant). Consider realm phasing to be nearly at an end. The EIL through much of the realm will be at the sub-critical level. The realm EIL at any point will be the sum of the original realm datum and the E-Flows of all proton decay. Since the realm does have a marginal expansion the realm EIL datum must gradually decrease. This is represented by the very gradual initial drop in the energy medium curve circle 6.

Let us for a moment evaluate the nature of the realm EIL that surrounds a proton in this phase of the energy medium cycle. Firstly the realm EIL datum is decreasing due to its marginal expansion. Secondly, far fewer (if any) realms will be in phasing sequence which will restrict any sudden EIL build-up. Thirdly, there will be no bursts of E-Flow from single Mps since these are non-existent at sub-critical EILs although this effect will be replaced by the more regular aspect of proton decay.

The net result is for a close balance to be maintained initially between these factors. As proton decay progresses, the proton will continue to lose mass and the resulting E-Flows will also be marginally diminished. The energy medium balance alters and the realm EIL will undergo a slightly greater rate of change. The energy medium curve 6 shows this increasing rate of EIL drop as a curve with an increasing downward dip. One must interpret this trend as the result of the gradual but total conversion of all realm mass back to the energy medium form. We shall discuss this aspect later in this section as at this stage it is vital that the reader obtain a concept of the importance and active roles played out during the initial part of the circle 6 energy medium curve representation. There are a myriad of phenomena that result from the proton structural decay at sub-critical realm EILs and which are in observation by us today. These observations and phenomena constitute our realm structure in very real terms and it is the intention here to give the reader a sample of some of the topics that we shall deal with in detail later on.

1. Proton Time. This is based upon the use of the overall event changes or sequence in the mass decay of the proton as a datum or clock from which the duration of all other events can be measured. The comparison of all events to the decay rate of the proton mass makes this measurement of a relative aspect since proton decay in itself is of a variable nature. This measurement of duration takes place automatically in units of Time that are directly related to the unit rate of proton decay. Time is therefore the rate at which events sequence. In realm phasing there was no even rate of event change and hence duration was an incomparable quantity. In proton decay the change of mass status is recognised as an event sequence that occurs at a uniform and consistent rate for a particular realm EIL. The proton event sequence is thus a base against which other events may be compared for establishing relative duration. This base thus used as an automatic event duration measurement device is termed as Proton Time. Proton time can be compared to pendulum time except the oscillation period is that of the Mp repetitive sequence. As the proton event sequence slows so will Proton Time and vice-versa. In order to attain an absolute sequence grading system one must establish a datum for the rate of decay of the proton. This is possible when we can specify set conditions around a proton such that a defined decay rate / Proton Time occurs. This can then be the datum rate of proton decay / Time upon which to base all duration measurements. This has been developed and presented in detail in the subject matter of chapter 5. The curious reader may find an inspection of that chapter of interest at this stage.

2. Gravity. In Fig 2.9 we showed the process by which linear momentum was absorbed by the proton surface Mps in a C_1^+ EIL thus causing it to acquire a motion away from the E-Flow source. When a proton is in a sub-critical EIL it is in decay and its response is quite different. In chapter 1 an explanation relating to the E-Flow(l) of Fig 1.13 was detailed. It was shown how the E-Flow(l) from all M-E/Xs at the proton surface resulted in an inward direction linear momentum of all Mps below the surface. Because of an evenly distributed pattern of M-E/X s over the entire proton surface the net result of all the E-Flow(l) linear momenta was to neutralise one another but at the expense of a compressive action towards the proton centre. The push by each Mp towards the proton centre being the result of the linear momentum absorbed during its E-M/Xs. This push thus being the basis of all push action (termed Force) in the realm. Since the push action is in balance there is no change in the overall linear momentum of the proton as a whole. That is assuming the proton to be in an area of uniform EIL all the surface Mps are thus subject to the same EIL conditions and so behave exactly alike.

Let us now alter this EIL uniformity by subjecting the proton to an E-Flow as shown in Fig 2.12.

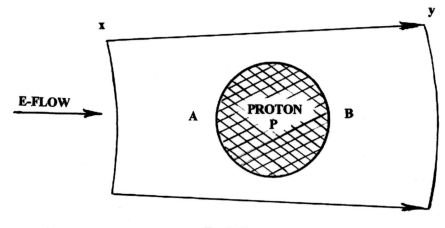

Fig 2.12

An E-Flow is an expanding energy medium entity and as such the EIL at a point in it is inversely proportional to the square of the distance from the source. For the moment let us ignore any effects of linear momentum absorption by the proton from this E-Flow. Thus when the proton P in the diagram finds itself within an E-Flow xy, the upstream A-side of the proton will be at a marginally higher EIL than the opposite B-side. The surface Mps at A will be in a region of higher EIL than those at B. Now the rate at which M-E/Xs occur at the surface of the proton depend upon the prevailing realm EIL. The higher the EIL the slower are the cyclic Mp sequence exchanges. In chapter 1 it was shown that the Fig 1.13 E-Flow(3) resulted in an energy medium grid around the proton that regulated and maintained a precisely distributed pattern of surface energy medium exchanges. The higher the energy medium grid intensity the slower the exchange action. The grid intensity must be regulated to a uniform level all around the proton. When an E-Flow is brought into play the energy medium grid intensity at one side 'A' is momentarily higher than the other side 'B'. The surface Mps immediately respond to the new EILs and a re-distribution of Mp duration is set up that equates to the changed conditions. The energy medium grid around the proton thus alters to suit the new conditions. This means that there are fewer Mps occurring on the A side of the proton than there are on the B side in a set period of proton time. This also means that there are a greater number of E-Flow(1)s on the B side of the proton than there are on the A side in a set period of

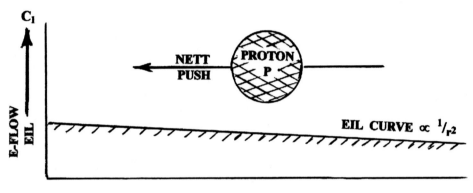

Fig 2.13

time. As such there will be an imbalance in the linear momenta of Mps pushing towards the proton centre. The net difference results in the proton as a whole acquiring a linear momentum in the direction B to A. This net pushing action upon the proton as a whole is observed to occur towards the E-Flow source and hence up the EIL gradient as shown in Fig 2.13. The greater the EIL gradient the greater is the difference in the Mps repetitive sequence rates on the opposite surfaces of the proton and hence the greater the directional push upon the proton as a whole. For the moment we shall ignore the effect of absorption of the linear momentum of the E-Flow.

In a sub-critical realm area an EIL gradient can be the consequence of several effects. The realm itself possesses a varying EIL across its expanse and hence will have a natural EIL gradient between any two points. It also has a finite though expanding boundary which may be logically assumed to be at EILs well below those more near its central regions. We may assume therefore that an overall EIL gradient for the realm is generally towards its centre. Protons will thus tend to remain within the confines of a realm. We shall discuss this aspect again in a later section.

Alternatively an EIL gradient is set up by the proton decay E-Flow. The EIL within that E-Flow decreases with distance according to the inverse square law. A proton that contacts such an E-Flow thus experiences the push that forces it towards the higher EIL. Since all protons are in decay then each produces an E-Flow that eventually will reach the others. If we may consider just two protons in proximity such that each is in the E-Flow path of the other, then we will observe that each receives a push up the EIL gradient in the direction of the other E-Flow source. This push that one proton experiences as the result of the E-Flow from the other is such as to force them towards each other and makes it appear as though an attraction exists between them. It must be understood that although each is influenced by the E-Flow EIL gradient of the other, the push action itself is independently contained within each proton body. We shall discuss this phenomenon in detail in chapter 4 on the gravitation effect.

3. Forcephoidal effects. In chapter 1 Fig 1.18 it was indicated that a series of peaked EIL spherical shells existed around the proton as a result of its decay E-Flows. The peak EILs were observed to diminish with distance from the proton surface in accordance with the inverse square law. These spherical EIL peaks travel away from the proton as part of the natural decay E-Flow and relate to the pattern of Mp* distribution on the proton surface. These peaked EIL spheres are termed as forcephoids and are to be found around all protons in decay. Since the EIL gradient in a realm area determines the pattern of Mp* on the proton surface then it follows that the forcephoids are not necessarily spherical in shape nor with the proton as their exact centre as shown by the configuration in Fig 2.14(b). We shall discuss these aspects again later on in another section. For

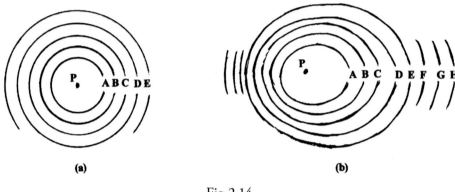

Fig 2.14

now let us obtain a graphical concept of the EIL profile of the forcephoidal peak. This is represented in Fig 2.15 for an E-Flow direction as shown.

Although the forcephoid traverses along with the E-Flow it may be considered as a semi permanent feature around the proton. This is because at a set radial distance from the proton the forcephoid peak is always the same. The spacing between forcephoids is small compared to the radial distance from source and so the space left by one is soon followed by another. In the Fig 2.15 forcephoid spacing is indicated by 'l'. When peak B traverses to the position D another forcephoid will have taken up B's former position and the set up will appear to be unchanged. We shall see later how this apparent forcephoid EIL permanence is basic to structuring a Mp shell around a proton to create a primary atom.

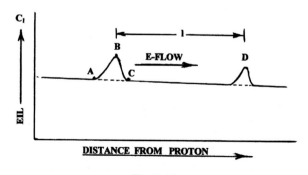

Fig 2.15

It is important to note the varying EIL gradients in and around the forcephoid. The EIL at 'A' is slightly higher than that at 'C' because of the inverse square law of diminishing EIL for an E-Flow. The EIL difference between peak point B and C is thus greater than that between B and A at any instant. Let us now see how a proton behaves when such a forcephoid phases through it. Let us assume that the forcephoid width AC is marginally greater than the proton diameter d as represented in Fig 2.16. Let us assume also that the proton is at absolute rest in a sub-critical realm EIL and that a single forcephoid approaches at velocity C_v from the left hand side. As the forcephoid side BC reaches the proton the steep EIL gradient must be sensed by the surface Mps. The push up the EIL gradient will occur within an event sequence and at the moment that the push actually develops the forcephoid will have traversed to a total envelopment of the proton. Nevertheless a delayed push upon the proton will occur in the direction C to A.

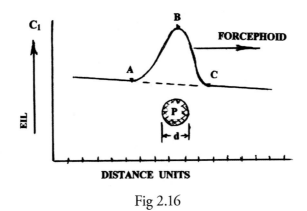

Fig 2.16

In Fig 2.7 we indicated how energy medium from a number of phasing realm quantities are partially absorbed into an E-M/X. The quantity of energy medium from each realm that flows into the E-M/X sphere of action depends upon the individual realm EIL and phasing velocity. The short event sequence duration of the E-M/X requires that for any additional energy medium absorption the phasing velocity must be very high. Now when a proton suddenly enters a higher, albeit sub-critical EIL that is nearly uniform, then the grid pattern of the surface Mps adjust to the new EIL. Mps that would have resulted in a one way M-E/X may now be able to reform into Mps. The additional energy medium that raises the proton surface energy medium grid to allow this to happen has a linear momentum in the direction of forcephoid traverse i.e. A to C. But the energy medium grid is self regulating and hence re-distributes the E-Flow(3) Mps which become spread further apart on the proton surface. The net difference in the proton before and after entering the forcephoid is for a marginal reduction of Mps at the surface. This means that there are more surface Mps in repetitive energy medium exchanges in the latter case which implies that some energy medium from the forcephoid peak is absorbed. In fact this is the case and it is apparent that all energy medium absorbed from the forcephoid must impart linear momentum to the proton body. At the relatively lower E-Flow EILs this imparted linear momentum is small compared to the surface Mp imbalance and so was not considered in the previous analysis. The proton within the forcephoid region however receives a much greater proportion of energy medium absorption and so acquires a linear momentum in the same direction as the forcephoid traverse. That is a push in the direction A to C. In chapter 8 we shall be discussing the case of a proton speeding through the realm and of its 'picking up' some of the energy medium in its path. Although the proton would still be in general decay the energies absorbed at its lead side would tend to slow it down along with its rate of decay. Fortunately, this also sets up a surface Mp imbalance with fewer Mps at the lead side as compared to the trail side. The result is a push in the direction of traverse. A balance in the requirement of these two effects is in mutual regulation and so no change of proton velocity occurs. However there is now a greater compressive action within the proton body which directly affects its moveability by an external force. The 'inertness' of the proton has simply been increased by its velocity through the realm.

Now in Fig 2.16 as the forcephoid traverse continues, the proton encounters the EIL gradient of side AB and by the differential Mp* process experiences a push in the direction A to C. This is in complete opposition to the earlier push caused by the EIL gradient BC. The combined result of these three effects is that the proton experiences a net push in the direction of forcephoid traverse

i.e. A to C. Since the forcephoid originates from another proton we may subsequently assume that protons must push away from each other. At larger distances the EIL peaks of forcephoids becomes insignificant and the gravitational type push caused by the E-Flow EIL gradient overrides. It must be noted that forcephoid widths and spacing as hypothesized above are extremely small albeit variable. We shall discuss these aspects more fully in chapter 7 on Magnetic and Electric type forces within the realm.

4. Atomic structure. In Fig 1.18 was shown a graphical representation of the relative positions of successive forcephoidal peaks traversing away from the proton body. These forcephoids must peak to very high EILs when close to the proton and diminish with distance according to the inverse square law (for EIL). When a large number of protons occur in a set area then the E-Flows and forcephoids from each must phase with one another. This would be similar in principle to realm phasing except that in this case the energy medium quantities originate from the decaying protons. At some defined maximum distance from the proton surface its forcephoid will phase with incoming E-Flows and other forcephoidal peaks and possibly attain an EIL that approaches C_1. Beyond this distance the maximum peak EIL would have to be sub-critical. At the present stage let us concern ourselves only with the achievement of peak EILs approaching C_1. Conditions within this forcephoidal peak are then conducive for the formation of Mps.

In the Fig 2.17 the decaying proton P_1 is shown with a forcephoid at the radial position R_1 from its surface. This forcephoid traverses radially away from P_1 and is bound to phase with the E-Flow and forcephoids from the neighbouring protons P_2 and P_3. At some defined distance from P_1 the phasing of all these energy medium peaks and E-Flows will result in an EIL build up that approaches C_1.

Let us denote this distance as the R_2 forcephoidal position. One would imagine that when the phasing approaches the R_2 forcephoidal position that a peak EIL even greater should be attained.

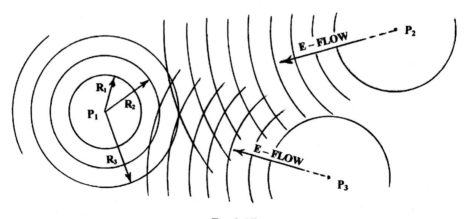

Fig 2.17

This is not so as there occurs an E-M/X phenomena at the R_2 forcephoidal position which reduces the quantities of energy medium reaching R_1 from P_2 and P_3. The forcephoid R_2 EIL as such is graphically represented in Fig 2.18.

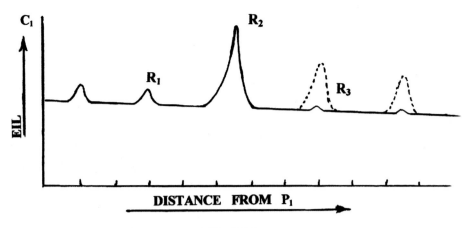

Fig 2.18

It will be noticed that no forcephoids of any consequence are indicated after the radial distance R_2 although these should logically have been as represented by the imaginary dotted peaks. The reason for this is that once a forcephoid achieves a C_1 zone EIL, by whatever energy medium is available, then E-M/Xs will occur within that peak. The quantities of Mps that form depend upon the total energy-medium concentrated at the forcephoidal peak. In Fig 2.17 for the sake of simplicity and ease of explanation we indicated the interaction between a limited number of protons. The protons that were formed within the realm during the circle 5 phase of the energy medium curve would be virtually countless. It would also be logical to assume that they would be fairly evenly dispersed within the realm. As such proton P_1 would be surrounded by a fairly even distribution of protons on all sides. The energies from the forcephoids and E-Flows of all these would therefore result in the R_2 position forcephoid from proton P_1 to be at a consistent high EIL over its entirety. Thus the Mps that form within the R_2 forcephoid would totally surround proton P_1. It would appear that for an instant there exists a shell of Mps around the proton P_1.

As mentioned earlier a Mp is a phase in the energy medium - mass medium exchange process and is extremely short lived (10^{-43} seconds in Planck time). As soon as the Mp-shell forms, the forcephoid R_2 becomes non-existent due to all its energy medium being absorbed into mass. There is thus no forcephoid to traverse to the outward position R_3 as represented in Fig 2.18. Also, since much of the energy medium has been absorbed at the R_2 position the forcephoid at R_1 remains very nearly at its normal decay level.

Let us now observe the events that affect the Mp-shell. The linear momenta of these Mps is the resultant of all the energy medium momenta absorbed. It would be logical to assume that at the R_2 position the energy medium from external sources greatly exceeds that within the proton P_1 forcephoid R_2. The net momenta of the Mp-shell must therefore be such as to cause it to contract towards P_1. Since these Mps are now at a sub-critical EIL then an early M-E/X must occur. The M-E/Xs that follow send out an E-Flow that phases with the next forcephoid R_1 that approaches. The spacing between the forcephoids as represented by 'l' in Fig 2.15 is of major importance in order that E-M/Xs can repeat. The forcephoid traverses out from P_1 at a set velocity C_v. The energy medium exchange cycle occurs within a set event sequence. The E-Flow from the Mp-shell therefore occurs at a defined moment. This E-Flow must phase with the next approaching forcephoid R_1 at an earliest in order that a C_1 EIL can be attained. This can only occur if the forcephoid spacing is small. This phasing will occur at a radial position somewhere between the former R_1 and R_2 (Fig 2.18) locations. Subsequently E-M/Xs will occur and a new Mp-shell comes into being around proton P_1 at a slightly closer-in radial position. As before this Mp-shell acquires a linear momentum that is the resultant of the absorbed energies. There is now a marginally higher proportion of the forcephoid energy medium making up the mass by virtue of the higher normal forcephoid peaks closer-in to the proton. Eventually the Mp-shell may form much closer-in to the proton P_1 and the considerably higher forcephoid peaks would contribute a larger proportion of energy medium to the masses. As such the shell will acquire a net outward linear momentum and so cause it to expand away from P_1. We can thus conclude that an Mp-shell will regenerate in a succession of E-M/Xs and M-E/Xs and will stabilise at a defined radial distance from the proton centre. This radial distance will depend upon the EIL conditions of that local realm area. The higher the EIL conditions then the greater will be the radial distance at which the Mp-shell stabilises and vice versa.

In the circle 6 energy medium curve most protons in the realm will encounter E-Flows from other neighbouring protons and many will acquire Mp-shells of their own. Once formed, this regenerating Mp-shell becomes a near permanent feature around the decaying proton. Also there are no forcephoids outside this shell as shown in Fig 2.18 and as such there will be no 'forcephoidal effects' as explained for the previous section. Although the proton is basic to the mass of the realm, the proton plus Mp-shell is however to be considered as the unitary structure from which combinations and complicated mass/shell assemblies will develop. This unitary structure is subsequently to be termed as the Primary atom and will be discussed at length in Chapter 6. We shall also then be dealing with some of the more complicated multi-proton nucleii and shell variations that account for the differing mass structures observed in and around our world today.

5. Galaxy formation. These have their origin in the very large Mp-congregations that formed near the peak of the circle 4 phase of our universe cycle. They were the result of the progressively increasing EILs with the resultant adding of Mps to the Mp-congregations that formed early in the circle 4 phase. There has to be a limit to the size of these Mp-congregations which would have subsequently absorbed a very large proportion of the realm energy-medium that existed initially. These large protons (for that is basically what they are) averaging in size close to the volume of our current solar system, became widely distributed across the realm zone. Let us refer to them as 'proton giants'. They remain independent and separate from one another. This is because they simply behave as our current ordinary protons except for their relative shear size and the very intense E-Flows emitted by them. Once the realm energy medium cycle enters the circle 6 phase the general realm EIL reduces to below C_1. However there is a zone around the proton giant that is always at or above C_1 EIL. One of the reasons for these protons giants to remain separate entities is because they repel each other at close distances even though they induce gravitational attraction for the farther out mass entities. This is represented in the graphical relationship shown in Fig 4.24 in chapter 4. The attraction factor is due to the natural realm grid EIL gradient while the repulsion factor results from the absorption of linear momentum of the very high EIL of the outward flowing energy medium of their E-Flows.

As the realm enters the circle 6 phase the realm EILs reduce to below C_1 and our current universe status prevails. EIL oscillations or light-waves can now begin to propagate (as explained in the next section) and the primary atom comes into existence with its Mp-shell surround. Light-waves can only operate in a realm zone that is below C_1 since they are essentially a plane of perfectly synchronised Mps that quantum-step in a set direction (10^{-35} metres approx.) and averaging a set velocity (3×10^8 metres/sec approx). Consider one such proton giant and assume it to be approximately the volume of our solar system. It is pure mass with countless Mps in a repetitive cyclic sequence. The E-Flow emitted by this proton giant is intense and within a surface zone around it (like the atmosphere is around the earth) the EILs are at C_1 and C_1^+ levels. The limit of this zone, being perhaps a few million miles, can be referred to as the 'event horizon' since light-waves dysfunction within this zone. Any Mp-congregations large or small that are formed within this zone will possess high linear momentum in the outward direction because of the high intensity outward E-Flows and so will be expelled from the zone. It is perhaps within this 'event horizon' zone that many of the hundreds of the hadron and lepton particles discovered in the laboratories get to be manufactured. Some of these would form close to the mass body and so evolve into larger Mp-configurations before they are expelled out through the millions of miles of event horizon and inter the open limits of outer space. Others would be formed nearer the outer limits of this event horizon and therefore have lesser time to construct before being expelled and so have a different/lighter quality of structure and perhaps even rather unusual in character. These are issues I shall leave to the interested reader to pursue. Outside the event horizon there is still a massive outpouring of energy medium E-Flows and repulsion effects (Fig 4.24) would dominate any other protons and primary atoms. At greater distances the E-Flow intensities are diminished and along with intersecting E-Flows from other bodies result in a gravitational attractive effect dominating. We shall discuss the setting up of a 'realm grid' in the next section. As primary atoms congregate to evolve into stars of varying masses and which follow their burning cycles (to form the more complex atoms) they would be attracted towards one or another of these proton giants. They would be drawn into an orbit around it and we would assume the orbital path to follow Newton's laws as for planetary motion in our solar system. This would in fact be incorrect. As we get closer in towards the proton giant there is an increasing component of repulsion action (Fig 4.24 again) causing a diminishment in the gravitational attraction component. As such the stable orbital paths need to be comparatively slower and are actually seen to match the velocity of the stars that are farther out. We shall refer to this as the 'cartwheel effect' of the rotation of stars around the proton giant galactic core. In the early age of galaxy formation there would have been many collisions of stars that were being drawn into orbit, some perhaps even orbiting in opposite directions. This would have been a turbulent period but in time stability is achieved as we see today. The cartwheel effect has actually been observed when scientists looked at distant spiral galaxies that happen to be orientated edge on to our view in order to measure their rotational velocities. Modern spectroscopic techniques are so sensitive that they can measure the Doppler shifts at distances out from the centre of the tiny image of the distant galaxy to give velocity measurements across its disc. These measurements have also been extended to measure the velocities of clouds of hydrogen gas (primary atoms) which form part of the disc of the distant galaxy using radio astronomy techniques. The results were a surprise to the scientist community. In almost every case it was found that the speed with which the stars are moving is the same all the way across the disc as far as measurements can be taken. It was confusing because it was assumed that stars farther out from the galactic centre should be moving more slowly in their orbits in exactly the same way that the outer planets of our solar system move more slowly than the inner planets. The explanation given above for the cartwheel effect is thus verified.

Let us now consider how close in towards the proton giant's surface is it possible for a star to progress. At some distance from it the gravitational attraction may be exactly balanced by the repulsion factor and as such the star does not require to orbit around it to maintain a stable distance. The star will simply float at this equilibrium position. However due to the attraction from other orbiting stars farther out, this stationary star will be drawn to follow a path around the proton giant in the same direction as the general galactic rotation. In doing so it will progress to an orbit farther out till it once again achieves an equilibrium condition.

Since our realm entered the circle 6 phase a considerable period of 'event-duration' has elapsed and most proton giants have evolved into these galaxies. The fact that there may be a few orphan proton giants lurking in the realm regions is no cause for concern as we just couldn't fall into one if we tried. Proton giants with or without their entourage of stars seem to be uniformly dispersed throughout the realm volume. Currently we see only a small proportion of it but we can give the reader an idea of this distribution in an exemplified geographical scale representation.

Consider our own Milky Way galaxy as the centre of our realm area. In size it is approximately 100,000 light-years in diameter and contains in the region of 10^{11} stars that make a complete revolution once every 200 million years. Now if we represent our milky way galaxy by an aspirin then our nearest galaxy Andromeda (M31) is only another aspirin 13 centimetres away. This plus a few other galaxies form our local group of galaxies. The next similar group of galaxies, the Sculptor group is then 60 centimetres away. Now the Virgo cluster, a huge collection of about 200 galaxies spread over the size of a basket ball, is only 3 metres away. The Virgo cluster is the centre of a loose swarm of galaxies called the local Super-cluster and of which the local group (us) and sculptor group are a part. On this scale just 20 metres away is another Super-cluster, the Coma cluster containing thousands of galaxies. Farther out there are even larger clusters of galaxies some 20 metres across. There are galaxies in clusters spread out in every direction from us and as far as we are able to view they are uniformly distributed across the sky. The brightest quasar in the night sky (30273) is on this scale only 130 metres distant. (Astronomers today believe there are perhaps 140 billion galaxies in the visible universe). Everything we can see (the entire visible universe) can be contained in a sphere roughly one kilometre across, on the scale where an aspirin represents our milky way galaxy As to the size of the realm one is currently unable to even estimate an opinion. Perhaps on this same scale it is the size of our currently visible universe! And so it all originated during the intense activity of the circle 4 phase of the realm energy medium cycle when EILs approached C_2.

6. The Realm EIL Grid. Let us suppose that in the circle-6 phase the mass in the universe has attained its distribution as indicated above. There is a general uniformity of galaxy distribution across the realm. Now in the circle 6 phase all mass is in a realm EIL that is below C_1 and as such will be in a state of general decay. This means that all mass entities emit energy medium in the form of an E-Flow in a continuous controlled and even manner. These E-Flows spread outwards in all directions from each mass and as such pervade the universe in time. Consider the E-Flows of all these masses spreading out and intersecting/phasing with one another. Since each mass is relatively located and does not alter location suddenly, then their E-Flows will also maintain a similar constancy. At any selected inter-galactic location all these phasing E-Flows will result in a defined EIL buildup. Different locations will attain differing EILs and these levels will remain static in the short term for each location. These points in space will thus have a characteristic EIL. Points in space at slightly different EILs will thus have an EIL gradient between them. If we draw successive imaginary lines or planes connecting all points that are at similar EILs then we should end up with a spatial grid pattern showing zonal EILs and their EIL gradients. If we were to perform this task for the entire realm then we would have established an interim Realm EIL Grid. Perhaps Einstein would have liked to have known of this spatial grid when he propounded his theory that all space was curved and that mass objects were caused to 'fall' in towards space dips caused by this curvature. We shall discuss this further in chapter 4 on gravitation.

7. Realm EIL Oscillations. Let us conduct a thought experiment on the energy medium exchanges of the Mp phenomena. Consider a single Mp to be contained within a restricted volume. Suppose that a barrier is set up around this volume that prevents energy medium from flowing past it. Thus no energy medium will be able to enter or exit from our restricted space. Now consider the Mp in this restricted space to be surrounded by realm energy medium at the circle 6 sub-critical datum EIL. By the process explained earlier the Mp will revert to energy medium as an E-Flow in the outward direction. Since no energy medium will be permitted to leave our restricted area then the EIL must increase to accommodate the additional energy medium. By defining the volume of this restricted space we can ensure that the EIL within it approaches C_1. The energy-medium in our thought experiment restricted space now attains C_1 EIL and hence converts to a mass-medium. This behaves as a highly stretch strained medium that immediately undergoes an implosive contraction. For details of this phenomenon see Fig 4.2. At its centre a Mp is created which having been overly compressed from the implosive action reacts rapidly and explodes to result in an outward E-Flow in all directions. The life of the Mp is estimated at the Planck time of 10^{-43} secs. As such this zone will continue in a hypothetically never-ending succession of E~M/Xs and M~E/Xs. Similarly the EIL in our thought experiment restricted space will alternate between an EIL approaching C_1 and a low EIL approaching zero. In reality however the picture is slightly different. Although the EIL oscillation does occur just like in our thought experiment, it does so at different locations each time. A single Mp on its own will undergo a M~E/X converting to energy medium in the form of an E-Flow and that will be an end to the oscillation. However if a whole plane of Mps (like in a Mp-Shell) with all its Mps in perfect synchronism were to undergo M-E/Xs at the same instant, then their E-Flows would intersect/phase at a forward position with the probability of attaining C_1 EIL. (See Fig 3.4). This would then result in E-M/Xs occurring all along the forward plane causing new Mps to form. This is then repeated progressively for the next forward plane position. In essence this stepping action of the EIL oscillations is the principle of light-wave propagation. Each step is referred to as a quantum-step equivalent to the Planck length of 10^{-35} metres and each successive Mp plane exists for the Planck time of 10^{-43} secs. The forward direction is the outward direction of the E-Flows from the atom nucleus. In general the wave-front will continue in one direction with a velocity that has been measured at 3×10^8 metres per sec in realm space. It is interesting to know that the E-Flow velocity may be deduced with the appropriate assumptions and has got to be somewhat greater than the wave front velocity V_C. Also it now becomes easier to comprehend the dual nature of the wave front as both particle and wave.

In order to further understand the concept of these travelling oscillations of wave-fronts it may be worthwhile to examine the analogy of water-wave propagation on the surface of a pond.

Imagine that a small pebble is dropped gently onto the still surface. The result is a minor splash at the point of entry and then a succession of waves (peaks and troughs) that 'appear' to traverse away across the surface at a constant velocity. Actually there is no flow of water in the wave direction as proven by the floating cork experiment. If one concentrates upon the cork it will be seen that when the water wave traverses, the cork simply bounces up and down. What really occurs is simply a bouncing action of the water surface at that point. The cork is alternately in a trough or on a crest of an oscillating pond surface. Yet why should this 'surface bounce' spread outwards from the impact point and at a set velocity?

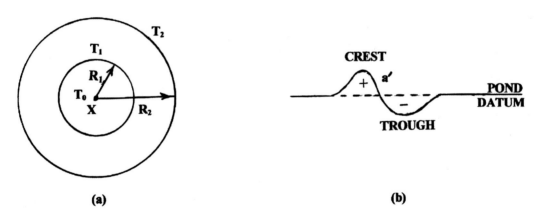

Fig 2.19

Now each surface peak is above the original pond datum level and therefore must contain some of the pebble momentum. However, as the crest traverses outward from the point of pebble impact it does so in the form of an increasingly larger wave circle. As the circle gets larger and the wave crest not much lower it would seem that the wave front acquires additional crest volume.

In Fig 2.19(a) the origin of the wave front is shown by the pebble impact point at X. Imagine the pebble to strike the pond surface at event sequence T_0. Assume that at the event moment T_1, the wave crest has traversed to the radial position R_1. If the cross-sectional area of the crest is a' then the total volume of water in the circular wave crest is given by,

$$T_1 \text{ Crest Volume} = 2\pi R_1 a_1'$$

Similarly the total volume of water in the circular wave crest at event moment T_2 and radial position R_2 is given by,

$$T_2 \text{ Crest Volume} = 2\,\pi\,R_2\,a_2{}'$$

In order that the crest energies at R_1 and R_2 positions be equal then the two crest volumes must also be equal. That is,

$$2\,\pi\,R_1\,a_1{}' = 2\,\pi\,R_2\,a_2{}'$$

let us assume that R_2 is twice the radial distance R_1, so that the above equation simplifies to,

$$a_1{}' = \square\; a_2{}'$$

This does not however relate to the observed phenomena which in fact shows a negligible change in the crest sectional area over such distances. Somehow additional energy is being absorbed by the crest as it traverses across the pond. This energy must therefore be obtained at the cost of some other source that must give up a proportional energy quantity. This transfer of energy takes places at the front of the travelling crest and results in a dip that is below the original pond datum. This dip develops as a trough that precedes the traversing peak. By virtue of this sharing of energies these are inter-dependent and must therefore always occur as a pair.

Let us now work out exactly how this energy transfer is brought about. There are two factors that hold the energy quantity of the water particles of the pond. These are the potential energy of the water by virtue of its relative position in the earth's gravitational pull and the surface tension (skin/diaphragm effect) of the pond surface. This surface tension acts like a stretched elastic skin which if displaced from the normal sets up a returning force. When the pebble drops into the pond it displaces a volume of water at its point of entry as shown in Fig 2.20(b).

The quantity of water displaced at X must re-locate elsewhere. This occurs as a movement of the displaced water particles into the area A and results in the water particles at A being displaced upwards above the normal pond datum. This upward movement of water must in its turn displace a similar volume of the air particles that are at the pond surface. The upward displacement of water particles at A occurs as a sequence that develops outwards from the pebble surface. The momentum

from the pebble's fall accelerates water particles in a lateral direction which then act like wedges to displace and accelerate water particles upwards as depicted in Fig 2.21.

The wave crest then develops as a successive lifting of water particles above pond datum. This uplift progresses outwards from the pebble and so gives the appearance of movement. The air particles that are subsequently displaced get compressed into the space ahead of the uplift and so result in that section of the pond surface being at a marginally higher air pressure. The water surface absorbs this pressure differential at that point by accelerating some of its water particles downwards which in turn must displace other particles laterally and by the "wedge principle" upwards (at an offset position of course). Subsequently a dip in the pond surface develops ahead of the water peak which also gives the appearance of movement.

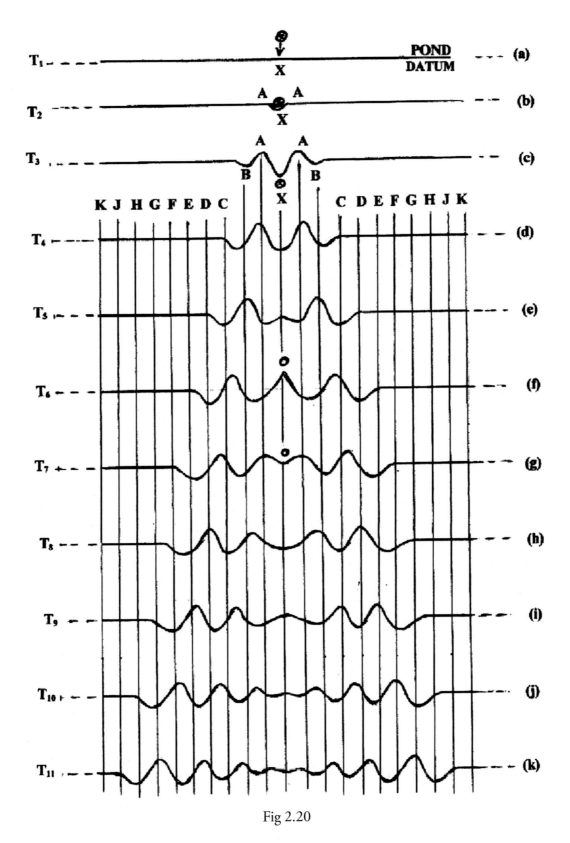

Fig 2.20

Now while the wedge principle for the peak acts in the outward direction, that for the dip acts inwards. Thus additional energy is pumped up into a progressively increasing wave peak as shown in Fig 2.22. When the surface of the water is displaced as either a prominence or a depression then the skin effect comes into play. This means that the surface acts as a stretched elastic diaphragm that sets up a returning force directly proportional to its displacement. The energy given up by the

65

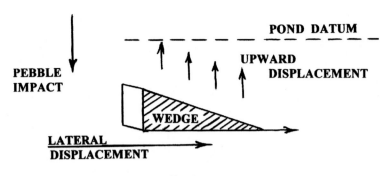

Fig 2.21

pebble has therefore not only raised a quantity of water against the earth's gravitational pull but it has also had to do so against an increasing downward force from the stretched skin. The same applies to the dip which receives a similar but opposing force upwards. When an equilibrium is

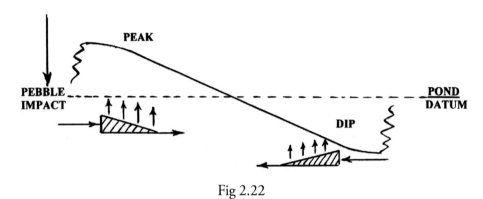

Fig 2.22

reached the peak ceases to enlarge as does also the dip. However, once the displacement ceases so does the wedge effect under it. The force due to the skin effect is still very much in action and this subsequently accelerates the particles of water back towards the pond datum level where the force ceases. Due to the inertia of motion the water particles continue beyond this level and result in a dip instead. This action would continue indefinitely as a bouncing water surface were it not for the internal friction in the water and air particles. In Fig 2.20 we have shown a sequence of this bouncing action caused by the initial displacement by a pebble at X and the resultant wave progression. The sequence shows how the oscillation at X is soon damped while the original Peak and Trough, although further out, appear virtually unchanged.

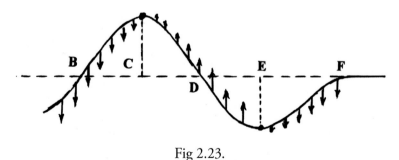

Fig 2.23.

In Fig 2.23 is indicated the direction of motion of this surface bounce at each point on the Peak -Trough combine. It is clear that each point on the water skin achieves a maximum displacement in each vertical direction that occurs successively with distance from epi-centre. The energy that maintains the peak despite its widening circumference is obtained from the trough that precedes it. It is worth mentioning that the Trough being on the outer side of the Peak must necessarily have the larger circumference. As such the Peak and Trough sectional profiles will be markedly different although each is part of the same unit.

From the view of a person looking at a fixed dot on the pond surface all that he will observe is a rise and fall at that point. For a person looking at a section of the wave he will observe a cyclic rolling action between the Peak and Trough. As explained the skin effect plays a major role in the surface level oscillation. A layer of oil on the surface destroys this skin effect and virtually ends wave propagation which bears out the fact that the energy imparted by the pebble at the epi-centre X does not support the wave entirely. An approaching tidal wave announces itself by the considerable drop in sea level before it strikes. Such waves have been known to travel thousands of miles building up an overall frontal power far in excess of the original energy expended in its formation. This is because the positive (crest) volume of the wave wreaks the havoc while the negative (trough) volume seems to pass unnoticed.

It is of paramount importance that the reader develop a working familiarity with water wave propagation in the manner just described. It is not intended to make any comparisons between EIL 'bounce' and the water wave. It is however intended to use the water wave sectional profile as a dynamic graph in the representation of the EIL oscillation / light-wave propagation within the realm. Although the next chapter deals with this topic at great length it is essential that an introduction be provided here as an aid to the conceptual visualisation of this realm characteristic. There can be no substitute for a mental model of this activity in the active circle 6 phase of our Realm.

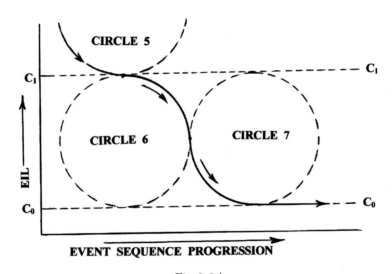

Fig 2.24

8. Radiation Background. In the Circle 5 phase the EIL is above C_1, but as it shifts down into the Circle 6 phase the EIL decreases to below C_1. Independent Mps will no longer form and those Mp-congregations that are very small, a fraction of proton size, will each eventually devolve into a series of E-Flow and spherical wave-front. This takes place across the entire realm zone that has just entered the Circle 6 phase. It is like a widespread crackling of mini-explosions that occur only once per particle. These now defunct particles would have been distributed widely and uniformly such that at any one observer location in the realm each wave-front / EIL oscillation will arrive from a progressively farther out spherical distance. It doesn't matter where in the realm the observer is located, the observations will be the same. As we progress well into the Circle 6 phase and time/event duration has now become a measurable quantity, the intensity and clarity of these EIL oscillations will become ever more faint. They will simply become a background effect to all the other phenomena occurring within the realm. By working back we should be able to establish the time duration when the Circle 6 phase actually came into effect. In essence our universe as we know it only commenced when the Big Crackle occurred at the start of Circle 6. This is when our Time began although much had been structured during the Circles 3, 4, and 5 phases as described earlier.

Let us return to the aspects of the energy medium curve circle 6 which in the early stages follows a trace that is very close to but below C_1. As the energy medium curve trace continues, the EIL progressively drops further and further from the C_1 zone as shown in Fig 2.24.

The initial part of the energy medium curve remains relatively close to the C_1 line indicating a very high realm EIL datum. Proton shell structures form easily in these conditions and the pace of the energy medium exchange sequence would be comparatively slow. Proton decay and subsequently Proton Time as defined earlier would also occur at a correspondingly slower event rate. EIL oscillations, which are basically a part of the energy medium exchange process, would also effect at the corresponding slower pace.

However, as the energy medium curve progresses the realm EIL drops away from the C_1 EIL line at an ever increasing rate of decline as indicated by the droop characteristic of the Circle 6 curve. The lower realm EIL results in an increase in the decay rate of the proton mass. Subsequently proton time and EIL oscillations are speeded up. Proton shells would also be effected as to position around the proton but this is complicated by other factors and so will be left for detailed discussion in a later chapter.

As proton decay continues the mass of the proton becomes gradually reduced. Subsequently, although the decay rate has increased at the lower realm datum, the E-Flows resulting therefrom must reduce with the diminishing proton mass. As mentioned earlier the realm EIL datum is the sum of the prevailing EIL conditions in that particular part of the realm. A reduction in any of these EIL factors will inevitably result in a reduced realm datum. Also since these factors are inter-dependent then each would affect the others. The end result is that the energy medium curve develops a droop characteristic for an increasing rate of EIL drop.

At some stage in the realm datum condition the EIL will be insufficient for the mass point exchanges in the EIL oscillation to operate beyond relatively short distances from the originating Mp-shell. The propagation of EIL oscillations will subsequently cease as a characteristic within the realm. The reduction in the EIL datum in conjunction with the much reduced proton masses will result in the gradual extinction of the Mp-shell around the proton for the more basic atom combinations. Finally, at the position on the graphical representation where the energy medium curve of circle 6 meets the energy curve of circle 7, we may assume that proton decay is total. The inter-play is quite complex especially where proton giants are concerned and the overall duration of the circle-6 phase would be large. As such in the end phase there would be no mass quantities left within the realm and the E-Flows that still proceed through it obeying the inverse square law for EIL with distance. This is partially represented by the circle 7 energy medium curve.

By means of the observable realm conditions around us today we may assume that the realm datum is currently at a position that is high up on the circle 6 energy medium curve. When we have established parameters for the energy medium exchange factors, the Mp and energy medium intensities etc., we should then be able to accurately establish our current realm status on this energy medium curve. For the present however, we have clues as to that position in the measurable rate of decrease in the gravitational constant G (indicating proton mass decay rate) and Hubble's Universal Red Shift Constant 'H' (indicating the speeding up of propagation of EIL oscillations between origin and earth). It would be too soon to enter into a discussion on this aspect at this stage, although some readers are bound to be ahead in their concepts and interpretations.

Energy Medium Sequence Circle 7.

The circle 7 energy medium curve as shown in the Fig 2.24 represents the dissipation of the E-Flows that resulted from the previous mass structures. Also represented is the natural expansion of the original realm by virtue of its increasing distance from the Mother Realm as suggested by Figs 2.3 and 2.11. This energy medium curve must therefore obey a law of a diminishing realm EIL that is a complex combination of the inverse square law for each E-Flow within it against the background of a marginally expanding realm. The realm EIL will thus tend towards C_0 and ultimately into the energy medium lines depicted in Fig 1.12.

The entire seven circle energy medium phases of the realm can therefore be inferred as occurring only once on the overall scale theorised. However there are bound to be recurrences on a minor scale within the realm where local phasing of E-Flows exist as indeed we shall theorise later. There must undoubtedly be similar events through multiple phasing of other realm groups emanating from the Mother Realm. The scope for speculation is endless. It would appear that the seventh phase of our realm energy medium cycle is a dying one in which all activity will eventually cease.

In the chapters to follow we shall be concerning ourselves with the more tangible structures and phenomena within the realm as we know it today and which were briefly described in the events of the circle 6 energy medium curve.

Chapter Summary

1. The seven circle energy medium sequence diagram represents the changes in EIL that occur as a result of multi-realm phasing. The event cycle is traced as an energy medium curve through a configuration of seven circles between the EILs of C_0, C_1 and C_2, on a graphical basis.
2. In the infinite cosmic void the probability of multi-realm phasing is greater in the proximity of a hypothetical 'Mother Realm' from which all other smaller realms are assumed to emanate.
3. By virtue of their expulsion from a grosser entity all realms must be considered in a state of expansion that is lateral to the direction of traverse away from the Mother Realm.
4. Energy medium concentrations that occur between C_1 and C_2 EILs result in the formation of complex Mp-congregations.
5. Energy medium concentrations above C_2 EILs simply do not exist. At this level all energy medium converts to a mass-medium structure.
6. The mass quantity level of Mps in a proton vary between an optimum level at the proton centre and a correspondingly reduced level at the proton surface. The differences depend upon the prevailing realm EIL datum.
7. When a proton is in a realm datum at below C_1 EIL it will undergo continual mass decay. Such is the case represented by the energy medium curve circle 6. Proton decay is the mechanism behind all the realm characteristics and action that we observe today.
8. The entire seven circle energy medium curve configuration points to the conclusion that all energy medium intensities must eventually approach C_0. The entire realm structure as we see it today is dependant for its actions upon the decaying process of the proton mass. This means that the realm functions as it does because it is simply working its way from a higher level energy medium status to a lower level energy medium status. The energy medium is portrayed as the basic constituent in a complete cycle of events from energy medium to mass and back to energy medium.

CHAPTER 3.

Propagation of EIL Oscillations

EIL Oscillation Events.

In the last chapter the reader was given a brief account of the nature of EIL oscillations within the realm. A graphical representation was made of EIL changes along a single radiant line from an originating source and a conceptual comparison made with the water wave profile spreading out from a central disturbance point. At that stage the intention was simply the formulation of a basic mental image of the nature of EIL oscillations. Since the graphs used in those descriptions were only representative of the spherical expanding characteristic of the EIL front, the reader was urged to imagine a similarity in the traverse of forcephoids. As such the peak in the EIL oscillation traverses through the realm as an ever widening sphere and successive oscillations from the same source follow at uniform intervals/spacing.

Of utmost importance is the source of the EIL oscillation. Let us for the moment return to a previous thought experiment in which we assumed a flow of the energy medium into a point position within the realm. As the energy medium flows into the point, the EIL within that point increases in proportion to the rate of fill. In Fig 3.1 we can represent the EIL changes within the imaginary point 'x' in relation to the duration of each event.

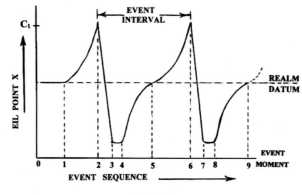

Fig 3.1.

Assume that the point x is initially at realm datum EIL as represented on the graph upto the event moment (1). At this instant the energy medium commences a steady flow into the point 'x'. The EIL within 'x' rises at an increasing rate towards the C_1 level based upon the theoretical condition that as 'x' fills with energy medium it behaves as though it were contracting in volume (like filling a conical vessel with water from an evenly flowing source; the water level rises faster as it nears the apex). We shall see later why this is so when we relate our imaginary point 'x' to the Mp-shell around the proton (Fig 2.18). Just before event moment (2) we observe that the EIL within 'x' attains C_1. Subsequently the energy concentration in x is sufficient for the initiation of the E~M/X sequence. When this finally commences at the event moment (2), the EIL in 'x' has progressed to a level marginally above C_1. The E~M/X results in the conversion of the energy-medium within 'x' into a Mp at the event moment (3). An energy build up has occurred from an E-Flow into a limited volume and a Mp has formed. The life of the Mp at 'x' is 10^{-43} seconds in Planck time, which is the event duration of its mass at the centre of the absorption sphere. This is shown on the graph as the interval between event moments (3) and (4) and represents the unstable/transitory nature of the Mp. At the event moment (4) the Mp commences the M~E/X sequence back to energy medium. This results as an explosion of that energy medium which flows outwards from the centre of 'x'. The expulsion of energy medium from the Mp is not instantaneous but occupies the event duration between event moments (4) and (5).

Now according to the conditions of our thought experiment, a continuous flow of the energy medium enters the point 'x' from an imaginary outside source. Thus there will be a continual tendency for the EIL in 'x' to rise. This trend must be superimposed upon the E~M/X and M~E/X effects for EILs within 'x'. At event moment (4) the Mp* E-Flow together with our imaginary source results in a very high rate of EIL rise. As we approach the event moment (5) the Mp* E-Flow depletes and the EIL rise is mainly due to our imaginary source. This is indicated by the reduced gradient of the EIL curve on the graph representing the interval for the M~E/X sequence. At the event moment (5) all the energy medium that had been involved in the earlier Mp formation is now outside the point 'x' and the build up of EIL in 'x' must once again depend more or less wholly upon our imaginary energy medium source.

We must now modify the EIL graph in Fig 3.1. to take account of the following additional energy medium conditions:

1. The E~M/X occurring near the surface of 'x' must absorb some energy medium from 'x'.
2. The M~E/X that follows will result in some of this energy medium flowing back into 'x'.

Since both of these effects are a direct result of the preceding Mp event within 'x', then they will be in perfect continuous synchronism with EIL changes in 'x'. That is to say, the absorption of energy medium from 'x' will always occur at the same relative event moment in 'x'.

In Fig 3.1 the E-Flow from event moment (4) sets up a C_1 EIL one step away from 'x'. We may assume that by the event moment (5) in 'x' that an E~M/X has commenced just outside 'x'. Energy medium will be absorbed from 'x' slowing down its own rate of EIL build up. Thus by the event moment (6) the EIL in 'x' may not as yet have attained C_1. However, slightly delayed event moment (8) in 'x' will receive the additional E-Flow from the M~E/X outside and thus attain the event moment (9) at a faster rate. It appears that the delay in EIL buildup to C_1 at event moment (6) is somewhat balanced by the faster rate of EIL buildup to event moment (9). Let us assume that the event interval has remained unchanged and that the EIL graph in Fig 3.1 fits the above conditions when the Mp events in 'x' are repetitive. The reader is reminded that the profile of this EIL curve is for conceptual purposes only and simply indicates the nature of the EIL changes within 'x'. If we may refer to the realm EIL datum as the base from which EIL changes are measured then a complete Mp sequence in 'x' occurs between the event moments (1) and (5). In future discussions we shall make frequent reference to this sequence.

So far we have assumed that the realm EIL datum remains unchanged. What effect would a changing realm EIL have upon the EIL curve of Fig 3.1? If the volume of point 'x' remains constant, then obviously at a lower realm EIL datum more energy medium will be required to flow in from our imaginary energy medium source. Subsequently the energy medium curve will take that much longer to attain C_1. The inverse applies for a higher realm datum. Once again the reader is asked to accept that the volume of 'x' at a high realm datum is larger than for a correspondingly lower realm datum. (This will become apparent when we relate 'x' to the Mp-shell of the proton). Subsequently the rate at which C_1 is achieved within 'x' remains unchanged. But what of the set quantity of energy medium within 'x' that must be available when an E~M/X sequence commences in order that a Mp may complete? To answer this we must relate point 'x' to the Mp-shell mentioned in chapter 2. We observed then that as a result of a uniformity of E-Flows and forcephoids from neighbouring protons, the centrally located proton P (Fig 2.17) developed a C_1 EIL in its R_2 forcephoid. Subsequently an E~M/X occurred within this forcephoid to set up (for that instant) what was a Mp-shell around the proton. This Mp-shell had a linear momentum that was either towards or away from the proton centre dependant upon the proportion of energy medium quantity absorbed from external sources. Each Mp-shell will dissipate and be replaced by another at the Mp event rate. We concluded then that an Mp-shell will regenerate in a succession of E~M/Xs and M~E/Xs and will appear to choose a definite radial distance from the proton centre. This radial distance will depend upon the EIL conditions of that realm area. The higher the EIL conditions then the greater will be the radial distance at which the Mp-shell recurs and vice versa.

It will be apparent to the reader that such a Mp-shell will contain a defined number of Mps dependant upon the total quantity of energy medium within the shell forcephoid. The farther the forcephoid position the larger will be its overall volume and the greater the number of Mps that will make up the shell, and vice versa. Now let us equate the point x of our thought experiment with this Mp-shell. Our imaginary energy medium source now becomes a reality and is a combination of the E-Flows and forcephoids from neighbouring protons. The Mp sequence is also a reality and occurs with a regular frequency. With changes in energy medium flow or realm EIL the Mp-shell retains its frequency at the expense of a marginal change in size and the quantity of Mps formed. This is discussed more fully in chapter 6 but for now let us briefly look at the mechanics of the Mp-shell.

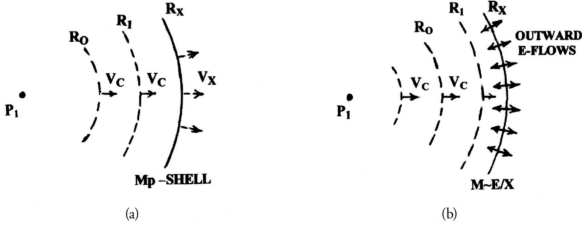

(a)

(b)

Fig 3.2

In Fig 3.2(a) is shown the instantaneous position of the Mp-shell relative to the proton P_1. The combination of linear momenta of all the absorbed energy medium results in a net outward momentum of the Mp-shell shown at the radial position Rx. Thus the forcephoid at position R_1 is approaching the Mp shell Rx at its normal traverse velocity Vc. Let the velocity of the Mp-shell be Vx in the same direction as Vc but small in comparison. Then forcephoid R_1 is bound to catch up with Rx quite rapidly. At (b) this is shown as nearly being achieved. However the Mp-shell has commenced an overall M-E/X with the resulting E-Flow as shown. The inward E-Flow meets the forcephoid R_1, phases with it to produce the next Mp-shell at approximately the same location as the previous one and with approximately the same net linear momentum. This sequence is repeated for each of the uniformly spaced forcephoids that successively approach each Mp-shell and the proton P_1 would appear to have a permanently structured shell surrounding it. We must leave this aspect for now and concentrate upon the nature of the EIL oscillations.

We can temporarily ignore the inward E-Flow and the Mp-shell (point 'x') as we did in chapter 2 and focus our attention upon the outward E-Flow as shown in Fig 3.2(b). In our earlier descriptions on the propagation of EIL oscillations we represented the EIL events as recurring successively at very small distances or steps from x. Let us look at the reason for these small steps and the factors that determine their size. In Fig 3.3 is shown a section of the Mp-shell at the M-E/X event. We have excluded the inward E-Flow detail from the diagram as this has been dealt with elsewhere

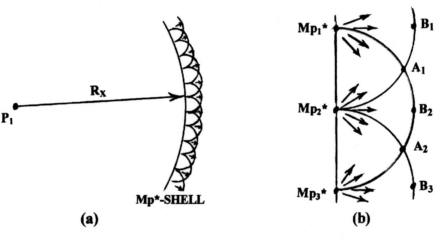

Fig 3.3

and would simply complicate the presentation here. The Mp-shell is represented at (a) by a uniform distribution of Mp events in perfect event synchronism on the surface of a sphere at the radial distance Rx from P_1. At (b) is an enlargement of a section of the same Mp-shell in a M-E/X sequence. The E-Flows from Mp_1^* Mp_2^* and Mp_3^* must be allowed to traverse outwards upto a radial distance that is greater than the minimum energy radius (Fig 1.10) before another E-M/X sequence is permitted. This is a rule of behaviour in which a second energy medium exchange cannot commence inside the sphere of influence of an existing exchange sequence. In Fig 3.3(b) we have represented the positions P_1, P_2 etc., as equilateral distances from two consecutive Mps. Thus A_1 and A_2 will definitely be outside the sphere of influence of the Mp-shell and if conditions are right may enter an energy-medium exchange sequence at the most convenient point. Now looking at Fig 3.3(b) the E-Flows from Mp_1^* and Mp_2^* arrive at A_1 at the same instant. If we assume that A_1 is at the edge of the minimum energy radius then the E-Flow from Mp_1^* and Mp_2^* will each be around the C_1 level when they approach this position. The phasing of both E-Flows will result in an EIL that is marginally above C_1 and so will initiate an E-M/X at A_1. The positions B_1, B_2 and B_3 could also be said to receive phasing energies that tend toward the C_1 EIL. However this would occur as a later sequence and may be discounted once an E-M/X has commenced at A_1, A_2 etc. It must be stated that A_1, A_2 etc., receive energies from an array of uniformly distributed Mp*s in the Mp-shell locality. This is very simply represented in Fig 3.4 for the equilateral configuration of three Mps on that shell.

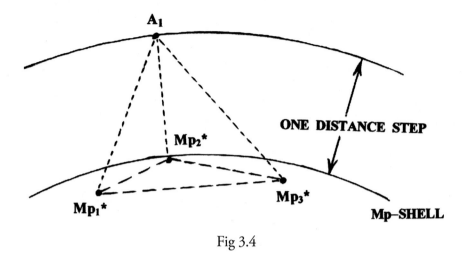

Fig 3.4

It must be stressed here that the prevalent realm datum EIL is the major factor in the attainment of C_1. The E-flows from the Mps* simply add the final amount of energy medium to the forward positions.

So far we have discussed (in general terms) the EIL changes that take place at a Mp location. We have also theorised upon the initiation of an E-M/X sequence one step from the Mp-shell structure. We must now consider the EIL events at these new step locations.

In Fig 3.1 we indicated the EIL changes that occur when a quantity of energy medium flows into an imaginary point x. We assumed that there was sufficient energy medium present for the E-M/X sequence to complete and for a Mp to exist for a set duration at the point x location. In Fig 1.10 we laid down the principles of Mp formation from energy medium at C_1^+ EIL. At this EIL the energy medium was presumed to develop a new characteristic that gave it a mass medium quality. This mass medium was considered to evolve as a highly stretch strained entity. This caused an implosive effect which resulted in the mass medium volume becoming sectionalised with each section imploding towards its section centre. These mass medium volume sections are assumed to be of an optimum size which depends upon the EIL of the former energy medium. These assumptions are a key feature in our theory and are based on the requirement for an intermediate functional entity in the conversion of our energy medium into mass as annunciated by Einstein's formula for energy/mass interchangeability, $E = MC^2$. This conversion aspect is treated more fully later (Figs 4.1 and 4.2).

Now in Fig 3.4 we have a slightly different energy medium aspect at the location A_1. Here we have a C_1^+ EIL occurring within a very localised zone. As such the conversion to mass medium will be on a limited scale within this zone. It is quite probable that this zone has a smaller volume than that of the optimum mass medium sections mentioned earlier. This means that a relatively smaller quantity of energy medium converts to mass medium which then implodes towards a theoretical mass centre called a mass point. Since this entity has considerably lesser energy medium substance than our former circle 4 and circle 5 Mps, then it follows that its existence as a mass point entity is also of a slightly shorter duration. Let us differentiate between these two levels of mass point and refer to the lighter mass process simply as a mass point event. Thus in the Mp-event we have the same sequence of E-M/X and M-E/X as in normal Mp formation. The difference lies in the event duration of each Mp structure. Fig 3.5 shows the event sequence of the Mp-event and should be compared directly with that of the Mp duration in Fig 3.1. In Fig 3.1 the event sequence shows the Mp with a finite duration indicated by the interval between event

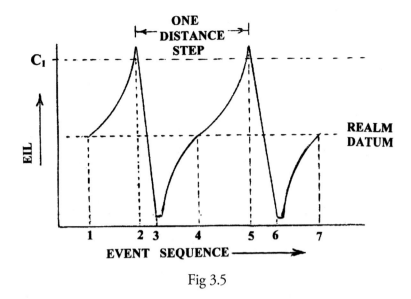

Fig 3.5

sequences 3 and 4. In Fig 3.5 the duration of the Mp is smaller at the event sequence 3. In all the other aspects the diagrammatic representation of the overall sequences are similar. From this we would conclude that the Mp-event is a viable theoretical event only with an overall cyclic duration that is somewhat lesser than that of a standard Mp sequence inside the proton.

The Propagation Front.

In the last chapter we presented a water wave profile analogy. At that stage the reader was requested to view the water wave as a sort of dynamic graph of the spherical EIL wave front. Although we know the mechanism of each to be totally different there are still comparisons in action that can be made.

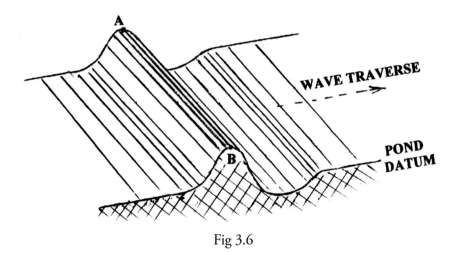

Fig 3.6

Consider the water wave on the surface of the pond. After the pebble has dropped in, a series of peaks and troughs spreads outwards from the point of entry. These waves spread outwards as perfect circles around that point and at a constant profile velocity. As the wave circle enlarges we can concentrate our analysis of its action upon a particular section. At large distances from the point of origin we can assume a small section of the wave front to be contained in a set of parallel planes that are transverse to the direction of propagation as shown in Fig 3.6. The line AB that represents the wave peak may be considered linear and transverse to the direction of wave propagation. We may now apply this principle to the spherically enlarging EIL oscillation front.

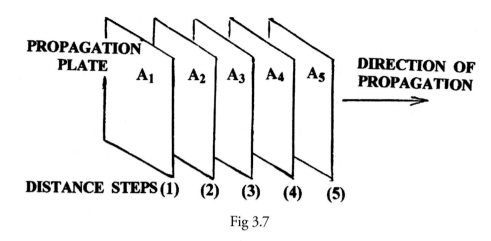

Fig 3.7

As the Mp event progresses in successive steps away from the Mp-shell, its sphere of action gets larger and larger. At large distances from its origin a small section of the Mp sphere may be looked upon as a flat section i.e. in the same plane. For this small section we would therefore observe the Mp-event progression in a series of parallel planes each spaced one distance step apart. Since we tend to observe and analyse effects on finite sections only we shall refer to each small section as a 'propagation-plate'. Thus, although we are in fact dealing with an expanding spherical EIL wave front, we can assume a small section of it to be flat and transverse to the direction of propagation steps. This is represented in Fig 3.7 by a succession of the propagation-plate at the distance steps (1), (2), (3), (4) etc. The concept of the propagation-plate as a section of the spherical EIL oscillation front plays an essential role in explaining the behaviour of these energy medium waves under varying realm conditions.

Propagation Velocity.

We indicated earlier that the Mp-event was simply an EIL oscillation over a spherical surface at a set radial distance from a centre of origin. Also that this EIL oscillation was set up or initiated in a sequence of small distance steps progressing away from that origin. The rate at which the EIL oscillation progresses from one step to the next is termed as the propagation velocity. In the previous section we ignored the spherical nature of the oscillation front at large distances from

origin and for purposes of analysis concentrated our attention upon a small section termed a propagation-plate (abbreviated p/p). We shall now look at the p/p in a bit more detail in order

to establish its characteristics. We must always remember that each EIL oscillation was initiated at the Mp-shell (our imaginary point x) and that its behaviour originated there. There are a large number of factors and peculiarities that control the p/p stepping action and in order not to confuse the reader we shall initially present a simplified event sequence. As we progress in our discussion we shall gradually develop a more complicated action sequence by including other factors as well.

Let us therefore commence our discussion with two of the basic factors that determine the propagation velocity of an EIL-oscillation. Before an EIL-oscillation can occur at the next forward step position the E-Flow from the current Mp-event sequence must traverse the step distance as indicated in Fig 3.8. Thus from the moment that the M-E/X sequence of the Mp-event at A

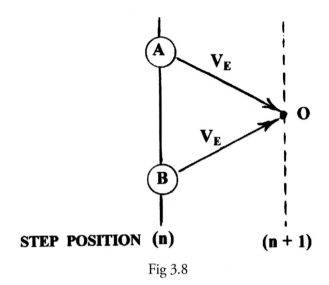

STEP POSITION (n)   **(n + 1)**

Fig 3.8

commences, a flow of energy medium traverses towards the imaginary point O. The position of the point O was selected as the earliest point (outside the sphere of influence of the Mp-events at A and B) that attained a C_1 EIL. So the first factor in determining the velocity of propagation is the rate at which the energy medium flows from position A to position O. Let us denote this E-Flow velocity as V_E. (N.B. Fig 1.14). Since this is a finite velocity then point O will attain a C_1 EIL after an event duration that is also finite. Thus the step position (n + 1) will be initiated as the next location for the Mp-event as part of an overall event sequence and occupying an event moment. Let the distance between the step positions (n) and (n + 1) be denoted by d. The traverse distance AO equals d/Sin 60 if we assume AOB is an equilateral triangle. Thus the earliest that a Mp-event can commence at point O is given by,

$$\text{E-Flow AO} = {}^{d}/.866 \, V_E \text{ event moments}$$

For the present let us assume that both d and V_E are quantities that do not vary.

Now consider the E-M/X sequence of the Mp-event at point O. All the energy medium flowing into the sphere of influence around point O is absorbed into the E-M/X occurring there. As such the E-Flow AO cannot proceed beyond the step position (n + 1). Referring to Fig 3.5, the duration of the E-M/X at the position (n + 1) is the interval between the event moment (6) and event moment (7) of the overall event sequence. Let Do denote the event duration of the E-M/X sequence at step position (n + 1). At the end of this the Mp-event sequence reverts to a M-E/X and a subsequent E-Flow to the (n + 2) step position. Since we have already accounted for this E-Flow as part of the earlier events we may assume that a complete propagation sequence is made up of two factors. The velocity of E-Flow between the step positions and the event duration of the ensuing E-M/X. Both of these are finite factors and added together are a simple measure of the EIL oscillation propagation velocity between the step positions p(n), (n + 1), (n + 2) etc., as given by,

$$\text{Step progression} = \text{Do} + {}^{d}/.866 \, V_E \text{ event moments / step}$$

At the present stage we are able to relate this to the 3 x 10^8 metres/sec velocity of propagation on the assumption that the step distance d is 10^{-35} metres, the shortest Planck length, and the Mp duration is 10^{-43} seconds in Planck time. We can infer from the above that if the mechanism of propagation is a stepping type sequence of events then VE the velocity of the E-Flow AO component can also be calculated and must certainly be greater than 3 x 10^8 metres/sec velocity, i.e. faster than light. More on this in a later chapter. See also Appendix 1.

Now consider the event sequence at the Mp-shell from where all EIL oscillations originate (Fig 3.1). The repetitiveness of the Mp-shell formation sends out one EIL oscillation for every completed Mp-event cycle. This EIL oscillation propagates outwards through the realm in successive steps as outlined above. Because of the shorter cyclic duration of Mp-events we may assume that a number of propagation steps will have been taken by the EIL oscillation before the Mp-shell emits another. This second EIL oscillation will trail behind the previous one at a distance of a set number of propagation steps. Let this number be denoted by N, then the spacing between successive oscillations will be N steps. Oscillations from the same source are considered to belong in the same event family. As such the distance between successive oscillations is termed the wave length (denoted by Lambda λ) of the series.

Thus, wave length λ = N steps.

This is shown in Fig 3.9 and it is important that the reader does not confuse propagation steps with the notion of spacing between trailing oscillations. An EIL oscillation events at each propagation

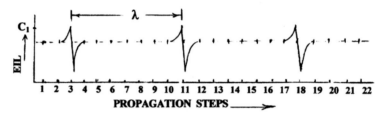

Fig 3.9

step in turn and at any instant must be assigned to one particular step position only. Thus the EIL oscillation is the same event entity at successive steps. A trailing oscillation is however a separate event entity and therefore may occupy a step position of its own at any instant. We now have two event entities in existence at the same event moment and are thus able to assign a finite spacing l between them. The example in Fig 3.9 shows three oscillation entities trailing one another at the finite spacing of eight propagation steps each.

Since all EIL oscillations from the same source and under the same realm conditions have the same propagation velocity, then the finite spacing between each will remain constant. The rate at which these oscillations pass a fixed position in the realm is a measure of their repetitive arrival at that position and is termed as the EIL oscillation frequency. Frequency therefore also reflects the repetitive event rate of the Mp-shell from which each oscillation originated. There is a simple relationship between the finite spacing λ, the frequency f and the propagation velocity Cv where,

$$f = {}^{Cv}\!/\lambda \text{ oscillations per second.}$$

The greater the finite spacing λ the longer will the next oscillation take to arrive and hence the frequency rate will be lower (and vice versa). Differences in frequency will depend upon the rate at which energy medium buildup occurs at the Mp-shell. As mentioned in the previous chapter this may depend upon two factors; firstly the rate at which the energy medium is supplied to the Mp-shell and secondly the energy medium capacity of that shell. We shall return to this at a much later stage in our discussions.

Propagation Events.

In the previous section we presented a simplified mechanism for the propagation of EIL oscillations through the realm. This was done in order that the reader may attain an understanding of the basics of the propagation characteristic. However we must now add to those basic features certain other events that are an essential part of the characteristics of propagating EIL oscillations. Our simplified view was that EIL oscillations propagated in quantum steps (10^{-35} metres) as a series of stop-start events. The Mp-event occurred at a p/p (stop) and then an E-Flow took place (start) towards the next p/p forward position. The velocity of propagation was therefore dependant upon the duration of the E-M/X sequence at the next p/p and the subsequent E-Flow velocity to the next p/p step. It was assumed that the p/p at a quantum step position was stationary within the realm. The law of conservation of linear momentum for all the energy medium involved in the E-M/X of the Mp-event indicates a net forward momentum of the p/p. In Fig 3.8 the high EIL at the point O is made up of several phasing components. The two main components are the E-Flow from the step position (n) to (n + 1) and the relatively high realm datum at O. Now if the E-M/X at O were to form a Mp, then that Mp would possess a linear momentum in the general direction of the next quantised step (n + 2). In a similar manner the Mp-event energies must possess that net linear momentum during the E-M/X sequence and prior to the Mp-event. Thus the imaginary point O in Fig 3.8 will possess a linear momentum in the direction of oscillation propagation. The p/ps as

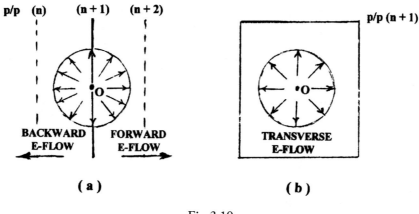

Fig 3.10

represented in Fig 3.7 must therefore move in the direction of propagation. This will be an additional quantity to be considered when computing the oscillation propagation velocity.

So far we have only considered the forward direction E-Flow from the M-E/X sequence of the Mp- event. We must now consider the effects of the remaining E-Flow. In Fig 3.10 the Mp-event is represented in the M-E/X sequence in the propagation plate (n + 1). At (a) the p/p (n + 1) is represented in a plane perpendicular to the plane of the paper and shows the E-Flow emanating from the imaginary point O. We have already accounted for the forward direction of the E-Flow; the forward direction being defined as the direction of propagation of the EIL oscillation.

In Fig 3.10(a) apart from the transverse E-Flow in the plane of the p/p it would appear that there is an equal distribution of the E-Flow between the forward and backward directions. However we must remember that the p/p has been ascribed a finite velocity in the forward direction of propagation and this velocity must be an additive component to all E-Flows emanating from the imaginary point O. The E-Flows represented in Fig 3.10(a) are relative to the moving p/p and a correction to a more absolute representation is shown in Fig 3.11.

Let Vp be the forward velocity of the p/p and although it is small in comparison to the E-Flow velocity V_E, it must nevertheless alter the overall distribution in the vector quantity Vp direction. In principle therefore we may conclude that a greater quantity of M-E/X energy medium is flowing in the forward direction and that the maximum differences between the forward and backward E-Flow velocities are plus and minus Vp.

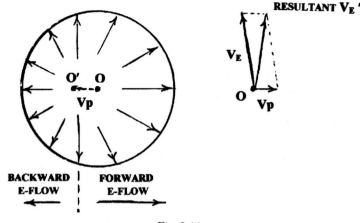

Fig 3.11

Maximum Forward Velocity = V_E + Vp
Minimum Backward Velocity = V_E – Vp

We must now consider the effects of this backward E-Flow and determine whether it results in high

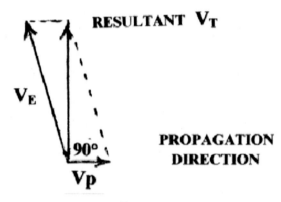

RESULTANT V_T

PROPAGATION DIRECTION

Fig 3.12

EIL points comparable to those forward of the p/p. Let us however first consider the effects of the E-Flow within the p/p as shown in Fig 3.10(b). Since we know that the p/p has forward momentum

then the actual transverse E-Flow will be in a parallel plane that is stationary within the realm datum and somewhat behind the p/p. This transverse E-Flow will be marginally less than Ve since it will be the resultant of the vectors V_E and Vp as represented in Fig 3.12.

If V_T is the absolute Transverse E-Flow velocity then,

$$V_T = \sqrt{(V_E{}^2 - Vp^2)} \text{ which indicates that } V_T \text{ is less than } V_E .$$

The plane of this E-Flow V_T remains stationary within the realm and parallel to the preceding p/p. In Fig 3.11 this plane would occur through the point O'at the commencement of the M-E/X event.

We know that there are a large number of Mp-events occurring in the p/p plane and that the imaginary centre points of each are spaced at equal distances within that plane. Fig 3.13 shows the Transverse E-Flows from a number of M-E/Xs across the plane. It is clear that within moments the E-Flows will phase

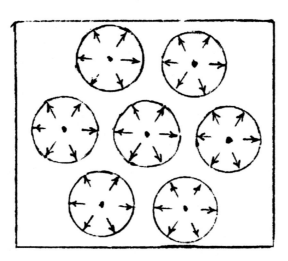

Fig 3.13

with one another and continue across the plane. At any point in this plane there will be virtually an endless arrival of E-Flows from more distant positions in that plane. After the preceding M-E/X, energy medium is dispersed away from each Mp-event centre resulting in a drop in the realm EIL datum at that point. This is more than compensated for by the arrival of the Transverse and Backward E-Flows from the preceding p/p events. As such a trailing p/p will be met by a reasonably well maintained realm EIL at every propagation step.

In Fig 3.9 the reader was asked to imagine an interval of eight steps between trailing oscillations. The propagation of the EIL oscillation requires that the Mp-event occurs at every step in turn. Thus from each step position a transverse and backward E-Flow will emanate. As each oscillation proceeds forwards it will meet the backward E-Flows from the previous p/p locations and absorb these into its own p/p E-M/Xs. By this means the backward E-Flows will not be allowed to proceed much further than the trailing oscillation p/p. The transverse E-Flow however may not necessarily coincide with the plane of the trailing p/p steps and as such may then proceed in the transverse direction for a longer duration than the p/p interval before it is absorbed in the sphere of influence of other trailing p/ps. The transverse nature of the EIL oscillation will therefore be more apparent than the forward or backward energy medium changes of the propagating wave.

In Fig 3.14 we have shown the p/p CD trailing p/p AB by eight propagation steps. We have also represented the backward E-Flows from a number of step positions behind p/p AB. It can be seen that the E-Flow from the step position (7) has not quite reached the previous step position (6) . However the E-Flow from the step position (4) has just about reached step position (1) having had the longer traverse duration since the p/p AB was at that position. If one were to continue representing the backward E-Flow from step positions (3), (2) and (1) we would see the E-Flows as having reached the trailing p/p CD and being absorbed into the E-M/Xs occurring there. During the stepping sequence of events from the position CD to step (1) a marginal backward E-Flow could pass beyond a trailing p/p. The probability of getting past the next p/p is however extremely remote and as such we may ignore any backward E-flows beyond the trailing p/p.

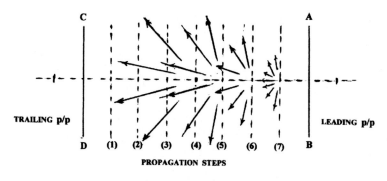

Fig 3.14

So far we have presented a rather straightforward and basic description of the type of events that are an integral part of the propagating EIL oscillation. The reader is now required to visualise the whole picture of events as a more complex phenomenon. The forward E-Flow from each M-E/X of the Mp-event phases with a complexity of backward direction E-Flows from the previous p/p step positions. There are also transverse E-Flows that contribute to the EIL datum within the E-M/X sphere of influence. There is a net forward momentum of the p/p that results from the conservation of linear momentum of all the phasing energy medium quantity. Thus a step position is not an absolute fixation within the realm. We shall now briefly consider secondary Mp-events that may occur if the phasing action of the backward E-Flow results in EIL points that attain C_1. The setting up of a plane of secondary Mp-events moving in the direction of the backward E-Flow would result in a diminishing series of EIL oscillations that occur at higher frequencies. Analysis of these oscillation harmonics as inter-step repetitions of the original Mp-shell exchanges becomes

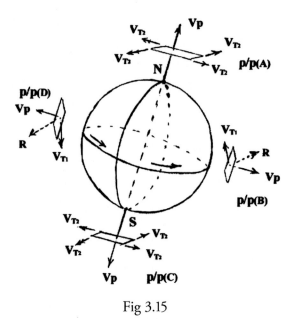

Fig 3.15

extensively complicated and until quantitative parameters for the energy medium EIL is brought into play we cannot detail the phenomenon here. At this stage the reader must simply be aware of such a phenomenon so that a complete mental picture is projected for possible future researches.

We know that the propagation plate has a linear momentum in the direction of propagation during its brief existence as the oscillation front. We must now assume that this motion is only a component of a more complex movement of the p/p. When an EIL oscillation sets out from the Mp-shell, the continued character of that oscillation will depend considerably upon events within that shell. Let us suppose that the Mp-shell as represented in Fig 3.3(a) rotates about the proton P_1. This is probable since the Mp-shell in its continued repetitions is quite likely to absorb energy medium flowing in from external sources. Thus in Fig 3.3(b) the Mp_1^*, Mp_2^* and Mp_3^* energy medium quantities may have a linear momentum in a lateral direction as well.

Consider a revolving sphere as shown in Fig 3.15 to represent the rotation of the Mp-shell. The axis of rotation is along NS and is shown inclined in a random position. The direction of rotation is as indicated. Since our intention is to establish the nature of the momentum of a propagating plate that issued from such a Mp-shell, let us select four such plates as represented by p/p(A), p/p(B), p/p(C) and p/p(D). Let Vp be the outward linear momentum which may be assumed the same for all four plates. The lateral momentum of each p/p will however depend upon the point of issue from the Mp-shell. Those points farthest from the axis of rotation will naturally possess maximum lateral motion while those near the axis centres NS a minimum. A section that centres on N or S such as p/p(A) and p/p(C) will possess an angular momentum that corresponds to the rotation of the Mp-shell. This angular momentum is transmitted to each p/p step in turn based on the conservation of linear momentum of each E-M/X event within the p/p. Fig 3.16 shows the changing attitude of this polar propagating plate at each successive quantum step.

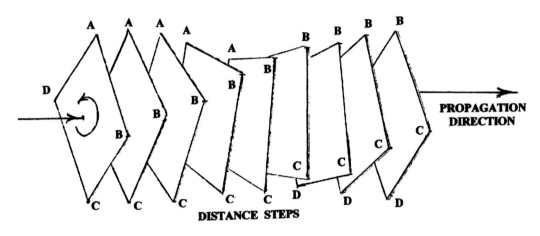

Fig 3.16

In the diagram ABCD represents the p/p stepping from left to right. In the eight quantum steps that are represented we show the p/p as having rotated through 90^0. This is purely an example of the rotation effect portrayed as a mental image and must not be related to the actual angular momentum rate of the p/p. Although the p/p will continue to exhibit the character imparted to it from its origin, nevertheless this could change and become vague as it traverses the realm and encounters varying energy medium conditions that get absorbed into it.

Let us now complicate this analysis by considering the Mp-events in the p/p(A). Each Mp-event centre will possess a linear momentum in a direction that relates to the spin of the Mp-shell. This means that the linear momentum of the energy medium at each Mp-event centre will cause it to traverse away from the centre of rotation as shown in Fig 3.17. The energy medium in each E-M/X will tend to traverse in a linear direction that is at first at right angles to its radial distance from the gyration centre N . However at successive propagation steps these E-M/X centres will be found at greater distances from the plate gyration centres and travelling in a direction more and more away from it. Consequently the propagation front along the axis NS will contain a depleting energy medium quantity and at some stage will be unable to progress to the next quantum step. The spherical propagation front will now contain a gap or hole along each polar axis and this will enlarge with each propagation step. In Fig 3.18 this is shown up as a conical gap in the propagation front along NS and will be termed

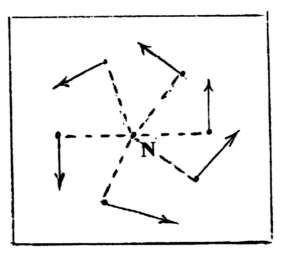

Fig 3.17

a polar cone. An observer within this region will not receive any EIL oscillations from the Mp-shell assuming of course that the proton centre is stationary with respect to the surrounding realm datum.

We now observe that the rotation of a p/p as represented in Fig 3.16 is no longer applicable and that in fact

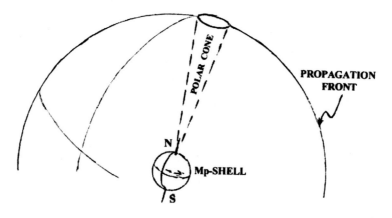

Fig 3.18

each plate will possess a component of velocity that is lateral to its direction of propagation. This simply means that a given quantity of energy medium in a particular E-M/X will occupy a laterally displaced position in the succeeding p/ps. The direction of propagation will still be at right angles to the plane of the plate and Fig.3.8 will need to be modified to take account of the effect of this lateral velocity vector on the E-Flow velocity V_E. This is shown in Fig 3.19 and simply shifts the position of O in the step position (n + 1). If V_T is the component of lateral velocity of the plate then the E-Flow velocity from A towards O will be slightly less than the former V_E and the E-Flow from B to O will have a velocity that is greater than V_E (by adding or subtracting vector V_T from V_E). Thus the E-Flows from A and B can reach the step position (n + 1) in exactly the same event duration as before. However they will both reach a point O' at the same moment if O' is closer to A than it is to B such that,

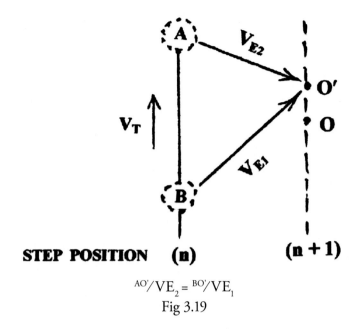

STEP POSITION (n) **(n + 1)**

$$^{AO'}/VE_2 = {^{BO'}}/VE_1$$
Fig 3.19

The energy medium quantities arriving at the point O in Fig 3.8 and at point O' in Fig 3.19 are found to be exactly the same and as such the size of quantum step has not had to alter. Since other EIL conditions are also the same then the velocity of propagation of the EIL oscillation will 'step' at an unchanged rate. In this section we have attempted a general characterisation of the event structure that makes up energy medium oscillations. We shall now vary some of the EIL conditions within the realm (in thought only) and examine the nature of the subsequent EIL oscillations.

Variations in Propagation Velocity.

Two motorists set out to cover a distance of 100 miles. They start at exactly the same instant and travel along at the maximum agreed speed of 50 mph. If motorist A makes a 5 minute stop every 15 miles and motorist B makes a similar stop every 20 miles, then which one completes the course first? The answer of course is the motorist who makes the least number of stops. The actual motoring duration of each is exactly the same i.e. 2 hours. However A made 6 stops and B only 4 stops of 5 minutes duration each. Hence B reaches the 100 mile position 10 minutes before A.

Consider the propagation of an EIL oscillation between two fixed points. As mentioned earlier, the oscillation progresses in successive quantum jumps. Let us assume that an oscillation B takes (n) steps to cover the distance between two points spaced one metre apart. Now consider another oscillation A that takes (n + 1) steps over the same distance. If we specify that the E-Flow velocity between steps is exactly the same and that the duration of the E-M/X and M-E/X sequences are also equal, then oscillation A will take longer to traverse the one metre distance than oscillation B. This is because oscillation A has had to include the duration of one additional Mp-event sequence in its overall journey events. From this example we would therefore conclude that the propagation velocity of EIL oscillation B is greater than the propagation velocity of oscillation A.

In both of the above examples we have gauged the average velocity of A and B over a set distance. We cannot say that at any particular instant, B is travelling faster than A. This would not be true since each is endowed with the same capacity for speed. The difference in average travelling rate over specified distances must therefore depend upon the size of step taken between each stop. As a further practical experiment the reader may attempt the following as represented in Fig 3.20.

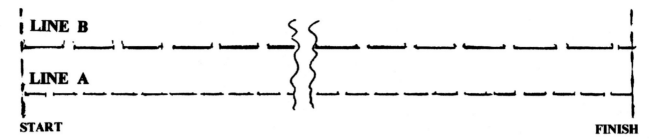

Fig 3.20

The requirement is to draw a series of dashes in two separate lines A and B over a distance of 20 centimetres. The length of dash in line A is to be 0.5 cms and that in line B is 1.0 cm. The inter-dash spacing is the same at about 0.25 cms. Using graph paper and working as consistently as possible for each case the reader is required to determine the line that is completed in the shortest duration. A stop watch may be used to establish the effective event duration of each. The author's own attempts involved a large number of 'runs' to establish a fluency of stroke which then resulted in the following best duration being recorded;

Line A (27 strokes) = 0.25 mins.
Line B (16 strokes) = 0.16 mins.

The indicated progressional velocity of Line A is thus 80 cms per minute and of line B is 125 cms per minute. We are now in a position to draw the conclusion that the progression or propagation of an EIL oscillation through the realm is directly proportional to the size of quantum steps taken. The study of the variation in propagation velocity is therefore a study of the structural make up of the quantum step dimension.

Earlier in Fig 3.3 we indicated the basic mechanics of transfer of energy medium from Mp-shell to the points A_1, A_2, etc., at which energy medium exchanges resulted in the Mp-event phenomena producing a spherical front at the new position - and so on and on. We must now review the mechanics of this stepping action and ascribe fundamental laws to explain its make-up and behaviour.

The quantum step.

This may be defined as the linear distance between the successive positions at which the EIL oscillation front is set up as a result of the recurring Mp-event phenomena. The interval in which the Mp-event recurs depends upon the earliest moment that a point outside the former sphere of influence attains a C_1 EIL. At a later stage we shall explain the reason behind a restricted volume (the sphere of influence) being retained as an out of bounds region for recurring Mp-events. This is the zone within the realm where linear momentum is created and imparted to the energy medium. The conditions within this region are therefore unique and although the conservation of linear momentum laws are not violated they must be looked at in a new light.

The location of an Mp-event centre is therefore a phasing position for the E-Flows from a number of centres in the previous p/p. For reasons given earlier we shall consider the quantum step as occurring in the forward propagation direction and normal to the plane of the previous p/p. Let us assume that the Mp-event centres are uniformly spaced within the p/p. The forward E-Flows from these must phase at some positions. The EIL contributed by each E-Flow diminishes with distance traversed and as such the spacing distance between the Mp-event centres would be an important consideration in determining the earliest point where a C_1 EIL occurs.

Consider the E-Flows from the three uniformly spaced centres A, B and C represented in Fig 3.21. Let the E-Flows from these intersect or phase at O such that the EIL at O approaches C_1. The position of O must be outside the influence of the absorption spheres of A, B, and C and the EIL will depend upon their distances from O. It may be necessary to include the phasing E-Flows from a greater number of Mp centres to attain a C_1 EIL at O. This will depend upon the spacing of the A, B and C centres and also on the prevailing realm grid datum.

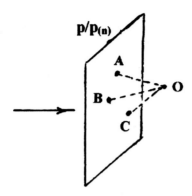

The Mp-shell is the origin of the propagation front and we see that this enlarges at each quantum step. The centres A, B and C will thus progressively become spaced farther and farther apart. The distribution or density of these Mp-event centres in the wave front would therefore be inversely proportional to the square of the distance propagated. This would indicate that the spacing of the Mp-event centres becomes infinitely large at infinite propagation distances. How then could the Mp-event recur and a quantum step be maintained over large distances? Actually the Mp event is a result of the mass medium status being achieved in the new plane of O. Only then will the Mp event centres result. This means that as the wavefront expands, the number of Mp centres Fig 3.21 increasingly adjust to the space and conditions that prevail. The Mp event density within the wavefront will therefore remain fairly constant. In Fig 3.18 we showed that the Mp-shell spin resulted in a gap in the propagation front which we termed as a polar cone. The rate of spin determined the size of cone and we now postulate that this polar cone region enlarges at an increasing rate as the wave front spreads. The theoretical aspects behind this surmise is that the velocity component of a section of wave front energy medium away from the rotation axis increases with distance. This is shown in Fig 3.22.

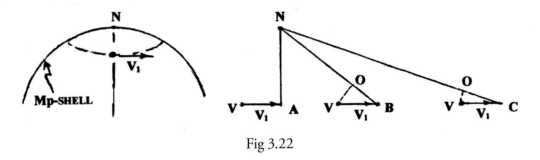

Fig 3.22

Let the Mp-shell impart a linear velocity V_1 to a Mp-event centre near the axis of rotation. Let this be represented vectorially in the successive positions VA, VB. and VC. Initially VA is at right angles to the rotation axis and has no velocity component along NA i.e. away from N. However as VA progresses to position VB a velocity component OB away from N is clearly existent. At the position VC this velocity component has increased farther as represented by OC. In the limit the away component approaches the value V_1. Now if the propagation velocity of the wave front is Vc and if we assume that V_1 is not negligible in comparison, then we may assume that the polar cone has an increasing lateral spread. The volume swept by the wave front may be represented by the revolution of a parabola about the Mp-shell spin axis as shown in Fig 3.23. It is clearly noticed that the propagation front is not as complete as we had earlier imagined. It is in fact a continually smaller proportion of the overall spherical surface at that radius and may be compared to an expanding band in a lateral direction from the spin axis.

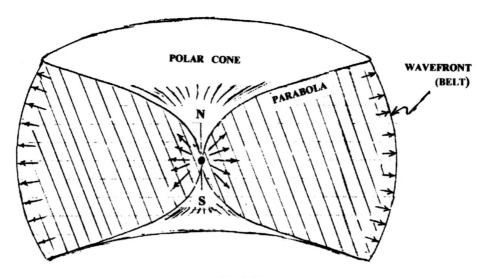

Fig 3.23

Earlier we made the assumption that a small section of the wave front may be considered to be flat and in the same plane at large distances from the Mp-shell. We may now state that this distance need not be so large as to approach infinity. In comparison with the Mp-shell diameter of approximately 10^{-10} metres, a distance of a few metres would be considered very large indeed. For these relatively large distances our analysis of a p/p would show a negligibly small wave front density change at successive steps and in all future considerations we shall accept this premise. However the change in wave front density is cumulative and must be taken into consideration only when a very large number of quantum steps are involved e.g. when computing the maximum distance that a wavefront may traverse.

Let us now consider the second factor that exerts some influence in the spacing of Mp-event centres within the wave front. Upto now we assumed that the Mp-shell was a simple distribution of uniformly spaced Mps over a spherical surface area. If we remember that the energy medium content of the Mp-shell is the result of an outward bound forcephoid phasing with the inward E-Flow from the previous shell M-E/X and that the forcephoid has a defined peak width (R2 in Fig 2.18) then we must assume that the Mp-shell retains some of this width in which to structure the Mp positions. We cannot compute this width of shell nor the pattern of Mp arrangement but we do know that all will behave exactly alike and in near perfect synchronism for their E-M/Xs and M-E/Xs. A section through the Mp-shell could be imagined as at Fig 3.24, the dots being

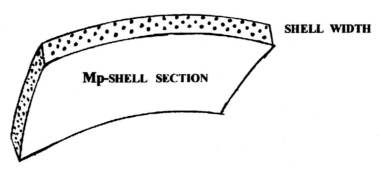

Fig 3.24

representative of the Mps in the shell width. The spacing of Mps are indirectly dependant upon the energy medium quantities and intensities that inter-phase with the forcephoids. That however is another topic and will be attempted in chapter 6. For the moment the reader is asked to accept the basic concept presented in Fig 3.21 for E-Flows with the difference that we are now dealing with a more complex combination from a larger depth of Mp-centres. In line with this principle we must assume that the wave front will possess a corresponding depth. The forward E-Flow now becomes a very complex assortment of interphasing energy medium and C_1 EIL will be achieved over a band width area. This produces the Mp-event at centres that nearly correspond to those in the previous wavefront. It is highly probable that a mixture of Mp sequences will form the initial few EIL oscillation fronts. However as the wave front expands the successive forward E-Flows set up a narrower C_1 EIL band resulting in a reduced width of Mp-event centres. Ultimately this front width is so reduced that the Mp-event centres occur one deep only. This sequence is represented in Fig 3.25 as a possible pattern of behaviour within the propagating front. At (a) we show the possible relative positions of the Mp-event sphere of influence in the wave front. At (b) the same amount of energy medium quantity is shown in a slightly more expansive though narrower wave front. Notice the trend for the two planes within which the Mp-event centres lie are brought closer together. At (c) this is more noticeable that the two rows of Mp-event centres are knitting together. At (d) there is virtually a single row of Mp-event centres. It is stressed that this is only a visual example of the nature of a wave front close to origin. The actual behaviour may be somewhat different in detail but the principle is the same which is that the energy medium density within the wave front in its early stages gets regulated to a certain extent by a narrowing of the front width. This compensates for reduced wave front density due to the spherical expansion.

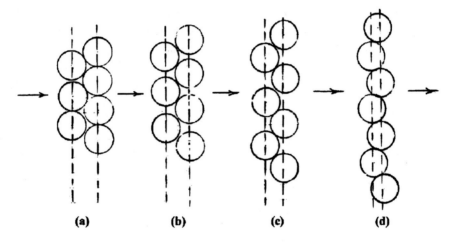

Fig 3.25

It must be re-stated that the sequence above only represents the wave front when its curvature is large, i.e. close to Mp-shell origin. The wave front at large distances however is represented by (d) above and a small section of wave front may be considered to have negligible curvature as defined for a p/p in Fig 3.7. As such we are able to return to our earlier concept of the p/p as a flat section transverse to the direction of propagation and ignoring plate thickness over large distances.

The above mentioned factors thus together ensure that a reasonable wave front energy medium density is set up at each forward propagation step. The reason that a wave front is able to traverse the phenomenal inter-realm distances is due to the negligible change in wave front density in the parallel propagating plate concept. There are of course additional energy medium factors that also assist in this expansive propagation and we shall come to those later. For now we have simply indicated why the wave front holds together in spite of its expansion. It will be noticed that the energy medium density of the wave front diminishes with distance traversed and no attempt was made to modify the inverse square law that still applies in all cases.

Let us return to Fig 3.21 in which the attainment of C_1 EIL at point O was dependant upon E-Flows from the Mp-event centres A, B and C. It will be obvious that as the spacing between these points changes so must the position at which C_1 EIL is attained. We must now attempt to establish the relative change in position of the point O (as representative of the next p/p) as the wave front proceeds. We must first consider the factors that cause a C_1 EIL to develop at the distance point O. So far we relied upon the E-Flow from just three points for the establishment of the required EIL at a forward p/p point. However with increasing propagation distances the energy medium density in the wave front must diminish and in our analogy we represent this by an increased spacing of the Mp-event centres. One of the properties of Mp formation principles is that in reduced energy medium density areas the sphere of influence is enlarged. This was defined in Fig 1.11 as a Minimum energy radius when the energy medium density approached C_2 levels. The Mp-event behaves in the same manner as far as the sphere of influence or energy medium absorption radius is concerned. The EILs set up in the forward position are at a marginally lower energy medium density and the Mp-event therefore operates over a marginally larger sphere of influence.

It may be remembered that when we presented our analogy of the water wave on the pond surface we mentioned that the crest of the travelling wave was maintained by kinetic forces given up by the preceding trough. In the case of the EIL oscillation the Mp-event survives not only upon the energy medium of the forward E-Flows but also upon the realm EIL datum that exists in its propagating path. It would be appropriate therefore at this moment to discuss the realm datum property and to describe its make up.

A consistent level of the energy medium at any point in the realm defines the realm EIL datum at that point. The basic realm existed as a cloud of the energy medium traversing the cosmic void in a direction away from the Mother Realm entity. At this event moment the realm datum would be considered as the EIL of the energy medium in its variable distribution across the realm. The marginal expansion of the realm would naturally result in a corresponding reduction in the realm EILs. The realm datum may subsequently be considered to be in a state of gradual reduction from a higher to a lower EIL. The earlier definition may thus be modified to include a "reasonably" consistent energy medium level as the essential nature of the realm datum.

With the advent of multiple realm phasing and the 7-circle energy medium curve phenomena, a very complex mixture of flowing energy medium may be described for the circle 6 status. The energy medium outflow from the proton mass may be considered to be reasonably continuous and obeying the inverse square law for intensity. As such the EIL of this E-Flow would appear constant at any set realm position. The continuous and consistent nature of this E-Flow intensity would thus contribute to the composition of the realm datum simply by its existence therein. The realm datum would therefore be at relatively higher levels when in close proximity to mass bodies.

For all realm positions there are a countless number of E-Flows prevailing from every direction. Some may have originated relatively nearby while others from near infinite distances. Each will contribute a small component of EIL towards the set up of a reasonably consistent energy medium datum. The existence of any phenomena that results in a flow of energy medium must itself add to the realm datum and in turn be affected by it. In the case of EIL oscillations, the E-Flow from each Mp-event M-E/X sequence becomes part of the realm datum even if only to be absorbed into a subsequent E-M/X . In the case where velocity may be ascribed to the proton mass then the component of E-Flow EIL would either increase or decrease at a set realm position depending on whether the source drew nearer or receded farther from it.

The realm EIL datum may thus be looked upon as a sort of three dimensional EIL spatial grid that defines the moment by moment energy medium quantity at each point within the realm. If we assume that mass bodies i.e. galaxies, are uniformly distributed throughout the realm then from simple mathematical analysis of E-Flows it would be fair to conclude that under similar conditions the realm datum near the realm centre was at a relatively higher level. There may also occur instances where local mass concentrations are so great that cumulative E-Flow EILs may result in a realm datum exceeding the C_1 level. There is scope for interesting speculation about these realm zones and their effects upon the circle 6 phenomena. Let us suppose that in the circle 6 phase the mass in the universe has attained its distribution as indicated for galaxies in our description in chapter 2. As we saw there is a general uniformity of galaxy distribution across the realm. Now in the circle 6 phase all mass is in a realm EIL that is below C_1 and as such will be in a state of general decay. This means that all mass entities emit energy medium in the form of an E-Flow in a continuous controlled and even manner. These E-Flows spread outwards in all directions from each galactic mass and as such pervade the universe in time. Consider the E-Flows of all these masses spreading out and intersecting/phasing with one another. Since each mass is relatively located and does not alter location suddenly, then their E-Flows will also maintain a similar constancy. At any selected inter-galactic location all these phasing E-Flows will result in a defined EIL buildup. Different locations will attain differing EILs and these levels will remain static in the short term for each location. These points in space will thus have a characteristic EIL. Points in space at slightly different EILs will thus have an EIL gradient between them. If we draw successive imaginary lines or planes connecting all points that are at similar EILs then we should end up with a spatial grid pattern showing us zonal EILs and their EIL gradients. If we were to perform this task for the entire realm then we would have established an interim Realm EIL Grid. The intention here is to give the reader a brief understanding of the character of the realm EIL datum in order that other related aspects i.e., wave front propagation, may be more easily explained. We shall discuss this further in chapter 4 and its effects on gravitation.

Let us now continue with our analysis of the wave front quantum step. One of our early assumptions was that the E-M/X phenomena could not repeat inside its sphere of influence. There is a very simple reason behind this conclusion. When a Mp-event occurs it has a tendency to absorb all the energy medium within a defined spherical volume. This volume is termed as the absorption sphere of the Mp or Mp-event as the case may be. This volume dictates the amount of energy medium that may be absorbed into a Mp. We mentioned the term minimum energy radius in the previous chapter and implied that the energy medium required to form a Mp at a realm EIL near C_2 levels would be wholly available in the smallest volume of energy medium sphere. As the realm EIL decreases towards C_1 levels, the Mp formation depends upon a proportionately larger spherical volume of energy-medium. Thus the absorption sphere increases in volume although a lesser total energy medium quantity is being absorbed into each Mp. A Mp formed when realm EILs are near to C_2 contains more energy medium than a Mp formed near C_1 realm EILs as stated earlier (Fig 2.6). When the realm datum drops below C_1 the trend for Mp formation ceases. However when certain local C_1 EIL peaks occur, this trend re-appears as in the case of the EIL oscillation. The absorption sphere initiated within those peaks may then extend marginally beyond the peak EIL regions. This occurs in the case of the Mp-event phenomena because of the lesser total energy medium within the absorption sphere. This also means that each Mp-event results in all energy medium within the absorption sphere acquiring a linear momentum towards the spherical centre in the E-M/X event stage. In the M-E/X sequence this linear momentum traces an outward path and must eventually exit beyond the absorption sphere boundary. This leaves a temporary void behind as represented in the diagrammatic sequence of Fig 3.26.

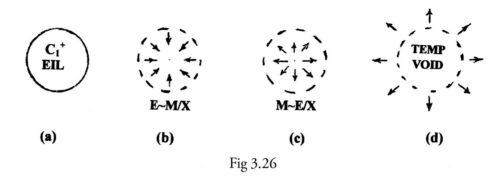

Fig 3.26

At (a) a C_1^+ EIL zone is attained locally and at (b) an E~M/X sequence is set up. At (c) this proceeds into a M~E/X sequence with the indicated outward E-Flow. At (d) this E-Flow resulted in all the energy medium flowing out of the absorption zone to leave a temporary void behind. Although the realm datum will rapidly re-establish in this zone from the pervading realm E-Flows, nevertheless there is a set event duration before this can normalise. This event duration prevents a C_1 EIL being immediately re-achieved with this zone. Hence we specify that when an E~M/X is to be set up from a preceding M~E/X, then it can only occur outside the absorption sphere of the earlier sequence. We must consider the energy medium requirement of the new E~M/X in that it requires an absorption sphere with sufficient energy medium to maintain a reasonable exchange sequence duration. Thus we would expect the next absorption sphere to set up such that it does not conflict with the low energy medium of the temporary void zone.

This is represented in Fig 3.27 where A and B are the Mp-event centres in the existing wave front YY. The absorption spheres around A and B meet at O. If OX is drawn normal to AB then it represents the line of peak EIL from the E-Flows of A and B. Consider the peak position C as a logical position for C_1 EIL attainment. However the absorption spheres around A and B are now temporary void zones and are so close to position C as to make it impossible for an absorption sphere to set up around C. The closest point around which an absorption sphere could be set up to contain similar quantities of energy medium as the former A and B absorption spheres is the geometric position D (ADB being an equilateral triangle). At this stage we may assume that the absorption spheres around A, B and D are quite similar.

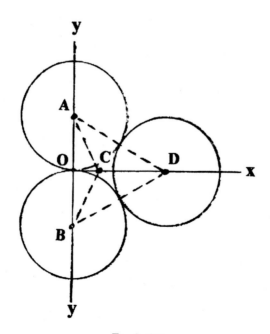

Fig 3. 27

92

Because all the Mp-events within a particular wave front are perfectly in phase with each other then the absorption spheres around A and B in Fig 3.27 will actually commence without sensing the other zone of action. As such the sphere of action of one will most certainly encroach upon the field of the other. This is shown in Fig 3.28. The energy medium in the intersecting zone being shared by the absorption spheres of A and B Mp-events.

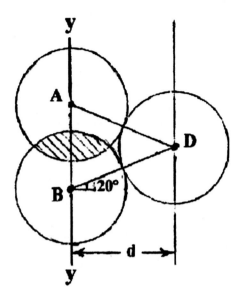

Fig 3.28

The position of the succeeding Mp-event sequence is shown at O under the condition of equal absorption sphere radius. The plane through O and parallel to the wave front 'yy' will then represent the next forward step position of the propagating EIL oscillation. The quantum step in the propagation of the wave front is then represented by the distance between these two adjacent parallel planes.

If we assume that the absorption spheres around A and B (Fig 3.28) intersect by up to half their radius, then the quantum step will be given by;

$$d = BD \times Cos\ 20^0\ \text{(approx.)}$$
$$= .94 \times \text{absorption sphere diameter.}$$

Or, quantum step d μ absorption sphere diameter.

We stated earlier that the size of the absorption sphere was inversely related to the realm EIL datum. Hence the quantum step size will also be inversely related to the realm EIL datum.

Thus the higher the realm EIL the smaller will be the quantum step size and vice versa. This means that the velocity of propagation of the wave front which also varies according to the quantum step size must therefore vary inversely as the realm EIL datum. The higher the realm EIL datum the lesser will be the velocity of wave front propagation and vice versa.

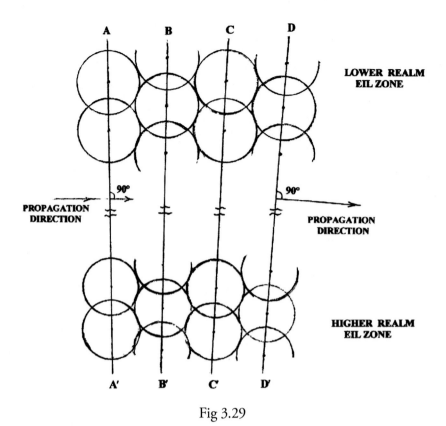

Fig 3.29

We declared that the realm datum was a three-dimensional energy medium grid at varying EILs. As such we may conclude that the EIL oscillation wave front propagates through the realm energy medium grid at a velocity that is variable and dependant upon local energy medium conditions. When the wave front traverses into regions of progressively higher realm EIL the quantum step will progressively shorten and the propagation velocity be correspondingly reduced. The inverse is also true for traverse into regions of progressively lower realm EIL. By the very nature of their effect upon the realm EIL spatial grid all mass bodies will therefore cause the propagation quantum step to progressively shorten when the wave front approaches the body and to lengthen on recession away. This is an important concept when explaining the changes that occur in the curvature of the wave front. This is reflected in a tilt of the p/p caused by the varying realm EIL datum along its plane. The direction of propagation of the wave front or p/p is always normal to the p/p plane. A variation of the quantum step size across the p/p will result in the successive p/p not being parallel to the previous one. This is highlighted diagrammatically in Fig 3.29 for a succession of propagation steps. These quantum steps are represented by the successive p/p positions AA' to BB' to CC' to DD'. The realm EIL datum at the ABCD side of the p/p is assumed marginally lower than that at the A' B ' C' D' side and that the change in EIL is uniform. Hence for the higher realm EIL zone the absorption sphere of the Mp-event will be smaller and the quantum step shorter than for the lower realm EIL zone. The Mp-events within each p/p remain in phase even though there are differences in absorption sphere size. In the diagram it is evident that the wave front along ABCD has traversed a greater distance than the wave front at A' B ' C' D' in the same event duration. As noticed in our experiment of Fig 3.20 we see a similar pattern between the two ends of the p/p. The effective propagation velocity of one is higher than that of the other. As such the p/p AA' undergoes a progressive tilt which becomes quite noticeable when it reaches the position DD'. The illustration in Fig 3.29 is a simple one although there are in existence a complexity of inter-related factors that affect the wave front propagation. It should be apparent to the reader that a mass body must exist somewhere in the direction of the lower part of the illustration. This indicates that when a wave front propagates past such a mass it must tilt in the direction as shown. Einstein's prediction of such an effect was a momentous one so long ago. It is considered however that by isolating each energy medium phenomenon the language of the text becomes more easily understood. At a later stage the reader may put together all the effects described to obtain an overall picture.

So far we have managed to relate propagation velocity to the quantum step and to the realm EIL datum. We must now set up a further relationship between the frequency of EIL oscillations and propagation velocity.

Oscillation Frequency Relationships.

In Fig 3.9 the reader was asked to imagine a set-up of about eight quantum jumps between successive EIL oscillations. The propagation of the wave front requires that the Mp-event sequence must event at each quantum step in turn. From each wave front position a transverse and a backward E-Flow emanate in addition to the forward E-Flow as indicated in Fig 3.10. EIL oscillations from the same Mp-shell source and under similar realm grid conditions will possess the same overall propagation velocity. Subsequently the finite spacing between successive oscillations will remain equal. The rate at which these oscillations / wave fronts pass a fixed realm position is the measure of their recurrence and is termed the oscillation frequency. The shortness or otherwise of the event duration between successive wave front arrivals indicates the spacing between trailing oscillations. The shorter the interval the higher is the repetitiveness or frequency of wave front arrivals.

In Fig 3.14 we had shown the backward E-Flow that emanates from each quantum step position for the series of events between a successive leading and trailing p/p. We also showed how this backward E-Flow was absorbed into the Mp-event absorption spheres of the trailing EIL oscillations. The larger the number of quantum steps between successive oscillations, the lesser will be the EIL of the backward E-Flow when absorbed by the trailing wave front. The quantity of backward linear momentum absorbed into each of the absorption spheres of the trailing p/p is therefore inversely proportional to the interval between successive EIL oscillations.

It is to be remembered that in Fig 3.11 the wave front p/p was ascribed a finite forward velocity Vp. This was the resultant of the linear momenta of all the energy medium quantity absorbed into the wave front absorption spheres. A simple measure of the propagation rate of the oscillation stepping sequence was earlier given by;

$$\text{Propagation rate} = D_o + {}^d\!/_{.866\ V_E}\ \text{duration / step.}$$

where, D_o is the duration of the E-M/X sequence of the Mp-event,

d is the quantum step linear distance, and

V_E is the E-Flow velocity in the M-E/X relative to the Mp-event centre.

We showed that the velocity of the wave front propagation through the realm was directly proportional to the quantum step size. Now in the above expression the term D_o assumes that the wave front propagation is at a dead stop while the E-M/X sequence completes. That is to say, the p/p at each quantum step position is assumed to be stationary. Due to the absorption of linear momentum into each E-M/X event we conclude that an overall forward linear momentum must be ascribed to the energy medium quantity within the p/p. Thus when the E-M/X is in progress the entire Mp-event energy maintains a net velocity Vp in the forward propagation direction. We may now conclude that there are two component factors in the wave front propagation velocity. Of primary importance is the quantum stepping sequence. This factor is responsible for contributing a major portion to the wave front velocity. Of secondary importance is the addition of the wave front velocity component Vp. Although Vp may be a minor part of the overall oscillation velocity its variations nevertheless add important effects to the wave front characteristics.

In order to understand how Vp may vary let us conduct a thought experiment as represented in Fig 3.30. Consider an absorption sphere around a central point A. Let the E-Flows entering this sphere denoted by the indicated arrow markers (1), (2), (3), (4) and (5) be as shown, such that each has a velocity component towards the right hand side only. The net linear momentum of the energy medium within the absorption sphere around A is indicated to scale by the vector quantity VA. Now let us consider an absorption sphere around another point B with the same directional E-Flows as at A but with additional E-Flows (6) and (7) as well. Since the velocity components of (6) and (7) are towards the left hand side then these will act in opposition to VA . The net linear momentum of the energy medium within the absorption sphere around B will be the difference between left and right hand velocity components. This is represented to scale by the vector quantity VB. It is clearly observed that VA is greater than VB. From this we may conclude that if A and B represent the Mp-event centres within a p/p, then the Vp component of the propagation velocity is greater for the wave front through A than through B.

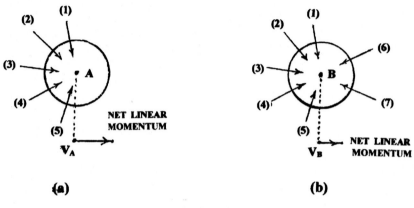

Fig 3.30

Let us extend our experiment of Fig 3.30(b) a bit farther. Let us assume that the E-Flows (1), (2), (3), (4) and (5) originate at fixed positions in a plane that is one quantum step to the left side of B. Thus each of these E-Flows will be at a reasonably constant EIL when entering the absorption sphere position around point B. Inversely let us assume that the E-Flows (6) and (7) originate in a plane that is about eight quantum steps to the right side of B but is variable in its choice of position. From this it will be apparent that the vector quantity VB will alter depending on whether E-Flows (6) and (7) enter the absorption sphere at a higher or a lower EIL. This will correspond inversely (inverse square law for EIL) on whether the E-Flows originated from a closer or a farther quantum step position. Fig 3.31 shows how this applies to the absorption of the forward as well as the backward E-Flows in the wave front actual set up.

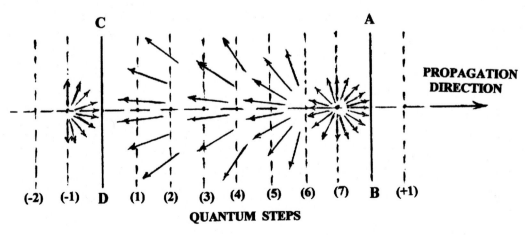

Fig 3.31

96

In the diagram, AB and CD are the relative positions of two consecutive EIL oscillations. AB leads CD by eight quantum steps. The E-Flows are indicated by arrows and are shown just as the Mp-event is about to occur at AB and CD positions. Let us concentrate our attention upon CD simply because both the forward and backward E-Flows entering this plane are clearly represented. The forward E-Flow entering CD is from the step position (-1) which is just one quantum jump away. The backward E-Flow entering CD had originated from the step position (4) which is 4 quantum jumps away. In the diagram only four sets of diverging backward E-Flows are shown from step positions (7), (6), (5) and (4). Those from step positions (3), (2) and (1) are not represented since these have already been absorbed into the wave fronts at the (-1), (-2), (-3), etc., step positions. Now the net linear momentum of the p/p CD depends upon the linear momentum of all the energy medium quantities absorbed there. If the linear momentum of the energy medium flowing in the forward direction is reasonably constant then we may say that the net linear momentum of the p/p CD will vary according to variations of the backward E-Flow linear momentum. We have already shown that the EIL of the backward E-Flows approaching CD is much less than the forward E-Flow to CD. In the diagram example Fig 3.31, the backward E-Flow has traversed a distance of four quantum jumps which is four times the distance traveled by the forward E-Flow to CD. Since the EIL obeys the inverse square law for distance then the EIL in the backward E-Flow will be 16 times less than the EIL of the forward E-Flow to CD.

If we allow the spacing between the consecutive wave fronts to vary, then by the same type of illustrative example we could show the backward E-Flow at CD to originate from a closer or a farther quantum step position. This would in turn result in a variation in linear momentum absorbed from the backward E-Flow. The greater the assumed spacing between p/p AB and p/p CD, the greater will be the distance from CD at which the backward E-Flow will have originated and the lesser will be the EIL when the E-Flow enters the absorption sphere at CD. Thus the backward linear momentum absorbed into the p/p will be relatively small and the net linear momentum VB in Fig 3.30(b) will be so much the greater. The smaller the assumed spacing between the p/p AB and CD, the lesser will be the distance from CD at which the backward E-Flow entering CD will have originated and the greater will be the EIL of the backward E-Flow entering the absorption sphere at CD. The backward linear momentum absorbed into the wave front CD will thus be relatively greater and the net linear momentum VB will by difference be relatively less.

From the above we may conclude that when the quantum step spacing between successive EIL oscillations is large then the wave front propagation velocity component Vp is proportionately greater; and vice versa. The spacing between the oscillations is termed the wave length of the propagating waves and this is inversely proportional to the frequency. The shorter the wave length the higher the frequency, and the longer the wave length the lower is the frequency.

Now, Wave length $\propto 1/$ frequency
and Wave length $\propto$ velocity component Vp
Therefore, velocity component $\propto 1/$ frequency

From this we are forced to conclude that, other factors being constant, the propagation velocity of an EIL oscillation is inversely proportional to its true frequency.

The reader is reminded that the illustration of E-Flows in Fig 3.31 is only a simple representation of a very complex series of events. Not only does the Mp-event recur at successive step positions but also shifts in the direction of propagation while in the energy medium exchange sequence. The absorption spheres are uniformly distributed across the p/p and the illustration has shown the effect from only one of these. The reader must visualise the rest.

E-Flows from more distant origins within the wave-length region must also be accounted for in the net linear momentum calculations.

Furthermore, the reader must not forget the Mp-shell origin of the EIL oscillation and that the energy medium events at that stage are mainly responsible for the repetition or frequency of occurrence of the wave front and that the realm datum is the major energy medium contributor to each p/p.

The measurement of true frequency is a virtual impossibility because of the variance pattern in the proton time rate (Chapter 5) by which we gauge all event duration. If proton time were to slow to half its present event change rate, then our measurements would show the velocity of wave front propagation to have doubled (even though it is actually unchanged). By the same virtue, the frequency of the oscillation will appear to have also doubled even though the wave length spacing is as before. Our concept of the velocity of light waves and its frequencies is therefore a relative one. However since frequencies may be observed only in relation to one another we would still maintain that under similar conditions a lower frequency wave has a greater propagation velocity than a higher frequency wave.

There are three factors upon which the observed frequency of an EIL oscillation is determined. Firstly, the oscillation rate of the Mp-shell origin determines the repetitive rate of wave front formation. The more rapid the Mp-shell oscillations the smaller the inter-wave spacing and hence the higher the frequency. Secondly, if the above oscillation rate is maintained but with the Mp-shell acquiring a finite velocity, then the actual wave-lengths will change. The wavelength will be shorter in the direction of the acquired velocity and greater in the opposite direction. This is termed the doppler effect and results in a defined wave-length change as shown in Fig 3.32.

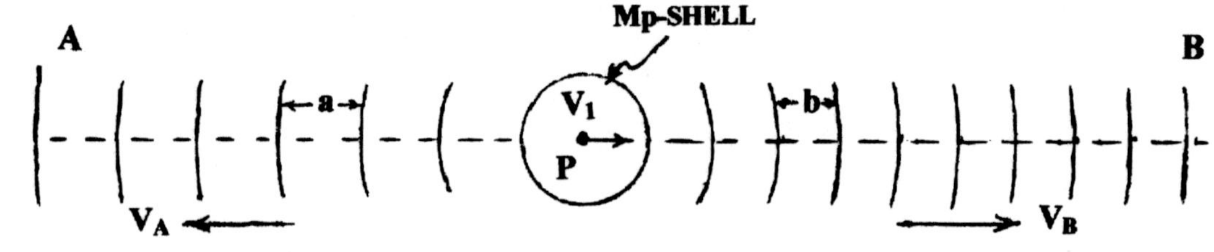

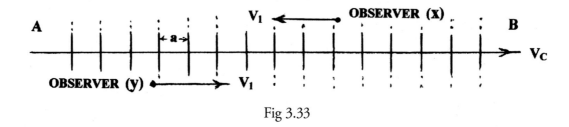

Fig 3.32

Let P be the proton centre of a Mp-shell and let it possess a velocity V_1 in the direction PB. By virtue of the proton velocity V_1 each successive wave front in the direction PB is effectively brought up to trail closer behind the previous one. Also each successive wave front in the direction PA trails farther behind the previous one. Thus the wave length in the direction PB represented by 'b' is less than the wave length in the direction PA represented by 'a'. This characterises the doppler effect as far as relative frequencies are concerned. There is however a more important resultant effect and that is that the velocity of the wave front propagation has marginally changed. If V_A and V_B represent the wave front velocities in the directions PA and PB respectively, then V_B is marginally less than V_A.

Fig 3.33

The third factor upon which an observed frequency is dependant is the relative velocity of the observer. This effect again is very similar to the doppler effect except that in this case it is an apparent effect witnessed by the observer only. The wave length remains unaltered but appears to meet the observer at a lower or higher repetitive rate depending upon whether the observers motion is in the same or opposite direction to the propagating waves. Fig 3.33 highlights this.

Let the velocity of propagation of the light wave be denoted by Vc in the direction AB. The wave length is denoted by 'a'. The true frequency is then given by,

$$F_T = \frac{Vc}{a}$$

Let the two observers (x) and (y) each travel at a velocity V_1 but in opposite directions as shown. Let V_1 be small compared to Vc. The observer (x) travels towards the wave fronts and therefore each wave front is met by (x) at a slightly earlier moment than if (x) had been stationary. The apparent frequency of the light wave observed by (x) will therefore be given by,

$$F_{(x)} = \frac{(Vc + V1)}{a} \text{ (greater than } F_T)$$

The observer (y) however is moving away from the approaching wave fronts and therefore each wave front passes (y) at a slightly later moment than if (y) had been stationary. The apparent frequency of the light wave observed by (y) will therefore be given by,

$$F_{(y)} = \frac{(Vc - V1)}{a} \text{ (lesser than } F_T)$$

The observers (x) and (y) each witness a different frequency although the light wave velocity and wave length have remained consistent. Actual observations show that $F_{(x)}$ is "blue shifted" while $F_{(y)}$ is 'red shifted'. It must be noted that the p/p velocity component Vp is unchanged and therefore frequency comparisons should only be made by one observer for any correlation of future experiment results.

A graphical representation of the relationship between true frequency and the velocity component Vp is given in Fig 3.34 and shows a drooping characteristic.

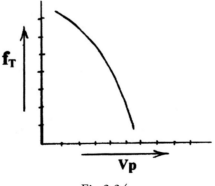

Fig 3.34

We end this discussion with the startling conclusion that differing light wave frequencies propagate through the same realm space at marginally differing velocities. The conglomeration of light waves reaching earth from some distant galaxy does so as a constant stream. We are able to separate each light wave frequency through velocity segregation in a prism and thereby confirm the above statement. We shall be discussing this effect in more detail in the section on refraction.

Other Propagation Relationships.

An interesting question is posed on the ultimate fate of the wave front. Whether the quantum jump mechanism is able to continue indefinitely or whether it ceases at some specific distance from origin. Figs 3.23 and 3.25 suggest that the answer is not a simple one. Firstly, the initial energy medium density within the wave front is variable and depends upon the complexity of the originating Mp-shell source (chapter 6). And secondly, the wave front curvature is subject to change as it propagates through a varying realm EIL grid / datum. The change in propagation direction as indicated in Fig 3.29 causes an overall change in the wave front curvature. With the wave front in the normal expanding mode i.e. convex curvature, the energy medium density will continually reduce and the wave front should eventually cease "stepping". If the wave front becomes truly flat in a particular direction then there is no reason to suppose a change in the wave front energy medium density over infinite realm distances. In practice however areas of the wave front are absorbed or deflected by other mass objects. As such the resulting section of wave front propagating through the realm will undergo a decay of energy medium density near its sectional edges even when the entire section is flat. The reader may find the mental examination of the stepping mechanism near the sectional edge of the wave front quite complex and interesting.

Another possibility is for the wave front curvature to become concave. This results in an increase in the energy medium density with progression because of a reduction in the spacing between absorption spheres within the wave front. In this case the energy medium density will increase upto the focal point and thereafter re-emerge as a wave front with the normal convex curvature. The wave front will subsequently appear to have originated at this focal point.

Let us now consider what happens to the propagation rate when the wave front density alters. One of the considerations to be remembered is that with reduced wave front density the absorption spheres enlarge and their spacing increases. This in turn results in an increased quantum jump length and hence a higher velocity of wave front propagation. Thus under normal expansion conditions the velocity of propagation of the wave front increases marginally with distance traversed.

So long as there is a high realm EIL datum the wave front is able to absorb large quantities of this prevalent realm energy medium to perpetuate its Mp-event stepping action. There are however certain extreme conditions of realm EIL which set a limitation upon the recurrence mechanism of the wave front steps. In the last chapter it was stated that EIL oscillations only occurred within the realm regions corresponding to the upper part of the circle 6 energy medium curve. This meant that EIL oscillations could not occur if the realm datum was C_1 or above. By definition a C_1^+ EIL is a realm zone where Mp formation occurs in profusion and at random. There is no synchronism of Mp formation as for the Mp-event in the wave front to achieve a uniform forward step. If a light wave were to propagate into such a zone then the wave front would de-synchronize after the first such step and subsequently lose its propagation characteristic. Also light waves could not originate from inside such a zone since protons would not possess a Mp-shell which is the source of all EIL oscillations. Incidentally this poses an interesting question regarding the fate of atomic structure in a realm EIL above C_1. The realm EIL datum or grid as we have referred to it increases in the proximity of mass objects. In cases of collapsed stars this mass object may result in the EIL grid exceeding the C_1 level and subsequently preventing the wave front from propagating upto it. Also no wave fronts originate from that object. We shall discuss this "black object" phenomenon in more detail in the next chapter.

Inversely, EIL oscillations do not occur when the realm EIL datum is so low that C_1 EILs are virtually unachievable by a wave front. The property whereby a C_1 EIL initiates the Mp-event is the essential factor in the quantum step mechanism. In the circle 6 energy medium curve the trend is one of continual reduction in the realm datum. At some point on this curve there will not be sufficient datum energy medium to permit the forward E-Flows in the wave front to attain a C_1 EIL at the required step position (Fig 3.27). Subsequently the Mp-event could not recur and the wave front energy medium would simply dissipate as an E-Flow. It will be noticed however that prior to this occurrence the propagation velocity would have increased somewhat (propagation velocity being inversely proportional to the realm EIL). By similar reasoning EIL oscillations cannot propagate through the cosmic void.

At this stage the reader should have acquired a consciousness of the character of EIL oscillations. It would not be possible to specify each and every aspect of that character against all the realm conditions that might be encountered by the wave front. It is expected of the reader to interpret those results for himself. So far we have shown that,

1. The velocity of propagation is directly proportional to the quantum step size and inversely proportional to the realm EIL.
2. The propagating velocity is made up of two components; the quantum step component and the p/p forward motion component Vp.
3. The p/p forward motion component Vp is directly proportional to the oscillation wave length or inversely proportional to the frequency. Subsequently the velocity of propagation becomes inversely proportional to the oscillation frequency.
4. The quantum step size increases marginally with the distance propagated.
5. The direction of propagation tilts towards the higher realm EIL datum.
6. The wave front propagation velocity is always less than the corresponding E-Flow velocity.
7. Changes in frequency are real only if they are the result of changes in the motion of the Mp-shell source.
8. The wave front cannot propagate in a realm EIL above C_1, nor if the realm EIL is below a defined minimum level.

From the above it would appear that there is an extremely variable situation for light waves within the realm and that all astronomical observations were subsequently prone to grossly erroneous interpretation. However all concepts are brought into parallel by the reasonably isotropic conditions around us that extends to a radius of over ten billion light years (our current limit of observation).

On the basis of our conclusions on the properties of EIL oscillations let us now attempt to explain some of the more commonly observed light wave phenomena.

Reflection of the Wave Front.

The physical laws governing the reflection of light waves should be well known to the reader. These laws were never derived from any wave theory but like most physical laws have become known through extensive observation. It is not the intention here to review such well established rules but simply to indicate a fit with our propagation model. A test of that model at this stage is essential if we are to progress further into the physical rules governing the realm energy medium. We have continually stated that there is a very complex pattern of action and reaction between energy-medium quantities within the realm. So far we are only able to present a simplified version of these actions in order to convey certain properties and characteristics of the energy medium. The reader must realise that a single p/p does not step through the realm on its own. It is accompanied by countless others issuing from every atom structure in the realm and at varying repetitive rates. To describe the inter-play of all these would simply confuse the overall picture which it is hoped the reader will eventually develop within his own mind. Differing views of the same aspect would certainly widen the objectivity within the theory.

There are two essential features in light wave reflections that we must now attempt to explain. The first is the reversibility of the propagation direction by a smooth object surface. The second is the direction of the reflected wave front when the incident light wave meets a plane object surface at an oblique angle. For the moment we shall ignore some of the other phenomena such as plane polarisation etc., until we approach that particular subject later.

In Fig 3.11 we first explained the motion Vp of the p/p. We showed that the backward E-Flow from the Mp-event was lesser in energy medium quantity than the corresponding forward E-Flow. Forward being the wave front propagation direction. Vp is essential to the division of the forward and backward E-Flows so that the former is greater. This may be looked upon as a fundamental rule for our propagation model. Thus in order to reverse the direction of propagation of the wave front we must somehow first reverse the p/p component Vp. The law of conservation will not permit this unless we can cause sufficient energy medium with opposite linear momentum to be absorbed into the E~M/X of the p/p.

In an earlier discussion we mentioned that the basic mass structure consisted of a recurring Mp-shell around a single proton. More complicated combinations subsequently formed from this basic atom such that multiple Mp-shells could exist around a multi-proton nucleus. The oscillating relationships between these multiple shells becomes more complex as their numbers increase. Basically each Mp-shell is a synchronism of Mps that must present an E-Flow to the realm during the M~E/X sequence. At the present stage we are only interested in the effect of this E-Flow outside the mass structure. This was the origin of our wave fronts as explained in Fig 3.3.

We also know that the circle 6 phase protons are in a general state of mass to energy medium decay. Subsequently all mass structures must therefore give out a seemingly continuous E-Flow that obeys the inverse square law for EIL with distance. At greater distances this decay E-Flow intensity will only make a marginal contribution to the realm datum and so have very little effect upon the wave front velocity component Vp.

In Fig 3.30 we indicated the effect of an opposing E-Flow upon Vp in order to show a relationship between propagating velocity and wave-length. The closer the p/p was to the source of a backward E-Flow the lesser was Vp. A similar situation now exists with regard to the wave front approaching a mass structure. Firstly the wave front enters a zone of progressively higher realm datum and secondly it must also encounter the repetitive energy medium releases of the Mp-shell.

A quantum step will eventually be reached when the net linear momentum of all energy medium quantities absorbed will be nil, i.e. Vp = 0. Let us assume that such a condition is attained when the wave front is within a few quantum steps of the mass structure.

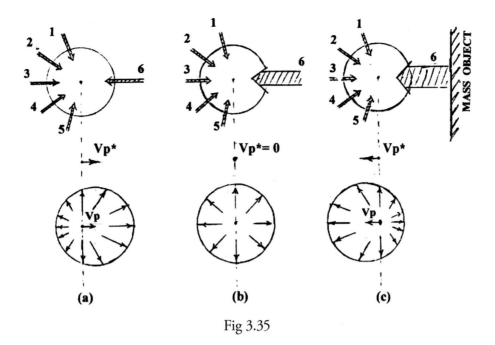

Fig 3.35

Let us now conduct a thought experiment to visualize the reversing process of wave front propagation. Assume the mass object to give out a continuous E-Flow. Fig 3.35 indicates the sequence of Vp change as the p/p approaches close to the mass object. The EIL of the E-Flow is represented by the width of the arrow vectors.

The upper diagram set represents the E~M/X sequence and the lower represents the subsequent M~E/X. Consider the mass object to the right to be of a solid compound mass structure so that we can for the moment ignore the behaviour of the Mp-shell phenomenon. The initial normal propagation is shown at (a) with Vp acting in the direction of propagation towards the mass object. The EIL of the E-Flow (6) arriving at the absorption sphere from the relatively distant mass object is low and so has very little effect upon Vp. At (b) however the E-Flow (6) is much greater and just balances the E-Flows (1), (2), (3), (4) and (5) such that Vp = 0. This situation results in an equal backward and forward E-Flow distribution from the subsequent M~E/X and should remind the reader of the E-Flow situation from the Mp-shell setup in Fig 3.2(b) . It must be noted that although the energy medium quantity available at each Mp-event is increased, the energy medium density of the wave front remains virtually unchanged. This is because the absorption centre spacing does not change but the absorption sphere size decreases marginally at the higher realm datum.

With Vp = 0 the subsequent division of E-Flows in the backward and forward directions must be equal. Because of the higher realm EIL there is the potential for achieving a C_1 EIL at both the backward and forward quantum step distances. The former however must progress into a relatively lower EIL region while the latter encounters an even greater energy-medium intensity. Although both may achieve a C_1 EIL the probability of sufficient energy medium being absorbed in the backward position for a strong wave front is relatively low. Any wave front propagating backwards from the Fig 3.35(b) position will subsequently be faint and barely noticeable. The forward E-Flow however progresses to the next forward quantum step position indicated at Fig 3.35(c) as a strong wave front. It will be noticed that the linear momentum absorbed at this location is such that Vp now acts in the direction away from the mass object. According to our fundamental rule a strong wave front will propagate in the same direction as Vp. Thus in our simplified thought experiment we have theoretically caused the wave front to be reversed or reflected by the mass object. Our experiment also indicated that it is theoretically possible for a single wave front to give rise to two separate wave fronts travelling in opposite directions. We shall observe the implications of this a bit later on in our discussion on refraction.

Keeping in mind the above simplified principle of wave front reversal let us now complicate the setup by considering the more intricate oscillating structure of the Mp-shell. The reader will remember that all our wave fronts originated from such a structure. However the Mp-shell shown in Figs 3.2 and 3.3 was that around the most primary of atoms. Here we indicated the repetitive formation of an Mp-shell around a single proton. As the quantity of centrally located protons increases so does the complexity of the surrounding Mp-shell structure.

For the moment let us keep this discussion as elementary as possible by assuming that for every proton entity within the nucleus there exists a separate Mp-shell section. (See Fig 6.33). Also that each Mp-shell has a repetitive beat or re-structuring oscillation that depends upon its relative concentric position around the nucleus. We shall see later that Mp-shell oscillations must have discrete frequencies in order that the shells may perpetuate (chapter 6).

Let us continue with our discussion by further assuming that each Mp-shell section results in a faint but very real EIL oscillation that spreads outwards from it. Therefore, the greater the number of shell sections the more complex will be the nature of the EIL oscillations emanating from the overall structure. This mixing of EIL oscillations that seem to originate from the same central source results in what is termed as a compound frequency. It may be defined as the apparent integration of two or more EIL oscillation frequencies ensuing from a structure of semi-spherical Mp-shells. An example of compound frequencies that we are generally able to observe is the light reaching us from the sun and stars. We term this as 'white light' but in fact it contains not only all the colours of the visible spectrum but other frequencies as well. Incidentally we are able to create white light by heating a mass object to incandescence. This indicates a relationship between the strong emitted compound frequencies and the temperature of a structure.

Now what is temperature? It is simply a measure of the realm energy medium that is absorbed by the Mp-shell structure over and above its basic repetitive requirement. This also represents the corresponding amount of energy medium that can be given up by the shell structure. The minimum energy medium level for Mp-shell structure existence is termed as the absolute zero of temperature. There would be a marginal variation of this absolute minimum quantity as the decay of the proton progressed and the shell structure gradually closed inwards.

In Fig 2.17 we indicated how the Mp-shell was formed and in subsequent reasoning showed that the radial distance at which the Mp-shell repetitions stabilised depended upon the EIL conditions of that realm area. The higher the EIL conditions then the greater would be the radial position at which the shell stabilised and vice versa. Thus at a higher energy medium level or temperature the Mp-shell will be diametrically larger (expansion phenomena). The oscillation frequency of a Mp-shell alters with change in its radial position but it would not be appropriate to detail such a characteristic at this stage. The reader is asked to simply consider a provisional concept of the increased shell size (and energy medium) in relation to the compound frequency setup as represented in Fig 3.36 for an imaginary shell structure. In the 'A' region at the relatively lower temperatures, the shell

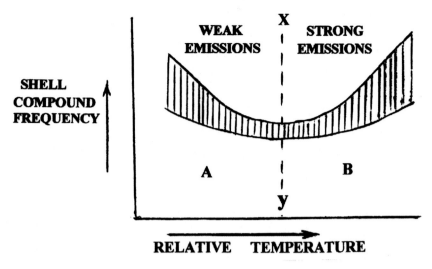

Fig 3.36

oscillations result only in very weak emissions and it is this characteristic that is so essential to the reflection process. The term "weak emission" refers to a wave front that is able to progress up to a limited number of quantum jumps only. In the 'B' region there is strong emission from a higher energy-medium shell structure. The separating indication 'xy' is simply a measure of the prevailing energy medium conditions and is therefore relative to such factors. The band AB will however always remain symmetrical about 'xy' at its varying locations.

Let us now, with the above theories in hand, return to our discussion of the reflection process. Consider the approach of a wave front towards a Mp-shell structure at the weak emission zone 'A' (Fig 3.36). The incident wave front mechanism must encounter the weak front emanating from this shell structure. In a weak emission the quantum jumps are initially very small but rapidly get larger with each successive step. When the step size enlarges out of all proportion to the available energy medium then the wave front will discontinue after a limited propagation as shown in Fig 3.37. Here we represent the outermost shell at XX around the nucleus N. The radiating lines A, B, C, D and E indicate the direction of the weak emission front from the Mp centres on the shell XX. The spacing between these lines increases with distance and successive quantum jumps get correspondingly larger. It is obvious that at some step position e.g. position 8, the spacing of the Mp-event centres

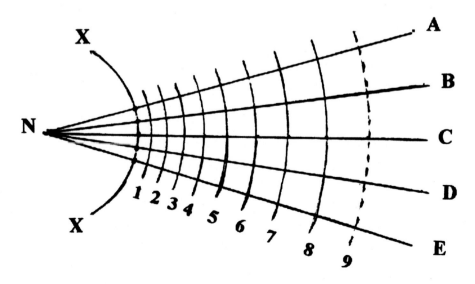

Fig 3.37

105

will have become so great that the energy medium quantities are unable to attain another C_1 EIL Mp-event. The propagation has ceased and the wave front energy-medium dissipated. However we are not really concerned with the fading of these wave fronts but rather with the propagation energies in the initial few quantum jumps. The reader should note the change in the energy-medium mode from our thought experiment of Fig 3.35, although the basic principle remains the same.

Now the incident wave front mechanism is unaffected at the hypothetical step 9 position. As it proceeds closer to XX however it must encounter an increasing intensity of opposite E-Flows within the weak emission wave front. The incident wave front will absorb some of this energy medium resulting in a marginal change in Vp the p/p velocity component. A position very close to XX will finally be attained as shown in Fig 3.38 where the incident p/p is represented by AA'. Let us assume that at this position the wave front AA' has absorbed a large proportion of the E-Flow from XX such that the velocity component Vp has reversed. It is very important that the E-Flow pulses from XX reach the AA' position just as the p/p at this position is in the E-M/X sequence in order that linear momenta may be absorbed and Vp reversed.

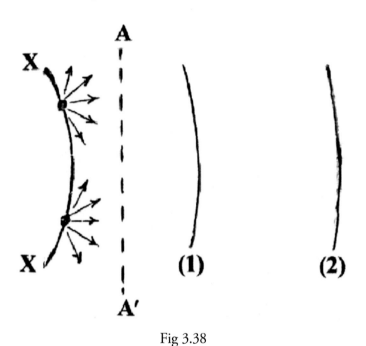

Fig 3.38

Now the incident wave front has a set frequency and so does the Mp-shell XX. In order that the situation depicted by Fig 3.38 may repeat for each incident wave front, the two frequencies must have a synchronous 'beat'. This does not mean that the two frequencies must be the same, but it does mean that the frequency of XX be a whole number multiple of the frequency of AA'.

Since the real frequencies being dealt with are of a compound nature on both sides it becomes apparent that the compound incident frequencies are not always totally reflected. In the case of compound white light reflection may only be for a particular frequency while the remainder propagate into the Mp-shell absorption cycle.

Another case would be when an alternating beat is set up and every second or third incident p/p gets to be reflected. In this case the reflected wave front frequency would correspondingly be half or one third of the incident frequency. The reader must also keep in mind that energy medium absorbed by the Mp-shells from frequencies not reflected result in increased energy medium levels of the shell. This would correspond to a change in the local shell temperature-frequency relationship (Fig 3.36) and could result in an independent strong wave front emission.

In order to progress with this chapter we must leave the reader to think deeply upon the above factors. We suggest he apply his own imagination and interpretation to the theories outlined above and develop our reflection model even further.

Oblique reflection.

In the previous section we assumed the direction of the incident wave front to be normal to the mass structure surface. The reflected wave front was subsequently reversed along this path and propagated away from the structure i.e. in a direction that was perpendicularly away from the reflecting surface. The exact point of reversal was indicated by the quantum step position at which the energy medium absorbed into the Mp-event sphere resulted in a reversal of the p/p velocity component Vp. The wave front comprises of centres of the Mp-event phenomenon distributed uniformly across the p/p plane. Thus each Mp-event sphere propagates to the next forward step position in a local energy medium coordinated manner. It would therefore be quite reasonable for us to deal with the principle of oblique reflection by the application of linear momentum laws to individual p/p absorption centres.

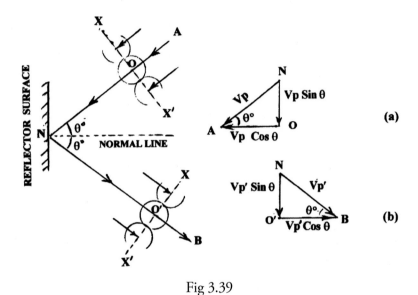

Fig 3.39

In Fig 3.39(a) consider O to be the centre of a Mp-event sphere in a wave front XX' propagating towards the reflecting surface but at the oblique angle $\theta°$ to the normal. The p/p velocity component Vp which is in the direction of propagation can be split into the components Vp Cos θ and Vp Sin θ such that the former is normal and the latter parallel to the reflector surface. As the absorption sphere steps closer in, the outward E-Flow emissions get absorbed into the wave front centres. Since these emissions are in an outward direction and symmetrical about the normal the Vp Sin θ component of the p/p linear momentum will not be affected. The Vp Cos θ component will however be initially reduced and then wholly reversed when large quantities of emission energy medium are absorbed (Fig 3.35) .

The p/p velocity component Vp is dependant mainly upon the frequency of the oscillation when other influences can be ignored. This was represented in Fig 3.31. Thus in Fig 3.39(b) we can determine the propagation direction of the reflected wave front by assuming that at large distances from the reflector surface the incident and reflected wave fronts propagate under the same realm EIL conditions. By large distances we mean several emission quantum steps. When the wave front has been reversed at the reflector surface and has propagated outside this minimum distance then the p/p velocity component Vp' will be the same as the incident value Vp. Now since the parallel components of this linear momentum vector has remained unchanged throughout the wave front reversal process we find that the vector diagrams NOA and NO'B are similar. This means that the net effect of the wave front reversal is to direct the propagating wave front in a direction such that

    (1)   Vp'   = Vp

    (2) Vp' Sin θ   = Vp Sin q

and   (3) Vp' Cos θ   = Vp Cos θ.

This is only possible if the reflected direction makes the same angle with the normal as the incident wave but on the opposite side of the normal line. The reader should note that we refer above only to ultimate reflected propagation direction at a position a large distance from the reflector surface. The close-in detailed reflection mechanics are a bit more complex and are represented in Fig 3.40 based upon the principles of energy medium linear momentum absorption at each propagation step.

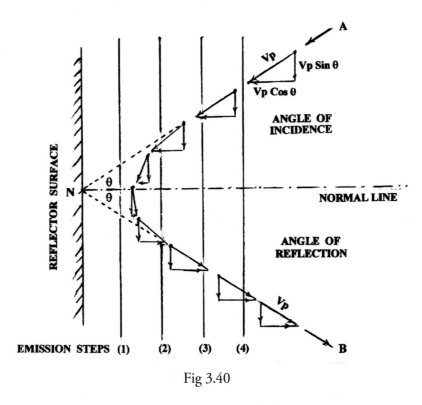

Fig 3.40

It must be remembered that the Mp-event absorption sphere is extremely minute compared to the Mp-shell of the atom structure. As such the wave front will always be reflected as a p/p and never as individual centres. An individual Mp-event cannot operate on its own but must be supported by a combined network of synchronised exchanges in order to achieve propagation to the next quantum step position.

Consider the incident wave front at A to approach the reflector surface at the angle θ to the normal line. As the wave front moves closer in the vector quantity Vp is reduced. The vector diagram also indicates a change in direction for Vp while its parallel component Vp Sin θ remains unchanged. The normal line component Vp Cos θ reduces to zero and then reverses its direction and rapidly increases to its former though reversed vectorial quantity. The reader may note that these close-in changes are not symmetrical about the normal line. This is because the changes to the component

Vp Cos θ are not the same for incident and reflected wave fronts. The backward E-Flow conditions get progressively worse for the incident Mp-event but which hardly applies to the reflected propagation. This however does not interfere with overall angled relationships and it is suggested that the reader conduct a personal research to develop this aspect further.

Now the smaller the incident angle θ the greater will be the normal component Vp Cos θ. In order that this energy medium linear momentum component may be reversed, a sufficient quantity of opposite linear momentum must be absorbed into each Mp-event. The greater the Vp Cos θ component then the greater will be the reversing energy medium requirements. The additional energy medium exists at positions closer in to the reflector surface by virtue of the higher emission wavefront densities nearer the source.

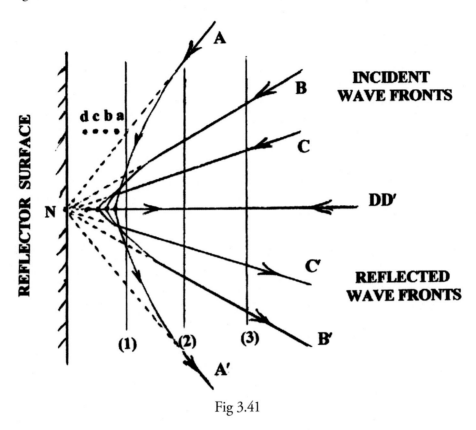

Fig 3.41

Thus an incident wave front propagating towards the reflector surface along the normal line will require the greatest quantity of energy medium for wave front reversal. This also means that this wave front will approach to the closest possible reflector surface position. A normal incident wave front is therefore reflected from a very close-in step position while other oblique angled incident wave fronts will get reflected from slightly farther out positions. The relative wave front paths for differing incident angles is represented in Fig 3.41. Propagation of wave front centres is represented by the incident paths A, B, C and D and the reversal positions by the normal line points a,b,c and d. It should be noticed that the curve leading from the incident path to the reflected path is not a smooth continuous curve but contains a sharp change of direction at the reversal point.

Let us now mentally observe a p/p or wave front section undergoing the reflection process. In Fig 3.42(a) we show the combined wave front centres A, B, C and D as a small wave front section approaching the reflector surface at some random oblique angle.

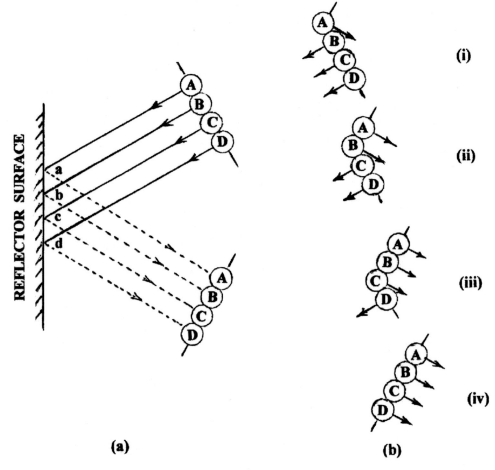

(i)

(ii)

(iii)

(iv)

(a)

(b)

Fig 3.42

From the diagram it is clearly seen that the wave front centre A will be the first to undergo a reversal of its propagation direction when meeting the reflector emissions at point 'a'. The other centres B, C and D will however still continue forwards towards their respective reflector contact points 'b', 'c' and 'd'. The directional status of all these Mp-event centres when each gets successively reversed is represented in the sequence at Fig 3.42(b). The reader is requested to study the diagram carefully and to establish a personal concept of this process in as much mental detail as possible by the application of the propagation model and linear momentum conservation principles.

The above represents an extremely small section of the p/p being reflected by an approximately flat Mp-shell surface section. For the more usual larger wave fronts the reflector surface may be made up from a combination of several Mp-shells of adjacent atoms. Subsequently this may present sections of the wave front with different surface angles. The same reflection laws will still apply but it will be seen that the reflected wave front will separate into sections that correspond to the size of each planar face of the reflecting surface. This is shown in Fig 3.43 and is known as reflective scatter of a wave front.

In this diagrammatic representation a large incident wave front A-K is split into three much smaller sections on reflection. Each section (A-C, D-G and H-K) is shown as propagating in a different

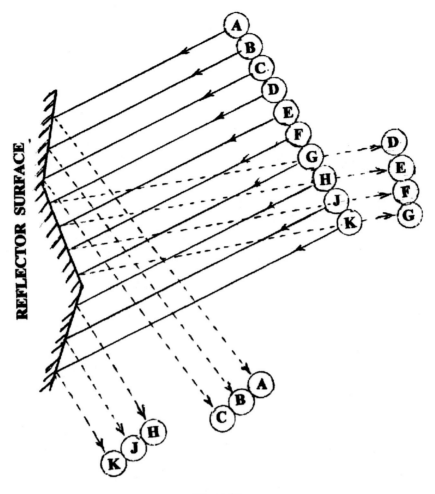

Fig 3.43

direction depending wholly upon the angle at which the reflection faces present themselves to the incident wave front. Here again each reflector face is very large in comparison to the size of the p/p absorption spheres and as such each reflected section of the wave front is able to propagate as a p/p unit.

There are mathematical formulae that could be derived for the energy medium changes within the Mp-event centres of a wave front undergoing reflection. These would however be subjective in character and not easy to simulate. They could however provide us with additional energy medium laws that extend into unknown areas. The reader is encouraged to attempt such a course for rationalising our propagation model and it may show that corrections are necessary to the reasoning presented earlier.

The reader must also realise that the issues within these models for propagation and reflection have been grossly over simplified. For instance we have ignored the decaying nature of all matter and the subsequent E-Flows resulting therefrom. This is not to say that the reasoning within our models is unjustified, but it does indicate that in reality a large number of external factors must also be taken into account. However at this stage of presenting a new theory it was not considered prudent to "bake the cake" entirely but to simply lay some of the ingredients upon the table.

Although we have no defined observable proof of the accuracy of our propagation model we have little cause to doubt its probability on the basis of its application so far. Let us now therefore consider some of the other commonly observed light wave characteristics in order to further test these energy medium exchange concepts.

Refraction of the Wave Front.

We have shown our propagating model to be subject to a set of energy medium physical laws. These were summarised earlier and indicated the variable nature in the stepping action / velocity of the wave front propagation. In Fig 3.29 we showed how the direction of propagation could be changed and it is this characteristic that we shall now look at a bit more closely.

In the previous section we showed that the p/p velocity component Vp could be reversed with a subsequent turn-around of the wave front stepping action. This reversibility in propagation direction was termed as the reflection characteristic of the wave front and depended upon the emission character of the reflecting body. We also indicated in Fig 3.35 that it was theoretically possible for a single wave front to give rise to two separate wave fronts travelling in opposite directions. 'Opposite' refers to the different directions initiated by the Forward and Backward E-Flows of the M-E/X in a p/p Mp-event sphere. In our propagation model we must therefore assume that such a dual set-up occurs during the reflection process but that we only notice the strong wave-front action.

In Fig 3.40 we showed the wave front propagation direction to change marginally just before reflection occurred. This marginal change in propagation direction is termed refraction and is much more apparent when the EIL differentials are high. We have already shown that the wave front propagates in quantum jumps or steps and that these always occur in a direction that is "forward" and at right angles to the p/p plane. In Fig 3.29 it was shown that this p/p plane could "tilt" and subsequently alter the propagation direction. This is the basic principle under which refraction occurs.

The tilt in the p/p can only occur when an EIL differential exists across the wave front plane. With the wave front entering normally into a region of higher EIL there may be no change in propagation direction but there is however a change in the quantum jump size and therefore a corresponding change in propagation velocity. This is shown in Fig 3.44 with the wave front and quantum jump grossly exaggerated.

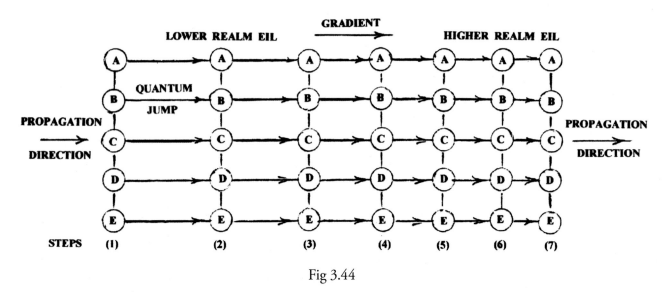

Fig 3.44

In the diagram, A,B,C,D,E, represents the wave front propagating from left to right from a region of lower realm EIL to a region of higher realm EIL. For simplicity we have shown this to occur over only seven quantum jumps. It is clearly seen that the quantum jump between steps (1) and (2) is much larger than the quantum jump between steps (6) and (7). As such the velocity of propagation has decreased with progression into the higher realm EIL zone although its direction has remained unchanged. Let us now consider the case where a wave front propagates through a region from a lower to a higher realm EIL but where the EIL change is oblique to the wave front plane. This is shown in Fig 3.45.

Consider the wave front ABCDE at the step (1) position. The absorption centre 'A' is within a lower realm EIL than the absorption centre 'E'. Corresponding realm EILs are indicated by point 'a' and point 'b' on the EIL gradient scale shown.

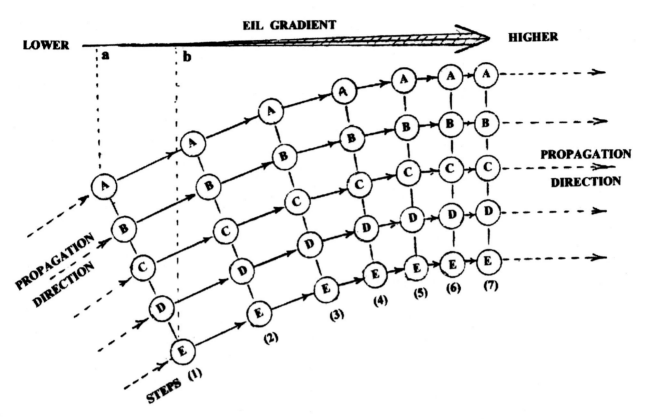

Fig 3.45

Subsequently the quantum jump for 'A' will be larger than the quantum jump for 'E'. As such the wave front plane at the step (2) position will have acquired a tilt towards the higher EILs. This tilting of the p/p plane continues at each successive propagation step until at the indicated step (7) position the EIL at 'A' and 'E' are the same. Thereafter the propagation will take place as represented by Fig 3.44 without any further change in propagation direction.

The reader will note that in both of the above illustrated cases we made no mention of the initial propagation velocity. This is because there was then no requirement for a comparison. However in the next case we must make such a comparison since we wish to establish the relationship between frequency of the wave front and refracting angle within a given realm EIL gradient.

We showed earlier in Fig 3.34 that the relationship between frequency and the p/p velocity component Vp was inverse. This means that under the same realm conditions a higher frequency EIL oscillation will have a lesser propagation velocity than a lower frequency oscillation. Thus when two different light wave frequencies approach an EIL gradient they will do so at marginally differing velocities. In Fig 3.44 and Fig 3.45 we did not represent the p/p velocity component Vp in the stepping set-up. Let us now correct this omission and represent Vp as a moving absorption centre as shown in Fig 3.46. Here we show the quantum jump just as before with the absorption centre E-M/X sequence commencing at the A' position. In the sequence duration from E-M/X to M-E/X the

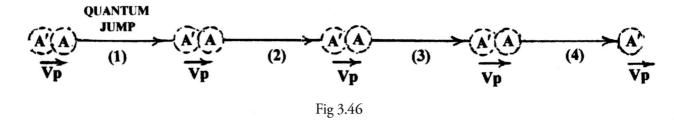

Fig 3.46

Mp-event centre will have moved forward to the A position. The rate at which this absorption centre moves forward is really the p/p velocity component Vp. The quantum jump size is determined by the prevailing realm EIL datum while the velocity component Vp is determined by the backward E-Flow absorbed from the preceding p/p. As such for a given realm EIL condition the quantum jump size may be assumed constant while Vp may not.

Let us now observe what happens when two differing light wave frequencies propagate through the same realm section. All realm conditions are identical for each frequency with the only exception being the E-Flows of each absorption centre. Let the light wave A be at a higher frequency than light wave B. If Vp_A and Vp_B represent the corresponding p/p velocity components then we have Vp_A as being less than Vp_B. This is shown in Fig 3.47 for a comparison between the propagation action of these two frequencies. The quantum jump size for both frequencies is shown to be dimensionally equal. In the illustrated example we have greatly amplified the difference in the

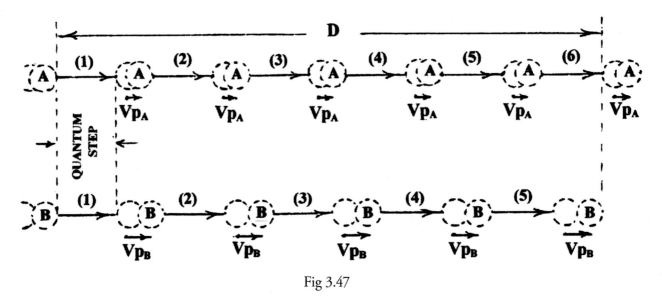

Fig 3.47

velocity components Vp_A and Vp_B such that one is nearly twice the other. We have also shown the quantum jump from edge to edge of the absorption circles rather than from their centres to simplify the discussion.

Consider the two light waves traversing the same realm distance D. Each wave front propagates forwards not only by a quantum step but also by a distance traversed by the p/p during the E-M/X and M-E/X sequence. The overlapping absorption circles represent the position of the p/p or wave front at the start and at the end of the energy medium exchange sequence. The reader must remember that the EIL oscillations or wave fronts are a large number of quantum steps apart and that the absorption centres above simply represent the stepping positions for a single wave front in its propagation from left to right.

In Fig 3.47 it is clear that the propagation rate of light wave B will be greater than that of light wave A by virtue of the difference in their p/p velocity component Vp. This results in a difference in the number of quantum steps that each light wave must execute to traverse the distance D. In our example we have shown that light wave A has had to make six quantum jumps to traverse the distance D, while light wave B required only five quantum jumps over the same distance. Since each quantum jump and each energy medium exchange sequence has the same duration in both cases, we may conclude that the propagation rates are in the proportion given below;

$$5 \times V_B = 6 \times V_A$$
$$\text{or, } V_B = 1.2 \, V_A$$

where V_A is the propagation velocity of light wave A and V_B is the propagation velocity of light-wave B.

Let us now establish a relationship between the frequency of an EIL oscillation and the refracting angle over a given realm EIL gradient distance. In Fig 3.45 we showed that the propagation direction of the p/p was changed at each quantum step position so long as an EIL gradient existed across the wave front plane. The change in direction or tilting of the p/p occurred successively at each new step position. Consider the example of light wave A and light wave B transposed into this layout. We know that light wave A will need to execute a greater number of quantum jumps to propagate over a given distance than light wave B. Since the degree of p/p tilt is a product of the variation in the quantum step size due to EIL differential across the wave front we may assume that both frequencies A and B are equally affected in their stepping action. That is to say the degree of tilt acquired per step by each frequency will be the same at the same EIL gradient.

Let the number of quantum jumps required by the light wave A to propagate over the distance D be large and denoted by N. We know that light wave B will propagate over the same distance with fewer quantum jumps. Let these be denoted by (N - 1).

We also know that the EIL gradient across the wave front diminishes as the wave front tilts towards the high EIL. Subsequently the p/p tilt per step gets less with each successive step.

If we denote the total refraction angle over the distance D by q then we have,

$$\theta = \phi_1 + \phi_2 + \phi_3 + \ldots\ldots\ldots\ldots \phi_{(N-1)} + \phi_N.$$

where $\phi_1, \phi_2, \phi_3,$ etc. represent the change in p/p tilt at each successive quantum step.

In our example let the refraction angles for the light wave fronts A and B be q_A and q_B respectively.

$$\text{Then, } \theta_A = \phi_{A1} + \phi_{A2} + \phi_{A3} + \ldots\ldots\ldots + \phi_{AN}.$$
$$\text{and } \theta_B = \phi_{B1} + \psi_{B2} + \phi_{B3} + \ldots\ldots\ldots + \phi_{B(N-1)}$$

Now both light waves are initially propagating in the same direction and therefore will enter the EIL gradient zone at the same oblique striking angle. Let us assume for the sake of comparison that the first step within the EIL gradient zone occurs from the same realm plane and at the same instant. The events are represented on a propagation diagram in Fig 3.48. The propagation diagram highlights the main features of our propagation model on a 2-dimensional geometric plan that takes into account the variation in quantum step size between adjacent absorption centres in the p/p and the distance that the newly formed p/p traverses before the next step sequence commences.

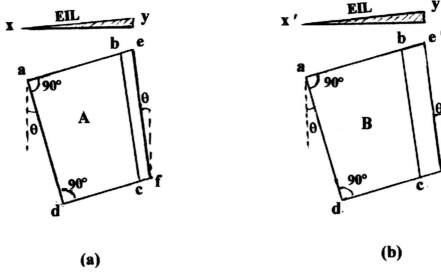

Fig 3.48

In the diagram ad represents the p/p at an angle θ to the vertical plane. The vertical plane has zero EIL gradient across it. The EIL gradient is along the horizontal direction xy as shown. The points 'a' and 'd' represent adjacent Mp-event centres and ab and dc are their respective quantum steps. Quite clearly the Mp-event sequence at 'd' occurs within a slightly higher realm EIL than that at 'a'. Subsequently dc will be less than ab. The new wave front therefore commences at the position bc. So far the propagation diagram is identical for both wave fronts A and B. As such the angle of tilt acquired at the new wave front position bc is the same in both cases. This means that the change in tilt angle is also the same, or,

$$\phi_{A1} = \phi_{B1}$$

However the p/p velocity components Vp_A and Vp_B cause each wave front to traverse a distance represented by be and be' respectively. Since Vp_A is less than Vp_B then be is less than be'. The new stepping off positions for the wave fronts is now represented by ef and e'f'.

Now ef and e'f', although at the same inclination to the EIL gradient, are at different EIL positions within that gradient. The Mp-event was characterised as being influenced by the EIL datum. The nature of the influence is reduced at higher EILs for a set amount of EIL difference and is best illustrated by the example in Fig 3.49.

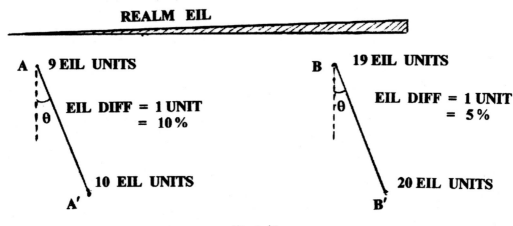

Fig 3.49

116

Here we have two wave fronts AA′ and BB′ inclined at the same angle to the vertical. The EIL difference between the respective wave front points is the same in both cases although each is in a differing realm EIL zone. As such the quantum step sizes will also differ, those from BB′ being smaller than those from AA′. At this stage we are more interested in the difference in the quantum step size within the same wave front. This step difference is dependant upon two factors. These are the realm EIL datum and the EIL difference between those centres. These factors may be combined and represented by the percentage EIL difference between the two centres. In the Fig 3.49 we have shown the centres A′ and B′ as being at 10 and 20 EIL units respectively. The EIL difference between the centres A and A′ is 1 EIL Unit as it is between B and B′.

The percentage EIL difference across the wave front AA′

$$= \frac{10 - 9}{10} \times 100, = 10\%$$

whereas the percentage EIL difference across wave-front BB′

$$= \frac{20 - 19}{20} \times 100, = 5\%$$

This means that the effective EIL difference between the centres A and A′ is greater than that between the centres B and B′. Subsequently the differences in quantum step size between adjacent centres in the wave front AA′ will be greater than the corresponding difference for wave front BB′. The positions of the new wave fronts from AA′ and BB′ will be such as to have caused the greater tilt change in the wave front from AA′.

If we apply this to our light wave A and light wave B comparison then,

ϕ_{A2} is greater than ϕ_{B2}

similarly $\quad\phi_{A3}$ is greater than ϕ_{B3}

and $\quad\phi_{A4}$ is greater than ϕ_{B4} and so on.

By mathematically summing up the two series above we could show that,

θ_A is greater than θ_B

From this we conclude that light wave A refracts through a larger angle than light wave B.

We may state this as a rule for our propagation model namely, that when wave fronts propagate into a region of increasing realm EIL then the angle of refraction of any wave is proportional to its frequency.

Let us now consider the inverse case whereby the light waves A and B propagate into a region of decreasing realm EIL. This is represented in the propagation diagram of Fig 3.50.

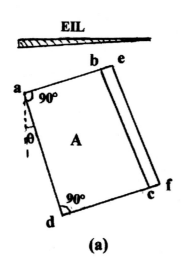

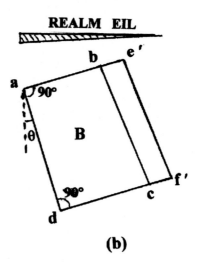

(a)   (b)

Fig 3.50

Let the number of quantum jumps required by light wave A to propagate over the distance D be denoted by N. We know that light wave B will propagate over the same distance with fewer quantum jumps from Fig 3.47. Let these be denoted by (N-1) .

In Fig 3.50 we see that the EIL gradient across <u>bc</u> has increased because of the increased tilt of the wave front. If we assume the same conditions as in the previous case then once again,

$$\phi_{A1} = \phi_{B1}$$

The p/p velocity components Vp_A and Vp_B each cause the wave-fronts to traverse the distance <u>be</u> and <u>be'</u> respectively. The effective EIL difference across <u>e'f'</u> will be greater than that for <u>ef</u> and subsequently light wave B will undergo the greater tilt at the next step. Then,

$$\phi_{A2} \text{ is greater than } \phi_{B2}$$
similarly $$\phi_{A3} \text{ is greater than } \phi_{B3}$$
and $$\phi_{A4} \text{ is greater than } \phi_{B4} \text{ and so on.}$$

We know that light wave B makes fewer steps than light wave A as represented in the series,

$$\theta_A = \phi_{A1} + \phi_{A2} + \phi_{A3} + \ldots\ldots\ldots + \phi_{A(N-1)} + \phi_{AN}$$
and, $$\theta_B = \phi_{B1} + \phi_{B2} + \phi_{B3} + \ldots\ldots\ldots + \phi_{B(N-1)}$$

This however does not indicate clearly which of the two series would provide the greater mathematical summation. We must therefore approach these from a different logical point of view. Assume that the difference in frequency between A and B is large. The difference in the number of steps taken over the distance D will be correspondingly great. The degree of tilting increases as the wave front propagates into the lower EIL zones and as such we have,

$$\phi_1 < \phi_2 < \phi_3 < \ldots\ldots\ldots\phi_{A(N-1)} < \phi_{AN}$$
where $$\phi_1, \phi_2, \phi_3, \ldots \text{ are small but increasing quantities.}$$

The difference between ϕ_1 and ϕ_2 is extremely small and subsequently the difference between ϕ_{A2} and ϕ_{B2} is infinitesimal.

Now, when we look at the two series for θ_A and θ_B again we note that greater importance must be given to the missing quantum jumps. These are enough to compensate for all the infinitesimal tilt additions to the θ_B series. When the frequency difference is large the θ_A series is clearly seen to contain the greater summation factors near the end of the series. As such θ_A is greater than θ_B. It can be subsequently shown that as the frequency range between light waves A and B diminishes, then θ_A reduces towards the θ_B level.

The reader may not find this approach convincing enough and may wish to resort to a more mathematical means to determine an answer. For now however let us accept that $\theta_A > \theta_B$. As such the higher frequency light wave A will refract to a greater extent than light wave B.

We may thus add to our propagation model rules by stating that when wave fronts propagate into a region of decreasing realm EIL then the angle of refraction of any wave is proportional to its frequency.

We may now show the direction that all wave fronts will follow when propagating through realm regions with EIL gradients that are both large and small. First however we must explain a little about the Realm EIL Grid. In present day land-mass geography we indicate terrain levels with contour lines upon a layout plan. In a similar manner we may indicate realm EILs with curved spatial lines. However the realm is a 3-dimensional entity and we would therefore need to represent realm EILs with a curved plane. A series of these equi-EIL planes is termed the Realm EIL Grid.

The steepest EIL gradient from any point on an equi-EIL plane is represented by the perpendicular line through that point. This is called the EIL Normal Line. Within the realm EIL grid the EIL Normal Line is not necessarily a straight line.

A sectional projection of the EIL grid on a layout plan would be represented by 2-dimensional EIL grid lines though these are misleading since they do not indicate the true 3-dimensional aspect of the EIL Normal line i.e. its true spatial direction.

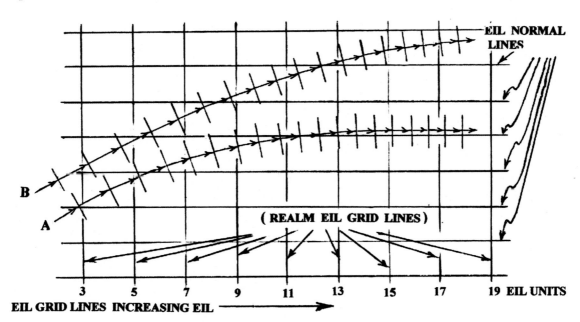

Fig 3.51

119

It is with reference to the EIL Normal line that we must indicate the refraction characteristics of a wave front. In order to illustrate these characteristics by a 2-dimensional diagram we will assume that the sectional projection includes the EIL Normal lines within its plane.

In Fig 3.51 are shown the relative propagation paths of light waves A and B of the previous example. The EIL grid lines are drawn vertical with an EIL gradient from left to right. The EIL Normal lines are the horizontal lines that intersect the grid lines at right angles. We see from the diagram that the change in the wave front propagation direction is always towards the direction of the EIL Normal lines. This was illustrated earlier in Fig 3.45.

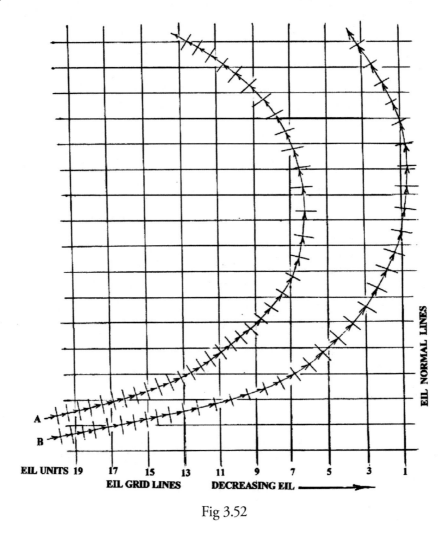

Fig 3.52

Fig 3.52 shows the relative propagating paths of the light waves A and B traversing towards the lower EIL grid lines. It is clearly seen that the light waves A and B undergo a change in propagation direction that is away from the direction of the EIL Normal lines. At some realm position the propagation direction will be at right angles to the grid normal line. Thereafter the propagation will continue with the characteristic represented in Fig 3.51. We see that the propagation direction has been changed from that of progressing into a lower realm EIL to that of progressing into a higher realm EIL.

Once again the conditions in Fig 3.51 and Fig 3.52 have been grossly exaggerated and assume that the EIL gradient is spread uniformly over infinite distances. Within the realm conditions are somewhat different. EIL gradients are extremely high for short distances only, i.e. close to mass objects. Thereafter the gradient diminishes according to the inverse square law but is easily upset by the presence of other mass body E-Flows. It was mentioned in a previous section that we considered the central regions of the realm to be at a higher EIL datum because of the focus effect from all the decaying mass within the realm. We would therefore assume that on average an EIL gradient exists between the centre and edges of the realm. We leave the reader to ponder upon this aspect and its effect upon long distance wave front propagation. The reader is invited to compare the wave front direction change in Fig 3.51 with that in Fig 3.41 and to note the differences.

Let us conduct a thought experiment and consider the propagation of a light wave through a zone of uniform realm EIL. We already know the characteristics of the quantum jumps and have concluded that the direction and velocity of propagation would remain unchanged. Let us now divide this zone along a set plane and permit the datum EIL of one zone to become greater than that of another. We now have two realm zones adjacent to one another at uniform but differing EILs. The EIL difference is clear cut and demarcated by the dividing plane as shown in Fig 5.53.

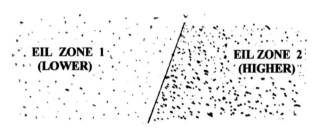

Fig 3.53

Let us further assume that these zones will forever remain separated and that the dividing plane is simply where one zone is in contact with the other. The dividing plane has no identity other than to indicate where one zone ends and the other begins. Consider a wave front propagating through zone 1, across the dividing plane and on into zone 2. We can ignore frequency and assume that the main difference between the propagating characteristics in each zone is caused by the difference in quantum jump size. The quantum jump size is larger in zone 1 than in zone 2 and subsequently the wave front propagation velocity is greater in zone 1 (zone 1 being at the lesser EIL).

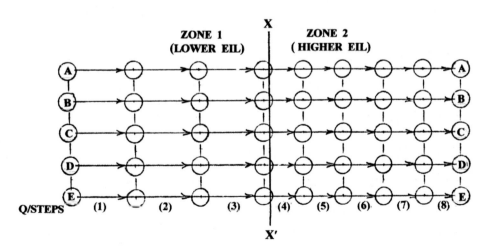

Fig 3.54

If the propagation from zone 1 into zone 2 is such that the wave front p/p is parallel to the dividing plane, then the entire wave front will enter zone 2 in the same quantum jump instant. Thus although the quantum jump size may alter, it does so with the entire wave front in unison and as such there will be no change in propagation direction as shown in Fig 3.54.

If however the propagation from zone 1 to zone 2 is such that the wave front p/p is not parallel to the dividing plane, then at some stage during zonal cross-over the wave front will span both zones as shown in Fig 3.55. The EIL differential between the two sections of the wave front p/p then results in a quantum step size differential. This in turn alters the plane of the wave front p/p and subsequently its direction of propagation.

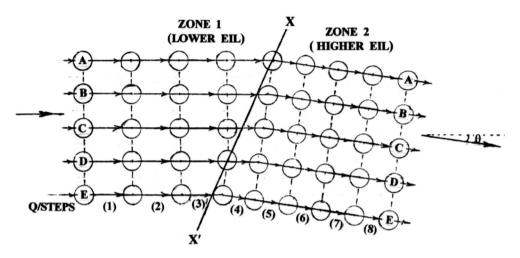

Fig 3.55

In the Fig 3.55 consider the quantum steps (1) and (2) of the wave front ABCDE. These are made entirely within zone 1 and therefore maintain a consistent step size and propagation rate for all wave front centres. As such propagation direction remains unchanged. At the step (3) however the wave front centre E enters within zone 2 while A, B, C and D centres may still be considered in zone 1. Now because of the relatively higher EIL of zone 2 the quantum jump size occurring within it is marginally reduced. This means that the step (3) size for the wave front centre is marginally less than that for A, B, C and D. As such the wave front portion between centres D and E acquires a tilt as shown while the wave front from A to D is unaffected. Since the propagation direction is normal to the wave front plane then in step (4) centre E will propagate at an angle q to the original direction. Centres A, B, C and D continue to step in the original direction although they also now begin to enter zone 2.

Let us assume (as part of our thought experiment) that the quantum step size in zone 2 is 2% less than that in zone 1. In order therefore to establish the position of the wave front centre E when the quantum jump spans the demarcating plane XX' it will be necessary first to show the full zone 1 quantum jump and then to reduce by 2% only that part which is inside zone 2. By applying this rule we are able to locate the position of all the wave front centres after step (4). As shown the entire wave front ABCDE is now at a tilt which gives the new direction of propagation. Since there are no other EIL differentials across the wave front then no further changes in propagating direction will occur. The wave front will therefore continue to propagate along this new direction while it is within zone 2.

In Fig 3.55 it will be especially noticed that the refraction is towards the normal on XX'. Let us now view a light wave propagating in the opposite direction from zone 2 into zone 1. This is shown in Fig 3.56.

Here we see that the wave front ABCDE makes the quantum steps (1), (2) and (3) entirely within zone 2. At quantum step (4) the wave front centres A and B both enter zone 1 but at varying distances. This results in a tilt to the wave front between centres A and C only. The tilt is in the direction shown because of the increasing size of the quantum step as the wave front propagates from zone 2 to zone 1 which is at the relatively lower uniform EIL. Continuing with our thought experiment and applying the same conditions as before, we can

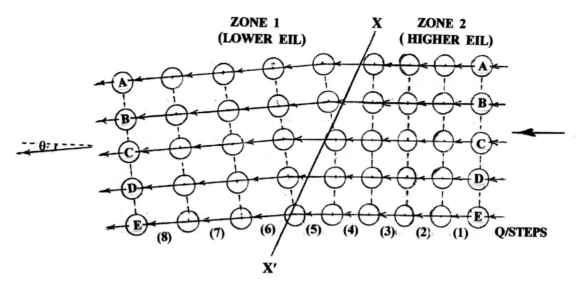

Fig 3.56

establish the positions of the new wave front centres by simply increasing by 2% only that part of the quantum step which is inside zone 1. Fig 3.56 shows that the refraction of the light wave is away from the normal on XX'. The reader is advised to compare Figs 3.55 and 3.56 and to note that essentially one is a reversal of the other. That is to say, a wave front centre traces the same path in its propagation across the demarcating plane between zones 1 and 2 when initiated from either direction as shown in Fig 3.57.

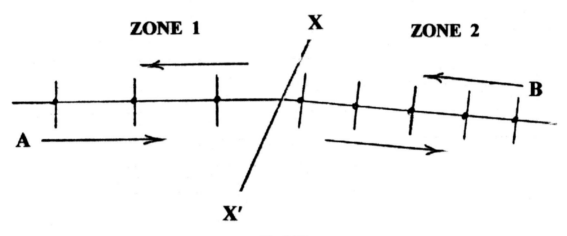

Fig 3.57

It is all very well in a thought experiment to assume the demarcating plane XX' to be infinitely fine. In reality no such situation can occur. Any such demarcating plane must offer an EIL gradient between zones 1 and 2 over a set band width. Relative to the smallness of the quantum step size this band width may be considered quite large. In order therefore to determine the refraction of the light wave between zones 1and 2 we must also take account of the EIL gradient set up between them. We must therefore add the conditions of Fig 3.45 to those of Fig 3.55 and obtain a composite as shown in Fig 3.58.

Here we represent X_1X_1' as being at zone 1 EIL and X_2X_2' at zone 2 EIL. As such the entire refraction will tend to occur within the EIL gradient zone as shown. If we assume that the EIL gradient is reasonably uniform then a light wave A traversing from zone 1 to zone 2 will trace the path indicated. Similarly by superimposing the Fig 3.51 conditions into the EIL gradient zone of

Fig 3.58 we may conclude that for differing light wave frequencies the higher frequency wave front will undergo the greater refraction. The reader should note that the term light wave has been used loosely and is meant to extend to the entire range of EIL oscillations including those invisible to the human eye.

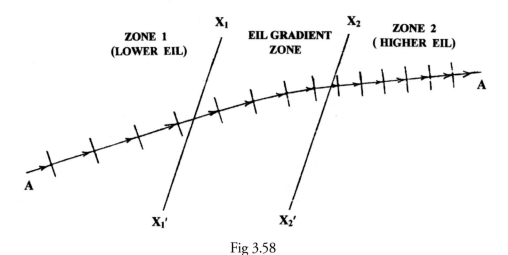

Fig 3.58

In chapter 2 (Fig 2.17) we briefly described the formation principles of the Mp-shell and the subsequent basic atom. Of the total volume enclosed within the atom sphere less than 0.1% is taken up by the central proton mass. The remaining 99.9% of the atom volume is occupied by realm energy medium at an EIL less than C_1 and related to the prevailing local E-Flows. As atoms pair up into molecules and then congregate together to form a larger mass body, this distribution of atomic volume remains reasonably constant. Although such mass bodies appear rigid and impervious, they are really quite empty as far as proton volume is concerned. The region between the protons of adjacent atoms is therefore mainly energy medium at the proportionate EIL datum. Due to the proximity of the decaying protons this EIL datum will be rather higher than the realm datum prevailing outside the mass body structure. If we assume the atomic structure within the mass body to be distributed in such a manner that their Mp-shell repetitions occur within a compound frequency band, then wave fronts at certain other frequencies may be able to propagate across those shells and into the atomic sphere. The mass body that allows this is considered to be transparent to those wave fronts. There are varying degrees of transparency offered by a mass structure but in general we consider an object to be totally transparent if the wave front emerges reasonably unaffected by its propagation through the structure.

By virtue of the oscillating nature of the Mp-shells the wave front is able to step through the shell zones without loss of wave front intensity. This can be compared to the school girl who steps cleanly through the rotation zone of a skipping rope operated by two school mates. It is also possible to achieve this when two skipping ropes are rotated in the 'Double-Dutch' fashion.

Assuming that the wave front has stepped through the Mp-shell let us now consider its propagation within the atomic volume. Apart from the central mass this volume is filled with the realm energy medium. The EIL is higher near the centre and diminishes with increasing distance as represented by Fig 1.18 (chapter 1). The existence of the energy medium is of course a consequence of the E-Flow out of the decaying central proton mass. The EIL gradient and the E-Flow EIL are therefore proportionately inter-related for any point within this zone.

Propagation across this volume will be considered in two parts as shown in Fig 3.59. Consider the wave front centre O propagating in the direction shown. It steps through the Mp-shell at A and propagates towards B which is on the directional centre line XX' of the atom. The EIL at B will be higher than near A by virtue of distance from proton P. As such propagation along AB will be into an upward EIL gradient. This will be the first part of the propagation. The second part will be the propagation from B to C along a downward EIL gradient. The EIL Normal Lines are along the radii from the centre P. Also EIL conditions are uniform around the proton P such that EIL gradients and E-Flow intensities are consistent at set distances from P.

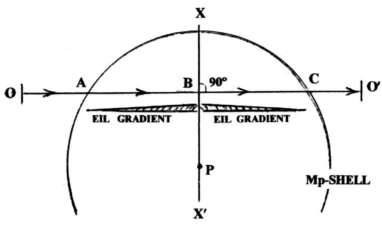

Fig 3.59

To consider the first part of the above propagation we note that an EIL gradient prevails across the wave front as it propagates from A to B. By virtue of our propagation diagram (see Fig 3.48), the change in propagation direction is towards the EIL Normal line. Thus theoretically the wave front O should refract towards the atom centre P. Such would certainly be the case if the EIL gradient was a static one as for a normal realm datum situation. However conditions inside the atom sphere are different from those outside. This is because the EIL gradient is caused essentially by the E-Flow from the proton centre. The EIL of the E-Flow overrides any other realm datum effects in this zone and as such must impart substantial directional linear momentum to the E~M/X of the wave front centres operating therein. We know that our propagation model comprises of two components. The forward E-Flow from the M-E/X of each wave front centre and the p/p velocity component Vp.

Earlier we conducted a thought experiment (Fig 3.35) to visualise the principle of wave front reversal. Let us now conduct a similar thought experiment but with the component Vp having to encounter a large E-Flow from an inclined direction as represented in Fig 3.60.

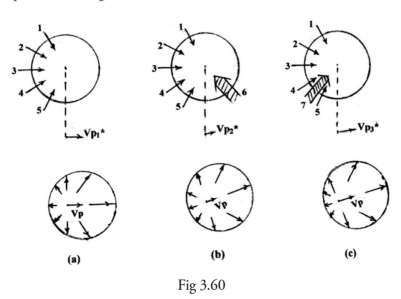

Fig 3.60

125

The upper diagrams represent the E-M/X sequence of the wave front centre while the lower represents the subsequent M-E/X. The initial normal propagation is shown at (a) with a p/p velocity component Vp represented by the vector Vp_1^*. At (b) the wave front centre absorbs a large external linear momentum from the indicated E-Flow 6. This additional linear momentum becomes a part of the p/p velocity component Vp with the new resultant vector quantity Vp_2^*. It will be noticed that Vp has deflected in the direction of the E-Flow 6. (Compare this with Fig 3.19 data). Since the E-Flow 6 acts in partial opposition to the propagation direction then the magnitude of Vp_2^* will be marginally less that Vp_1^*. At (c) the wave front centre again absorbs a large external linear momentum from the indicated E-Flow 7. The additional linear momentum absorbed into the velocity component Vp results in the net vector Vp_3^*. Once again it will be noticed that Vp has deflected in the direction of E-Flow 7. Since E-Flow 7 acts partially in the direction of propagation then the magnitude of Vp_3^* will be marginally greater than Vp_1^*.

The reader should note that this phenomena of Vp deflection is only applicable in circumstances where normal realm datum influence becomes secondary to an E-Flow situation. Such a zone of influence is certainly to be found within the confines of the atom between the proton centre and the Mp-shell. We shall now apply the above principles of linear momentum absorption and Vp deflection to the wave front propagation within this zone.

For the purpose of explanatory simplification let us assume separate propagation paths to account individually for each of the following effects;

   (a) refraction due to EIL gradient,
   (b) Vp deflection due to large E-Flow intensity.

The mean deviation computed from these two paths would determine the resultant propagation direction. This is represented in Fig 3.61 in which the path OAB_1 is that solely due to the Vp deflection trend and path OAB_2 the refraction due to EIL gradient. Refraction along the path AB_2 conforms to the behaviour of our propagation model in the EIL gradient environment of Fig 3.51. Refraction is thus towards the EIL normal line. As the wave front centre progresses along AB_2 the EIL differential across the wave front increases resulting in a greater refraction trend. Maximum refraction will occur theoretically at the path position B_2 when the wave front path is transverse to the EIL normal line i.e., maximum EIL gradient. Vp deflection along path AB_1 represents the acquisition of additional linear momentum outwards from the atom centre. As the wave front centre propagates along AB_1 it

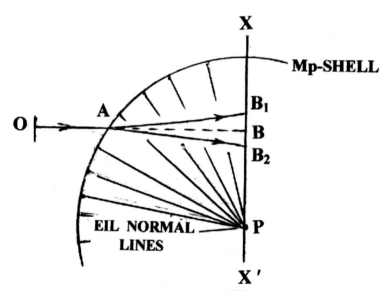

Fig 3.61

is in effect progressing closer in towards the proton centre and thus into a region of higher E-Flow intensity. At the path position B_1 the wave front centre will be at its closest position to P and will absorb a maximum outward linear momentum (relative to other positions on AB_1). As such the Vp deflection trend will be a maximum at the B_1 position.

Now in reality we are looking at the same wave front centre propagating along one path somewhere between AB_1 and AB_2. From the above we notice that a relationship exists between the two effects. Each exerts a maximum influence at the directional centre line XX' position. The absorption of linear momentum depends upon the E-Flow intensity and the refraction is dependant upon the EIL rate of change at that position. Now the EIL rate of change or gradient is simply the dissipation of the proton decay E-Flow according to the inverse square law for distance. Thus the E-Flow EIL and the gradient EIL are one and the same quantity at any position within the atom volume. The two effects are related by the same EIL condition. This means that the refraction in our propagation model for this zone will be proportional to the quantity of linear momentum absorbed from the

E-Flow. When the refraction is large the linear momentum absorbed will also be large and vice versa.

Consider the path position A. Let us assume that at this point Vp deflection and refraction are both nil. The propagation path will therefore continue beyond A in an extension of the straight line path OA. At some point beyond A linear momentum will be absorbed by the wave front centre as will a slowing down of the propagation rate due to a smaller quantum step size in the higher EIL. A Vp- deflection trend outwards will therefore come into effect the same moment as the refraction trend inwards. At this stage we may assume both these effects to be extremely small and since they are of opposite nature the difference is negligible and the propagation path continues in a straight line. This balance is only to be found within the atomic volume zone because of the over-riding E-Flow conditions. The balance is one where the refraction due to the inward EIL gradient is always exactly equaled by the Vp deflection in the opposite direction. It does not matter whether the path OAB passes nearer to or farther out from the atom centre the balance is maintained and the propagation path remains in a straight line.

Let us consider the second part of the wave front propagation within the atom zone i.e. from the directional centre line XX' to Mp-shell in Fig 3.62.

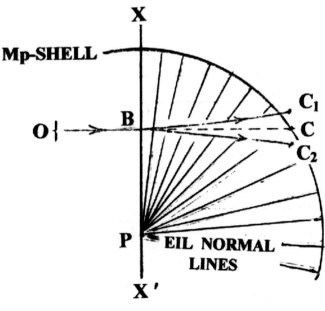

Fig 3.62

The E-Flow and resulting EIL gradient conditions within the atom volume are consistent all around the proton centre. The wave front centre O however propagates along the path BC which is from a higher to a lower EIL. In other words the propagation is along a decreasing EIL gradient with a refraction trend as shown in Fig 3.52. The refraction is seen as being away from the EIL normal line. As such refraction along the theoretical path BC_2 conforms to the behaviour of our propagation model. As the wavefront centre progresses along BC_2 the EIL differential across the wave front decreases resulting in a reduced refraction trend.

The linear momentum absorbed by the wave front centre deflects the p/p velocity component Vp in the direction of the E-Flow. The Vp deflection trend is along the path BC_1. Maximum Vp deflection trend is at B while the minimum trend is at the position C_1. Once again we notice a corresponding opposite balance between these two effects. We may assume therefore that the two opposing trends nullify each other and no resultant path deviation occurs. We may therefore draw the conclusion that a wave front propagates through the inner volume region of an atom without any change in propagation direction. This would apply to the stepping action of all frequencies.

Now although the frequency of the wave front may enable it to step through the Mp-shell it is not able to do the same upon reaching the central proton. Since the proton is a very large congregation of mass points there is no common frequency within. All E-flows entering the proton sphere will be absorbed into one or more of the E-M/Xs occurring therein and no corresponding E-Flow will exit the opposite end. The wave front section that steps into the proton structure is therefore wholly absorbed by it. Fig 3.63 illustrates this as a circular gap in the p/p that continues beyond the proton sphere. The reader may follow this through in fine detail if so desired and conduct a logical analysis of the propagation characteristics close to the proton. The Fig 1.13 conditions must of course be taken into account.

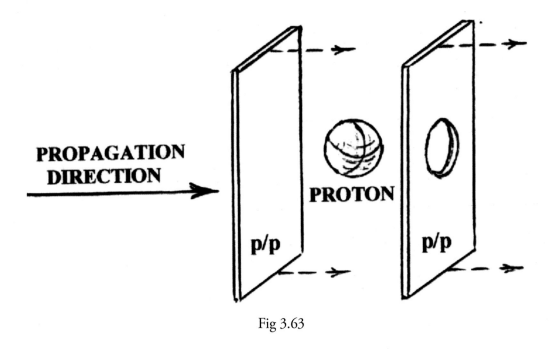

Fig 3.63

We must now return to our refraction topic and discuss the overall characteristics of a wave front propagating through a transparent body. Consider a cube of this transparent material to be situated within a uniform realm datum represented by the zone 1 of Fig 3.58. Let us assume that the structuring within the material cube is such that the EIL within is reasonably uniform and represented by the zone 2 EIL. Between the two EIL zones there must exist a transition zone with an EIL gradient that blends one zone with the other. This is presented diagrammatically in Fig 3.64.

Let us for the moment ignore the reflection phenomena while assuming the surfaces of the cube to be perfectly smooth. Let a wave front centre propagate along the path AB as indicated. There is no change in propagation direction so long as the propagation is wholly within the uniform EIL of the zone 1. At an assumed position B the E-Flows from the cube M protons can no longer be ignored and a build up of EIL commences. A maximum zone 2 EIL is achieved at the surface of the cube. The wave front propagating towards the surface of the cube will therefore progress up an EIL gradient and so will refract towards the EIL normal line.

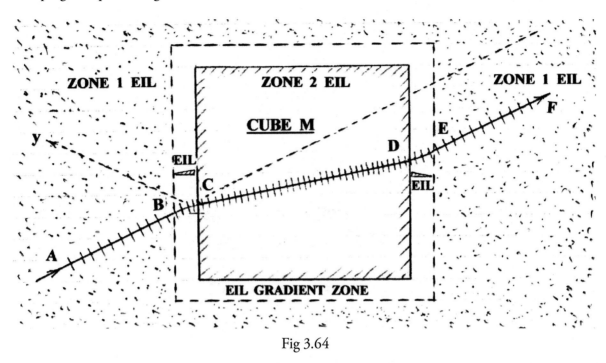

Fig 3.64

The EIL normal being perpendicular to the cube face. The wave front propagation path BC in this transition zone will thus be curved as high lighted by Fig 3.51.

Propagation within the cube M is at a uniform zone 2 EIL and as such no path deviation occurs. CD is thus a straight line path. At D the wave front must once again enter the EIL transition zone. The wave front propagating outwards away from the cube surface will therefore progress down an EIL gradient and so will refract away from the EIL normal line. The wave front propagation path DE in this zone will also be curved but in the direction high-lighted by Fig 3.52.

Since the EIL gradient conditions on each face of the cube are similar then the path deviation along BC will be of the same magnitude as that along DE but opposite in directional sense. The propagation path EF will thus have returned to the original direction but with a lateral displacement as shown. An extension of the principles in Fig 3.51 and Fig 3.52 indicate that propagation paths vary according to frequency. For the above configuration however the emerging path will always be in the same direction as the incident path though the lateral displacement will differ.

In the thought experiment of Fig 3.35 we drew the conclusion that under certain conditions of E-Flow a single wave front could give rise to another wave front that would propagate in an opposite direction. This was a logical conclusion whereby the backward E-Flows from the wave front centres are sufficiently boosted by the local energy medium conditions to set up the additional backward position wave front. From this it becomes apparent that the wave front propagating towards the surface of the transparent cube M acquires this additional wave front which is then observed as a reflection along 'B-y' in Fig 3.64.

In Fig 3.10 and Fig 3.41 we illustrated the phenomenon of reflection. We have also just explained the refraction phenomenon. There is no logical reason for either phenomenon to be preferred when a wave front approaches a mass object. There is also no logical reason for one or the other to be eliminated from physical reality under the same conditions. We may therefore conclude that in every case where a wave front approaches a mass structure that at some position very close to that structure a dual wave front is developed from both backward and forward E-Flows of the initial wave front. The backward direction wave front is observed as the reflected wave front. The resultant intensities of the reflecting or refracting wave fronts may differ and as such will depend upon the properties of the mass object.

In the case of transparent mass objects the refracting wave front is easily observed. When the mass object is not transparent then we must assume that the refracting wave front does still occur but is ultimately absorbed within the overall structure of the object. The reader must realise that transparency is relative to the frequency of the wave front as our theory indicated earlier. Incidentally, when the refracting wave front is absorbed by a mass object there must also be an absorption of linear momentum from the wave front centres. As such the mass object should receive a "push action" in the wave front propagating direction.

Fig 3.65 gives a combination of the reflection and refraction processes at the surface of a transparent object. The wave front A propagates towards the surface XX' of the transparent object. There is a correlation between the emission step positioning as represented by Fig 3.37 and the EIL

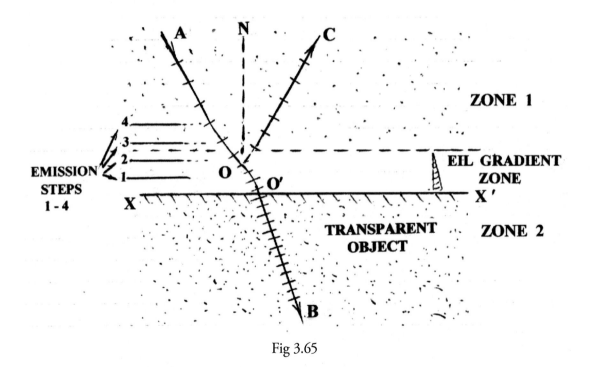

Fig 3.65

gradient zone near XX'. The wave front approaches the EIL gradient zone and is initially influenced by the surface wave emissions as explained in Fig 3.40. At the position O in Fig 3.65 the p/p velocity component Vp becomes small due to the absorption of large quantities of energy medium linear momentum directed outwards from XX'. Subsequently the backward and forward E-Flows from the M-E/Xs of the wave front centres at O are both able to set up corresponding wave fronts in their respective positions. The backward propagating wave front progresses in the direction OC having acquired a tremendous boost to its p/p velocity component Vp. The wave front at C is considered to be the wave front A that is reflected off the surface XX' and obeying the rules of reflection as set out earlier. The forward propagating wave front progresses along OO' and acquires additional energy medium into its wave front centres by virtue of the increasing realm EIL datum as it approaches zone 2. As such there is an increase in this wave front intensity rather than the expected decrease.

The wave fronts B and C are now totally separate and independent of each other and are clearly observable as such. However each bears the same frequency characteristic as the incident wave front A. The reader should note that the direction AO is not in a straight line but curves as shown in Fig 3.41. This is only a marginal deflection since it ceases at the point O which is relatively far out from XX'. The propagation along OO' corresponds of course to our propagation model theory for refraction. Further logical extensions may be undertaken by the reader.

Let us once again ignore the reflection phenomenon and concentrate upon the propagation path of a wave front through a transparent object. We have already observed the wave front path through the cube shaped transparent object of Fig 3.64 and shall now observe the wave front path through a transparent prism as shown in Fig 3.66.

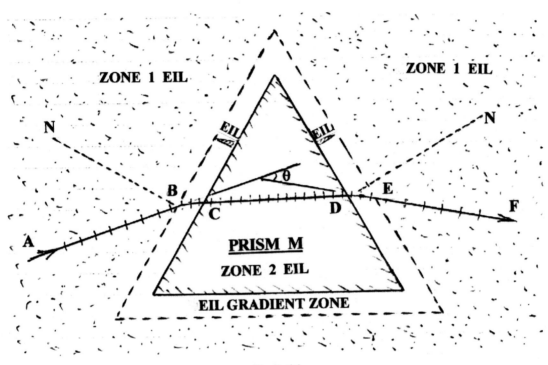

Fig 3.66

Using our propagation model theory the wave front A propagates towards the prism M along the indicated path AB. At B the wave front enters the EIL gradient zone. Propagation is then along the curve BC which is a refraction of the wave front path towards the EIL normal line. The EIL normal. line being that perpendicular to the prism face at C. Propagation through the transparent prism is within the uniform EIL zone 2. As such the path CD is a straight path. At D the wave front enters the EIL gradient zone again and the path DE is along a decreasing EIL gradient. Refraction is therefore away from the EIL normal line DN. Propagation along EF is within the uniform zone 1 EIL and is a straight path. The propagation directions of the incident wave front A and the emerging wave front E differ by the indicated angle q.

If we consider wave fronts at differing frequencies to propagate through the above prism then according to our propagation model principles for the same incident path the indicated angle q will be greater for the higher frequency. Compound frequency wave fronts propagating along the same path AB will therefore propagate through the prism and emerge along different wave front paths as shown in Fig 3.67.

Newton used this principle to separate the various frequencies within the compound frequency band emitted by the stars. Because much of the separated band was within the human eye response range the effect was quite spectacular. Each separated frequency brought forth a different response on the retina that appeared as visible colours

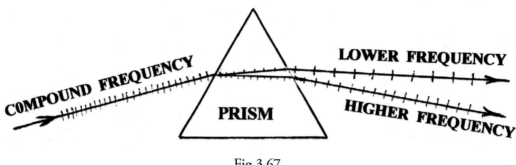

Fig 3.67

when interpreted in the brain centres. The colour grades varied from red for the lowest visible frequency to violet for the highest. The reader may of course obtain all this from any standard text on the subject and therefore we shall skip over such detail. Incidentally, any detailed study of compound frequencies must take into account the rule that individual frequencies within such a frequency band each propagate at a velocity that is inversely proportional to its frequency. This means that lower frequency wave fronts propagate faster than higher frequency wave fronts. As such the frequencies of Newton's spectrum captured on photographic plate were not all emitted at the same instant. The red frequency wave front must have left the star source much after the violet frequency wave front in order to reach us at the same instant. Refer to Fig 3.31.

Let us digress for a moment. The spectrum of frequencies from a Star 200 million light years distant that we may receive today were emitted under realm datum conditions that prevailed 200 million years ago. According to the circle 6 energy medium curve in our seven circle theory, the realm EIL datum follows a diminishing trend. As such an elapsing period of 200 million years must result in a change in the realm datum EIL. This change is graphically represented in Fig 3.68 by the relative positions of A and B on the circle 6 energy medium curve.

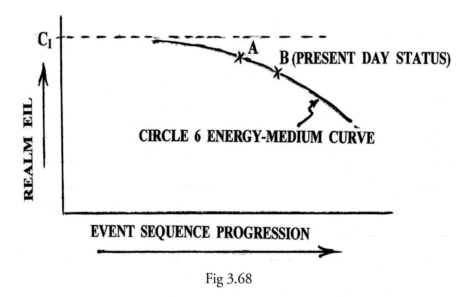

Fig 3.68

Let us temporarily ignore the variance in proton decay rate in order to simplify this discussion. The realm EIL datum at A is higher than at B. The emission of wave fronts depends upon the repetition frequency of the Mp-shells of the atom structures. The repetition cycle of the Mp-shell in turn depends upon the E~M/X and M-E/X overall sequence duration. This cyclic sequence is slower at higher realm EIL conditions due to the marginally improved stability or life of the Mp formations. Each successive wave front emitted from the Mp-shell (Fig 3.3) will be correspondingly later. As such a larger duration gap will exist between successive EIL oscillations. The wavelength of emissions for realm status A will thus be greater than those for the status B, other factors being the same. All emissions received by an observer must originate in the past. An observer at B status will therefore receive emissions at correspondingly longer wavelengths from stars that are proportionally more distant. This has been identified in the red shift phenomena for the known emission and absorption lines within the spectrum. Sir Edwin Hubble produced a linear relationship between the distance of a star/galaxy from earth and the red shifting of its spectrum of wavelengths. More recently however astronomers using more powerful devices have probed further into the realm and recorded some very large wavelength shifts, always towards the longer wavelengths. They extended Hubble's relationship formula and found it to be a curve. Perhaps this may compare with the circle 6 portion of our energy medium curve and permit us to establish a true position for B. We shall discuss these aspects in detail later on in the text.

We shall now pursue another interesting refraction phenomena. One which relates refraction with the internal reflection within a transparent body. Let the wave front approach the surface of the transparent body as shown in Fig 3.69(a). If the EIL of the transparent body is at zone 2 and the surrounding EIL being at zone l, then an EIL gradient zone will exist at the interface as shown. Zone 2 EIL is greater than zone 1 EIL as for the previous descriptions.

Let the wave front path AO make an angle θ_1 to the EIL normal line ON. The wave front enters the EIL gradient zone and propagates along OO' i.e., into a decreasing EIL zone. As such refraction is away from the EIL normal line resulting in the zone 1 propagation path O'B. The refraction angle ϕ_1 is greater than the incident angle θ_1.

In Fig 3.69(b) we have increased the incident path angle to θ_2 such that the refraction angle ϕ_2 is very nearly 90°. Our theory indicates that the wave front path along O'B cannot be maintained at right angles to the EIL normal lines (Fig 3.52). As such a marginal increase in the incident path angle from θ_2 to θ_3 results in the new propagation path OO' of Fig 3.69(c). This path causes the wave front to refract back into the transparent body. Since the wave front path OO' through the EIL gradient zone is symmetrical about an imaginary centre line then the angle of re-entry at O' will be the same as the angle of exit at O. The wave front would appear to have undergone reflection at the

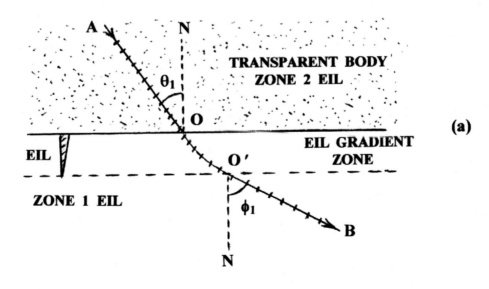

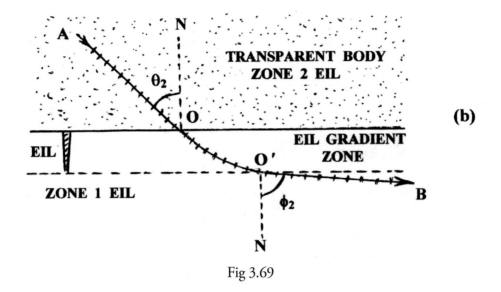

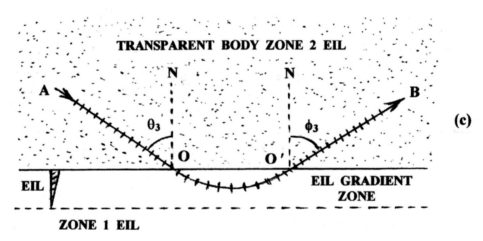

Fig 3.69

internal surface of the transparent body. Observation shows the physical characteristics of this refraction to be the same as those for reflection by the wave front reversal principle of Fig 3.40. This internal reflection can only occur if the propagation is from a higher EIL zone to a lower EIL zone The angle between q_2 and q_3 at which the above phenomena of total internal reflection is seen to occur is termed the Critical angle of internal reflection. The critical angle varies inversely as the difference in EIL of the two zones. The greater the difference in EIL the greater is the EIL gradient between the two zones and the smaller will be the Critical Angle θ.

In Fig 3.69(c) the refraction of the wave front was represented by a simplified propagation path dictated by the propagation diagrams of Fig 3.50. The actual events are far more complex. Because of the Critical Angle θ adjacent wave front centres enter the EIL gradient zone in a staggered sequence. The quantum step of the earlier entrants will therefore be marginally larger (because of the decreasing EIL). The subsequent sequence of quantum steps is shown in Fig 3.70.

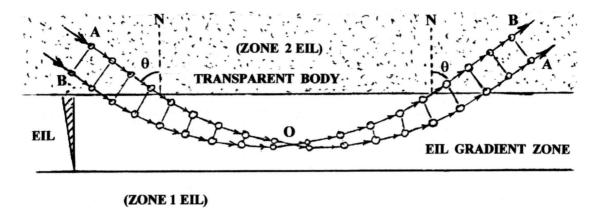

Fig 3.70

The wave front centres A and B are shown in successive quantum step positions for the Critical Angle θ. The reader will notice that the wave front centres draw closer together as propagation proceeds into the EIL gradient zone. At the position O the wave front centre B has already achieved the 90° parallel propagation direction while wave front centre A has not. As such the next propagation step for B takes it back towards zone 2 while A continues towards zone 1. The position O is therefore a cross over point for the wave front centres A and B as shown. The position of O will vary according to the propagation path of each adjacent pair of wave front centres. This is

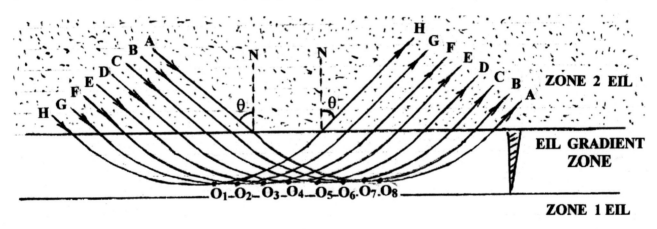

Fig 3.71

represented on Fig 3.71 by the positions of O_1 for the paths G and H, O_2 for the paths F and G, O_3 for the paths E and F, and so on. These are of course the focal points for the respective wave centre pairs and the reader is asked to appreciate the fact that propagation continues thereafter in the normal manner. Different frequencies will refract differently within the EIL gradient zone and once the critical angle for that frequency has been exceeded the principles of internal reflection are the same. In actual experiment, prior to the attainment of the critical angle and total internal reflection, the bright refracted wave front is accompanied by a very weak internal reflection. This is due to the set-up of a weak backward propagating wave front due to the EIL conditions at the zone 2 inter-face. The reader may analyse the logical detail of this event as a personal project using the principles of our standard propagation model. The reader is also advised to continually maintain a parallel with the established rules of refraction through the refractive index ideas in conventional physics.

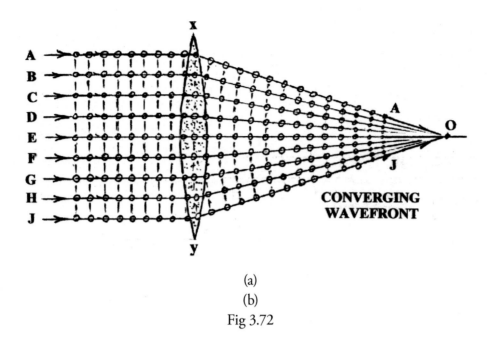

(a)

(b)

Fig 3.72

So far we observed wave front propagation through a prism and a cube of transparent matter. We shall now extend that investigation to propagation through a lens shaped transparent object. The lens in profile may be considered as either convex or concave and with which the reader must be wholly familiar. Each may be structurally defined as a composite of a large number of prism elements. If the theoretical apex of the elemental prisms all point away from the lens centre then the lens is the convex type. The inverse of this gives us a concave lens. It is a simple exercise to ascertain the wave front path through these multiple prism elements. This is detailed in Fig 3.72 for both types of lenses. At (a) a flat wave front A - J approaches the convex lens 'xy' in the direction shown. By applying the procedure in Fig 3.66 to each wave front centre the position and direction of successive propagation points may be determined. It will be noticed that after propagation through the lens 'xy' the plane of the wave front has become curved and the propagation paths converge towards the position O. An important phenomenon occurs at this wave front focus point which we shall discuss shortly.

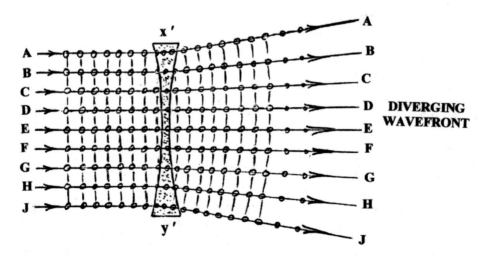

At Fig 3.72(b) a similar wave front is shown to propagate through a concave lens x'y ' resulting in a diverging wave front. The physical effects of both types of lens are analysed in considerable detail in most standard texts on the subject so that further detail here is unnecessary. It will be noticed however that the curve of the wave front after passing through the lenses is also dependant upon the number of quantum steps that were taken by each wave front centre within the lens structure.

Let us consider the wave front when it converges to the theoretical point O (Fig 3.72a) . When a wave front is emitted from its Mp-shell source it has a curvature that represents a true spheroid surface. When this wave front is allowed to propagate through a material structure such as a transparent convex lens then the curvature of the resulting wave front would be as imperfect as the quality of the lens surface profile. Although this surface may appear smooth and perfect to the naked eye, under magnification a totally differing view can be obtained. The result is that convergence of this imperfect wave front does not propagate towards a single point O but rather to a profusion of such points around the position O.

If the wave front curvature had been perfect then the theoretical point O would have been a reality and the wave front would have resulted in the formation of a Mp at O. The emergence of any front from the ensuing M-E/X would not have had the identity or character of the in-going wave front and so recognition of aspects the other side of the lens would be impossible. However we are fortunate that this is not the case and that as a result of an imperfect wave front approaching the position O a cross over of wave front centres occurs around this position as illustrated in Fig 3.73.

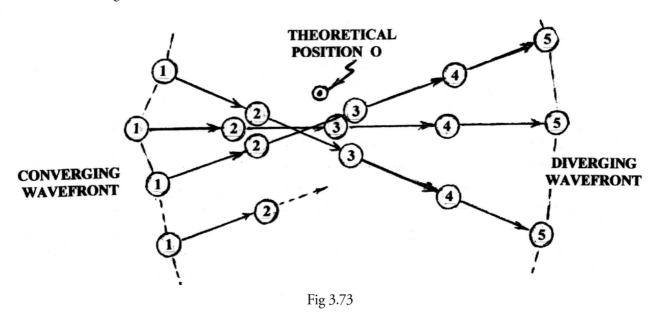

Fig 3.73

In the diagram the event progression is represented by the numbering sequence inscribed within each wave front centre. Each numeral represents the same event moment for all centres and as such indicates the nature of the wave front cross over. The reader is not bound to this aspect in our theory and may pursue an alternative view point where this becomes available. Perhaps the wave front centres, through the process of the quantum step, do align themselves automatically into a perfect curvature and so result in a Mp at O. Wave fronts from different directions would obviously have differing focal positions then perhaps the new diverging wave fronts issuing from the Mps would be a simple sequential re-issue of all the earlier converging wave fronts. We shall not dwell on this any longer as we have left it open to further investigation.

Let us now progress to an aspect that resembles a concave lens but is wholly without mass. We know that close-up to the surface of any mass object there exists a fine zone of relatively higher EIL. This EIL is compounded from the mass decay E-Flow and the natural wave front emissions. Fig 3.74 shows this as an EIL gradient dropping off rapidly with distance from the object surface.

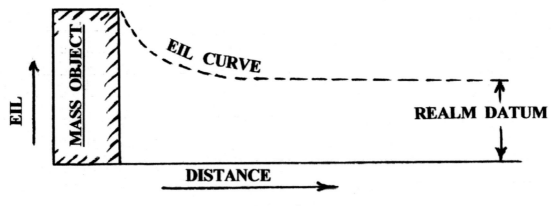

Fig 3.74

Now consider a wave front that propagates past the object. According to the rules of our propagation model the rate of quantum stepping or wave front traverse is inversely proportional to the realm EIL. As such we may represent wave front propagation past the object by showing a refraction of the wave front relative to traversing position of its centres as represented in Fig 3.75. The reader will notice that the EIL gradient behaves somewhat like one half of a concave lens. The wave front tends to work its way past and marginally around the mass object. The propagation directions alter as shown.

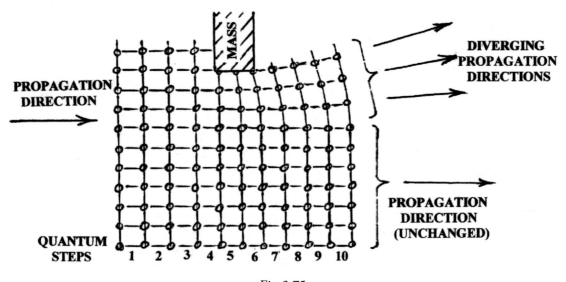

Fig 3.75

To expand on this theme consider a second mass object positioned near the first. Let the distance between them be small so that the EIL gradient caused by one is opposite to that of the other. Let the EIL curves blend together as a smooth continuous curve as represented in Fig 3.76.

Now let us observe the propagation path of a flat wave front traversing the gap between the masses A and B. This is represented in Fig 3.77 and clearly indicates the similarity of action with that of a concave lens. The presentation is a simplified one since we have ignored the inversion of the wave front centres that are reflected off the inner faces of A and B.

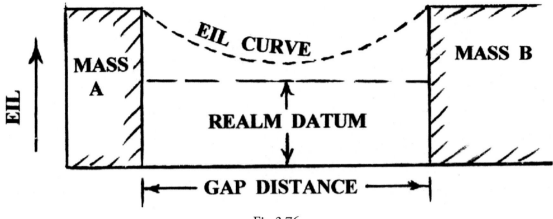

Fig 3.76

These reflected wave fronts are still a part of the main wave front but appear to originate at the surfaces of A and B. The reader may compare this aspect with that of the principles connected with diffraction and so draw their own conclusions. There is much detail that can be built up on this topic but it is not our intention to do so but rather to recommend it as a personal project.

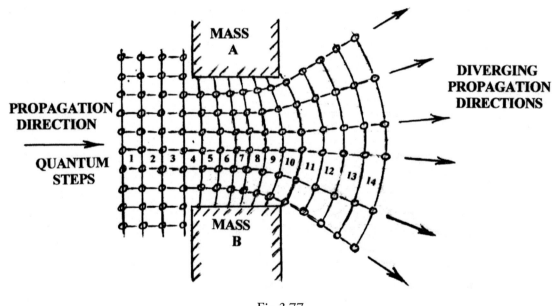

Fig 3.77

Another consideration is when a wave front passes on either side of a mass object as shown in Fig 3.78. Here we notice separated portions of the same original wave front being refracted towards each other's path such that an over-lapping phenomena is bound to result. These are shown as the upper and lower diverging wave fronts and it must be remembered that they are like identical twins having complete similarity in character. The wave fronts are coherent.

When such wave fronts cross each other's propagation paths there are bound to exist locations where the quantum step positions for a wave centre do coincide. Also at these locations the coinciding wave centres may or may not be at the same Mp sequence event. That is to say the E~M/X or M~E/X event of one may or may not correspond to that of the other. We may reason however that when the coinciding wave centres are in the same Mp sequence phase then a strong or bright wave front section is propagated. When the coinciding wave centres are in opposite Mp

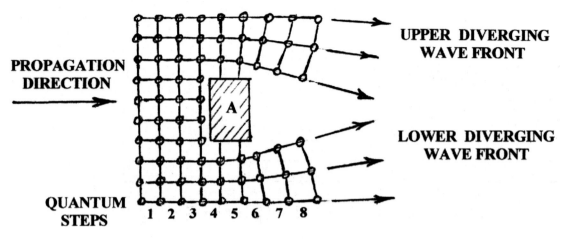

Fig 3.78

sequence then each absorbs energy medium from the other and that section of wave front loses synchronism with the rest and is not observed to propagate in its original form. This is the phenomenon of interference of coherent wave fronts and is dealt with in detail in texts on the subject. It would be impossible to discuss all the known aspects of light wave propagation in relation to our propagation model as this is not the purpose of this text. However we do recommend that the reader re-examine these phenomena with the energy medium theory in mind and perhaps present a related explanation.

Before concluding this chapter we should like to speculate (perhaps rashly) upon the phenomenon known as polarisation of light waves. The speculation is rash but necessary since it could pave the way for other more logical explanations of the phenomenon.

Consider the propagation plate as presented in the Fig 3.7 and Fig 3.16. Now assume that the interference phenomenon similar to that mentioned above is somehow brought into play in the plane of the p/p. Further assume that the interference is along parallel planes in the p/p and these planes have a very small spacing between them. The result is a very fine slicing up of the original p/p as represented in Fig 3.79. The p/p may then be said to be polarised in the direction of the sliced

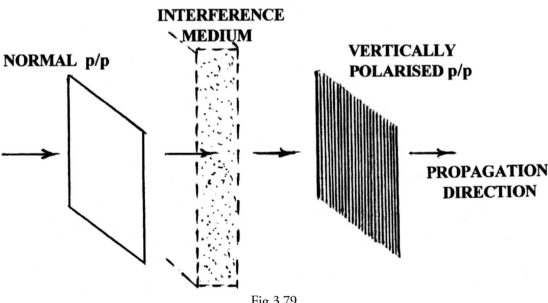

Fig 3.79

lengths. Each slice (white strip) propagates as an individual plate and is separated from adjacent slices by a gap (dark strip) containing no wave centres. The p/p slice maintains its propagation as it may be considered as a plate that is several wave centres wide on the one hand and infinitely long on the other. Over an extended propagation distance the wave centres spread out and so reduce the gap strip. Eventually this gap is reduced to zero and the polarisation effect disappears. The p/p is once again a complete plate albeit reduced in intensity. This principle also applies to the case when the p/p has a rotating characteristic as shown in Fig 3.16.

Now consider the effect of a second polarisation shortly after that above. If the slicing up is in the same direction i.e. vertically, then the already polarised p/p strips will not be interfered with and will continue unchanged. However if the re-slicing effect is transverse to the strip lengths then the overall result is very small individual sections of wave front remaining. These are represented in Fig 3.80. Unfortunately these wave-centre sections are too small to contain sufficient wave centre energy medium for the necessary achievement of C_1 EILs at the next quantum step position. As such the wave front is not renewed and simply dissipates as an errant E-Flow to the realm. In the diagram the 'light bits' indicate the wave centre sections and these are seen to be totally separated at the second transverse polarisation.

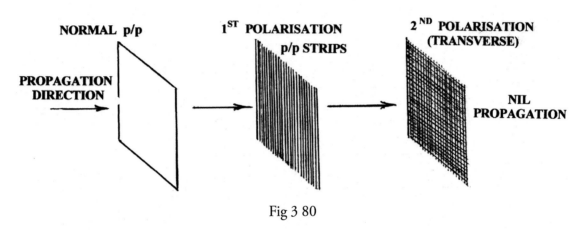

Fig 3 80

Certain crystalline atom structures may cause these polarisation conditions to be brought about and they usually polarise in a single plane only. Reflection and refraction in certain incident directions may also cause polarisation in defined planes. The reader is asked to specially consider the phenomenon of wave centre interference as a chief cause behind polarisation and is recommended to further the application of propagation model principles in the pursuit of greater detail.

In the next chapter on gravitation we shall be discussing impulse in masses arising from a general pattern of E-Flow across the realm and the resulting 'realm datum' setup. We would ask the reader to continually consider all the aspects of wave front propagation in such a complex energy medium environment. Perhaps the detail in this chapter could then be extended by those so inclined.

Chapter Summary

1. The Mp-shell of the atom is the source of all propagating wave fronts.
2. The wave front is made up of perfectly synchronised Mp-event centres.
3. The Mp-event is the sequence of events in the formation of a Mp in the wave front. It provides a theoretical explanation for wave front mechanics.

4. The wave front is represented as a propagation plate which makes a stepping action in the same forward direction. These steps are known as quantum jumps or steps.

5. The propagation plate may also have the property of rotation about its propagation direction axis.

6. The velocity of propagation is directly proportional to the quantum jump size and inversely proportional to the prevailing realm EIL.

7. The propagation velocity is made up of two components: the quantum step component and the p/p forward motion component Vp.

8. The p/p forward motion component Vp is directly proportional to the oscillation wave length or inversely as the frequency. Subsequently the propagation velocity becomes inversely proportional to the oscillation frequency.

9. The quantum jump size increases marginally with the distance propagated.

10. The direction of propagation is always normal to the p/p which tilts towards the higher realm EIL datum. Wave front refraction is a resulting phenomenon.

11. The wave front overall propagation velocity is always less than its corresponding E-Flow velocity.

12. Changes in oscillation frequency are real only when they are the result of a change in the motion or status of the Mp-shell source.

13. The wave front cannot propagate in a realm EIL above C_1 nor in a realm EIL that is below a certain (as yet undetermined) minimum.

14. The Mp-shell structures of all mass objects produce wave front emissions in a compound frequency band. The reflection process is a resulting effect.

15. The Red Shift phenomena is a result of the diminishing realm EIL as represented by the circle 6 energy medium curve. Mp-shell repetitions are slower at the higher realm EILs.

16. Propagation plates under coherent conditions may interfere with one another. Diffraction and polarisation are phenomena resulting from such an interference action.

CHAPTER 4.

Gravitation

Newton's first law of mechanics states that each body continues in its state of rest or of uniform motion in a straight line unless it is acted upon by an external resultant force. Newton defined the force acting on an object as the rate of change of its linear momentum, the momentum being the product of its mass and velocity. This may be expressed more explicitly in the following manner. The linear momentum of an entity remains constant until linear momentum from another entity is added to it. The linear momentum of an entity is defined as the product of its energy medium quantity and its velocity. Linear momentum is a vector quantity, the momentum being in the same direction as the velocity.

The most basic form of linear momentum change is that which occurs during the E-M/X sequence of the Mp event. In Fig 1.7 the phasing energy medium of a number of realms was represented by vectors. Each vector component had a linear momentum that would be absorbed into the Mp. As such the linear momentum of the Mp was the resultant momentum of all the absorbed energy medium components. The linear momentum of each component has been added to that of all the other components during the E-M/X sequence to produce a single resultant linear momentum in the new Mp entity.

The reader is aware of the phasing property of the energy medium that rules out the transfer of linear momentum through a non-existent collision process. This means that the linear momentum of an energy medium quantity can only be added to that of another energy medium quantity by going through the Mp formation sequence. We are also aware that although we have not as yet evolved a parameter for the energy medium we do know that its quantity is proportional to the product of its EIL and volume. The linear momentum in a small volume of high intensity energy medium may therefore be the same as in a large volume of relatively lower intensity energy medium. The reader must also remember that we defined the energy medium as being a perfectly homogeneous entity that had no structure or individual particles to make up the whole. It was indivisible in its volume and yet an infinitesimal part of it could be randomly absorbed into a Mp. The medium itself had no inertia and was unaffected by any kind of 'forces'. It had no constituent other than itself and would travel in a perfectly straight direction until absorbed into a mass particle.

The mass particle that so absorbs an energy medium quantity also acquires the corresponding linear momentum from that energy medium quantity. The rate at which such linear momentum is acquired by a mass entity (like the proton) represents the rate of change in its linear momentum. This is defined as linear impulse and results in change to the velocity of the mass entity. Linear impulse is an event that occurs only during the E-M/X phase of the Mp formation sequence. A succession of linear impulse events within a mass entity results in continued change in the linear momentum of that entity. This also means a continued change in its velocity. The make up of a mass entity is a congregation of a very large number of Mps acting in a defined event sequence. Therefore, although the linear momentum absorbed by each Mp is of a discrete value, the resultant linear impulse on the whole mass entity has an overlapping continuous effect. The mass entity appears to be in a state of continued linear momentum and velocity change.

The rate of velocity change is termed acceleration of the mass entity and the linear impulse is considered to be the force that is causing this to happen. The reader should please note that from now on the terms Force and Linear impulse are synonymous and will be used flexibly and as necessary.

When an object A is moving, it is said to have an amount of linear momentum given by,

$$\text{Linear momentum} = (\text{mass of A}) \times (\text{velocity of A})$$

Newton's second law of mechanics states that the rate of change of momentum is proportional to the impressed Force and takes place in the direction of the straight line along which the force acts. Thus for object A we may write,

$$\text{Rate of change in Linear Momentum} = (\text{mass of A}) \times (\text{Rate of change in velocity})$$
$$\text{or, Linear Impulse} = (\text{mass of A}) \times (\text{Acceleration of A}).$$

This may be expressed more explicitly in realm energy medium terms as follows. The rate of acquisition of energy medium momentum by an object is proportional to the rate of change of its velocity. Also, the acceleration of an object is proportional to the resultant linear impulse acting upon that object.

Newton's third law of mechanics states that for every action in one direction there is an equal action in an opposite direction. Essentially this reflects in a law of conservation of linear momentum which states that if no external forces act upon a system (of colliding objects) then the total momentum (of the objects) in any given direction remains constant.

It was important that we started this chapter with Newton's laws of mechanics. Since these laws are observational laws, they are important clues to consider in the make up and behaviour of all mass structures. We have already indicated the very basic nature of linear momentum change at the Mp level and it is important that the reader note this as the origin of all impulse / force actions.

The law of conservation of Linear Momentum implies that momentum cannot be created. It is possible in practice to initiate momentum when none was previously in existence. Such momentum changes are produced by explosive means. An example is a bullet of mass m of 50 grams say, fired from a rifle of mass M of 2 kg with a velocity v of 100 m/sec. Initially the total momentum of the bullet and rifle is zero. From the principle of conservation of linear momentum when the bullet is

fired the total momentum of bullet and rifle is still zero, since no external force has acted on them. Thus if V is the velocity of recoil of the rifle then,

$$mv \text{ (bullet)} + MV \text{ (rifle)} = \text{zero}$$
$$\text{or} \qquad mv = - MV$$

The momentum of the rifle is equal and opposite to that of the bullet. Solving the equation we see that,

$$V = -\frac{mv}{M} = -\frac{(50 \times 100)}{2000} = -2.5 \text{ m/s}$$

This means that the rifle moves back or recoils with a velocity only about 1/40th that of the bullet. The high velocity of the bullet exactly compensates for its small mass relative to that of the rifle and as such the explosive force produces the same numerical momentum change in the bullet as in the rifle.

If however we consider each mass individually and totally distanced from the other, we may conclude that they have both acquired a measurable linear momentum. These masses may in turn pass on that acquired momentum to other mass objects through collision. The relevance of this analogy to the creation of similarly balanced linear momentum in the energy medium of the Mp will be discussed shortly.

Let us now look at the inverse of the above example. Consider two masses A and B to be joined by a stretched spring. If we hold A and B apart and stationary, then the linear momentum of both will be realistically and individually equal to zero. When we release both masses at the same instant they will commence moving towards one another with individual velocities that fit the equation,

$$M_A V_A + M_B V_B = 0$$
$$\text{or} \qquad M_A V_A = -M_B V_B$$

In other words each mass acquires a linear momentum that is equal and opposite to that of the other. The law of conservation of linear momentum has not been violated. The acquiring of equal amounts of inward direction linear momentum is termed implosive action, while that of acquiring equal amounts of outward direction momentum is explosive action. Both of these types of action are present in the sequence of events leading to and resulting from Mp formation. They form part of the most basic phenomenon within the realm that imparts linear momentum to the energy medium quantities so involved.

Let us now integrate these principles with the phenomenon of Mp formation. There are three phases which characterise the Mp event. The first is when the energy medium becomes a mass-medium that undergoes implosive action. The second is the duration of existence of the Mp entity at the spherical centre of the implosive region. The third phase is when the Mp reverts to the energy medium under a rebound or explosive action set up within the Mp. We shall now examine each phase in a bit more detail.

The first phase is initiated when the realm EIL exceeds C_1. The energy medium concentration is then sufficiently dense for it to acquire a new property. This relates the energy medium character very closely to that of a mass medium at the same energy medium density. Physically each is indistinguishable from the other as far as physical make up is concerned. Yet each is different in pattern of behaviour. The EIL for the energy medium state is a very high one. However the same medium density for the mass medium state is considered extremely low. The mass medium therefore behaves like a highly stretched or expanded version of a mass unit. As in our spring example there now occurs a tremendous tendency within the mass medium for a volumetric contraction or implosive action. This characteristic results in the mass block being split up into defined spheroid volumes each of which contains a fixed energy/mass medium quantity. The size of spheroid is inversely proportional to the EIL attained earlier. This means that the largest spheroid volumes occur at EILs just above C_1 while the smallest occur at EILs approaching C_2. Yet within this relationship the mass quantities are greater for the smaller spheroids at the higher EILs. This is because the EIL range between C_1 and C_2 is hugely out of proportion to the spheroid size range.

This principle is illustrated in Fig 4.1 which shows the relative size between spheroid A at the EIL near C_1 and the spheroid B at an EIL near C_2. The energy/mass medium density of spheroid B is many times that of A yet

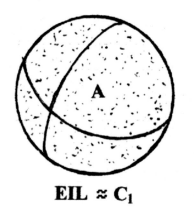

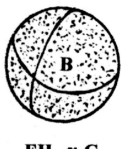

EIL $\approx$ C$_1$

EIL $\approx$ C$_2$

Fig 4.1

the volume difference is marginal. Therefore the energy medium quantity in the smaller spheroid B is greater than that in the larger spheroid A. Subsequently, Mp_A will contain less mass quantity than Mp_B. We must therefore re-define unit mass as a quantity that is relative to a particular C_1^+ EIL. The reader is asked to visualize the afore

mentioned split up of the mass medium block into defined spheroids as an event that takes place at the commencement of the implosive action. The implosive action commences as soon as the energy medium achieves mass medium status. The split up is therefore random by location. This also means that although a majority of the spheroid volumes will conform to the above energy medium density size relationship, a negligible minority of much smaller volume sections will inevitably result. For the moment we shall ignore these exceptions.

Now these spheroid mass medium entities have all formed simultaneously from the same mass medium block. The implosive action causes each to contract in volume towards the spheroid centre. This contraction occurs at a very high rate and continues until mass of a certain optimum density is attained. This central mass entity is termed a Mass-Point (Mp) and becomes the basic unit of mass in our realm energy theory. This unit of mass is however relative to a particular C_1^+ EIL as mentioned above and so must be quantified as such. This will only be possible when an energy medium parameter has been established. The achievement of optimum density completes the first phase of the Mp.

The second phase concerns the duration of the existence of the Mp entity. The character of the Mp is very different from the energy medium from which it is formed. The Mp is a defined energy medium particle with a physical structure that is unable to phase with other Mps. It would therefore be physically possible for one Mp to collide with another. Since the Mp is a result of a contracting mass medium, it is logical to assume that it does not achieve rigidity at the optimum density. The Mp is therefore an elastic and compressible entity. Because of this the duration of existence of the Mp is limited.

The implosive action ends when optimum density is attained. However the linear momentum of the spheroid contraction continues inwards even after optimum mass density has been reached. Subsequently there is a brief but intense centrally directed impact action upon the Mp that compresses it further to a condition well in excess of optimum density. This sets up an opposing explosive action within the Mp that is a reaction of its over compressed material. It is important that we understand the exact nature of the impacting event before continuing to the third phase sequence.

Consider the Mp as a sphere in structural shape and assume it to be made up of a large number of infinitely fine concentric shells as represented by its cross-section in Fig 4.2(a). An enlargement of a section near the surface is given at (b) and shows the successive layers A,B,C,D,E etc. The reader should note that we have layered the Mp for explanatory reasons only and that the true structure should be considered as uniform and homogeneous. Compare it to a piece of rubber material. If you stretch it beyond its normal volume an implosive action is set up and if it is compressed below its normal volume an explosive reaction is set up.

In Fig 4.2(b) consider each layer to possess a defined linear momentum towards the Mp centre. Now consider the layers A,B, C, D, E etc., to continue in sequence all the way to the centre spot of the Mp. At this spot the momentum of the innermost layer will be neutralised by collision with its own opposite face. Let this be the nth layer from the surface layer A. Just prior to this event instant the material making up the Mp will have attained optimum density and the implosive action will have ceased to exist. However the momentum of each layer will still be in existence and therefore the Mp will undergo further change.

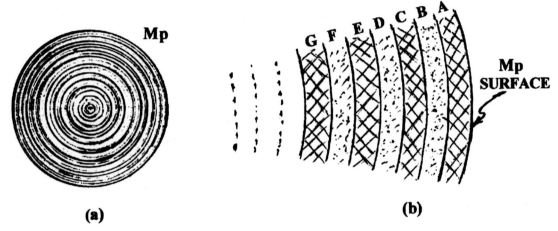

Fig 4.2

Consider the nth layer which has just had its momentum neutralised. This means that any section of this layer has zero momentum. Now the (n - 1)th layer and all of the others still have an inward momentum. The next event is therefore the hypothetical collision of the (n - 1)th layer with the stationary nth layer material. The result is that the momentum of the (n - 1)th layer is neutralised while the nth layer material undergoes further compression. This sets up an explosive action within the nth layer material which has now been compressed to above optimum density. The action of the nth layer material would therefore be to explosively expel the (n - 1)th layer as soon as its momentum is wholly neutralised. This cannot happen because the next event is another hypothetical collision of the (n - 2)th layer with the stationary (n - 1)th layer. The result is a compression of the (n - 1)th layer material which in turn transmits further compression upon the nth layer material. The explosive action builds up. The next events continue in a similar manner for each successive material layer. The compression action is transmitted progressively to all the underlying layers and therefore the innermost layers will be under the greatest compression. As such they are also under the greatest potential explosive action.

The hypothetical collisions or impacting of each layer upon the others below maintains a balance with the build-up of the internal explosive action. The impacting events however must end when the momentum of layer A of Fig 4.2(b) is finally neutralised. The uppermost layer A gives up its momentum by setting up compression in layer B. Layer B is under explosive action which now acts upon the stationary layer A. This event marks the end of the second phase i.e. the duration of existence of the Mp entity.

The explosive action in the layer B now acts outward upon the material of layer A. This causes the layer A to be pushed outward with a finite linear momentum. Under this momentum the material in layer A separates from layer B and is forced to expand in volume. The outward momentum causes the rate of expansion of the mass medium to far exceed the opposite rate of contraction during the earlier implosive sequence. This is because the contraction related to the entire energy medium quantity of the Mp while the current phase expansion is for a small fraction of that quantity. The expansion of the spherical layer A is such that its density reduces rapidly to the C_1 level. The mass medium then reverts to the energy medium state as the explosive action of the mass medium layer A becomes fully effective. The layer A is now in the energy medium state with a finite linear momentum spherically outward from the Mp centre.

The expulsion of layer A now results in layer B becoming the outermost material layer on the Mp. In the process of pushing layer A outwards the layer B has expanded to optimum density and relieved the compression upon layer C. The explosive action in the layer C now acts upon the material of layer B. This causes the layer B to be pushed outwards and to follow very closely behind layer A. This sequence of layer by layer emission continues all the way down to the nth layer when the entire Mp has reverted to the energy medium state. The effect of this layer by layer explosive exchange sequence is to produce a fairly uniform E-Flow from the region of the former

Mp. The velocity of this E-Flow is also fairly constant. Now the explosive action within the innermost layers is greater than that within the more outer layers. Subsequently the momentum imparted to the layer (n - 1) will be much greater than that imparted to the layer B. However the layer B will have a much greater rate of expansion by virtue of its location on the Mp and so will come under the least duration of implosive action. The layer (n - 1) however must undergo a greater duration of implosive action before its expanding mass medium reverts to realm energy medium. The balance is such that the same resultant E-Flow velocity is attained from each Mp

material layer. The third phase or M-E/X sequence is completed when the nth layer is expelled. The reader may draw his own conclusions as to the relative duration of each Mp phase.

By comparing the energy medium state before and after the Mp event we note that the difference is in the momentum of that energy medium. The law of conservation of linear momentum has not been violated since the resultant momentum was always zero. We must note however, that the end result is an E-Flow emanating in all directions from a central point. Since this E-Flow is in a continual state of expansion then its EIL must continually decrease. In chapter 1 Fig 1.3 we showed this EIL to obey the inverse square law for distance. In Fig 1.8 we showed that phasing realm energy medium quantities were absorbed into the Mp structure and that the linear momentum of the Mp was the vectorial resultant of all the absorbed momentum. In Fig 1.14 we discussed the effect of this Mp linear momentum upon the third phase E-Flow velocity distribution. The reader may perhaps wish to refresh his memory of that discussion by referring back.

In chapter 2 we explained the principle Mp groups that led to the congregation of Mps into hyper-complex synchronised groups which behaved as relatively permanent mass entities. All this occurred during the circle 4 phase of the energy medium curve and resulted in the formation of these semi-permanent Mp congregations called protons and proton giants.

With the large scale absorption of the realm energy medium into these protons a permanent energy medium quantity is lost to the realm. This causes a lowering of the realm EILs. Although realm phasing may continue, the resultant EIL will continue to drop as more Mps congregate. At some stage such large quantities of the energy medium will be locked in protons that the realm EIL decreases to below C_1. Subsequently individual Mps will cease to form but protons will continue as Mp-congregations. It is this below C_1 zone existence of the proton which is of the greatest importance and relevance to the physical phenomena of our world today.

Let us visualise a specific Mp well inside the proton body. Although the realm EIL is below C_1, the energy medium exchanges and E-Flows inside the proton results in EILs well above C_1. As M-E/Xs occur, energy medium flows into the surrounding area. These E-Flows phase with each other and result in EILs well in excess of C_1. When Mps reform inside the proton complex they absorb all energy medium entering their zones of action. As such all the E-Flow from a M~E/X gets to be re-absorbed into other E~M/Xs well before the surface of the proton can be reached. There is thus a kind of balance between the M~E/Xs and E~M/Xs within the body of the proton. But what of the acquired linear momentum of a re-formed Mp inside the proton? We may assume that within the body of the proton each E-M/X is surrounded by a directional uniformity of E-Flows such that the resultant momentum of each newly formed Mp remains zero relative to the proton.

Let us now consider a Mp on the surface of the proton. By definition a Mp is short-lived and exchanges to energy medium in a sequence of events resulting in a spherically outward E-Flow. In Fig 4.3 this third phase is depicted for a Mp at the proton surface. The resulting E-Flow may be divided into three directional sections as follows:

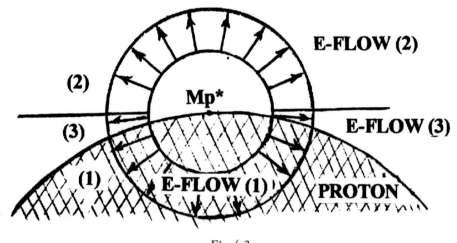

Fig 4.3

1. The E-Flow (1) is the section that flows into the body of the proton.
2. The E-Flow (2) is the section that flows outward away from the proton surface.
3. The E-Flow (3) is the section that flows along a plane that is tangential to the proton surface.

The reader should be aware that the present discussion refers to conditions in the realm that relate to the circle 6 phase of the 7 circle energy medium curve. The proton is therefore in a state of decay which is reflected in a permanent loss of some of its surface Mps. Fig 4.3 shows the exchange of a surface Mp into a spherical E-Flow which we have conveniently divided up into three sections. Let us now consider each section individually and in detail.

The E-Flow (1) phases into the body of the proton and must therefore be absorbed into the E~M/Xs of the Mps re-forming there. The absorption of this energy medium quantity is also to have absorbed its linear momentum. It must be noted that the E-Flow (1) is being treated as a separate entity and has a net linear momentum that acts towards the centre of the proton body. The E-Flow (1) contains approximately half of the Mp* energy medium quantity since it represents nearly half the total spherical E-Flow. (Mp* denotes the location of a previously existing Mp). We can assume therefore that proton decay is the loss of only half a Mp's energy medium quantity for each Mp permanently exchanged at the proton surface.

Consider a small part of the E-Flow (1) from the surface Mp* centre O as represented in Fig 4.4. The E-Flow is in the general direction of the proton centre. We have shown a random arrangement of Mps near the proton

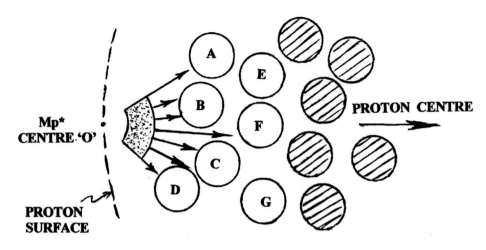

Fig 4.4

surface to a grossly exaggerated scale. As the E-Flow proceeds inwards, some of it is absorbed in the E-M/Xs of Mps at A, B, C and D. The rest of the E-Flow continues on and gets absorbed in E,F and G. Let us for a moment suppose that there was no E-Flow from the Mp centre O. The Mps A, B, C, D, etc would still have occurred in the positions shown but from a uniformity of surrounding E-Flows. As such the linear momentum of each would remain unchanged.

Now the absorption of any energy quantity from the E-Flow (1) by the Mps A, B, C, D, etc., means that their linear momentum must change and that the impulse acts in the inward direction. The Mps that gain this additional momentum must therefore also acquire a finite motion. This motion is in an inward direction and causes them to collide with other internal Mps. These then result in a transfer of linear momentum from Mp to Mp through E-Flow absorption and result in a general momentum of Mps towards the proton centre location.

In Fig 2.6 we showed that Mps within the proton and progressively nearer to its centre contained a proportionately greater mass quantity. The general EIL inside the proton would subsequently result directly from E-Flows arising from the M-E/Xs of these Mps. It can be shown that Fig 2.6 also represents the pattern of EILs within the proton since Mps at greater mass levels also produce E-Flows at correspondingly greater EILs. This means that E-Flows from Mps nearer the proton centre have a greater energy medium content than those from Mps nearer its surface. Logic therefore dictates that in a Fig 1.7 situation of energy medium absorption in the formation of a Mp, that for any E-M/X event within the proton an isotropic absorption of energy medium and hence linear momentum cannot occur and that Mps should in general acquire an impulse in a direction away from the proton centre. This however is not the case. The cyclic duration of the full Mp event sequence is directly proportional to the EIL of the environment within which it occurs.

As such the frequency of full cycle Mp repetitions will be lesser the closer these are to the proton centre. This would indicate that the M-E/X cyclic event occurs with greater frequency near the proton surface than at its centre. At this stage we must once again assume that the energy medium difference in the M-E/X contents from optimum and minimum mass point levels is a smaller variation than that resulting from a Mp's cyclic frequency at near C_2 and C_1 EILs. This means that although Mps near the proton centre produce E-Flows with a greater linear momentum Mp for Mp, it is however the Mps at higher operating frequencies that impart a greater impulse to their surroundings over a set period of time. Thus the Mps in a proton acquire greater linear momentum from the surface zone E-Flows than from those that emanate from its more central regions. As such the proton is maintained in a state of compression with all Mps possessing a net linear momentum directed towards the proton centre.

There are a large number of Mps upon the proton surface that undergo this M-E/X sequence. In theory therefore proton decay should occur through this very large number of M-E/Xs occurring uniformly about the proton surface. Since all Mps just below the surface acquire an inward linear momentum, there is a general impulse towards the proton centre. The inward linear momentum of Mps upon one side of the proton is found equal to the inward linear momentum of Mps upon the opposite side. The net result of all the E-Flow (1) acquired linear momentum is a general neutralisation against one another but causing a compacting action between the proton internal Mps. Each internal Mp is therefore being pushed towards the proton centre and is the main action that holds the proton together. It also causes the proton to behave as an extremely compacted Mp-congregation unit.

Let us now suppose that the number of Mps distributed uniformly about the proton surface that undergo the M-E/X can be varied. This means that we can also vary the proton compacting action. A greater number of distributed M-E/Xs will result in a greater E-Flow (1) quantity. This would result in greater impulse of the internal Mps and so a more compacted proton. For a lesser number of surface distributed M-E/Xs the proton compaction would be correspondingly less. The reader should note that here for the first time in our discussions we have come across a mass entity in continued existence with considerable internal action. Although we have just observed it in its balanced state, the internal impulses causing compaction are together of relatively immense importance and significance.

So much for the principle of action within the proton. There is considerably more to the behaviour of Mps at its surface and this is where the E-Flow (3) of Fig 4.3 plays an important role. The main effect of this E-Flow is that a region of high EIL is set up close to the surface of the proton especially when combined with the E-Flow (3) of adjacent surface Mp*s. Since M-E/Xs occur uniformly about the surface this results in a high EIL grid surrounding the proton and close to its surface. This energy medium grid may be compared to the atmosphere that surrounds our planet Earth (to aid the readers conception). The EIL within this region may exceed C_1 and as such give rise to new Mp formation. It is the subsequent M-E/X action of these Mps that prevents the otherwise rapid decay of the proton surface Mps.

In Fig 4.5 is shown a section near the proton surface at a particular event moment. The surface Mps are represented by the row A and successive inner Mps by the rows B,C,D,E etc. These Mps formed from the M-E/X action of previous Mp*s that existed a moment earlier in the rows (a), (b), (c), (d), (e), etc,. The reader will notice that the internal rows will easily reform from the M-E/X E-Flows of the surrounding Mp*s. All the Mps of a proton are surrounded by other Mps except for those at the surface. These surface Mps cannot therefore receive the same amounts of reforming energy-medium as the others. The surface Mps should therefore in theory be lost to the proton.

Subsequently in a matter of a few Mp event cycles the inner rows B,C,D,E, would each represent the surface row in turn and so be lost. This sequence would result in a very rapid demise of the proton mass.

In practice however this does not occur. The E-Flow (3) causes a region of C_1^+ EIL at the proton surface that functions as a barrier to widespread surface Mp losses. This region or EIL grid (as we shall term it) provides the proton with an EIL zone that corresponds to the Circle 5 realm energy medium curve conditions. As such independent Mps will form on the outer side of the surface Mps. These EIL grid Mps are not a part of the synchronised proton Mp-congregation but nevertheless they do position the surface Mp in a more uniform E-Flow situation. More surface Mps will subsequently

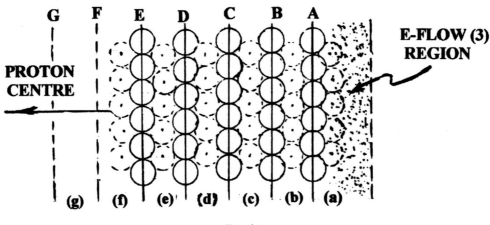

Fig 4.5

be able to undergo cyclic repetition. The proton surface EIL grid was initially set up from the E-Flow (3) of its surface Mps. This EIL atmosphere was then found to be conducive to the formation of other secondary Mps . The E-Flow from these secondary Mps then not only added to the EIL atmosphere but also provided many of the proton surface Mps with enough energy medium and momentum to maintain them in a zero impulse situation. As such the number of proton surface Mps that are lost is reduced considerably. When a large proportion of the proton surface Mps are protected in this manner the result is a reduction of the E-Flow (3) atmosphere. We draw the conclusion that a certain proportion of surface Mps must be lost by the proton in order to maintain the minimum necessary EIL atmosphere. For the surface Mps to maintain this minimum atmosphere balance a very high rate of proton decay would need to be maintained. This does not fit in with reality and we find that this is because we ignored the presence of other adjacent protons and the Realm datum itself.

In chapter 3 we defined the Realm datum at a point as being the intensity of the energy medium resulting from a countless number of E-Flows from all directions arriving at and phasing through that realm position. The existence of any phenomena that results in a flow of energy medium must itself add to the realm datum and in turn be affected by it. The energy medium outflow from a proton may be considered to be reasonably continuous and obeying the inverse square law for intensity. As such the EIL of this E-Flow would appear constant at a set realm position. The continuous and consistent nature of this E-Flow intensity contributes to the realm datum simply by its presence therein. In the case where velocity may be ascribed to the proton mass then the component of E-Flow EIL added to the realm datum at a set position would either increase or decrease depending on whether the proton approached or receded from it.

The realm EIL datum may thus be looked upon as a three dimensional EIL spatial grid that defines the moment by moment energy medium intensity or EIL at any point within the realm.

The E-Flow (2) shown in Fig 4.3 is thus a major contribution to the realm datum when taking into account all mass within the realm. In chapter 1 we considered this entity in some detail and a revision to Fig 1.15 is advised at this stage.

The proton surface EIL grid is thus reinforced to some considerable extent by E-Flows from adjacent protons. This subsequently supports the action of the secondary Mps in the surface zone with the net result of a further reduction in proton decay. It must be appreciated that Mp decay at the proton surface is essential to the support of the surface EIL grids within the proton group. Although some individual protons may receive more energy medium than that lost, this would only be a short term feature and would be at the expense of other protons.

Logic dictates that the EIL of this proton surface 'energy medium atmosphere' is at a consistent level all around the proton. This is because of a built-in cause and effect feature which operates to regulate this energy medium level. Consider an increase in the external E-Flow intensity from adjacent protons. The proton surface grid will undergo a rise in EIL and the secondary Mps will form at a proportionally increased mass quantity level. Subsequent exchanges cause the proton surface Mps to also receive a proportionally increased E-Flow on their outward side. This reduces the imbalance of E-Flows that surrounds the surface Mps and causes fewer to be lost in the decay process. This in turn reduces the E-Flow (3) quantity and thereby reduces the surface grid EIL. We observe that an increased input of energy medium from external sources has resulted in a decreased E-Flow from the surface Mps. The result is a minimal change in the surface 'energy medium atmosphere' EIL.

Now consider the effect of a decrease in the external E-Flow energy medium. The proton surface EIL grid will undergo an immediate drop in EIL. The secondary Mps will form at a proportionately reduced mass quantity level. Subsequent exchange events will result in the proton surface Mps receiving a proportionately reduced E-Flow from the 'atmosphere zone'. This increases the imbalance in the directional E-Flows to these Mps which causes a greater number to be lost in the decay process. The E-Flow (3) quantity thereby increases and raises the surface grid EIL. Once again we observe a minimal change in the surface 'energy medium atmosphere' EIL.

In future discussions we shall only make reference to the Realm EIL datum as the main external source supplying energy medium to the proton surface grid. This brings us to the conclusion that the proton decay rate is inversely proportional to the Realm EIL datum at that location. At this stage the reader should have a fairly clear mental picture of the proton and its behaviour in decay. However, let us summarise its main features with the aid of the diagrams in Fig 4.6.

1. The proton is a Mp-congregation formed during the Circle 4 and 5 phases of the 7-circle realm energy medium curve.
2. During the present circle 6 phase the proton enters a process of decay through the loss of energy medium from its surface Mps.
3. The E-Flows from these surface Mps results in the internal Mps receiving an inward impulse causing the Mp-congregation as a whole to be in a continual state of compression.
4. The E-Flows also result in a supply of energy medium to other protons and the realm datum.
5. These E-Flows at the proton surface result in a high EIL 'energy medium atmosphere' around the proton.
6. The energy medium atmosphere represented in Fig 4.6(a) causes random secondary Mps to form within it which in turn help to suppress the decay of the proton surface Mps.
7. The realm datum supplies energy medium to the proton's atmosphere zone.
8. Some proton surface Mps leap outward into the surface zone because of impulse from internal E-Flows. These are represented by the Mps A, B and C in Fig 4.6(b).
9. The action within the surface zone is one of furious activity. Secondary Mps are in a state of continual mass and energy medium exchanges. E-Flows are generated in all directions causing most proton surface Mps to remain at zero linear momentum.
10. The EIL of the surface zone is maintained at a reasonably consistent level by an automatic variation in the proton decay rate.
11. Finally the proton may be viewed as a highly compacted spherical Mp-congregation unit enveloped in a protective surface layer of secondary Mps in a high EIL energy medium atmosphere.

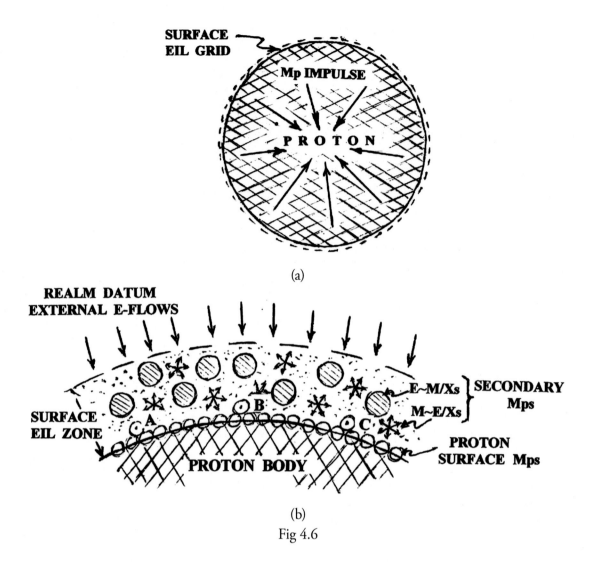

SURFACE
EIL GRID

Mp IMPULSE

P R O T O N

(a)

REALM DATUM
EXTERNAL E-FLOWS

E~M/Xs
M~E/Xs

} SECONDARY
Mps

SURFACE
EIL ZONE

A   B   C

PROTON BODY

PROTON
SURFACE Mps

(b)

Fig 4.6

Let us consider the distribution pattern around the proton surface for those Mps that are permanently lost. We know the surface grid remains at a uniform EIL all around the proton. If the realm datum is uniform and unchanging then the quantity of external energy medium acquired by the surface grid will also be uniform for all positions around the proton. This also means that energy medium from Mps lost at the proton surface must be supplied to the surface grid in a similarly uniform manner. This uniformity of energy medium supply from the proton surface can be achieved only by an even distribution of lost Mps across that surface. Thus Mps causing decay of the proton will be evenly spaced about the surface.

If the realm datum remains uniform but increases in EIL then the Mps lost at the proton surface will need to supply less overall energy medium to the surface grid. Because each Mp must lose all or none of its energy-medium quantity the above reduction is made by fewer Mps being lost. Since this energy-medium must still be supplied to the surface grid in a uniform distribution the fewer Mps being lost acquire a proportionately enlarged though even spacing about the proton surface.

A decrease in the uniform realm datum EIL will subsequently mean a greater supply of energy medium from the proton surface is required to maintain the surface grid at its consistent EIL. This is achieved by a greater number of Mps being lost. Since the increased energy medium must be supplied to the surface grid in a uniformly distributed manner, then the Mps being lost simply acquire a proportionately tighter (though even) spacing about the proton surface.

From the above it will be obvious to the reader that the decay rate of the proton is inversely proportional to the surface spacing of lost Mps. It must be mentioned however that it is possible for the proton to be in a state of nil decay and yet continue with a pattern of surface Mp losses. The above varied spacing is figuratively represented in Fig 4.7.

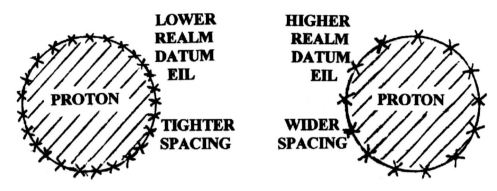

Fig 4.7

When an EIL parameter does evolve we shall be in a position to equate the losses and gains in energy-medium quantity terms for the proton mass. This could lead to a determination of the rate of proton decay and also proton time as discussed in the next chapter. For the moment it is sufficient that we continue with our conception and logical analysis of these realm events.

So far we have permitted the compacted Mp-congregation or proton to remain in a state of unchanged net linear momentum. This is because we assumed the proton to exist in a region of absolutely uniform realm datum. In the realm such regions are few and far between. As such it would be more appropriate to assume the proton to be in a region of varied EIL. We shall now determine the effects of such a realm datum upon the Mp events within and upon the surface of the proton. This situation is represented in Fig 4.8. Each half of the proton is shown to be located within a differing realm EIL datum. The surface EIL grid is depicted as the broad zone around the

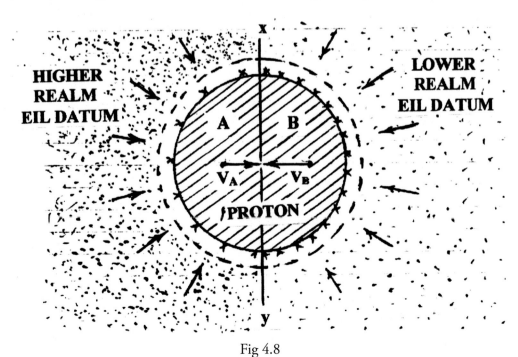

Fig 4.8

proton. Let the plane xy, passing through the centre of the proton, be considered as the plane separating a region of higher realm EIL from a region of lower realm EIL. The realm datum by definition is an accumulation of the EILs of E-Flows from the other protons/mass bodies in the realm. We mentioned the uniform distribution of galaxies across the realm in an earlier chapter and its short term constant positioning. As such the realm datum is shown in Fig 4.8 as a flow of external energy medium into the proton surface zone. We have already described the behaviour and events of the surface grid zone. We know that the EIL of this zone remains at a consistent level all around the proton and that variations are quickly adjusted by corresponding surface Mp action. Now if we apply the conclusions represented in Fig 4.7 to the present case we find that surface Mp losses on the proton half 'A' are fewer than those on the other half 'B'. That is, the surface spacing of lost Mps on side B is more closely knit than those on the side A.

In Fig 4.3 we saw that each Mp lost at the proton surface caused an unbalanced E-Flow (1) to enter the Mp-congregation. The linear momentum of this E-Flow was acquired by the internal Mps as shown in Fig 4.4. This set up a general Mp impulse towards the proton centre which was neutralised by a similar action on the opposite side of the proton. Now in Fig 4.8 consider the E-Flow (1) quantity absorbed by the internal Mps. The proton section A has the lesser E-Flow (1) quantity than does the proton section B. Therefore the net linear momentum of the E-Flow (1) towards the proton centre is greater within the section B. Since the entire E-Flow (1) quantity is absorbed into the E-M/Xs of the internal Mps then the linear momentum acquired in that absorption will be the total linear momentum of the E-Flow (1). This means that the linear momentum acquired by the internal Mps in section B is greater than that acquired in section A. The resultant linear momentum of all the Mps in section A may be represented by the indicated vector quantity V_A as shown. The resultant linear momentum of Mps in section B is correspondingly represented by the vector V_B.

Now V_B is greater than V_A. The section B impulse is in the direction of section A while the section A impulse is in the direction of the section B. That is to say the sections A and B are pushing against each other. Since the push of section B is greater than the push of section A, the proton as a whole will move in the direction of B to A. The gain of linear momentum by the proton is therefore the difference in the vector quantities V_A and V_B and will be in the direction of V_B. The proton as a whole will obey Newton's second law and accelerate in the direction BA.

$$\text{Impulse } V_B - V_A = \text{proton mass} \times \text{acceleration.}$$

The proton is seemingly under an external push action or force that acts in the direction of the higher realm datum. A net impulse has been acquired by the proton as a direct result of its position in a varied realm datum. The reader should note that the impulse (i.e., rate of change of linear momentum) of the proton is a result of variations in the absorption of energy medium by its Mps and not because of some external collision effect. Each Mp-event responds to changes in realm EIL and the proton as a Mp-congregation will abide by the same rules. The difference in the realm EIL across the width of the proton may therefore be considered to be the source of the push effect within the proton. This push effect is termed the Gravitation effect and its level of action is directly proportional to the EIL gradient of the surrounding realm datum.

The reader will no doubt be concerned about the observed gravitation effect that appears to operate according to Newton's formula,

$$F = G \cdot M_1 M_2 / r^2 \text{ (the usual notations)}$$

in which two masses M_1 and M_2 are pushed towards each other by a force F which is directly proportional to the product of their mass quantities and inversely proportional to the square of the distance r between them. G being assumed as a constant in the short term and called the gravitational constant. This is a law that was a direct result of observation and we have no intention of altering the formula (for now). However we do intend to refute the idea of mutual attraction as a direct action between the two masses.

In the present day realm (circle 6 curve) all mass is in a state of normal decay. Which means that each mass entity is continually giving up some of its mass to the realm as a flow of energy medium. This E-Flow has an intensity that is inversely proportional to the square of the distance from source. The E-Flows all phase with each other causing an EIL that is the arithmetical sum of the individual intensities. This is the general make up of the Realm EIL datum.

Now the E-Flow (2) from a proton is an external E-Flow that has no event with the proton itself but only with other protons. This means that the proton does not feel its own E-Flow (2) as an integral part of the Realm datum surrounding it. This is a logical pattern of behaviour and is based upon the fact that the proton Mps can only respond to conditions within its surface zone caused by energy medium entering there from the outer side. The E-Flow (2) does not contribute sufficiently to its own proton surface EIL grid to affect the behaviour of its surface Mps to any great extent. When considering the influence of the realm datum upon a proton we must therefore ignore that particular proton's outward E-Flow (2) .

Our contention is that masses do not exert any direct influence upon one another. They add energy medium to the realm datum which then undergoes a local EIL change resulting in a local realm EIL gradient. Any proton that happens to be within this realm EIL gradient zone will then respond in the manner described in Fig 4.8. Namely, it experiences an impulse in the direction of the increasing EIL gradient.

The realm datum may be considered as a 3-dimensional grid defining realm EIL from point to point at any event instant. Points of equal EIL will be represented by a curved plane passing through those points in a manner similar in principle to geographical map contour lines that indicate points of equal height above sea level. It is essential that the reader develop a detailed mental picture of this realm EIL grid concept in order to fully understand its character and effect upon the proton. Since mass objects like galaxies and proton giants tend to occupy fairly stable realm positioning (in the short term), then this spatial EIL grid also has a similar consistency of location (in the short term). Initially the realm existed as a cloud-like energy medium entity traversing the cosmic void in a direction away from a much grosser 'Mother Realm' entity. At this stage the realm datum would be considered as the EIL of the energy medium in its variable distribution within that realm. The marginal expansion of the realm would cause a gradual lowering of the datum EIL with traverse duration.

With the advent of multiple realm phasing and the subsequent 7-circle energy medium curve phenomena, a complex pattern of flowing energy medium has evolved within the realm. Today we are at the circle 6 stage in the life of the realm cycle and a very large number of E-Flow sources are in existence. The original realm EIL is small compared to the current summation of individual E-Flow EILs. For every realm position there are a countless number of E-Flows arriving from all directions. The major players in the source of these E-Flows are the galaxies that are uniformly distributed throughout the realm as we described in chapter 2 (when we exemplified their distances from us using an aspro to represent the size of our galaxy on page 63.).

The E-Flow from the each proton /galaxy mass may be considered to be reasonably consistent and fairly continuous. The EIL within this E-Flow obeys the inverse square law for distance and since mass objects are relatively stationary within the realm their E-Flows contribute a reasonably constant EIL to a set realm position. Some E-Flows may have originated from relatively nearby masses while others may have traversed vast distances. Each will contribute a proportional component of EIL to the overall realm EIL grid (in time). It is obvious that nearby protons will contribute a larger component of EIL to a grid point than those farther away. The realm datum grid would therefore have a higher EIL at positions close to mass bodies. As such a 3-dimensional plan of the realm EIL grid would show an increasing EIL gradient for positions approaching a mass object. This is represented in Fig 4.9 for a realm section around the mass object A. The lower shaded area shows the realm datum EIL in that area if we

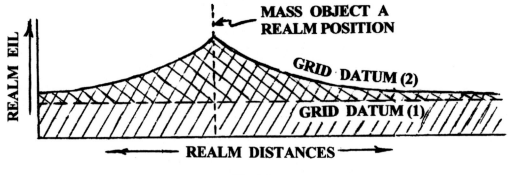

Fig 4.9

exclude the EIL component in the E-Flow from mass object A. The mass object A will consider itself to be at this realm grid datum(1) EIL and its protons will behave accordingly. Since there appears to be no EIL gradient in the grid datum(1), the mass A would receive no impulse and so remain in equilibrium.

The E-Flow from the mass A however would contribute its component of EIL to the realm grid and if another mass object were to approach mass A then it would experience the realm EIL datum(2). The closer its approach to mass A the greater the EIL gradient of the grid datum and hence the greater the impulse towards A. The approach of another mass towards mass A would cause a change in the grid datum(1) by virtue of another EIL component being added to it. The grid datum(1) would gradually increase towards the approaching mass and at any instant the grid EIL relative to each mass would be as represented in Fig 4.10. A and B are the two masses at the realm

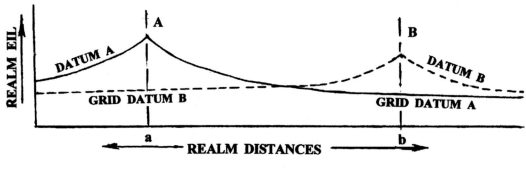

Fig 4.10

positions a and b. Let mass A be greater than mass B. The E-Flow quantity from A will then be greater than that from B. The mass B will experience the grid datum A EIL while mass A will experience the grid datum B EIL. Each grid datum being the cumulative EIL of all E-Flows other than its own.

At the mass A position the grid datum B has an EIL gradient sloping upwards in the direction of B. As such the protons in mass A receive an impulse towards mass B. Similarly at the mass B position the grid datum A has an

EIL gradient that slopes upwards in the direction of mass A. The protons in mass B therefore receive an impulse in the direction of mass A.

Now the impulse received by each proton in mass B is dependant upon the EIL gradient of grid datum A at position b. The total impulse received by B will depend upon its total proton quantity. We may therefore conclude that the impulse pushing mass B in the direction of mass A is also directly proportional to the mass of B. Also the amount of the EIL gradient at position b is dependant upon the E-Flow quantity from mass A. This in turn depends upon the mass of A. Therefore the impulse received by mass B is also directly proportional to the mass of A. Since the EIL gradient of grid datum A decreases with increasing distance from A according to the inverse square law then we may also conclude that the impulse received by B is inversely proportional to the square of its distance from A. As such we have,

$$\text{Total Impulse in A (in all its protons)} \propto {}^{(M_A \times M_B)}/R^2$$

(where M_A and M_B are the masses of A and B and R is the realm distance between them).

Let us now consider the realm datum around A and B that would be prevalent for a third mass C. The grid datum EIL that mass C would respond to would be the arithmetic sum of all the E-Flow EILs around it other than its own. In Fig 4.11 this is done by graphical summation of E-Flow EILs from masses A and B on top of the 'Realm Datum Base' (due to the isotropic E-Flows from all the other mass of the realm). If C is a single proton its effect upon the Grid Datum AB would be negligible. To determine the impulse in C for any position between the masses

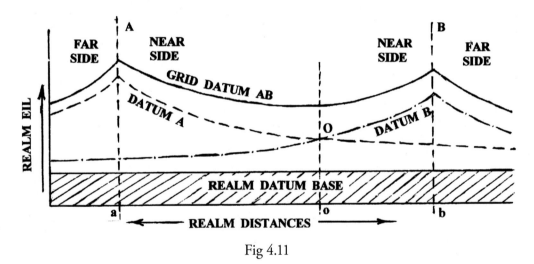

Fig 4.11

A and B, one would simply need to ascertain the grid EIL gradient at that position. At the realm position O indicated by the intersection point of datum A and datum B on the straight line between masses A and B, we note that the EIL gradient of the Grid Datum AB is nil. As such if the proton C were to be located at O it would undergo no impulse and would remain in equilibrium. It should be remembered that Fig.4.11 only represents the Grid Datum AB for a particular instant since both masses A and B are in a state of impulse towards each other.

A closer examination of the Grid Datum AB line highlights the fact that the EIL gradient on the far side of each mass is greater than that on the near side. Thus if C was to be positioned at a set close-up distance from A or B, then it would receive a greater impulse towards that mass if the positioning was on the far side of it. The difference in gradient arises because the EIL gradients of Datum A and Datum B are of opposite slope on the near side of the masses. An addition of their EILs will therefore reduce the overall gradient. On the far side however the EIL gradients of Datum A and B slope in the same direction so that the addition of their EILs increases the overall gradient.

The realm datum base represented in Fig 4.11 may be assumed to be reasonably constant and with an EIL gradient approaching zero. It is the sum of the EILs in a pattern of isotropic E-Flows from very distant galaxies. The measure of distance is relative and normally means outside the realm zone under investigation such that the E-Flows enter the zone equally from all directions. A part of the realm datum base is also the EIL of the original realm energy medium which may be related to the expanding realm prior to phasing.

The gradual but continual decay of the proton mass within the circle 6 phase means that an overall depletion of mass within the realm is taking place. This depletion of mass occurs mainly by a reduction in the quantity of Mps that event within the proton. This in turn results in a decrease in the E-Flow (2) EIL that effectively emanates from each proton to the realm. The isotropic E-Flows subsequently arriving at some isolated distant point in the realm will therefore progressively diminish in EIL. Since this is the Realm Datum Base at that point we may conclude that it does not remain constant as considered earlier but undergoes a very gradual EIL diminishment.

In a similar manner with progressive loss of mass the E-Flows from A and B in Fig 4.11 will also gradually undergo a reduction in EIL. The local grid datum AB will therefore prevail at a reduced EIL. Protons that enter a reduced grid datum EIL tend to undergo an increased rate of decay. This can only slow down the rate of general grid datum decline temporarily. It must be noted that although protons may decay at a higher rate they do not necessarily produce an E-Flow (2) of higher intensity. The intensity of the E-Flow (2) depends upon both the mass of the proton and its rate of decay. The rate of decay is defined as the rate of loss of mass as a percentage of the total mass of the proton. It is therefore clear to see that an earlier proton may produce a greater E-Flow quantity at a low rate of decay than that which a later (and lighter) proton produces at a higher rate of decay.

Let us for a moment consider the Grid Datum of the Realm as a whole. If we assume that masses exist reasonably uniformly across the realm volume then E-Flows from these masses will result in a higher Grid Datum near the centre of that volume. As such an increasing EIL gradient will exist across the realm and in the general direction of its volumetric centre as shown in Fig 4.12.

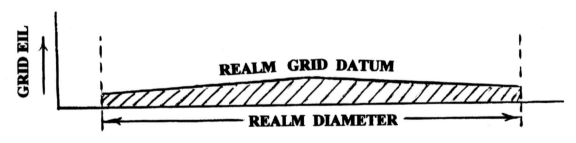

Fig 4.12

According to our theory all masses will receive a net impulse directed towards the centre of the realm. All the E-Flows will however proceed beyond the original realm diameter and so cause its extension. A sort of growth by default. This extension will progress initially at the E-Flow velocity and will reach a dimension limit when the E-Flow EIL approaches C_0. Now this dimension limit will tend to decrease as the decaying mass quantities within the realm gravitate up the EIL gradient towards the realm centre. This is because the E-Flows have a traverse limit at which their EILs reduce to C_0 (See Fig 1.12).

The realm covers a very large expanse and it may not be possible for protons to traverse the distance to the realm centre before complete decay occurs. It is therefore purely academic to consider the EIL changes in Grid Datum as more mass approaches the realm centre. It may be that the 7 circle principles are repeated in part if and when the Grid EIL is pushed towards the C_2 level.

We mentioned earlier that as proton decay progressed, the mass of the proton became less and its E-Flow intensity reduced. This factor must therefore be taken into account in the formula for impulse received by adjacent mass objects. In Fig 4.10 we showed that,

Total Impulse in A (in all its protons) $\propto {}^{(M_A \times M_B)}/R^2$
(where M_A and M_B are the masses of A and B and R is the realm distance between them), and, Total Impulse in B (in all its protons) $\propto {}^{(M_A \times M_B)}/R^2$

Therefore, Total Impulse in A = Total Impulse in B.

Now because the mass object A has a greater quantity of protons than mass object B, we may conclude that the impulse in each of B's protons is somewhat greater than the impulse in each of A's protons. That is to say, object A has more influence on each proton in object B than B has upon each proton of A.

Fig 4.13 represents a proton that is in a region of EIL gradient. The gravitational push is in the direction of the EIL 'Normal line' and up the EIL gradient (see Fig 3.51). The Grid datum EIL lines shown are the contour

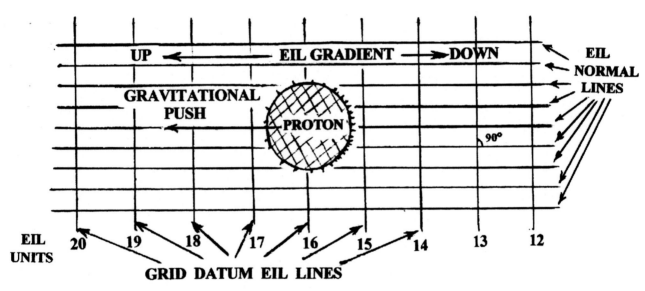

Fig 4.13

lines that represent the curved grid plane through points of equal EIL. The EIL normal line is that which indicates the maximum gradient between the curved grid planes (as described in chapter 3). The dependence of gravitational push upon the grid datum alone is clearly illustrated by the absence of other protons in Fig 4.13.

Let us now attempt an evaluation of the gravitational impulse due to grid datum gradient. The net impulse of each proton is the resultant of all the Mp impulses directed at its centre. In Fig 4.8 we showed this resultant to be in the same direction as vector V_B. It must be realised that if we summed up the centrally directed impulses for all the Mps in the proton, we should find ourselves

with a very large impulse quantity indeed However the impulses of the Mps in one half of the proton are approximately balanced by opposing impulses in the other half. The slight difference in impulse balance appears as the resultant proton impulse (i.e. a rate of change in proton linear momentum). The balancing of opposing Mp impulses results in the extremely compacted Mp-congregation as described earlier.

The overall proton impulse is in effect a very small proportion of the total impulse that compresses the proton structure. In order to develop a formula for the gravitation push upon a mass object we must include this "small proportion" as a factor in the impulse equation. This small proportion factor is the percentage imbalance of impulses within the Mp-congregation that is directly dependant upon the slope of the realm datum EIL gradient. Since the realm datum is reasonably consistent in its EILs, it stands to reason that the above factor will also correspond. Let us denote this percentage imbalance of impulses within the proton as the Factor X and written X^* since it is without dimension (being a percent figure).

We must recognise X^* as being variable and dependant upon grid datum gradient. In its turn the grid EIL datum depends upon E-Flow intensity from mass objects near and far. Consider a proton at a defined location in such a realm grid. If Ic is the total impulse quantity of its Mps then the resultant impulse is given by;

$$\text{Impulse per proton} = X^* \times Ic$$

Alas, this equation is not a very practical one since we are unable to measure or determine the factors on the right. We can however link them to other known factors and so form a quantifiable equation.

Consider the two objects A and B in Fig 4.10. The impulse per proton of B is proportional to the grid datum A gradient at B's location

$$(X^* \times Ic)_B \propto \text{grid datum A EIL gradient.}$$

Now, grid A gradient $\propto$ E-Flow quantity from A

$$(") \propto M_A$$

Also, grid A gradient $\propto {}^1/R^2$ (Inverse square law)

Therefore $(X^* \times Ic)_B \propto {}^{MA}/R^2$

Impulse per B proton = Constant B x ${}^{MA}/R^2$

Total impulse in B is given by,

$$M_B \times (X^* \times Ic)_B = \text{constant B} \times {}^{MA}/R^2 \times M_B$$

if we consider each proton as a numerical mass unit.

Similarly, total impulse in A is given by,

$$M_A \times (X^* \times Ic)_A = \text{constant A} \times {}^{MB}/R^2 \times M_A$$

Experiment shows the impulses in A and B to be equal and opposite.

Therefore we have, Constant A = Constant B.

This is a common gravitational constant and is denoted by the letter G.

Therefore total impulses in A and in B = $G \cdot M_A M_B / _R 2$

We now have an equation that is a practical one with quantities that are measurable by current techniques. The constant G has been determined experimentally and careful measurement shows,

$G = 6.67 \times 10^{-11} N m^2 Kg^{-2}$ in SI units.

The dimensions of G are,

$[G] = (MLT^{-2} \times L^{2)} / _M 2 = L^3 M^{-1} T^{-2}$

This would indicate that G is a rather complex quantity and performs a specific function in the above equation. It is obvious that G is a sort of conversion factor in a formula to correlate the response of a mass object to its EIL gradient environment.

We have already shown that there is a defined relationship between the mass of an object and its effect in raising the grid datum around it. Consider the grid datum curves from two separate masses A and B, and represented together for purposes of comparison as in Fig.4.14.

Let us ignore the grid datum base for the moment in order to simplify our discussion. Also let us assume that a peak grid EIL occurs at the surface of each mass object. Fig 4.14 shows the main characteristics of each grid datum curve. These may be summarised as follows:

1. The change in EIL obeys the inverse square law in all cases.
2. At the same realm grid EIL corresponding curves have the same slope.
3. Each curve commences at the surface of the mass object and at a realm grid EIL that is directly proportional to its mass. (Here we assume the make up of the masses to conform to the same structural pattern. We shall consider variations in proton concentration later on),

In relation to Fig 4.14 these characteristics may be interpreted as under:
EIL at A is greater than the EIL at B in the proportion that M_A is greater than M_B.

That is, $^{EILA}/EIL_B = ^{MA}/M_B$

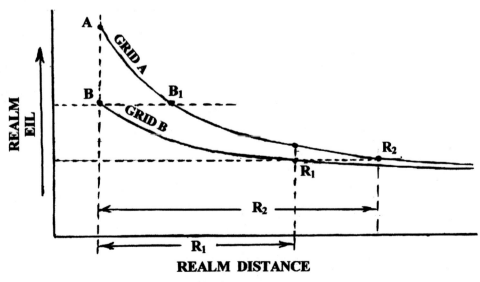

Fig 4.14

Point B on grid datum B is at the same EIL as the point B_1 on grid datum A. At these points both curves have the same slope. A proton at each of these positions would therefore be under similar impulse conditions.

Protons at the realm distance R_1 from A or B would be under different impulse conditions. The impulse towards A being greater than the impulse towards B. However a proton at the realm distance R_2 from A is at the same EIL as the proton at R_1 from B. As such both protons will be under similar impulse conditions and therefore we have,

$$^{MA}/(R_2)^2 = {}^{MB}/(R_1)^2$$

The curve sections BR_1 of grid datum B and B_1R_2 of grid datum A are similar in all respects. If we graphically shift the mass B to position B_1 we get a single grid datum curve that can represent both A and B masses. By extending this curve in either direction we achieve a grid datum curve that is applicable to all mass objects on an individual basis. This is represented in Fig 4.15. The reader will note that on the lower side the curve tends towards C_0 EIL, which is a realistic and acceptable trend. The other end of the curve indicates a trend towards infinite EIL which is absurd since C_2 is the upper limit for EIL. In practice mass and volume are proportional to one another such that very large masses also tend to occupy very large volumes. This reduces the E-Flow intensity from the various protons so that internal distances become a major consideration. At the very high EILs the curve trend changes due to influence of other factors which we shall detail a bit later on. For the moment consider the grid EIL curve in the diagram as representative of our known realm conditions.

Let P be the realm position of a proton such that it is at the distance R_B, R_A and R_X respectively from M_B, M_A and M_X. The grid datum curve for each mass is that portion of curve from its surface EIL to y '. If we consider each mass separately we see that the proton P is at exactly the same EIL and slope for all the masses. This means that the impulse of P towards each mass will be the same.

$$\text{Impulse of P} = G \cdot {}^{MB}/(R_B)^2 = G \cdot {}^{MA}/(R_A)^2 = G \cdot {}^{MX}/(R_X)^2$$

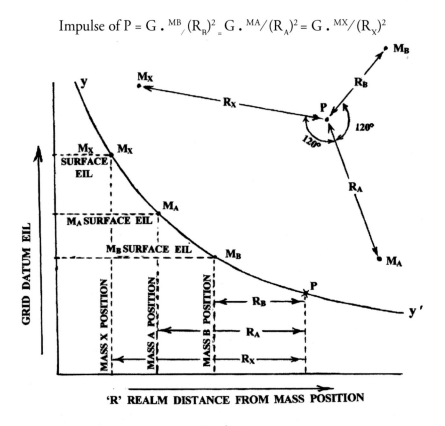

Fig 4.15

If we position P as shown in the spatial configuration with respect to M_B, M_A and M_X then the net impulse of P is zero. If a diagram similar to that in Fig 4.11 is drawn up the reader will observe P to be at a realm position of zero EIL gradient.

Before we proceed further let us take a detailed look at the gravitational constant G in an attempt to determine its function and character.

We know that,

$$\text{Proton impulse} = X^* \times Ic,$$

where X^* is the small proportion factor that increases with EIL gradient. At zero gradient $X^* = 0$ and therefore the proton impulse is nil. On the other hand Ic, which is the total centrally directed internal impulse, has a marginally inverse relationship to grid EIL. In Fig 4.8 we showed a difference in Mp events for the two halves of the proton. The varied spacing of surface Mp action was presented as the chief cause of a net resultant impulse. We did not then consider the duration of the Mp-event cycle as we wished to present the reader with a physical model describing the principle of the gravitational push. Referring to Fig 4.8 consider that the distribution of surface Mps is entirely uniform all around the proton. Now in a set duration consider the surface Mps on the B side of the proton to undergo 'N' exchange cycles. In the same duration the surface Mps on the A side will undergo fewer exchange cycles since they are at a higher realm EIL. Let the A side undergo 'n' exchange cycles. Therefore in a set duration the E-Flow quantity from the surface area B to the centre will occur more frequently than from the surface area A. If the impulse in the two halves is the same for each exchange cycle and is denoted by $Ic/_2$ then,

$$V_B - V_A = Ic/_2 \times N - Ic/_2 \times n \text{ (for the same period)}$$
$$\text{Net Proton impulse} = Ic/_2 (N - n) \text{ (towards the higher EIL)}$$
$$\text{Therefore we have, } X^* = (N - n)/_2 \text{ (our small proportion factor)}$$

Our earlier presentation was of course easier to visualise and so served an important purpose in the overall description of the gravitation phenomena. For the short term duration of the circle 6 energy medium curve the EIL variation is small and subsequently we may assume Ic to be constant per Mp exchange cycle of the proton Mp-congregation. It should be noted that the EIL in the seven circle representation is to a logarithmic scale while that in Fig 4.15 is to an arithmetic linear scale.

Now we know that the proton decay phenomena results in a gradual decrease in the mass of the proton. This means progressively fewer Mps in the proton and a subsequent decrease of total impulse Ic. For the present however we shall ignore the ageing of the proton since we are concerned with the impulse action for masses as they exist at a particular event moment. We shall assume that the protons in all masses are approximately the same age and so have near equal mass.

$$\text{Proton impulse} = Ic \times X^* \text{ where Ic is constant.}$$

Consider the proton P as being at a distance R_A from the Mass M_A. Let us ignore all other masses for the moment. The E-Flow from M_A extends spherically outwards and its EIL obeys the inverse square law. At the proton P we have,

$$EIL \propto {}^{MA}/(R_A)^2$$
or $$EIL = C_1 \times {}^{MA}/(R_A)^2 \text{-----------------------(a)}$$

where C_1 is a constant factor that relates E-Flow EIL to the mass of the emitting source.
In the Fig 4.15 EIL curve for all masses, the curve gradient is directly proportional to the grid EIL. We have,

EIL gradient $\propto$ EIL

or   EIL gradient = C_2 x EIL --------------------(b)

where C_2 is a constant factor that relates the EIL gradient to the grid EIL.
Also we have,

$X^* \propto$ EIL gradient

or $X^* = C_3$ x EIL gradient -----------------------(c)

where C_3 is a constant factor that relates the small proportion factor X^* to EIL gradients. From (a), (b) and (c) above

$$X^* = C_1 \times C_2 \times C_3 \times {}^{MA}/(R_A)^2$$

subsequently,
Proton P impulse towards $M_A = Ic \times C_1 \times C_2 \times C_3 \times {}^{MA}/(R_A)^2$
But P impulse = $G \times {}^{MA}/(R_A)^2$
where G is the experimentally determined gravitational constant.
Therefore, $G = Ic \times C_1 \times C_2 \times C_3$

We see therefore that the experiment factor G is really made up of a number of different constants which perform their own relative conversion function with the exception of Ic. It is important to note that Ic had been assumed constant and is currently considered as part of G. We assumed Ic as constant by ignoring the effect of proton decay upon the mass of the proton.

Over a longer event duration period however proton decay cannot be ignored. Over such a period the proton would have lost some of its mass, and with fewer Mps at its surface Ic must inevitably decrease in quantity per exchange cycle. This means that the E-Flow from each proton is gradually and continually decreasing in quantity which results in the general droop characteristic of the realm datum indicated by the circle 6 energy medium curve. The droop occurs in spite of a marginal increase in the Mp cyclic rate at the lower realm EIL.

If the reader will refer back to Fig 3.68 and our brief explanation for the Red Shift phenomena it will be noted that the decreasing EIL characteristic of the Realm datum with event sequence was responsible for the increasing rate in Mp-shell repetition. Now the Red Shift in frequencies and Ic are both directly related to the circle 6 energy medium curve that indicates changes in realm datum EIL. We may therefore consider a direct connection between a change in Ic and a change in frequency of Mp-shell repetitions. As such a change in Ic equates to a change in Mp-shell frequency. A measurement of this change is effected by the observing and measuring of Mp-shell frequencies from past moments in the emissions from distant stars. Hubble's Constant H indicates this change. According to the circle 6 curvature the rate of change in realm EIL is not constant but is indicated at an increased droop rate. In Fig 3.68 we note that the section AB is very nearly linear in profile when AB is small. Hubble's constant will be true under these short term conditions. Over a longer period however we cannot ignore the true character of the circle 6 energy medium curve. Hubble's constant H must therefore be exchanged for a similar rule or equation for H that conforms to the true realm EILs for past and future conditions. The curvilinear aspect of this trend should correspond to all our Red Shift observations. Since our observation platform of today will be different from our observation platform of tomorrow by the change in realm EIL, the above observations must be judged with respect to their date of observation.

Now Ic is an integral part of the gravitational constant G. We must draw the conclusion that if Ic is decreasing then G will also decrease at the same rate. Since the rate of change of Ic is indicated by a link up with Hubble's constant H (or a revised equation for H), then G will also change accordingly. In all cases therefore the rate of change of G as a fraction of G must be of the order of Hubble's constant for to-day. This is estimated to be a decrease of 1.5×10^{-18} of G per second. As such in a human lifetime G will change by only a few parts per billion (1000 million). The reader will appreciate that we can effectively ignore these changes for purposes of calculating the gravitational action upon masses as they exist within the realm to-day and at reasonable distances from one another. For very great distances we would need to take account of the duration of E-Flow traverse to evaluate G at the date of emission.

In our discussion so far we have concentrated upon the aspect of gravitation that causes masses to be pushed towards each other. We may define this phenomenon as gravitational attraction. There is however another related effect that we have not yet taken into consideration. This effect arises from the linear momentum within the E-Flow. Any absorption of energy medium from this E-Flow means that a proportionate quantity of linear momentum is also absorbed. The absorbing entity will therefore acquire an impulse in the same direction as the E-Flow. Since E-Flows originate from the proton and flow in a radial outward direction then the impulse in an adjacent proton will be along the same outward path. The conclusion is that by the absorption of a part of the E-Flow from adjacent masses an impulse action is set up that tends to push those masses away from each other. This action is defined as gravitational repulsion.

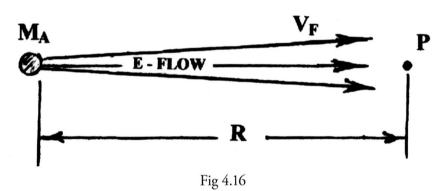

Fig 4.16

It must be pointed out that the gravitational attraction and repulsion effects are both the result of proton decay but are entirely different in their modes of action. However, it is essential that the reader realise that both effects are in continuous action at the same realm location, and it is the resultant of the two actions that determines the net impulse acquired by a test proton. In order to illustrate this concept we shall imagine a test particle of mass in the form of an isolated proton being gradually brought into the sphere of influence of a large mass object.

Let us assume the E-Flow velocity to be constant relative to the decaying mass object and so denoted by V_F. Let a test-proton P be positioned at a distance R from an object of mass M_A as shown in Fig 4.16.

If d denotes the diameter of the test-proton P,

then its x-sectional area $= \frac{\pi}{4} d^2$

or, $= d^2$ (if d is very small)

Now the E-Flow from the mass M_A that enters this sectional area of P during a complete Mp-event cycle represents the total energy medium quantity that will be absorbed by P in that duration. The volume of energy medium absorbed is given by,

$$\text{E-Volume} = L \ ' \ d^2$$

where, $L = {}^{VF}/_{\text{(Event Duration)}}$ (a distance quantity)

The energy medium quantity is defined as the product of its volume and EIL. Its linear momentum is the product of its energy medium quantity and velocity. Therefore linear momentum absorbed by P per Mp-event duration is given by,

$$\text{Impulse } P_R = (\text{E-Flow EIL}) \times (\text{E-volume} \times V_F)$$

Now, $\quad$ E-Flow EIL $\propto {}^{MA}/R^2$ (inverse square law)

or, $\qquad\qquad$ EIL $= C_1 . {}^{MA}/R^2$

where C_1 is a constant factor that relates the maximum surface EIL to the mass of the decaying source.

Therefore, Impulse $P_R = C_1 \times {}^{MA}/R^2 \times L \times d^2 \times V_F$

Since L, d and V_F are reasonably constant then we have,

$$\text{Impulse } P_R = \text{Constant } C_R \times {}^{MA}/R^2$$

where C_R is the Repulsion constant in this instance.

This impulse acts in the radial direction away from the mass M_A. Now we also have an impulse quantity acting directly towards M_A due to the EIL gradient across the test-proton P which is given by,

$$\text{Impulse } P_A = G \times {}^{MA}/R^2$$

The net impulse is then given by,

$$\text{Net impulse} = G \times {}^{MA}/R^2 - C_R \times {}^{MA}/R^2$$
$$= (G - C_{R)} . {}^{MA}/R^2$$

This indicates that the repulsion impulse diminishes with distance according to the inverse square rule. Since the quantity of M_A E-Flow absorbed by the test proton is a very small proportion of the whole, then the linear momentum acquired from it is also very small. A comparison made between this quantity and those involved in the compression of the proton entity show the latter to contain much greater impulse potential. A very small imbalance in the proton compressive action results in

a relatively large net impulse. This means that for most of the observed positions around a mass body the gravitational attraction impulse far exceeds that of repulsion.

Because the net effect between masses has been observed as one of attraction, we tend to ignore the very small repulsion effect altogether. We must somehow correct this by including the repulsion constant C_R as a negative quantity beside G. As such in the experimental evaluation of G we have in reality obtained a value for $(G - C_R)$. Fig.4.17 represents these impulses in graphical form relative to distances from M_A and shows attraction as a positive impulse quantity.

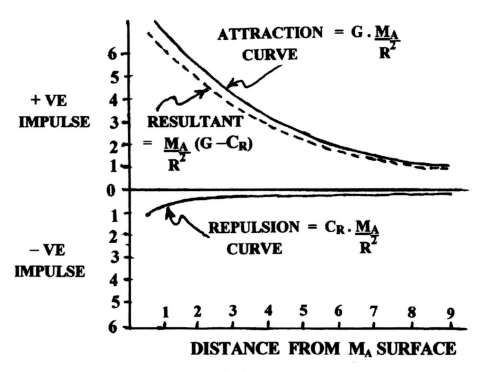

Fig 4.17

The reader will notice a similarity in the character of the positive and negative impulse quantities with distance. The resultant impulse is represented by the dotted curve and is seen to be wholly positive in its characteristic. Previous experiments and calculations on the attractive nature of gravity have in fact been based upon the character of this resultant curve.

In order that we may explain the full nature of both repulsion and attraction effects between masses, let us conduct another of our simple thought experiments. Consider two protons A and B as the only mass entities within the realm and that they are separated by a realm distance that is very much greater than the maximum energy radius as represented by Fig 1.12. That is, the EIL of the E-Flow from proton A will have decreased to C_0 before reaching proton B. Consider B as the moveable test proton that responds to the influence of the E-Flow from A. The reader will remember that we defined C_0 as the absolute minimum EIL achieved by expansion of the energy medium. Since the linear momentum of the E-Flow must carry the energy medium beyond the distance at which this occurs, and since further expansion of the energy medium is impossible, then we must assume that the E-Flow separates into lines or very narrow beams (Fig 1.12).

Earlier in this chapter we described the process at C_1^+ EILs when the energy medium changed to a mass-medium property which then underwent violent volumetric contraction. The contraction process caused the mass medium to separate into small contracting spheroids which subsequently imploded into Mps. Perhaps the separation of the energy medium at C_0 EIL and radial velocity V_F may be compared to that of the mass medium entity prior to Mp implosion. If so, then the narrow beam x-section may be similar in dimension to that of the mass medium spheroids. We have no way of confirming this but we can however assume that the C_0 beam section must be very small indeed.

Initially these C_0 energy medium beams flow very close together in a beam-bundle that is practically indistinguishable from a normal E-Flow. However since each C_0 beam emanates from the same imaginary sphere centre, they must gradually separate with continued propagation. We must now talk of closeness or population density of the beams in the bundle. This density will reduce with propagation distances although each beam has a constant EIL of C_0 and velocity V_F. The travel range of each beam is infinite and it can propagate through the cosmic void as easily as through the realm. Any section of the beam that enters the sphere of an E-M/X sequence will however be absorbed into the resulting Mp.

Proceeding with our thought experiment let us position the test proton B at a nearly infinite distance from A. The E-Flow beam bundle will be at a very low population density such that the beams are separated by a large distance from one another. This situation is represented in Fig 4.18 where adjacent beams in the bundle are spaced a distance D apart.

Let the proton diameter be 'd' such that at this stage it is smaller than the bundle spacing D. Consider the test proton in the position B_1. There is no EIL gradient across it and therefore no impulse exists towards A.

$$\text{Impulse } G \times {}^{MA}/R^2 = 0 \text{ (i.e. no impulse towards A)}$$

Also, there is no linear momentum from any beam entering or being absorbed by B_1. As such

$$\text{Impulse } C_R \times {}^{MA}/R^2 = 0 \text{ (i.e. no impulse away from A)}$$

The test proton B_1 is in perfect equilibrium.

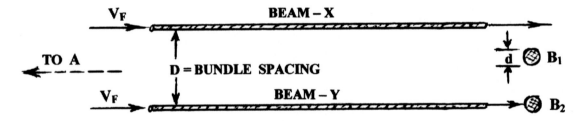

Fig 4.18

Now consider the test proton in the position B_2 which is in the propagation path of beam-Y. The beam-Y is intercepted and its C_0 energy medium absorbed into the E-M/Xs within B_2. This means that the linear momentum of the energy medium in a section of beam-Y is added to B_2. The test proton B_2 therefore acquires an impulse away from A. Since there is still no EIL gradient across B_2 then there can be no impulse towards A. The net impulse is therefore a repulsion of B_2 away from A and in the direction of V_F. However, as B_2 begins an acceleration along the beam-Y it absorbs less linear momentum from the beam. This is because its relative velocity to the beam diminishes and so reduces the length of beam section absorbed. In the limit when B_2 attains the velocity V_F no further linear momentum will be absorbed. It will be noticed that in this instance the impulse in B_2 is directly proportional to its velocity relative to that of beam-Y.

Let us continue with our thought experiment and now re-position the test proton B much closer to A but still outside the E-Flow maximum energy (medium) radius. This situation is represented in Fig 4.19 with a much enlarged view of the proton B. The proton diameter d is unchanged but the C_0 beams are shown bundled more closely.

Subsequently the number of C_0 beams intercepted by B are high. The linear momentum absorbed is proportional to the population density of the beam bundle and will be N times that of the previous

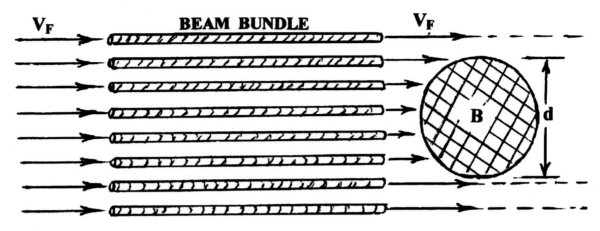

Fig 4.19

B_2. N being the number of C_0 beams within a sectional area of proton diameter d. The test proton B, through absorption of this linear momentum therefore acquires an impulse away from A. There is still no EIL gradient across B and therefore no impulse towards A. The net effect is thus a repulsion of B away from A and once again as B accelerates along the beam bundle it absorbs less of its linear momentum. This is because not only does its relative velocity to the beams diminish but also because of a decrease in the population density of the beam bundle.

As we experimentally position the test proton B outside the maximum energy radius (medium) but even closer towards A the population density of the beam bundle will increase greatly such that an EIL difference of C_0 will gradually develop across B. The situation is now one in which the repulsion impulse in B caused by the absorption of linear momentum from the beam bundle begins to be counteracted by a very slight attractive impulse. The repulsion impulse continues to increase as B is brought closer to A. The attractive impulse however is also building up at an even greater rate of increase. At some position near the maximum energy (medium) radius distance from A these opposing impulses within B will be in approximate balance. The resultant impulse is zero and B will remain in equilibrium. Fig 4.20 indicates the attraction and repulsion impulse trend as the test proton B is moved from infinity to near the maximum energy(medium) radius position. The impulse units marked are purely arbitrary and to a grossly exaggerated scale. They are for comparative purposes only.

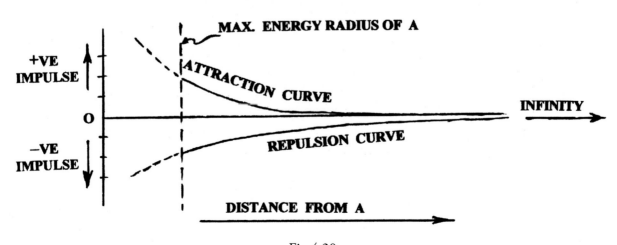

Fig 4.20

Continuing with our thought experiment let us position the test proton B just inside the maximum energy radius distance of A. B is now in a region of E-Flow and increasing EIL gradient. The attraction impulse is now greater than the repulsion impulse as is clearly observed if we extend the curves of Fig 4.20 into the E-Flow zone. The resultant is thus an attractive impulse towards A.

As we proceed to closer-in positions, the former gravitation and repulsion constants become applicable such that,

$$\text{Resultant impulse} = G \cdot \frac{1}{R}2 - C_R \cdot \frac{1}{R2}$$

where the masses of A and B are considered as unity. The conditions now are those represented earlier in Fig 4.17. The attraction impulse is closely related to the grid datum gradient while the repulsion impulse is simply a linear momentum quantity being absorbed from the E-Flow of A. This relationship prevails for distances right up to several proton diameters from A. Let us define this minimum distance by Nd, where d is the proton diameter and N is a positive integer.

It has already been explained that the attraction impulse caused by the EIL gradient across the proton is really due to the differing Mp-cyclic rates of its surface Mps. This caused an unbalance in the proton body's centralised (compressive) impulses with a resultant that then acted in the direction of the higher grid datum EIL. Although the EIL gradient is the essential factor causing this resultant impulse, we must also consider the relationship between the Mp cyclic rate and the grid datum EIL.

We showed in a previous section that changes in grid EIL resulted in an inverse change in the Mp-cyclic rate, i.e. as the grid EIL increases the Mp-cyclic rate slows down. Now at grid EILs substantially below C_1 this relationship is an inversely linear curve as shown in Fig 4 21.

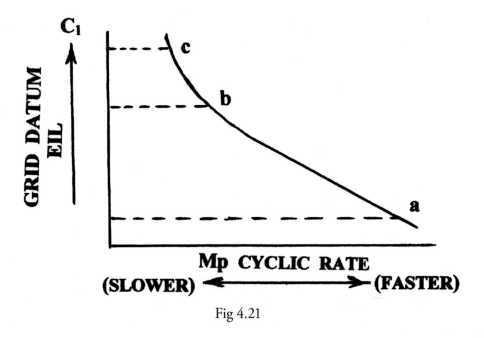

Fig 4.21

This is true only upto a certain level of EIL represented by the 'ab' section of the relationship. At the higher grid EILs the correlation curve takes on a curvilinear aspect as indicated by the curve section 'bc'. We observe therefore that the rate of change in the Mp-cyclic rate progressively diminishes at the higher grid EILs. As a direct consequence the relationship between grid gradient and attraction impulse will also change accordingly.

In our thought experiment we shall assume that this change in the attraction impulse characteristic becomes significant when the test proton B is less than the distance 'Nd' from A as defined above. As such a modified attraction impulse curve will be as represented in Fig 4.22. It should be noted that the curve section 'bc' also indicates that the former gravitational constant G is undergoing a continued devaluation as B approaches A. At the distance (Nd - m) from A in the graph we see that attraction impulse has reached a peak and thereafter at lesser distances undergoes a decrease in value. It is important at this stage that we fully understand the significance and correlation between the characteristics represented in Fig 4.21 and Fig 4.22 in order to follow the behaviour of our

test proton B with regard to proton A. Incidentally in Fig 4.22 we have shown the probable trend 'bd' for the impulse if G had remained unchanged. This will correct any misconception that may have been formed earlier.

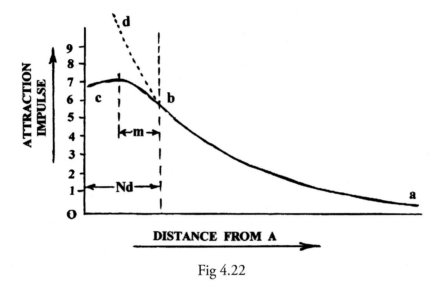

Fig 4.22

As such in the limit when the grid EIL approaches C_1 the value of G is very small compared to its previous level. In fact it is considered that G reduces to a value less than the repulsion constant C_R.

Let us now consider the repulsion impulse set up in the test proton B. Since this is simply the result of an absorption of linear momentum from the E-Flow of A then the equation,

$$\text{Repulsion Impulse} = C_R \cdot {}^1\!/R2$$

remains unchanged even for distances less than 'Nd'. As such the repulsion impulse trend with distance from A is represented by the curve of Fig 4.23. The energy medium absorbed by B from the E-Flow of A becomes very great as this E-Flow EIL increases at distances less than 'Nd'.

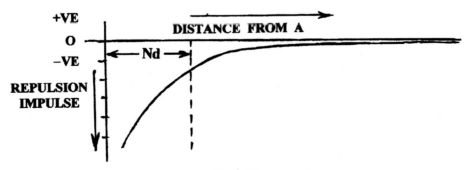

Fig 4.23

This is reflected in the massive increase of repulsion impulse as B approaches A and R^2 becomes very small indeed. In the limit when A and B are in surface to surface contact then the minimum distance between their centres will be given by 'd'. This contact however gives rise to a totally different set of impulse conditions which we shall consider shortly. For now however let us show up the resultant impulse experienced by our test proton B over the distance 'Nd' by combining Fig 4.22 and Fig 4.23 into a single characteristic. This is represented in Fig 4.24. The attraction and repulsion impulses in B act along the same path but in opposition to one another. The resultant impulse is therefore a straight forward arithmetical difference given by,

$$\text{Resultant impulse} = (\text{Attraction impulse}) \text{ minus } (\text{Repulsion impulse})$$

If the resultant impulse has a positive value then the proton B will be attracted towards A. If the resultant impulse has a negative value then B will be repulsed away from A. In Fig 4.24 we note that for distances greater than x from A, the test proton B experiences an attraction towards A. While at distances less than 'x' the proton B is repulsed away from A. At the distance 'x' the resultant impulse is zero.

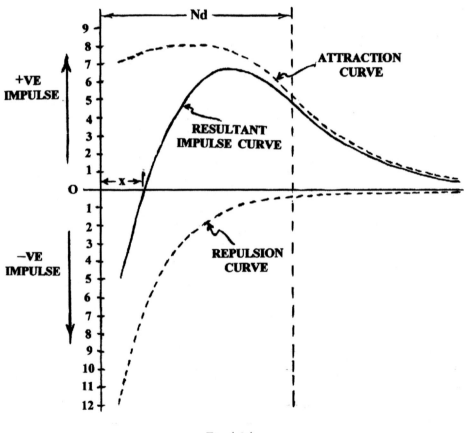

Fig 4.24

The reader should remember that as part of our thought experiment we are not only looking at the effects of A upon B in isolation but have also for the time being excluded other phenomena such as Mp-shell and forcephoidal effects from our current discussions (see Figs 2.16 and 2.17). We are thus able to evaluate the wholly individual characteristics of each particular phenomenon in isolation. Later we may consider the net result of the actions of differing phenomena with a much clearer interpretation of the facts.

To continue with our thought experiment let us locate the test proton B a few 'd' away from A. The net impulse in B will be negative and large. As such B will be repulsed away from A. Because of this action we would find it extremely difficult to hold B very close to A. However, as part of our thought experiment let us induce B with a linear momentum that takes it directly towards A. Let this induced momentum be such that B is caused to collide with A. Consider the situation at the moment of physical contact. In Fig 4.25 the protons A and B are shown in contact along the surface section 'xy'. Within this section the former surface Mps are now covered by those

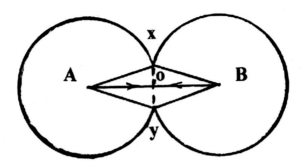

Fig 4.25

of the other proton and as such will behave just like internal Mps. Consequently there will be no linear momentum imparted towards the proton centre from the surface section 'xy'. In proton A therefore the compressive action Ic directed at its centre (Fig 4.6a) will be out of balance by a quantity of linear momentum that should have originated at its surface section 'xy'. This unbalance results in a net proton impulse in the direction A to O which is a push directly towards proton B. A similar action also takes place within the proton B with the result of a net proton impulse in the direction B to O which is a push directly towards proton A. Protons A and B subsequently appear to be bonded to one another by this mutual impulse action. We should note that this bond is very strong because it is part of the same action that maintains the Mp-congregation in compression. Since the section 'xy' contains no surface Mps then there is no external E-Flow from one proton to the other through this area. As a result the absorption of E-Flow linear momentum by A and B is also diminished. As such the repulsion action is considerably reduced and no longer obeys the characteristic portrayed in Fig 4.24. Subsequently the net impulse will be a mutual attraction impulse between A and B. Now this compressive bonding between A and B is dependant upon the area of the contact section 'xy'. The greater the contact area the greater will be the compressive action of A towards B and vice versa. Since the proton is a congregation of Mps and as such is not a rigid entity, we may assume that a certain amount of deformation will occur at the contact area. We must assume also that the Mp-congregation, by virtue of its own compression, will offer an increasing resistance to progressive deformation such that a balance against the joining action is finally attained.

We consider that this balance may be achieved at some pre-determined geometric conformity. Perhaps this equilibrium is reached when the section area of 'xy' is part of the cone with apex at the proton centre and which just envelopes the other proton as represented in Fig 4.26. The reasoning here is rather abstract and assumes that the repulsion component approaches a minimum level. However there are other assumptions that the reader might make to establish a different balance

position. Whatever the final position the two protons are now in a state of equilibrium with regard to one another and are to be considered as a bonded unit. It would require a very large pull action to separate A from B once this equilibrium state had been achieved. It will be noticed that their centres are now less than 'd' apart due to the deformation at 'xy'. In this bonded state the protons A and B will behave as a single Mp-congregation which causes a single though oval Mp-shell structure to occur around it (see Fig 3.2). We shall be discussing this aspect more fully in a later chapter. It is important to note that the characteristics represented in Fig 4.24 only apply when

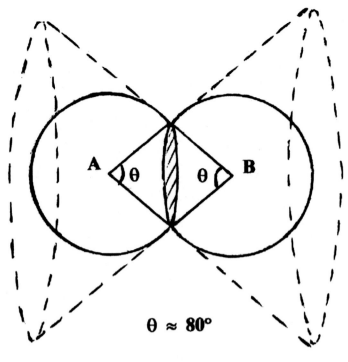

$$\theta \approx 80°$$

Fig 4.26

A and B are not in contact with each other. Once contact is made the conditions change to those just described.

In order to physically separate the bonded protons we find that a large pull is required and over a much greater distance than expected. As part of our thought experiment let us assume that we are able to cause A and B to be magically pulled apart. The sequence of events would be as represented in Fig 4.27. At (a) the protons A and B have a very strong bonding impulse pushing them firmly against one another. This is the state of equilibrium between A and B. At (b) we have applied our magical pull and have caused the centres of A and B to be moved slightly more apart. The bonded protons behave as a single Mp-congregation so that our magical pull causes it to deform at its weakest section. This occurs at the 'xy' position and subsequently results in a decrease in the contact area. The attraction impulse bonding A and B together is somewhat reduced though still large in comparison to any repulsion effect. If we were to remove our magical pull the two protons would simply return to the situation at (a).

At (c) both A and B have been pulled farther apart and although the contact section 'xy' is very much reduced there would still be a resultant attractive impulse between them. At this stage we may reduce our magical pull considerably in order to just balance that resultant. We must also remember that each proton is now absorbing linear momentum from the outward E-Flow of the other. This causes a repulsion impulse build-up which is an aid to our magical pull. At the present spacing of the protons and prior to physical contact there was a large repulsion action between them. Obviously once contact was made the impulse characteristics altered. At (d) we show the deformation in the combined (A + B) Mp-congregation as approaching a dimension limit. Even if

we were to hold the protons at this spacing the E-M/X and M-E/X cyclic repetitions within this waist zone C would not contain sufficient energy medium to prevent a rapid and total decay of its Mps. This means that there are too few Mps in the thinly stretched zone C for mutual repetitive E-M/Xs to continue as in the main body mass of the proton. As such the mass in the stretched zone C will rapidly convert to energy medium leaving the protons A and B once again as two separate entities at shown at (e). Both A and B will now behave according to the characteristics

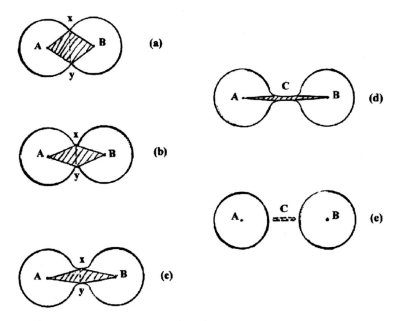

Fig 4.27

represented in Fig 4.24 i.e. a strong repulsion impulse pushing A and B away from each other. The reader should note that the energy medium released by the M-E/Xs of the Mps in zone C may be considered as a supplementary flow of energy medium that occurs only when bonded protons are pulled apart. (Perhaps the splitting of the Uranium atom is brought to mind here).

It must be stressed however that the above close-in distance impulse conditions follow different laws depending on whether the protons existing as separate entities are made to approach one another, or whether they have bonded together and are in the process of being pulled apart. In the former case Fig 4.24 applies right upto the contact distance of 'd'. In the latter case repulsion comes into play as the net impulse only after separation of the protons, i.e. at a distance 'nd' when 'n' is greater than unity.

(a)　　　　　　　　　　　　　　　　　　　(b)

Fig 4.28

The reader may investigate the conditions that prevail when three or even four protons come into mutual contact. We would do well to bear in mind that some multiple contact positions develop overall stronger impulse bonds than others. In Fig 4.28 four protons are shown in contact. It is clear that the positioning at (a) is not as compact as the positioning at (b). Also at (a) each proton is in contact with only two of the others while at (b) each proton is in contact with three other protons simultaneously. Therefore the multiple contact position shown at (b) develops a stronger inward impulse bond than for the positions at (a). We shall look into this aspect of natural positioning of protons in the atom nucleus in our discussions on the primary atom in Chapter 6.

So far in our thought experiment we have determined the impulses set up in the test proton B when it was at a defined distance from proton A. We must however now consider the impulses that are generated when a relative velocity exists between them. In an earlier description (Fig 4.2) on the three phases of the Mp sequence we showed that a flow of energy medium radiated outwards off the Mp surface. This energy medium was expelled with a velocity V_F that was constant relative to the Mp centre. Now if the Mp itself has an initial realm velocity then the E-Flow will also possess this additional velocity as represented by Fig 4.29. Let the Mp velocity be V_1 in the direction shown. Then the absolute E-Flow velocity in the same direction is $(V_F + V_1)$ and in the opposite direction $(V_F - V_1)$. The E-Flow velocity in any direction is thus given by,

E-Flow velocity = $V_F \pm V_1 \cos \theta$ (where q is the angle made with the vector V_1).

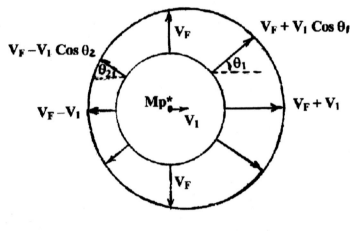

Fig 4.29

If the proton A (in our thought experiment) is permitted a realm velocity V_A then the E-Flow reaching proton B will possess a velocity $V_F \pm V_A \cos \theta$. The angle q giving the direction of V_A relative to the position of B as shown in Fig 4.30.

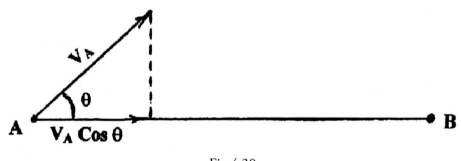

Fig 4.30

The E-Flow velocity intercepted by B is either greater or lesser than V_F by the vector component $V_A \cos q$ depending upon whether $V_A \cos \theta$ points towards or away from B. Let us now determine the effect of a changed E-Flow velocity upon the attraction and repulsion impulses induced in B. If the velocity of the E-Flow reaching B is greater than V_F, then the linear momentum absorbed by B will be correspondingly greater. This means that the repulsion impulse in B will have increased. If however the velocity of the E-Flow from A that reaches B is less than

V_F, then the net effect is a decrease in the repulsion impulse of B. This example may be inverted by assuming A to be stationary and B to be in motion. The result is the same and indicates that the repulsion impulse induced in B is dependant upon their relative velocity. As such the repulsion impulse curve of Fig 4.23 would alter marginally. This means that the resultant impulse curve would also change accordingly. We shall return to this aspect shortly.

The EIL gradient within an E-Flow does not depend upon its velocity of flow V_F. The EIL at any position is dependant upon its distance from A. The EIL difference between two such positions at any instant determines the EIL gradient. The EIL differential across B therefore depends entirely upon its distance from A. This determines the unbalance in Ic (the total centrally directed impulses) which results in the net attraction impulse. The relative velocity of B simply means that its position with regard to A (and hence its attraction impulse) is in a state of continuous change. As such the attraction impulse curve of Fig 4.22 will not require any modification.

From the above we observe that the resultant impulse curve of Fig 4.17 will alter only by the change caused in the repulsion impulse. This means that the net attraction impulse between two objects is marginally dependant upon the relative velocity between them. In the expression,

$$\text{Net impulse} = (G - C_{R)} \cdot {}^{(MA \times MB)}/R2 \ldots \ldots [\text{usual notations}]$$

for the attraction between the masses M_A and M_B we note that C_R is variable with velocity and is the only cause of a variation in the net impulse. From this we may conclude that the net attraction impulse operating between two masses is marginally different and depends upon whether they possess a relative velocity towards or away from one another. Compared with relatively stationary masses we shall find that the net attraction impulse between two masses at any instant is lesser when a relative velocity of approach exists between them and greater when a relative velocity of recession exists between them.

So far we have attempted an explanation of the basic mechanism of the gravitational effect. Because of the vastness and bulk of the realm, individual protons come under a complex pattern of push and pull in all directions. However using the basic concepts of proton impulse presented above the reader should easily be able to establish a resultant. The concept of EIL gradient and a general realm EIL grid will always indicate the direction in which a proton will gravitate. Our aim here was not to quantify impulses between masses but to lay down the principles of action for those impulses. We shall now consider an extension of these principles of impulse in areas of action that relate to the more practical situation.

We have already indicated that the primary atom consists of a Mp-shell around a solitary proton. Approximate dimensions are as follows:-

Proton diameter = $0 \times 5 \times 10^{-15}$ metres.
Mp-shell diameter – $1 \times 1 \times 10^{-10}$ metres.

By earlier definition the Mp-shell is a fine layer of perfectly synchronised Mps. Their formation results from the EIL build-up in the forcephoids emanating from the proton as described in Fig 2.17 and Fig 2.18. At this stage a revision of that section is advised. See also Fig 3.2.

It will be apparent to the reader that the Mp-shell is a product of the central proton energy medium and not vice versa. The Mp-shell is also dependant on the external E-Flow arriving from other nearby proton structures. In chapter 3 we saw that the Mp-shell was fundamentally the origin for EIL oscillations and we then concerned ourselves primarily with oscillations that propagated outwards into the realm. Let us now consider the events for the Mp*-shell E-Flow that courses back towards the atom centre or nucleus.

If one considers the Mp-shell position to be the commencement of all wave fronts then we must also know that a backward E-Flow will emanate from the M-E/X sequence at this position. This backward E-Flow must be the cause of a backward stepping wave front. Under equilibrium and uniform realm energy medium conditions, the position of the first backward quantum step is found to coincide with the instantaneous location of the next outward bound forcephoid. Subsequently a buildup of EIL occurs and there is then enough energy medium present for normal Mps to form. Therefore an Mp-shell will form instead of just another wave front. Due to the quantities of linear momentum absorbed this Mp-shell will possess an expanding feature i.e. a resultant velocity of its Mps that is in the radial outward direction. It is considered that the forcephoid energy medium quantity will be the second greatest single contribution to the Mp-shell make up after the realm EIL datum. Under high realm EIL conditions the cyclic duration of the Mps in each shell is prolonged, resulting in a lengthier outward traverse. Hence we have a larger atom structure. However a positional balance is maintained because the further out (from the proton) that the forcephoid must traverse to mate with the first backward quantum step position, the lesser is its energy medium contribution to the next Mp-shell. This means that the resultant Mp-shell velocity will be correspondingly diminished. Under reduced realm EIL conditions the cyclic duration of Mps is reduced resulting in a shorter outward traverse. Here we have a smaller atom structure. Again a Mp-shell positional balance is attained with the forcephoid energy medium and the resulting increased outward velocity. However, we are not about to discuss this aspect in any detail here since it is dealt with fully in chapter 6. What we are really concerned with now is whether any part of the backward E-Flow actually penetrates back to the proton surface. If it does, then we must evaluate its influence upon that proton.

Under equilibrium and uniform EIL conditions the forcephoids emanating from a single proton nucleus are perfect spheres in an outward expanding mode. The resulting Mp-shells are then also perfectly spherical at their prospective locations. The decay of one shell (M-E/X) gives rise to the formation of the next shell (E-M/X) by virtue of the phasing of forcephoidal energy medium and Mp*-shell backward E-Flows. The synchronised and co-symmetrical nature of both these features results in most of the backward E-Flow being absorbed into the succeeding Mp-shell.

Earlier in the chapter we indicated that if all factors were equal then the E-M/X phase of the Mp-event was of a shorter duration that the corresponding M-E/X phase. This factor may now be applied to that of the Mp-shell's repetitive formation. Since the backward E-Flow is an issue of the M-E/X of the previous Mp*-shell and the absorption of energy medium into the E-M/X of the new Mp-shell occurs within a relatively shorter event duration, then it follows that this E-Flow cannot be totally absorbed. This means that a proportion of the above backward E-Flow passes beyond the critical first quantum step position after an Mp-shell was initiated there. Since this E-Flow remnant and the newly formed Mp-shell are traversing in opposite directions there is no effect of one upon the other. Let us now concentrate our attention wholly upon this inward traversing energy-medium quantity.

Although the E-Flow from a specific Mp* of the shell obeys the inverse square law for EIL with distance, nevertheless when we consider the very great quantity of synchronised Mps in the shell and if we add or phase together much of the inward flowing energy medium 'remnants' then a progressive increase in EIL must occur as they approach the centre. Fig 4.31 demonstrates this in principle.

This build-up in EIL, coupled with the normal E-Flow velocity V_F, results in the linear momentum of the inward flowing energy medium being focused and thereby being concentrated along a reducing x-sectional zone. This trend continues until the energy medium phases with the proton surface grid when all of its linear momentum gets absorbed. The absorption of such an amount of linear momentum by the surface grid must cause a repulsion action within the proton. However, since the inward E-Flow emitted from the Mp-shell position reaches the proton equally (and at the same instant from all directions then there is no net repulsion action to cause any change in the

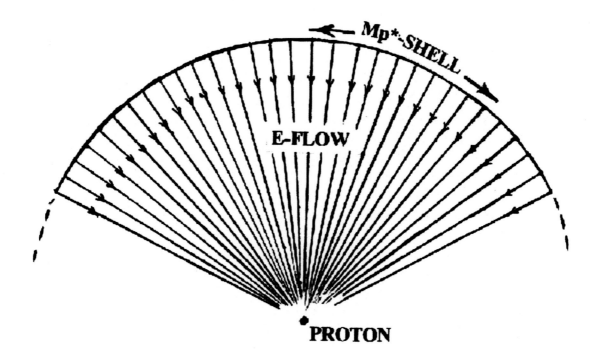

Fig 4.31

linear momentum of that proton. The absorption of all this inward directed linear momentum does however result in the further compaction of the proton (see Fig 4.6) .

Let us for a moment suppose the Mp-shell to become displaced with regard to the proton centre. The proton surface and the Mp-shell as such will not be equidistant from one another. This is represented in Fig 4.32.

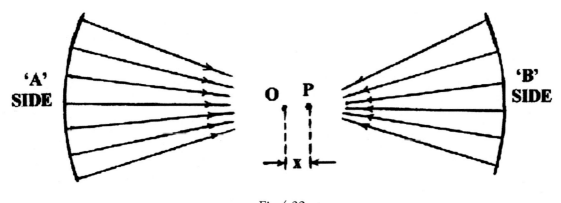

Fig 4.32

Consider the two sides of the Mp-shell indicated and let the displacement be towards the left hand side. As such the 'B' side of the Mp-shell will be closer to the proton than the 'A' side. To all intents and purposes it would appear as though the proton had been displaced from a central position 'O' by an amount 'x' towards the 'B' side to take up its new position at P.

181

It is important that we realise that the inward E-Flow is not restricted to the directional lines indicated in the diagram. We have simply represented that part of the E-Flow which is relevant to the argument at hand and have ignored all other directional flows. This is because in our present argument we are only concerned with the repulsion impulse resulting from absorption of the inward flowing energy medium by the proton. We must also remember that this inward E-Flow occurs intermittently and at the frequency of the Mp-shell repetitions. Its effect upon the proton is therefore to be considered as quite separate and in addition to the gravitation phenomena described earlier.

In Fig 4.32 the E-Flow reaching the position P from the B side of the Mp-shell will arrive earlier and at a slightly higher EIL than the E-Flow from the A side of the Mp-shell. Simple mechanics indicates that P will receive a net impulse towards the central position O. The reader should conduct a more comprehensive analysis of this aspect in order to draw the conclusion that the Mp-shell does indeed exert a centralising impulse upon its proton. This is a most important phenomenon and is responsible for the transfer of linear momentum between colliding atoms and subsequently between larger mass bodies.

A brief explanation of the means by which this transfer of linear momentum actually occurs will now be put forward, although the more aware reader must (most certainly) have already intuitively interpreted the facts. In Fig 4.33 are represented two primary atoms in collision. Consider both to be in equilibrium with A being stationary while B has a uniform velocity (and hence linear momentum) in the direction BA. It will be noted that the protons P_1 and P_2 are both in the central

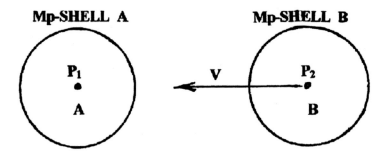

Fig 4.33

position relative to their Mp-shells. (Remember that the Mp-shell is an outwardly traversing and highly repetitive Mp structure with a life duration of 10^{-43} seconds in Planck time). When the Mp-shell B gets close to Mp-shell A then the outward E-Flow from the M-E/X of each shell is successively absorbed into that of the other. This means that the outward traverse of the portion of the Mp-shell nearest the other atom is reduced relative to the remainder of shell. This results in a displacement of the Mp-shell positioning with regard to its proton centre just as indicated by Fig 4.32. The closer that one shell gets to the other the greater is this displacement effect and the greater is the impulse upon the proton in the opposite direction. Incidentally this also means that when Mp-shells happen to be in proximity and initially at rest then a natural repulsion effect will exist between them. In the case of B colliding with A the end result is that the initial linear momentum of B is shared equally by both A and B. This means that the linear momentum of B is reduced to half while A gains the same amount. The total linear momentum of the set remains unchanged. We see therefore that linear momentum is transferred from one atom to the other not by a mere physical collision between their shell structures but by a combination action of energy medium absorption by both shell and proton. Remember that the shell must follow its proton since forcephoids issue from the proton itself. In turn the shell exerts a centralising impulse upon the same proton although at a different event moment. The sequence of collision is therefore as follows,

1. Mp-shell distorts.
2. Proton is centralised by impulse from shell.
3. Forcephoids contain new linear momentum of proton.
4. Mp-shell contains the new linear momentum of proton.

Let us consider two adjacent primary atoms at rest with regard to one another. Now even if the shells made contact their protons would still be a relatively large distance apart as far as Fig 4.24 was concerned. The distance between the protons would certainly be greater than 'x' and even possibly greater than 'Nd'. As such the gravitation effect between the two protons would be one of an attraction impulse as represented by the resultant impulse curve. The two adjacent atoms would therefore receive an impulse towards each other.

However we have already shown that the Mp-shell of one atom has a distortion effect upon the Mp-shell of the other causing a net repulsion between the two atoms. The Mp-shells cause the adjacent atoms to receive an impulse away from one another. At the contact position it is considered that the repelling action is the greater impulse and so the atoms will be caused to move away from each other. At larger distances the gravitation impulse is the greater impulse and so the atoms will be caused to move towards each other. At some relative positioning these impulses will be equal and opposite and the two atoms will remain at rest.

If the above case were to be repeated in a zone of higher realm EIL (or temperature), then the final rest positions would occur at a slightly greater distance apart. This is because there is a marginally greater quantity of energy medium in the Mp-shell causing a slightly greater distortion effect upon the Mp-shell of its neighbour. These impulses which exist between atoms and molecules of matter can be linked with the bulk properties of the so-called solids, liquids and gases.

Let us now consider the action of a large mass structure (such as the earth) upon a single test atom. When the test atom is outside the earth, the gravitational impulse is attractive and obeys the inverse square law. To determine the gravitational effect upon the atom it is legitimate to assume that the whole of the mass of the earth is concentrated at its centre. Inside the earth however this assumption must change since there is mass to either side of the test atom. The impulse towards the earth's centre is represented in Fig 4.34. If we assume the test particle to be a certain distance inside the earth then we can establish its impulse response by dividing the earth into two portions and considering the effect that each portion of mass has upon the particle. In Fig 4.35 we represent the test particle at the position 'y' inside the earth. We then divide the earth along an imaginary plane 'xx' which passes through 'y' and is normal to the line joining 'y' to the earth's centre O. We may now consider the earth to be in two mass sections A and B as shown. If we assume that all of the mass M_A of section A is concentrated at a point A and that of Mass M_B at a point B, then the test particle will behave as if it were in the situation represented in Fig 4.11. It must be remembered

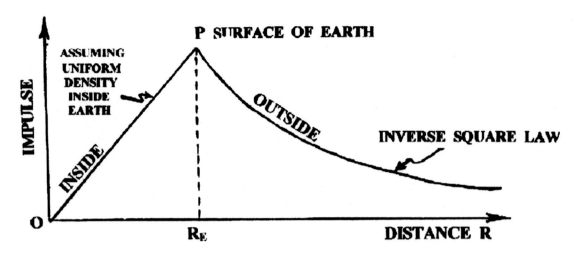

Fig 4.34

that 'xx' is simply an imaginary plane and the points A and B represent the mass centres on either side. As our test particle moves closer to the earth's centre the theoretical mass M_A will increase and M_B will decrease. When 'y' is at the earth's centre O then M_A and M_B are equal and the grid

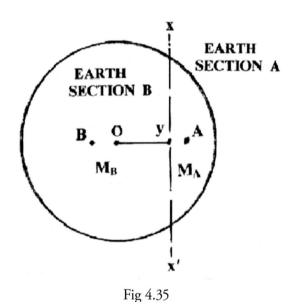

Fig 4.35

datum line will be evenly distributed such that the EIL gradient at O is nil. Thus when at the earth's centre the resultant impulse action in the test particle is nil. The reader may consider that this is logical since the test particle is at the mass centre of a very large congregation of other similar particles and therefore the attractive impulse towards them is equally balanced in all directions. This is an important consideration in that perhaps we can ignore all of the mass of the earth except for those few atoms that are adjacent to the test particle. A non uniformity in these relatively few atoms could be of more significance to the test particle than the entire remaining mass of the earth.

Let us expand upon this theme by considering a similar positioning of a body (such as the earth or sun) within a much larger mass assembly (such as our milky-way galaxy). The earth (or sun) is surrounded by billions of other similar masses and may therefore be compared to our earlier test particle. Yet the earth's main impulse responses are towards the sun and neighbouring planets. We can realistically ignore the effect of the mass of the galaxy as far as impulses set up on earth are concerned since these are relatively uniform and in balance from all directions.

The galaxy itself is part of a local cluster of similar bodies which in turn are part of a super cluster. Here again our galaxy is only considered as impulse responsive to those of its own local group. Response to the super cluster being balanced by a uniformity of action in all directions. These theoretical considerations could continue ad-infinitum but for the limiting factor of the Realm dimension. And it does indicate that so far as the grid datum is concerned the EIL gradient from the more distant masses approaches zero. This also means that the E-Flows from these distant masses arrive as a uniform EIL from all directions. We may therefore consider the realm to be isotropic in its mass distribution such that individual bodies, although not at the centre, may be deemed to be as such for purposes of overall action. We need therefore only concern ourselves with the realm datum effects caused by neighbouring masses and can safely ignore those more distant ones as being part of the isotropic conditions around us. (Neighbouring is used in a broad and relative sense). It is extremely important for the reader to visualise the realm datum as an energy medium structure that is a buildup of the E-Flows from all directions. We have already stated this earlier and perhaps a revision of that topic would reinforce the readers' concept of its make up (Fig 4.9).

Let us once again return to the concepts of realm grid gradient and realm EIL and recognise a similarity with Einstein's popular spatial grid. We consider that he had a strong notion of the reality of an EIL grid but lacked the necessary terminology and background to effectively phrase his

ideas. He ultimately presented his concept of gravitation in conventional grammatical terms by indicating that space was curved. Mass objects were then 'assumed' to move downhill into the space dips created by this curvature. Einstein did not indicate the means whereby a mass received impulse towards these dips because he considered gravity as an ever present and invisible action and portrayed it in space by the terrestrial principle of downhill acceleration.

EIL oscillations i.e. light-waves in space also apparently followed this downhill path and we have already discussed this in detail in chapter 3. It was extremely difficult to represent the 3-dimensional aspect of space by this 'downhill' example and as such there were many who were critical of this theory. Energy medium was only a vague ether to the laws of physics and there was no room for such an inconclusive and intangible entity in its regal enclosure. It is hoped that we are now considerably removed from that stage and that realm energy medium is accepted as a basic constituent in the physical sense since it possesses form, density (EIL) and linear momentum. Meaning that it has property and form.

The grid datum EILs and gradients are the slopes and curvature that Einstein mentally visualised but was unable to describe in the appropriate context. The main conceptual difference is that masses do not exist within space-dips but at grid datum EIL peaks as represented in Fig 4.9 and Fig 4.11. Also masses do not accelerate 'downhill' but develop a net impulse in the upward EIL gradient or 'uphill' direction. Perhaps Einstein was wise to present his theory in a conventional form and trust to later more accurate development.

Let us extend our concept of mass beyond the primary atom and consider the theoretical achievement of a much greater mass object. We have already discussed the distortion aspects of the Mp-shell. Consider this distortion as a uniform effect all around the shell. The shell would then be reduced in its spherical size (diameter) but would remain in equilibrium as far as a net linear momentum was concerned. When atoms congregate they exert this effect upon one another, especially since a mutual attractive impulse exists between them. The Mp-shells now form at close-in distances and must therefore contain a higher concentration of energy medium. Higher realm EILs within these zones are indicated by our conventional temperature factor. As more atoms congregate the Mp-shells undergo a further uniform distortion and the EIL must rise sharply. At some progressive stage of events the outward linear momentum of the shell will not be able to withstand the distortion from its neighbours and two or more protons may then crash together. A new proton combination occurs which may be as represented in Fig 4.26 and Fig 4.28. This would then set up its own Mp-shell configuration with nearly twice the outward linear momentum of the primary shell. Incidentally each proton pair generates a single though compound Mp-shell as we shall observe in chapter 6.

Larger masses (or atom congregations) cause increased attractive impulses resulting in an ever increasing compressive action inside its structure. More Mp-shells must eventually give way to these pressures and more atom nuclei would crash together. It will be remembered from Fig 4.24 that a very large net repulsive impulse is effective when protons or nuclei approach within a distance 'x' of one another. It is this effect which subsequently cushions the impact effect of nuclei crashing into one another so that disintegration of a Mp-congregation does not occur.

Let us now assume that a very large quantity of protons have been pushed into contact with one another and that individual proton deformation is minimal. One may imagine the protons to be relatively positioned just like marbles tightly filling a pouch bag. Note that although the marbles are all in close contact with each other there are interstitial air spaces between them. The overall mass density of the marble congregation must therefore be somewhat less than that of an individual marble. Similarly, when a very large atom structure (such as a star) collapses and becomes a proton congregation then we get a structure with an overall mass density that is extremely high but somewhat less than that of the individual proton. Such bodies have been studied by astronomers and have been named as 'neutron stars'.

Let us proceed an evolution step further in the cyclic events of such a congregation. With a structure that is sufficiently large the protons nearer its centre come under considerable compressive impulses which causes the individual protons to distort. This deformation reduces the interstitial spaces resulting in the structure becoming more and more dense. Ultimately there are no interstitial spaces whatsoever to separate individual protons. We now have a body at the centre of the proton congregation that is much larger than a proton but with the same mass density. Suppose that a large proportion of the congregation were to acquire this condition then we should find ourselves with a very large (perhaps a few miles in diameter) proton type object indeed. This object would possess the same mass density and Mp concentration as the proton. As such it ought to behave in exactly the same manner as the proton so far as attraction and repulsion action was concerned (relative to its size of course). However because of the reduced surface curvature, the E-Flow(2) (see Fig 1.13 and Fig 1.15) would build-up to much higher EILs. Depending upon the body size this EIL could reach levels in excess of C_1 upto considerable distances from its surface. EIL oscillations could not occur normally as a wave front within this C_1^+ zone and as such a space traveler could not view its surface. The impulse curve of Fig 4.24 would however still apply for a test particle approaching this body but with a new dimensional aspect. Any particles formed within the C_1^+ zone should in general also possess a linear momentum that would expel it away from the body and out of the zone. Our fictitious space traveler waiting outside the C_1^+ zone should be able to monitor these particles. Further speculation is left to the imagination of the reader. Perhaps the reader would prefer to parallel our proton giant with the proverbial Black Hole concept. In chapter 2 we indicated the formation of much larger proton giants during the circle 4 phase of the 7-circle theory. These giants are approximately the volume of our solar system and are at the heart of the galaxies as they exist today as discussed on page 61. Any extension of the theories presented so far should always be made with the 7-circle theory principles in mind.

There are many other aspects of gravitation that could be discussed here but at the cost of a considerable extension to this chapter. Since there are other zones of basic interest we feel it best to temporarily curtail the present topic.

Chapter Summary.

1. The linear momentum of an entity is the product of its energy medium quantity and velocity.

2. The linear momentum from one energy medium entity can be added to that of another energy medium entity only by the process of a full Mp-event sequence.

3. Mass particles that absorb an energy medium quantity also acquire its linear momentum. The rate at which linear momentum is acquired is defined as linear impulse.

4. The law of conservation of linear momentum specifies that linear momentum cannot be created. However explosive or implosive actions produce linear momentum in opposing directions but with a zero resultant.

5. The E-M/X sequence represents the implosive phase in the formation of a Mp while the M-E/X is the explosive phase resulting in the outward flow of energy medium.

6. The current proton is a Mp-congregation formed during the circle 5 phase of the 7-circles curve representation. In the present circle 6 status the proton is in a gradual process of energy-medium decay through a loss of some of its surface Mps.

7. Proton internal Mps receive a net inward impulse that results in compression of the Mp-congregation.

8. E-Flows at the proton surface result in a high energy medium grid or 'atmosphere' around the proton which aids in the suppression of surface Mp decay.

9. So long as the proton is in a region of uniform realm EIL it will remain in equilibrium. When the proton is in an EIL gradient zone it receives a net linear impulse in the direction of the higher EIL gradient.

10. The net linear impulse received is termed the gravitation effect and its extent is directly proportional to the local EIL gradient.

11. The realm datum is a 3-dimensional grid defining realm EIL from point to point. It is higher at positions close to mass bodies.

12. Masses in proximity find themselves in the EIL gradient arising from each other. As such each receives a net impulse towards the other.

13. The EIL gradient is inversely proportional to the distance from a mass body.

14. Masses in proximity also absorb linear momentum from the E-Flow of the other body. This results in a repulsion impulse in each body.

15. Gravitational attraction and repulsion effects are both the result of proton decay but are entirely different in their mode of action.

16. The resultant impulse curve is a combination of the two forms of action. For distances greater than 'x' the net impulse is one of attraction between two masses. For distances less than 'x' the net impulse is repulsion between them.

17. Protons that are in contact are bonded together by the same action that causes compression of its internal Mps.

18. Protons that are about to contact and bonded protons that are being separated do not obey the same laws for a given distance between their centres.

19. Relative velocity between mass bodies results in a change to the repulsion impulse curve. The attraction impulse is unaffected.

20. The Mp-shell of an atom exerts a centralising impulse upon its proton nucleus.

21. Transmission of linear momentum by collision of two atoms events through a chain reaction comprising of Mp-shell distortion and a subsequent proton centralising action.

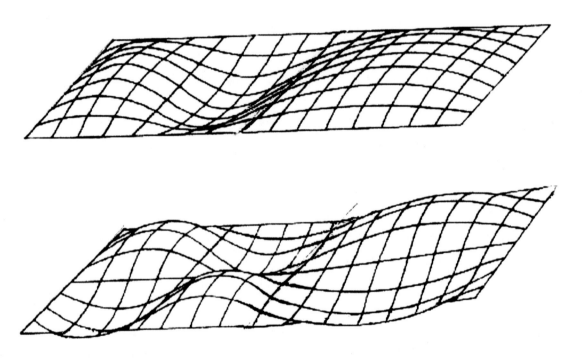

2-Dimension representation of examples of a section of realm grid.

CHAPTER 5.

Proton Time

In an earlier chapter we mentioned that the phenomenon of proton decay could be used as a comparison tool upon which the measurement of all event duration could be based. In this proton decay we observed a consistent decrease in the total energy medium quantity of the proton structure which is recognised here as an absolute condition of energy medium status change. It is the rate of change in energy medium content within the proton that can become a 'yardstick' for duration measurements by which we may precisely locate the 'time position' of any event moment. Since proton decay is a phenomenon characteristic within the circle 6 stage of the energy medium curve only, then the measurements based upon this 'yardstick' must be strictly limited to this phase of the realm life cycle. The concept of elapsing event moments has never before been the subject of a serious consideration and it is the intention here to remedy that omission by developing interrelated physical rules that characterise and define its measurement.

Let us first introduce a concept of elapsing events and then develop a basic notion for duration measurements along lines that are (at first) conventional and familiar. To do this we must commence another of our thought experiments. Let us set up a simple pendulum and wholly enclose it within a glass box. Let there be a total vacuum within the box and allow the pivot to be completely frictionless. Now let the pendulum be induced with an oscillation. Under these conditions the pendulum should continue oscillating indefinitely without change in amplitude or 'beat'. This could present us with a device to measure the duration of other events. For instance, we could measure the duration in which a man runs 100 metres simply by counting the number of oscillations that the pendulum makes between the start and the end of the run. This principle is indeed used by all conventional timing devices. Unfortunately each oscillation is exactly like all the others and therefore we must maintain a count of each 'beat' to be able to locate events and measure their duration. This count is maintained in terms of seconds, minutes, hours, days, weeks, months and years. Of course if for some magical reason every clock around the world was stopped for a few weeks or months we could easily reset them at the end of that period by simply relating to some known astronomical cycle. As such we had not really lost our pendulum oscillation count because of this external reference point. If however we placed ourselves in an enclosure without reference to any such external features then we should be unable to determine an event duration once the count had been lost. This indicates quite clearly that the oscillations of the pendulum are not sufficient on their own as an independent duration measurement device. There must be some other additional factor put into the pendulum oscillations that would permit us to determine, at a glance, the exact number of elapsed oscillations since a previous observation.

Let us change the operating conditions for the pendulum in its glass box vacuum by imparting an extremely small factor of friction to the previously frictionless pivot. Under these new conditions the amplitude of oscillation must diminish with each successive 'beat' due to absorption of a kinetic energy quantity at the pivot. As such the pendulum could not oscillate indefinitely and therefore must come to rest after a finite number of oscillations. The pendulum now has a finite operational' life' with each oscillation being unique by virtue of its displacement amplitude. For a set location on earth the oscillation period of this pendulum will be constant and independent of the changing amplitude of swing (when the amplitude is reasonably small).

As part of our continuing thought experiment let us assume that the operational life cycle of this second pendulum is about 3 billion oscillations. If the period of each oscillation is the equivalent of one second then our pendulum should oscillate independently for approximately 100 years. At any instant during this period we could locate precisely any event moment and duration simply by observing the amplitude of oscillation. The amplitude of the oscillations at any particular instant is represented on a graph Fig 5.1 showing a gradual decay of amplitude with elapsing oscillations. We now have a comparison tool that can not only measure other event durations but which can also locate precisely any event moment occurring within the next 100 years.

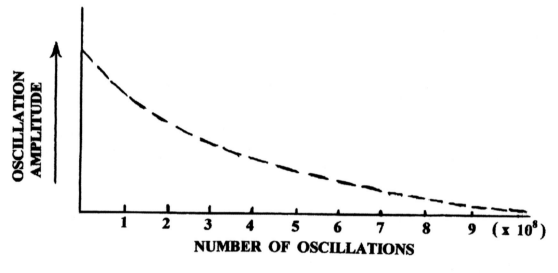

Fig 5.1

In this set up we therefore have a system for sequential grading of other events. This system however cannot be considered as an absolute one since the period of oscillation is in turn dependant upon other factors such as changes in the pull due to gravity or length of pendulum rod. In our thought experiment let us maintain the length of pendulum rod as constant but vary the pull due to gravity. Let us cause gravity to increase marginally. The pendulum will 'beat' quicker, meaning that the oscillation period is shorter. The graph of Fig 5.1 will still apply to the oscillations which will decay at the same rate per oscillation. However this means that the pendulum will still make the same total number of oscillations (3 billion) but in a shorter overall duration (say 95 years). Does this mean that the yardstick of duration measurement is inconsistent and therefore of limited application?

In order to determine whether or not this is the case let us broaden our thought experiment and suppose that everything we observe around us is basically constructed from countless little oscillating pendulums. Include also the basic structural make up of the human body and that all phenomena occurring around us have their functioning somehow inter-linked to those of an oscillating pendulum. Let us now consider two of our boxed pendulums that are identical in all respects. Place each at a separate location such that the gravitational pull on one is twice that on the other. Obviously we must locate each pendulum box on a different planet in order to achieve this condition. Let us refer to these separate locations as planet Earth and planet X, where Earth's gravity is twice that of planet X.

The period of oscillation of a simple pendulum is inversely proportional to the square root of the gravitational acceleration 'g' as given by,

$$\text{Period } T = 2\pi\sqrt{L/g}$$

where L is the length of pendulum and in our case is deemed constant. Then,

$$\text{Period } T \propto \sqrt{1/g}$$

From this we calculate that the period of oscillation for the pendulum on planet X is 1.4 times longer in duration than its counterpart on Earth. Let us transport ourselves to planet X and actually observe the pendulum oscillations there. However, according to our supposition that all things including ourselves are interconnected in function to mini-pendulum type oscillations, as soon as we arrived there the reduced gravity would immediately affect all our internal pendulums as well. Everything that we observed would similarly be functioning in relation to the gravity of planet X and as such we would not notice any change in the period of the pendulum. Also, our measurement base would not appear to have altered the least bit from that on Earth. Our concept of event duration based upon the boxed pendulum 'yardstick' would appear constant and we would conclude that there had been no change in oscillating period with change in gravity. Of course we know that this is incorrect but the reader is made aware by this example that relativities must change before we are able to realise that a change has taken place.

Let us further illustrate this principle by extending the above experiment with the help of some more transportation magic. Consider both box pendulums to be initially located on Earth. Let us start up the oscillations of both pendulums at the same event instant and with exactly the same displacement amplitude as per the conditions of Fig 5.1. Now let one of the pendulums be magically relocated onto planet X without any upset being caused to its oscillations. Let us allow 10 years to elapse on Earth and then we magically fetch back the box pendulum from planet X. Let us now determine the total number of oscillations made by each pendulum. The amplitude of the Earth-bound pendulum indicates a total of 315,360,000 elapsed oscillations equivalent to 10 years on Earth. The planet X pendulum however indicates a total of 220,752,000 oscillations in the same period. This corresponds to only 7 years of oscillations on Earth. This shows that our box pendulum 'yardstick' is not an absolute constant.

Although we did not realise it our concept of duration had actually slowed when we were on the planet X. Our concept of duration however seemed constant at both locations simply because there was no relative change in the periods of our box pendulum and all the other countless structural mini-pendulum events around us. We did not recognise this change in oscillating status since its 'beat' relative to the 'beat' of all other phenomena around it remained constant. Our measurement of elapsing moments and event duration must therefore remain with this 'yardstick' feature even when we know it to vary. We refer to this yardstick or relative system of duration measurement as Pendulum-Time and in the above example we could state that Pendulum-Time is slower on the planet X than it is on Earth.

The above is a very crude example of a general principle that does apply (though in essence only) to the conditions that prevail within the current circle 6 phase of our Realm. The reader must by now be quite familiar with the action within the Mp-event cycle and the factors that influence its sequence duration. Indeed we may compare these energy medium and mass medium exchanges to the oscillating sequence of the pendulum. Would it not therefore be preferable to use the Mp-event as a real duration measurement yardstick? Unfortunately each Mp is a one-off occurrence and has a life of only one exchange sequence. This rather limits its use as a measurement base. However if a sufficient quantity of Mps were grouped closely together such that a continuing series of Mp-events took place, then perhaps we could recognise a similarity with the frictionless pendulum. This would present us with the earlier prospect of having to maintain a continuous count of the succession in Mp-events in order to locate event moments. We must therefore recognise a second parameter within this system before we can consider it for use as a yardstick for duration measurement.

To achieve this let us propose a more realistic set of operating conditions for our group of Mps. Assume that the Mps in this congregation operate in a partially synchronised manner such that the M~E/X E-Flow from one Mp is totally absorbed into E~M/Xs of newly forming Mps surrounding it. Each new Mp is therefore built from energy medium released by the preceding Mps. Since the Mp-group or congregation must contain a limited quantity of Mps then those at its boundary positions will not be as completely surrounded as those nearer its centre. We must therefore assume that part of the M~E/X E-Flow from Mps at these outer locations escapes re-absorption into the group. This means that a finite quantity of energy medium is lost by the group whenever a M~E/X sequence occurs at or near its boundary. The Mp events within the congregation occur as a regularly repetitive cyclic sequence with a 'beat' that is inversely proportional to the prevailing realm EIL. The energy medium loss must occur at the end of each beat sequence, and hence that loss will also be regularly repetitive (and seemingly continuous) until no further Mps remain.

Let us assume that over a cyclic duration of one 'beat' the total loss of energy medium quantity is the equivalent to that of a single Mp. Consider all E-Flows leaving and entering the congregation as having been taken into account. This Mp-congregation with its regular 'beat' and energy medium decaying condition is somewhat similar in function to that of our earlier decaying pendulum oscillations and may therefore also be employed in keeping 'time' i.e. event duration. The reader by now must have surely made the connection between our example and the reality of the proton structure, and so arrived at the conclusion that Mp oscillations within the proton coupled to its energy medium decaying characteristic is the most basic 'yardstick' for duration measurement. Since all local phenomena are directly related to the proton's characteristics, then as was annunciated in the principles for Pendulum-Time so may we state all duration measurements in reality as being relative to the proton phenomenon. As such this automatic yardstick of duration measurement will from now on be referred to as Proton-Time.

The reader must realise that as in the Pendulum-Time analogy, Proton-Time is also a variable. In the case of the proton phenomenon variation in its functional 'beat' is caused by differing realm EILs. The phenomena of gravitation and wave front propagation have been reasonably explained in previous chapters and as such the reader must be familiar with their properties and dependant variability. We would ask the reader to mentally review those properties and thence to relate them directly with the beat variation of Mps within the proton. We may draw the conclusion that the proton 'beat' is variable and that its subsequent rate of energy medium decay is indirectly related. This would mean that the rate of proton decay is greater at lower realm EILs and lesser at higher realm EILs.

Our measurement and concept of event duration is linked to the uniform change in the energy medium status of the proton. As such the beat aspect is not directly taken into account when Proton-Time is under consideration. Changes in the overall energy medium status are sufficient for the evaluation of event duration. The functioning of the proton is therefore an automatic pace indicator for all local phenomena and it is this fact that permits us to relate proton decay to Proton-Time. Once again we would stress that this applies only to conditions in the circle

6 phase of the Realm energy medium cycle. As such we could say that 'Time' only commenced at the start of the circle 6 phase. Before that 'time' as we now know it had no relative meaning.

Let us illustrate our Proton-Time concept with the aid of a conventional application. Consider the measurement of the velocity of light. We are in fact measuring the duration that a wave front takes to propagate over a set distance. This would basically be like timing a man running in a race. The duration of the run depends upon the timing device and its functional pace of operation. A fast clock would measure a slow race and vice versa. Suppose that two athletes race over a set distance and one outruns the other. Also suppose that each athlete was timed in the run by their personal coach. Suppose also that when the two coaches compared stopwatch times they found them to be exactly the same. We would obviously conclude that either the winning athlete was timed with a stopwatch that ran fast or else that the other athlete was timed with a watch that ran slow. Unfortunately our measurement of the velocity of light is similarly handicapped by a rather variable timing device and the results do not therefore show up true differences of velocity.

We know that the velocity of wave front propagation varies inversely as the prevailing realm EIL. Now let us first consider measurement of wave front velocity in a higher EIL zone. The velocity of light will be somewhat reduced within this zone. However the higher realm EIL will also cause the proton Mps to function or beat at a marginally slower cyclic rate. This results in a correspondingly slower rate of overall energy medium loss by the proton and hence Proton-Time will also function at a correspondingly slower pace. Since our concept of duration remains unchanged then our proportionately slowed down internal clock will indicate a higher than actual velocity of light. In fact the correlated conditions are such that no change is detected in the measured velocity of a particular light wave under varying EIL conditions. As such it has been erroneously assumed that the velocity of wave front propagation was absolutely constant.

The measurement of wave front velocity in a lower EIL zone will produce the same result since the actual propagation velocity and Proton-Time rates are both correspondingly increased. It is like measuring a faster running athlete with a fast running stop-watch. As all factors remain constant relative to one another no change is detectable.

Let us now consider the impulse which is gravity. This is also affected by changes in realm EIL but unfortunately we must rely upon measurements of acceleration in falling objects to determine its value. Since acceleration is linked to velocity measurement we are again unable to detect a realistic impulse change (refer to Fig 4.21 and Fig 4.22). Let us suppose that our solar system just happened to traverse into a realm grid zone at a substantially higher EIL than at present. All protons of our system would then functionally beat at a somewhat reduced pace. Proton-Time will have slowed correspondingly but we would not appreciate this since our concept of duration would remain unchanged. The velocity of light would in reality be reduced but our measurements would not indicate any change. The impulse action within each proton body would be marginally less because of the slower rate of Mp events. This means that acceleration due to gravity was lessened but we would not notice a change in its measured value (just as in the case of the slower athlete being timed by a slow running stop-watch). However linear momentum of moving or rotating objects would remain unchanged and therefore objects would appear to gain momentum (measurement by a slow running clock).

The earth would appear to complete one revolution about its own axis in less than 24 hours. The velocity of orbiting satellites and planets would remain unchanged and as such the reduced actual gravitational impulses will cause them to move into more distant orbits. This would result in us observing that the duration of a particular orbit as being approximately constant (caused by measuring a longer duration with a slow running clock). Measurement of these physical distances would however show up as real and without our Proton-Time theories we could never expect to explain such apparently massive linear momentum changes. We would probably need to speculate upon some vast unknown force being secretly in action. Indeed, when Proton-Time is slowed then the linear momentum of all masses would appear (on measurement) to have increased. Yet in reality the only change that has occurred is in our measurement 'Yardstick'. Fortunately for us realm EIL changes are extremely gradual and therefore we may never have to deal with such an occurrence.

At the higher realm grid EIL, transfer of linear momentum between colliding masses will also have slowed in line with the proton's reduced Mp functional beat. The reader may analyse this aspect by considering that a slower Mp event cycle results in a diminishment of implosive and explosive actions (Fig 4.2). The subsequent E-Flow from a Mp would thus be at a marginally reduced flow momentum i.e. reduced actual velocity.

Subsequently the forcephoids of Fig 3.2(a) would take longer to arrive at the Mp-shell position. The Mp-shell would also possess a greater 'life' and the frequency of its repetitive existence would reduce. As such the observance of the frequencies of all wave fronts would appear unchanged to us (who are now measuring all duration with a slow running Proton-Time clock). However, frequencies that are initiated in a high Realm EIL zone and which then propagate across vast distances to a lower EIL zone will be observed there as being at a reduced frequency (as we are now measuring that same frequency with a slightly faster running Proton-Time clock). This particular feature is applicable to the phenomenon known as frequency red shifting and has already been discussed (Fig 3.68).

Our concept of duration is therefore wholly related to the functional 'beat' or energy medium decay of the protons that make up our material selves and surroundings. Although protons in grouped assembly within an atom nucleus may vary in quantity and thus in 'beat', it is the overall average within our structural locality which determines for us a pace for Proton-Time. It would be extremely time consuming for us to extend this discussion to any more examples relating Proton decay to other functional duration measurements and we propose that the reader conduct a personal meditation on this aspect. For the moment we would therefore conclude with the statement that proton decay, because of its uniformity, is the invisible yardstick by which we are forced to gauge all duration measurements for all the phenomena occurring around us. Because the make up of all realm material is structured with protons, this measurement yardstick is defined as Proton-Time. We shall now proceed with the development of physical rules that apply to and adequately describe the characteristics of Proton-Time in the case of the primary atom. The more complex atoms would follow the same rules but would make our discussion far too complicated at this stage.

We have thus far indicated that for every proton mass within an atom - primary or otherwise - events may be traced along a defined sequence order related to proton decay. To be able to correlate the relative 'life' time positions of simultaneously occurring other proton events, we need to evolve an event graded system based upon a common datum that will satisfy all conditions in a varying Proton-Time situation. Since we are not able to produce dimensional parameters for an absolute Proton-Time pace, we must resort to relativity with a theoretical proton decay situation. We shall achieve this through a relative comparison of individual proton decay energy medium curves in a graphical representation.

The total decaying duration of a proton is termed as its 'life'. The life of a proton is not necessarily the same as that of another since the duration depends upon the rate of decay, which is variable with realm EIL. Generally, protons have a longer life in the higher EIL zones such as those that prevail near large mass bodies (see Fig 4.9 to Fig 4.15).   Proton decay for purposes of duration measurement commences when the surrounding EIL just enters the circle 6 phase of the realm life cycle. At this condition the proton is assumed to be at the C_1 optimum mass level denoted by T_{Max} where T represents the total energy medium quantity in a proton. Now proton decay is uniformly progressive so long as the proton structure is in a stable condition. For this there must be a sufficiently large number of Mps in existence within the Mp-congregation for structural compaction to be effective (Fig 4.6a). The proton mass at the limit of this stability is termed the minimum decay mass and is denoted by T_{Min}.

The total energy medium that can therefore be released by a decaying proton during its life time is given by,

$$\text{Total decay energy medium} = T_{Max} - T_{Min}$$

We assume that instability of the T_{Min} proton results in the total disintegration of its structure through a rapid release of all its remaining energy medium in E-M/Xs. The overall aspect of proton decay is graphically represented in Fig 5.2.

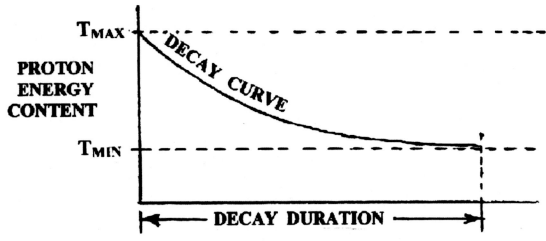

Fig 5. 2

Let us now consider the age aspects of the decaying proton. The age of the proton is defined as the quantity of energy medium that it has lost since it was at T_{Max} and is given by,

$$\text{Proton Age} = T_{Max} - T_{Current}$$

Since the decay rate is a variable according to realm conditions then it is probable that protons of varying age co-exist near one another. This means that protons of varying mass should be available to us. As protons continue to disintegrate when they achieve maximum age there will be other younger protons to continue the function for duration measurements.

We must now define a standard for the decay curve which we shall consider as the absolute maximum rate for proton decay. From now this total decay energy medium is to be referred to as the total normal decay energy medium and denoted by TN. In Fig 5.2 above, the decay curve will therefore be referred to as the TN-line. The slope of the TN-line indicates the rate of decay of the proton mass. Consider a single proton placed entirely on its own in a void region for its total decay duration. This means that the proton decay will not be affected by any realm EIL datum or by any E-Flow from other protons for its entire decay duration. The only influence on its structure would be that from its own internal Mps. As such the proton decay rate would be at an optimum and the decay duration a set minimum. Let us set this up as our standard rate of decay and refer to this decay curve as the 'absolute TN-line'. The duration in which this total decay occurs is a minimum and will be the standard or normal period of decay denoted by PN.

However in an actual realm situation there are considerable amounts of E-Flow reaching the proton from practically every direction. This makes up the realm datum grid as described in Chapter 4 and results in a substantial quantity of energy medium being absorbed into the E~M/X sequences of the Mps of the proton. The net decay loss is thus considerably reduced and the proton thereby has a

longer 'life'. As such in reality the slope of the decay curve will always be less than that of the absolute TN-line and the elapsing period for its total decay of a greater duration than PN. Fig 5.3 compares an actual proton decay line, or TA-line, with the absolute standard TN-line. Each decay line is an exponential curve between T_{Max} and T_{Min}. The energy medium loss diminishes as the proton mass depletes and is a constant percentage of the remaining mass.

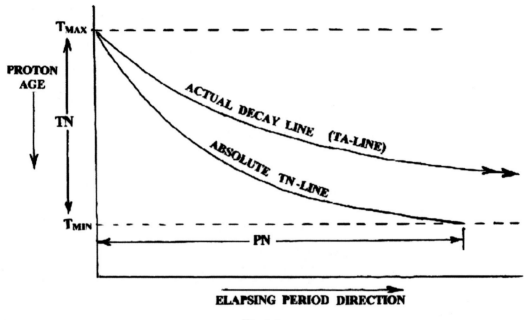

Fig 5.3

For purposes of simplicity we shall assume all protons to have commenced decay with the same mass/energy medium content i.e. T_{Max}. The reader may remember that protons formed in the Circle 4 and Circle 5 phases and that their mass quantity depended upon the realm EILs at which they formed. If we are to use the absolute TN-line as a duration measurement base we must then specify its overall decay rate as a fixed standard. We can do this by ascribing a set rate of decay over its entire decay period PN. All TA-line decay rates must then be considered in relation to this standard. Comparison of decay rates must always be made at the same proton age i.e. energy/mass medium content. This is an essential condition for comparison with the TN-line standard because of the diminishing rate of energy loss with progressive decay. In Fig 5.3 therefore comparisons between standard and the TA-line may only occur for positional points on the same horizontal line.

The reader may compare the absolute TN-line with the curve for the decaying oscillations of the pendulum in Fig 5.1. In that case we compared the oscillating rhythms of two pendulums under different conditions of gravity. We could have obtained an absolute oscillation rhythm by defining preset conditions and then comparing all other oscillations against that norm. In the present real life situation of proton energy medium decay this is exactly what we have tried to achieve. By definition we have secured the highest possible natural rate of decay as the absolute standard. All other decay rates may then be individually compared to the standard and thence correlated to one another. It would be by this process of inter-correlation of decay rates that we may determine a relative functional rate for our concept of duration and thereby the ageing process in different realm locations.

For the present we must look upon the ageing rate as being directly proportional to our concept of duration and thence its measurement. We (as proton structures) can only relate to these positive changes in proton make-up. Thus our concept of duration and thereby the elapsing of event moments would appear at a standstill if the proton entered an environment where its decay was zero. Even in this situation the Mps within the proton continue their cyclic formation sequence at only a marginally reduced rate. The elapsing of these most basic of events (E-M/Xs and M-E/Xs) must then surely be the true gauge for the measurement of duration. Unfortunately we are unable to relate to these events directly and therefore must somehow relate our concept of duration to the Mp cyclic rate in order to establish a unit for 'Time' i.e. elapsing event moments. The reader must realise that Time is not a physical

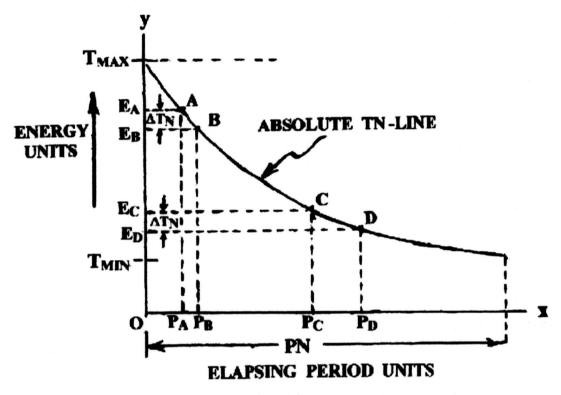

Fig 5.4

entity within our realm but rather is a process for measuring the duration of physical events in multiples of a chosen basic event duration. We shall return to a discussion of this aspect later in the chapter.

Let us now define the parameters that operate with the TN-line. In Fig 5.4 the TN-line is shown represented as a curve on a graph with ordinate (o-y axis) calibrated in energy-medium units and the abscissa (o-x axis) in elapsing period units. Consider the age positions A and B on the TN-line.

Assume these to be very close together and separated by an energy medium difference of ΔTN. Let the energy medium contents at A and B be E_A and E_B energy medium units respectively. Then,

$$Decay_{AB} = \Delta TN = E_A - E_B \text{ energy medium units.}$$

Since the total TN decay of our standard proton takes place over a fixed duration PN, then each age position on the TN-line (graph) corresponds to a particular event moment in its life period. Let each age position be represented on the horizontal axis according to its elapsed period from T_{Max} status. Let the elapsed period for the A and B age conditions be P_A and P_B period units respectively. Then,

$$\text{Duration of Decay}_{AB} = P_B - P_A \text{ period units.}$$

Subsequently we have,

$$\text{Rate of Decay}_{AB} = \frac{\text{Decay in energy medium units}}{\text{Duration in period units}}$$

$$= (E_A - E_B) / (P_B - P_A)$$

Since we have described this TN-line as having a standard rate of decay then,

$$\text{Rate of Decay}_{AB} = (E_A - E_B) / (P_B - P_A) = 1 \text{ (or unity)}$$

When we apply the above formulation to the C and D age positions of Fig 5.4, we find that,

$$\text{Rate of Decay}_{CD} = (E_C - E_D) / (P_D - P_C)$$

Since we have already expressed the desire to use this TN-line as an absolute decay standard, then the rate of decay all along the TN-line must be unity. This means that,

$$(E_A - E_B) / (P_B - P_A) = (E_C - E_D) / (P_D - P_C) = 1$$

In Fig 5.4 we have represented the decay quantities in both cases as being DTN. Therefore in order that the above equation may be satisfied we must have,

$$P_B - P_A = P_D - P_C$$

In the diagram it is clearly observed that the duration $P_D - P_C$ is the greater and as such the equation cannot be fulfilled unless we have a continuously changing period parameter to suit the equation. To achieve this the duration or period units must be stretched with the age of the proton. Since comparison of decaying rates must always be made at the same age conditions the relativities will remain unaffected. In Fig 5.4 therefore the elapsing period parameter must be drawn close to an inverse logarithmic scale to correspond with the TN-line function. Only then may the rate of decay all along the TN-line be expressed as unity and for it to be an absolute proton decay standard.

Since the actual rate of decay of protons in our realm is always less than the optimum TN-line rate, then so will our concept of duration also always be slower than optimum. According to our earlier example we could say that our stop-watch was running slower than standard. Our concept of elapsing events or Event-Time is inextricably linked with proton decay and as such in practice we observe no physical changes in our duration measurements.

At this stage it is essential that we briefly discuss the meaning of Time in the context of a parameter that quantifies the location and duration of all events in our realm locality. Actually, Time itself is a non-entity. By this we mean that it has no substance or form and therefore does not exist as a physical structure or thing. It is therefore simply a term that is used for evaluating duration in multiples of a most basic event. The units of Time are thus based upon the duration of these most basic physical events within the realm. We have already shown that the Mp-event sequence is the most basic event cycle within the realm and that this can be further sub-divided into its separate event phases. However it would be simpler to consider a time unit as the duration of a complete cycle of events for reasons that are obvious. Shorter events may be quantified in fractions of this unit. Let us for now consider a unit of Time as the duration of the Mp-event cycle. We could term this unit as the Mp-Time unit or Mtu. Since the duration of the Mp-event cycle is inversely proportional to the realm EIL, our time unit will also vary accordingly. If desired, we could set the EIL conditions for an absolute Mtu standard. It is considered that this would not be of any particular practical importance since our duration concept is not related to the 'beat' of Mp-events but rather to the overall decay rate of the proton.

We must recognise however that the Mtu is variable by only a marginal percentage across the EIL range and as such is an excellent measurement parameter for the progressive elapsing of event moments. We must accept therefore that the true measure of event duration, or Time as we know it, is really based upon the action of a physical entity, i.e. the Mp.

In a void region there are simply no events taking place and as such there could be no measure for duration or Time. In a massless realm there would probably be an isotropic E-Flow condition from previous M-E/Xs. As such a duration measurement unit would have to be based upon the traverse of an E-Flow section over a set distance. However since we would not be in existence to appreciate this realm condition such a speculation is of academic relevance only.

In our earlier pendulum example we would have considered each complete oscillation as a time unit. However we were only able to gauge the actual elapsing of event moments by the oscillation decay process. The extent of this decay indicated the elapsed oscillation period and this gave us a concept of duration for elapsing event moments. It is this concept of duration that indicates to us the progression of events around the pendulum. In the case of the proton, its decay governs our concept of duration and also what we consider the progression of Time. All of our observations on duration measurements are therefore a biased view and subject to considerable deviation from the truth. In order to determine the actual pace of time or Mtus we must first establish our decaying status relative to the TN-line of Fig 5.4.

To establish the relative decay or Event-Time rate for a specified realm situation we would need to make a theoretical assessment of all the influence factors and then plot an actual decay line or TA-line. However because we are now dealing with undefined parameters we are unable to do this. There is however an alternative method for achieving a close approximation to the TA-line. This is based upon the two extreme conditions of decay. The TN-line is the one extreme for maximum decay rate and the other is a zero decay line for C_1 EIL conditions. These are indicated in Fig 5.5. The zero decay line is the horizontal line through T_{Max} which we shall call the Tzero or T0-line. The zero decay status does not necessarily have to occur at this proton energy level as it may be induced upon a proton at a more advanced age because of special conditions of velocity and/or realm EIL. We shall discuss this aspect later. In diagram Fig 5.5 it should be possible to draw a complete set of geometrical curves between the position of the T0-line and TN-line so as to represent every proton decay condition imaginable. It would then simply be a matter of intelligent choice for the appropriate TA-line to fit a particular realm condition. However all known factors pertaining to realm EIL and E-Flows from mass objects must be carefully evaluated prior to choice of TA-line.

Let us consider some of these factors briefly. Firstly, our realm EIL will be quite high up in the circle 6 configuration (Fig 3.68) so that we would consider our TA-line to be well away from the absolute TN-line position. Secondly, E-Flows from large masses would increase the realm datum level so that proximity to such masses would further inhibit proton decay. Velocity through the realm datum may have further decay inhibiting or rather age inhibiting effects and so will also have to be considered. In the case of the massive proton giants (formed during the circle

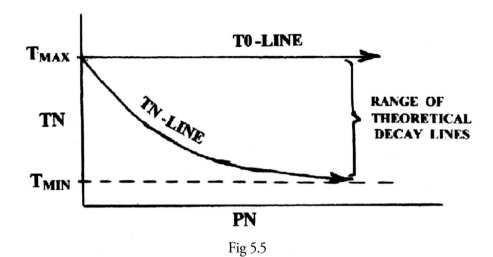

Fig 5.5

4 phase when EILs approached C_2, and also the 'black hole' resulting from the total collapse of a large neutron star), then the TA-line would approach a C_1 EIL condition as we got nearer to it. When the TA-line coincides with the T0-line we must assume that our concept of duration (i. e. event-time) ceases to function. Time is at a standstill.

Now in order to differentiate between the various TA-lines we must locate them within our Fig 5.5 graphical representation by reference to their positional distance from the TN-line. If we define the T0-line as being at the 100% away position, then we may reference the T0-line also as the (TN+1) -line. Let us illustrate this principle by considering the TA-line for a point on the Earth's surface. We shall use a comparative rating method for estimating the various EIL factors that inhibit proton decay. It does not matter too much if we are inaccurate in our initial estimates for a TA-line position since a relativity with other TA-lines will always be maintained. For a point on the Earth's surface let us commence our estimate with the assumption that the realm datum accounts for a TA-line at a 60% position i.e. (TN + 0.6)-line. The next consideration would be to set a numerical scale for the influence that various mass bodies exert in raising the local realm EIL. The size and mass of the body is of major importance since this would indicate the proton concentration/spacing within its structure. In the case of the Earth we shall assume all its protons to be located near its core centre so that computations for EIL at its surface may be simplified. We must further assume that all mass body assemblies have a uniform compactness within the overall realm structure.

A point on the Earth's surface is approximately 4000 miles from its core centre. The E-Flow from the assumed proton concentration at this centre will be at a considerably reduced EIL when it approaches the surface positions. If we set this influence as a standard for a 1% increase in realm datum EIL, then we have a TA-line given by the (TN + 0.61)-line. We must next consider the EIL influence of the E-Flow from the moon which is approximately 200,000 miles away. We estimate a marginal influence at the earth surface of 0.1%. The sun and planets total influence is 0.1%. For our entire galaxy we estimate a local influence of 0.3% since the earth is rather distant from the central core. We could not exclude the influence of our cluster of galaxies nor the super-cluster to which these belong. Rather than becoming involved in laborious considerations let us agree upon a blanket allowance to cover all these distant masses. Let us assume 0.1% of EIL influence for all these masses external to our galaxy. As such we now have,

$$\text{TA-line} = (\text{TN} + 0.616)\text{-line}$$

The reader should keep it in mind that this is only a very rough estimate for a TA-line at the earth's surface but which serves to illustrate the estimating process. We could establish this TA-line as a bench mark for the comparative estimating of other TA-line positions. In the above we did not consider the influence due to velocity through the realm as this aspect would be premature at this stage. A separate section has been allocated to this topic in Chapter 8 as considerable theoretical detail needs to be explained prior to any current statement of facts. However we can indicate that the velocity of a proton through the realm does cause additional realm energy medium to enter within the proton body to be absorbed into its internal Mp E-M/Xs and so reduces the actual decay rate. This factor is negligible for velocities that are small in comparison with the wave front propagation velocity. As proton velocity increases and approaches that of wave front propagation velocity i.e. speed of light, so does the TA-line approach the T0-line in decay characteristics.

As already stated it does not matter too much that our earthbound TA-line is inaccurately positioned, provided that all other TA-lines are positioned in corresponding relativity. The important function of the TA-line is to indicate relative event time rates for differing realm locations. Fig 5.6 shows a set of decay curves at 10% location intervals from the absolute TN-line. The percentage figure indicates the reduction in the slope of the TA-line as compared with the TN-line datum. It also refers to the increase in the overall period of decay. It will be noticed that the logarithmic scale applies to the spacing between TA-lines.

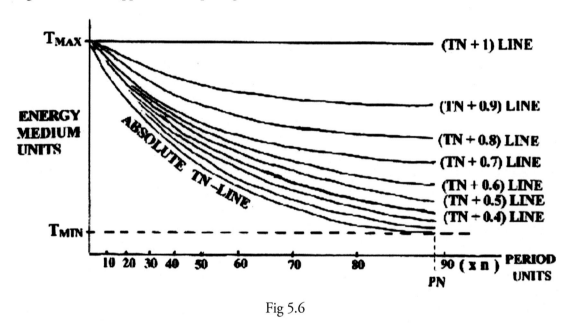

Fig 5.6

We shall show that extension of the TA-lines beyond PN is unnecessary when evaluating actual decay duration. The scale of the percentage change in the sloping of successive TA-lines is based upon the principle of a uniformly increasing quantity when calculated on a successively larger base. Perhaps the mathematician in the reader could extend the development of these TA-lines into a more accurately defined set of curves.

Let us now show how Event-Time in an actual location may be determined in relation to the TN-line standard. Consider the TA-line evaluated above for a position on the Earth's surface i.e. the (TN + 0.616)-line. This is represented in Fig 5.7. Assume the age of the protons at the earth's

surface to be represented by the position A on the TA-line as shown. Let these protons decay by a small amount DTN so that they gradually age to the subsequent position B. The corresponding age positions on the absolute TN-line are shown by C and D respectively.

In order to determine our decay rate (or Event-Time rate) we must compare the ageing period from A to B with that from C to D using the same start point on the PN scale. We must therefore project the curve section AB into the position CE as shown. Let us project CD and CE onto the period unit base and let CD occur over ΔPN units

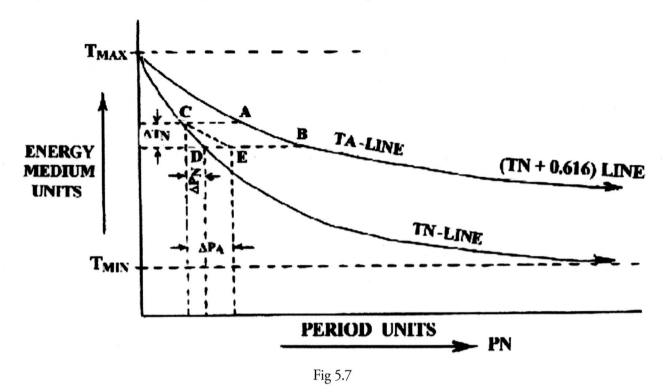

Fig 5.7

and CE over ΔPA units. Clearly ΔPA is much greater than ΔPN. We know that for the TN-line the decay rate is unity and given by,

$$\text{Decay}_{TN} = {}^{DTN}/\Delta PN = 1$$

For the TA-line the decay rate is given by

$$\text{Decay}_{TA} = {}^{DTN}/\Delta PA$$

A comparison of the actual decay rate with the absolute standard shows that,

$$^{\text{Decay}}TA/\text{Decay}_{TN} = {}^{DPN}/\Delta PA \text{ which is less than 1.}$$

This indicates the amount by which decay on the earth is slower than the optimum decay rate. This should work out at 61.6% slower. Also, from this we assume that our concept of duration has correspondingly slowed by the same amount. We do not observe any sluggishness in the passage of Event-Times since functional relativities remain unchanged. According to the example in which we measured the duration of a distance run by two athletes we could now state that our stop-watch is running 61.6% slower than standard. In the case of conditions that approximate the (TN + 1)-line, we would assume this stop-watch to be at a standstill. At this point our concept of duration or Event-Time becomes so distorted that we are unable to measure duration. Now although the proton may have ceased to decay this does not mean that its Mp-functional beat has ceased. Far from it. The T0-line situation may be compared to the pendulum example which has a friction-free pivot. The oscillations continue indefinitely at a set pace but without change in amplitude.

We see therefore that the Mp beat within the proton does not relate to our concept of duration. We know that the cyclic rate of Mp formation inside the proton varies between a maximum and minimum rate only and that both are high order quantities (Fig 4.21). The mechanics of the Mp does not permit its cyclic action to slow by very much (Fig 4.2). The characteristic behaviour of our concept of duration (Event-Time) and also the Mp-time unit (Mtu) changes are both graphically represented in Fig 5.8 for purposes of comparison. It is clearly observed that there is only a marginal correlation between the two graphs which simply indicates a decreasing trend when the realm EIL increases. So far as we are concerned we may consider the Mtu to be a constant.

The problem before us is to somehow relate the aspect of proton decay in its role as a counting device for elapsed beats with the true number of such beats taking place. In Fig 5.8 we have shown a condition on the right hand

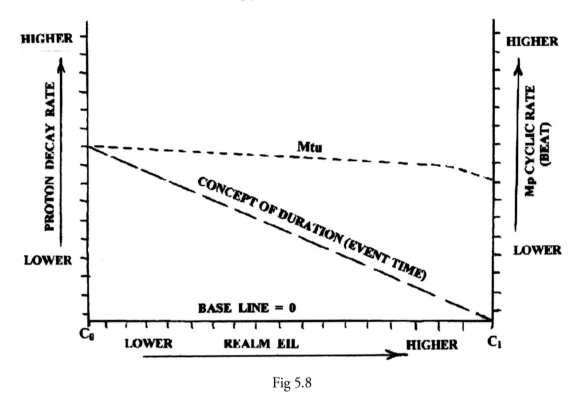

Fig 5.8

side of the graph wherein this counting device provides an accurate figure of elapsed beats. This means that the curve of Fig 5.1 may be transformed to provide an accurate relationship between proton decay and elapsing Mtus. Unfortunately as the proton encounters varying realm conditions this relationship becomes distorted. This distortion is the difference between the number of beats that we presume to have elapsed and the actual reality. In Fig 5.8 the widening gap between the two lines indicates the extent of the disparity that occurs at increasing realm EIL conditions. Our concept of duration or Event-Time will always seem the same, yet we must accept that this is not so. If we may evolve an Event-Time unit (or Etu) based upon the absolute TN line conditions, then perhaps we could progressively relate our concept of duration to the Mtu.

Let us assume that at the TN-line condition our concept of duration gives a true measure of duration. Then,

$$^{Etu}/Mtu = 1 \text{ (for Fig 5.8 left hand side)}$$

This means that our stop-watch is running at a correct and standard rate (for the earlier athletes example). All duration measurements will thus be true.

With increasing realm EIL conditions however the proton decay rate slows below the TN-line optimum and thus we have,

$$^{Etu}/Mtu > 1$$

This means that the original proton decay curve is no longer representative as a counting device for elapsed Mtus. In our duration measurements however we continue to equate the Etu and Mtu as being equal. As stated earlier our concept of duration is related entirely with proton decay as this is a feature that is measurable physically through the interactive processes of the atom structure. On the other hand the Mtu is quite elusive and unavailable to any interactive measurements and therefore is not drawn into our duration concept. As the difference between the Etu and Mtu gets larger our concept of duration gets slower and slower. Ultimately for a zero decay condition of the proton we have,

$$^{Etu}/Mtu = \text{infinity (for Fig 5.8 right hand side)}$$

The reader should note that although the basic unit for duration (the Mtu) is virtually unchanged here, it is the Etu which has become so large that it is immeasurable. Our concept of duration therefore becomes non-existent and duration measurements impossible. In conventional terms we would of course incorrectly state that under these conditions Time was at a standstill. What we would really mean was that our duration measuring ability had ceased to operate and that Event-Time was at a standstill. It should be clearly understood that it is the ageing process for the proton that is at a standstill and that the elapsing of event moments still have accountability in the basic duration unit of the Mp cycle.

We would repeat once again that Time is a non-entity. It does not exist since it has neither property nor substance. It is therefore simply a concept by which we evaluate the elapsing moments of a physical / mass structure. This concept is now set up as a physical identity by the definition of a unit of time as the duration of the most basic mass structural sequence. As such we must consider the time factor not as a new dimension or duration entity but rather as a facility for announcing duration. Since our conception of time is grossly prone to a distortion of event times we must correct for this by defining each Etu in multiple Mtu terms. We can achieve this by evaluating the realm conditions at a given realm location as we did for the set of curves in Fig 5.6. However on this occasion we shall simply indicate the increase in the Etu with progressively slower rates of proton decay. Fig 5.9 shows a curve giving Etu values in terms of multiple Mtus for realm conditions between the TN and T0-lines.

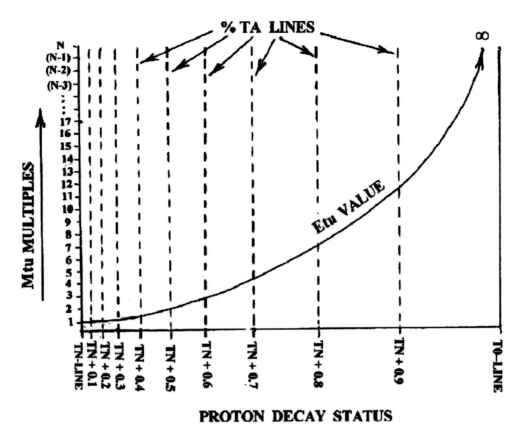

Fig 5.9

The proton decay status is indicated on the horizontal axis by a series of parallel TA-lines spaced similar to those in Fig 5.6. The Etu value increases in an exponential fashion and approaches infinity for the T0-line realm condition. Let us assume that Fig 5.9 is an accurate representation of the given parameters. Then for every TA-line condition we could determine the true quantity of Mtus that occur within the duration of each Etu. Subsequently, having first evaluated our TA-line position by the method described earlier, we could then override our concept of duration measurements and obtain a standard value for the elapsing of moments. This standard must remain on an Mtu basis for an overall realm application.

The reader must remember that Event-Time will always remain as our duration measurement parameter and that the Mtu is simply a theoretical base for the absolute correlation with other event moments. Event-Time is a variable feature by realm location and it is therefore essential that we inter-relate event moments across the realm according to some form of absolute measurement. Without this feature we could not expect to evaluate the true velocity of wave front propagation under differing realm EIL conditions nor could we accurately indicate the impulse that is caused in masses as a result of EIL gradient and E-Flow.

When we measure the velocity of wave front propagation we express the result as a distance traversed per Etu. Since this is a distorted result based upon our current concept of duration, then measurements conducted at different realm locations (under differing EIL conditions) would yield different results. We have already stated that wave front propagation has a velocity inversely proportional to the realm EIL (see chapter 3). It is considered that measurements by physical observation experiments would obscure this relationship by considerably inflating the measured velocity. In order to correct the results we must convert the measurement Etu to an absolute Mtu basis by using the data provided in Fig 5.9. Only then may the propagation velocities in differing EIL conditions be compared and a true relationship obtained.

It must be remembered also that the Mtu is not an absolute quantity but is marginally variable with EIL as shown in Fig 5.8. We could however consider applying a progressively increasing correction factor to the Mtu in order to achieve an absolute 'Time' quantity (in the EIL range from C_0 to C_1). Perhaps the reader would care to extend such speculation by an analysis of the practical applications in connection with the impulse due to gravity (Chapter 4). The marginal variation in the Mtu is bound to have an effect upon the total internal impulse quantity Ic within the proton (Fig.4.6) with subsequent effect upon the gravitational push (Fig 4.13) up an EIL gradient. Distortion in our concept of duration prevents us from achieving direct experimental proof of such hypothetical conclusions. Any attempt to do so must therefore be through proving indirect relationships and then applying correction factors presented by the data of Figs 5.8 and 5.9. Perhaps the resulting data could be used on a trial and error basis to achieve a more accurate estimate of our TA-line positioning.

Our chief concern in all of this is to know the relative ageing in protons at various realm locations and the subsequent evaluation of the actual elapse of 'Time' in terms of a standard unit. Since this basic unit has already been defined as the rate of cyclic events within the proton then we could justifiably refer to this system of duration measurement as Proton-Time.

There are many hypothetical cases that we could discuss but of very great importance to us is the Event-time variance that exists across the realm. Although most of the realm's current protons were formed at random event moments within the period of the circle 5 phase, their emergence into the circle 6 phase is very likely to have been very nearly simultaneous. As such we could assume that most protons started off their current decaying status at approximately the same optimum energy level (age). Their current age would of course depend upon several factors. Those that existed near the edge of the realm would have had marginal influence exerted upon their decay and so will possess a TA-line close to the optimum TN-line. Others more near the centre would have had a slower decay rate. Let us consider two such extreme positions and represent their TA-lines as in Fig 5.10.

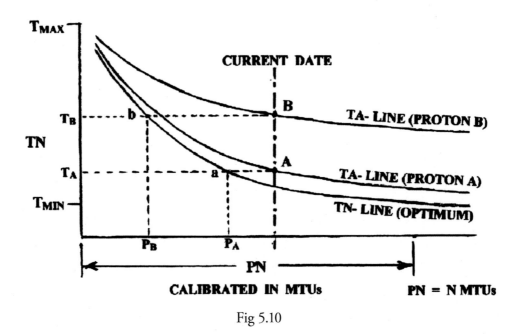

Fig 5.10

Earlier in connection with Fig 5.6 we mentioned that extension of the TA-lines beyond the PN position was not essential when evaluating age or decay duration. We meant of course that calibration of the period co-ordinate was unnecessary beyond the PN quantity. In Fig 5.10 let this co-ordinate be calibrated in Mtus such that,

$$PN = N \text{ Mtus.}$$

Consider the TA-lines for the proton A (realm edge) and the proton B (realm centre). The current date position for each is shown by the points A and B on the respective TA-lines. However this does not indicate the true aged condition of each proton. To determine this we must relate the A and B positions on the TA-lines to their respective energy medium content positions on the TN-line. This is achieved by projecting A and B horizontally across to their respective positions 'a' and 'b' on the TN-line. Earlier in the chapter we defined proton age by the quantity of energy medium it had lost since being at T_{Max}. As such the ages of protons A and B would be represented by the energy medium values T_A and T_B as shown. However now that we have evolved a theoretical time unit perhaps it would be more appropriate to express the age level of a proton in terms of equivalent elapsed Mtus. P_A and P_B in Fig 5.10 represent this equivalent age status for the protons A and B respectively on the PN co-ordinate. Since the equivalent age of a proton cannot exceed PN then clearly extension of this co-ordinate beyond the value for PN is unnecessary. Also in Fig 5.7 we could now more accurately compare the decay rates of protons through comparison of their decay periods in the expression,

$$\text{Decay rate proton A} \Big/ \text{Decay rate proton B} = {}^{\Delta PA}/\Delta P_B \text{ (using Mtus)}$$

In Fig 5.10 the equivalent age of each proton was indicated for a 'current date' position on the TA-line. We specify the current date as being the total number of elapsed Mtus since the proton commenced its decay in the circle 6 phase of the realm energy medium cycle. A more appropriate term for this dating feature would of course be 'Realm-Date' (abbreviated to R-D). However in order to have a consistent dating system throughout the realm we must permit the assumption that all protons did not commence their decay at the same instant. For our purpose, however, we shall assume realm dating as having commenced with the earliest decaying protons. Using a standard TN-line Mtu we may then ascribe positive or negative Realm-Dates to the events within the realm. Positive R-Ds would indicate events occurring after proton decay had commenced i.e. events within the circles 6 and 7 phases of the realm energy medium cycle. A negative R-D would concern those events that occurred prior to the commencement of proton decay i.e. in the circles 1 to 5 of the realm energy medium cycle. The principle of realm dating will need to be given much thought and careful analysis before it can actually be fully applied. We believe however that as space travellers traverse the realm, such a Time system would be invaluable as a true duration measurement tool.

Consider the case of a realm traveller leaving Earth and settling on a planet X near the centre of our galaxy where the realm EIL datum is at a higher level. Suppose that after a period of a few years (according to his atomic clock) he wished to return to Earth. What relative elapse of time will he find on Earth? Let us initially ignore the duration of the journey between the two planets. We may now simply concentrate upon the elapsing of Event-Time on planet X as compared to Earth.

Let us assume that protons on planet X are influenced by the realm datum to such an extent that their decay TA line coincides with the (TN + 0.700)-line. The TA-line for a point on the Earth has already been estimated as the (TN + 0.616)line. Initially the traveller is on Earth. At the Realm-Date T_1 he is magically transported to the planet X. His proton decay rate and concept of duration must now follow the TA-line for planet X. This is represented in Fig 5.11 by the line AC. At the Realm-Date T_2 the traveller is transported back to Earth. If the traveller and an earthbound companion were the same proton age at the R-D = T_1 and are represented by point A

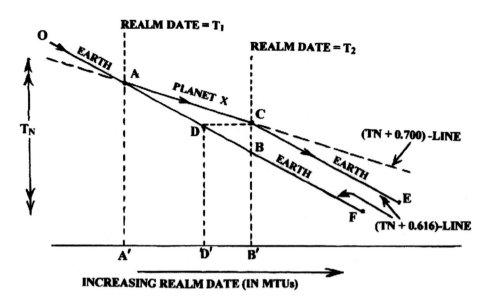

Fig 5.11

on the vertical TN ordinate, then their respective proton ages at R-D = T_2 are given by points C and B respectively. Thus when the traveller returns to Earth he will find that a greater number of years had elapsed on Earth than on Planet X. The equivalent age of the Traveller as compared to that of his companion would be represented by the position D' in relation to the position B' on the R-D co-ordinate. We could thus determine the true number of years that had elapsed on Earth over the Realm Date period from T_1 to T_2 from the following equation,

$$\frac{\text{Elapsed years on Planet x}}{\text{Elapsed years on Earth}} = \frac{AD'}{AB'}$$

The decay line for the Traveller is given by OACE while that for his companion is given by OABF. We shall discuss the effect of high velocity realm travel upon the decay characteristic of a proton in a later chapter.

Correlation between all the factors influencing the rate of proton decay, the setting up of TN and TA-lines, evaluating the age of protons and Realm-Dates extends into a large and complex study. This is too massive a subject to be given a fully detailed coverage at this stage and so we shall for the moment postpone any further analysis of this topic.

It is hoped that at some future date an exact science for the absolute grading of Time will develop. Perhaps the foregoing discussion and hypothetical Time values will cause a debating of the matter and so at least result in more thought being given to a virtually obscure science. Further development in the Laws of Time is therefore left for future generations when more 'clues' are uncovered.

• •

Chapter Summary.

1. Circle 6 phase protons undergo a continuous and uniform loss of energy medium. The rate of energy medium loss is inversely proportional to the realm EIL. This rate of change in the energy medium content of the proton may be used as a gauge for duration measurement.

2. The total decay duration of the proton is defined as its 'Life' and occurs between set levels of energy medium content. These are Tmax and Tmin. The total normal decay energy medium between these levels is denoted by TN where,

$$TN = Tmax - Tmin..$$

3. The age of a proton is usually indicated by its current energy medium state represented on the decay curve. It is the quantity of energy medium that has been lost in decay since the Tmax status. Age was also defined in terms of equivalent Mtus by representing it as the duration of decay to the present level if that decay had been at the optimum rate i.e. according to the absolute TN-line.

4. Time is to be considered as a non-entity because it has neither substance nor form. It simply does not exist as an entity. It is a relative concept whereby elapsing events may be compared in sequence and in duration to that of some other very basic physical event. A unit of duration or Time will therefore be defined as the duration of that basic event cycle. In the case of a simple pendulum the duration of each complete swing could be taken as the unit for Pendulum-Time. Evaluation of the duration of other pendulum events would then be determined in multiples of this Unit.

5. Units of Proton-Time are based upon the duration of the Mp-event cycle. The optimum decay curve indicates the actual number of Time Units i.e. Mtus that have elapsed since the Tmax state of the proton. When the decay rate decreases, our concept of duration remains with the optimum curve relationship (age to Mtus). As such our concept of event duration / Event-Time indicates to us fewer Mtus than have actually elapsed. Our concept of duration thus slows in line with the reducing rates of proton decay.

6. Event-Time as a distorted duration concept may be corrected and converted to absolute units of Proton-Time by means of a graph correlating Event-Time and actual elapsed Mtus under varying realm EIL and proton decay conditions.

7. A Realm-Date system may subsequently be established from the above to provide an accurate and consistent historical sequence for the physical events within and across the realm.

8. Protons of different ages may occur across our realm due to variation in rates of decay. Assuming that most protons entered the circle 6 phase at approximately the same instant and at Tmax status, then our Realm-Date and age relationships could indicate the pace of Event-Time in differing parts of the realm.

CHAPTER 6.

The Primary Atom

In the foregoing chapters of this text we presented the reader with a descriptive formation of the mass point, the Mp-congregation/proton and finally the Mp-shell. A special combination of events and conditions within the Realm led to these formations culminating in the Primary Atom structure. In this chapter we shall briefly review the general concept of mass structure formation before proceeding with a more detailed look at the primary atom and its behaviour in our realm EIL environment. The reader is reminded to be aware of all the other realm phenomena and conditions that are operative within the realm but which are not brought into focus when discussing a particular aspect. The characteristics of the activity within the realm are a complexity of interrelated behaviour. As such a complete understanding of a particular aspect can only be realised through an awareness of the many other events that may have a related effect upon it.

Consider the gravitational impulse phenomena. This requires a full understanding not only of the internal functioning of the Mp and of the proton but of proton decay, the realm datum grid and the set-up of EIL gradients. These in turn require a basic 'realm phasing' and '7 circle theory' knowledge for a proper appreciation. Quantification of the impulse cannot be accurately determined without first taking account of proton time at that location, - and so on and on. It would be nearly impossible to explain the aspect of any one of these phenomena without reference to another.

We assumed that realm phasing commenced at a location quite near to the Mother Realm boundary (Fig 2.3). The buildup of energy medium concentrations subsequently took place resulting in EILs of C_1 and above. At C_1 energy medium levels the energy medium properties altered marginally to become an energy/mass medium. The EILs were similar but the mass medium status was one in which it had evolved as an extremely stretch-strained entity. Since there was no way in which this status could be maintained the mass medium underwent rapid implosion. This results in the large mass medium block being split up into defined spheroid volumes each of which contain a set amount of energy/mass medium quantity (Fig 4.1). The spheroid volumes implode upon themselves causing each to contract in volume towards the spheroid centre point. This contraction occurs at a phenomenal rate and continues until mass of a certain optimum density is attained. This central mass entity is a Mass Point (Mp) and is the unit of mass in our energy medium theory.

The implosive action ends when optimum density is attained. However the linear momentum of the spheroid contraction continues inwards even after optimum mass achievement. Subsequently there is a brief but intense centrally directed impact action upon the Mp that compresses it to beyond optimum density. This sets up an opposing explosive action inside the Mp resulting in reversion of the Mp to energy medium in the form of an energy medium flow outwards (Fig 4.2). As such the Mp is an extremely short-lived mass medium entity. By comparing the energy medium state before and after the Mp event we note that the difference is in the momentum of that energy medium. The law of conservation of linear momentum was never violated since the resultant momentum was always zero. We must however note that the end result is an E-Flow emanating equally in all directions from a central point. Since this E-Flow is in a state of continual expansion its EIL will continually decrease. We have shown that this EIL decrease obeys the inverse square law for distance.

When large quantities of Mps revert to the energy medium status there will be a considerable quantity of energy medium in motion. These E-Flows will phase with one another causing even further buildup of realm EIL. Subsequently in the circle 4 phase of the realm energy medium curve the EIL approaches C_2. Individual Mps now form and sequence in great profusion and in closer proximity to one another. Their repetitive formation is at random yet also extremely rapid. The population density of Mps at any instant is also very high. It is probable that some locations will have a higher population density than others. It is at these locations that Mp-congregations evolve. Near the centre of these Mp-group locations Mps that formed would have absorbed energy medium from a more or less isotropic E-Flow situation. These Mps would thus acquire a stationary or zero linear momentum status. The Mps occurring adjacent to these stationary Mps would be considered as part of the same Mp group.

Now all Mps when they exchange to the energy medium do so in the form of an E-Flow. With Mps forming continuously and at total random all over the realm we may imagine a situation of intense E-Flows and phenomenal impulse action when those E-Flows were re-absorbed into other Mps. The E-Flow from one Mp* tends to get absorbed into the Mps that form in its surround. In an Mp-group situation we assume that external E-Flows cannot pass through the group location because eventually they will be wholly absorbed into the Mps making up that group. This means that very high intensity E-Flows enter these Mp-groups but that a similar high intensity E-Flow does not exit from the group boundary. At this stage the group boundary is a vague and indistinct zone but nevertheless of tremendous significance. E-Flows that are emitted from a group depend upon the group size and boundary Mp spacing. The groups' emitted E-Flows are at a much lesser intensity than those prevailing in the surrounding realm. The Mps in a group would thus come under action of a net inward E-Flow causing re-formed Mps to develop a linear momentum in the direction of each group centre. As a result these Mp groupings tend to become more compacted and so acquire a relatively small but distinct identity (Fig 4.6).

So long as the realm EIL remains high (near C_2), the E-Flows entering each tiny Mp-congregation would cause all adjacent Mps to acquire a net linear momentum towards that mass body. Since there is a profusion of Mp formation in this circle 4 phase of the realm energy medium cycle, then the Mp-congregation would buildup its Mp content extremely rapidly. It would seem that there ought to be no limit to the growth of a proton structure if the EIL remained sufficiently high. This is precisely the case and it is at this stage that proton giants are formed. These can attain a size approaching that of our entire solar system in volume. Proton giants would form in such large numbers that the realm EIL would diminish to well below C_2. Thereafter there would not be sufficient realm energy medium EILs to produce these giants again. In the circle 5 conditions the realm EILs are somewhat less and the Mp congregations that form are much smaller. This is when our current day protons were formed and maintained.

The Mp is an extremely short-lived particle of 10^{-43} secs in Planck time, and therefore all Mp-congregations must produce the sort of E-Flows shown in Fig 4.3, even at near C_2 realm EILs. The intensity of this outward E-Flow depends largely upon the curvature of the group's surface. The flatter the surface, i.e. the larger the congregation, the greater will be the outward E-Flow intensity. This means that at some critical Mp-congregation size the intensity of the E-Flows being expelled in a generally outward direction is equal to or marginally greater than the E-Flows being directed onto it from the Realm. In such a situation, all Mps forming close to its surface would in general acquire linear momentum in an away direction and so restrict further growth of the congregation. At this stage we would consider the proton to have attained its optimum size (or mass quantity) for that particular realm EIL. Therefore we may consider it as a general rule that the size of the proton today can be traced back and related directly to the realm EIL of the circle 5 phase at which it was formed.

Conditions for a proton in a realm with EILs ranging somewhere between C_1 and C_2 would be very different from those that we observe today. Physical laws would not respond as they do now. EIL oscillations would not propagate as wavefronts, since Mp centres could not maintain the necessary synchronism for their stepping action.. Although a proton giant exists at the heart of every galaxy today in the circle 6 phase of the realm, conditions within its C_1^+ surface event horizon zone would correspond to those of the earlier circle 5 energy medium phase. EIL oscillations would not occur therein and therefore a space traveller could not view its surface. However this does not mean that our physical laws have altered. It simply means that new conditions have been imposed causing different phenomena. The laws pertaining to linear momentum and Mp formation must however remain as the basic concepts that prevail under all realm conditions. By applying these correctly we could extend our physical laws to cater to the new conditions. For example, the impulse curves of Fig 4.24 would be quite different in the circle 4 or 5 phases. Since the proton would not be the sole source of E-Flows, the grid datum would not necessarily peak only at mass locations. Mp formation would still be prolific and as such also a source of E-Flows. As such there would be additional factors included in the gravitation impulse curves which would then operate under a completely different format. Since all of this would occur prior to the commencement of Proton-Time (as we know it today) then duration measurements would of course be impossible.

The large scale absorption of realm energy medium by the proton formations causes a gradual lowering of the realm EIL. At a stage in the circle 5 phase when vast numbers of optimum size protons have formed, the realm energy medium EIL diminishes in intensity to a level that must reduce to below C_1 and so to enter the circle 6 phase. Subsequently individual Mp formation will cease, though the proton entities will each remain as a stable mass entity. It is the prevalence of the proton in this sub C_1 EIL environment which is of the greatest relevance to the physical phenomena of today's world. Since the proton is merely a congregation of Mps then we may wonder at its survival in the present lower realm EIL conditions.

Consider a specific Mp well inside the proton. Although the general realm EIL is below C_1 the energy medium and mass medium exchanges inside the proton structure result in EILs there that are well above C_1. When a M-E/X occurs, the resulting E-Flows are absorbed into the adjacent E-M/X sequences. It is only at the surface of the proton that M-E/Xs occurring there allow energy medium to flow out to the realm. The E-Flow from a surface M-E/X is directed in three classified directions. Firstly, about half is directed back into the proton body. This imparts a linear momentum to adjacent Mps which then causes a general compaction of the proton structure. All Mps in a proton thus tend to possess an impulse towards its geometrical centre (Fig 4.3). Secondly, about half of the E-Flow is directed outward into the realm. This results in a net loss of energy medium by the proton and is classified as a gradual decay. It results not only in EIL gradients within the realm but also concentric peaks of energy medium around the proton called forcephoids. The decay characteristic is reasonably uniform and thus is useful as an automatic event duration measurement device. Thirdly, a marginal proportion of E-Flow is tangential to the proton surface resulting in a high intensity energy medium grid around the proton (Fig 4.6). This surface zone performs an important function in regulating the decay process. The net loss of proton energy medium is thus restricted to a small proportion of the basic Mp energy medium for each M-E/X at the proton surface. Energy medium flowing into this surface zone from other neighbouring protons helps to reduce the energy medium loss even further.

The decay of the proton results in a flow of energy medium away from the proton surface (Fig 1.15). Each M-E/X at the proton surface results in a defined quantity of E-Flow (AB) with an EIL that obeys the inverse square law with distance (Fig 1.16). Because of a fairly uniform distribution of M-E/Xs at the surface, the E-Flows tend to partially overlap causing EIL peaks to develop. The net result is a uniform E-Flow from all around the proton with EIL peaks at regular intervals (Fig 1.18). The series of spherical EIL peaks traverse outwards at E-Flow velocity and are termed Forcephoids. The spacing between forcephoids is dependant upon the Mp sequence cyclic duration and the velocity of E-Flow (Fig 2.15). It is this relationship that permits the Mp-shell not only to be initially set up around the proton but also to perpetuate itself through repetitive formation.

When a large number of protons occur adjacent to one another, the E-Flows and forcephoids that are emitted phase with each other and set up localised zones of high EIL. When a forcephoid traverses through such a zone it may acquire an EIL peak that exceeds C_1. Fig 2.17 represents such a situation. It is important to note that forcephoids do not originate from a particular source or at the surface of the proton. A forcephoid or traversing EIL peak only occurs at some distance from the proton surface when the E-Flow pattern from several adjacent M-E/Xs overlap. Now when a forcephoid peak achieves a C_1 EIL then Mps may form there. When the local realm EIL is enhanced by additional decay E-Flows from other adjacent protons then a spherical forcephoid peak exceeding C_1 may be achieved around that proton. Mps will form that are all in perfect synchronism and as such for a very brief instant a shell of Mps comes into existence around the proton. The net linear momentum of this brief mass shell is in the radially outward direction.

Since the Mps of this shell are in perfect synchronism by virtue of their simultaneous formation conditions, then their M-E/Xs will also continue in synchronism. This means that an E-Flow emanates from each Mp within the shell location. The E-Flow that propagates 'away' from the proton results in the formation of EIL oscillations discussed in chapter 3. The E-Flow that propagates 'towards' the proton phases with the next traversing forcephoid and causes EILs in excess of C_1 to be attained. Once again a Mp-shell surrounds the proton (Fig 2.18). The linear momentum of this shell is still outwards and when its M-E/Xs commence it will be at approximately the same location as the previous shell. This process of shell formation is repetitive and continuous in a realm location where the grid datum is reasonably high. Since the repetitive formation of this Mp-shell is very rapid it would appear to be a permanent feature around the proton. This status of a proton plus Mp-shell is structurally defined as the Primary Atom. For the moment we shall not concern ourselves with the phenomenon of atomic fusion as this becomes a complicated issue. Let us first discuss the separate functions of the single proton and its Mp-shell.

Consider the effect that the proton nucleus has upon the realm environment. Essentially, the proton is in continuous decay and therefore it is the main source of energy medium to the realm environment (in the circle 6 energy medium curve state). As a supplier of energy medium the proton must be looked upon as being at a high energy medium condition. Perhaps in comparison with other realm zones we may consider the proton to be a positive energy medium source. The notions of positive and negative energy medium functions are thus purely relative to whether energy medium emanates from or is absorbed into a particular mass particle or structure. In the case of the Mp-shell there is no E-Flow to the realm since energy medium is dissipated outwards in a compound series of EIL oscillations (Fig 3.3). Since the Mp-shell is in a continuously repetitive state, the E-M/Xs of the newly forming Mp-shell is the structure that perpetuates the atom. Thus the Mp-shell's repetitive structuring seems to absorb energy medium continuously. By definition we may therefore consider the Mp-shell as an energy medium absorber and performing a negative energy function within the realm.

When a proton exists without an Mp-shell then a very strong forcephoidal environment will be present and will influence other protons nearby. The Mp-shell however, by virtue of its repeated energy medium absorptions prevents any forcephoid from traversing out into the realm beyond the radius of the atom structure. Because we have defined the proton as a positive energy medium function then we would also have considered its forcephoidal effects within the environment as a positive effect. When the Mp-shell causes this effect to disappear (Fig 2.18) we conclude that it must have an opposite-to-proton effect upon the environment. We consider the Mp-shell to have counter-balanced the effects of the proton so that the Primary Atom (or PA) is an apparently neutral entity. This is an incorrect assumption based purely on the lack of forcephoids emanating from the PA. An E-Flow is still in existence beyond the sphere of the atom as are the EIL oscillations indicated by the shell compound frequency layout in Fig 3.36. We shall return to this concept of positive and negative classified behaviour a bit later on.

Let us consider a forcephoidal environment within the realm. This has got to be the result of free protons in existence around that location. There are two main reasons for the absence of a shell structure around a proton. Firstly the protons are widely separated from other protons across an expansive realm volume. Here, there is an insufficient concentration of phasing energy medium from the forcephoids of even the closest protons. As such forcephoidal peaks do not get to attain the uniformly concentric C_1 EIL around any proton as indicated for Fig 2.17 and so there is no Mp-shell formation. Since proton formation in the circles 4 and 5 phases of the realm energy medium cycle occurred across a very large realm volume, then there are bound to be vast quantities of these protons still in existence. It is considered that widely expansive clouds of these protons fill the realm environment such that much of the prevailing realm mass is in this state. We include proton giants in this setup though their function has been mainly in the creation of galaxies. Never the less, their contribution to realm environment is similar. The concentric forcephoids that then emanate outward from each such proton subsequently set up a very complex and extensive forcephoidal environment in that part of the realm. The pattern of forcephoidal peaks that develop within this environment may produce some interesting impulse effects on other protons (and atoms which happen to come by).

The second reason for the absence of the Mp-shell around protons is when these are packed together so tightly within an atomic grouping that a continuous repetition of the Mp-shell is not achieved for some structures. The reason behind this occurrence is that the EIL within the compacted atomic grouping has localised high spots far in excess of C_1 and synchronism among Mps in the same Mp-shell is lost. If Mps reform without perfect synchronism then there is no continuous shell and forcephoids will emanate beyond it. In the more complex atomic structures of solid objects this level of energy medium concentration is always present inside the mass to some degree and as such forcephoids will be emitted. The pattern of complex forcephoidal peaks that occur around such a mass structure usually conform to some approximate internal structural pattern. Approximate because the proportion of missing shells is very small compared to the overall atom quantities therein. In the more complex atoms we have not only a complex nucleus of a large proton grouping but also a complexity of multiple Mp-shell structures with variable individual mass content. Another source of shell-less protons is when a large quantity of PAs are compacted through gravitational impulse. The mass of such a compact body becomes so great that the EIL near its central regions exceeds C_1 by a large factor. Synchronism between Mps is impossible and no Mp-shell can exist. It is in these regions that the more complex atoms are manufactured. At this stage we are however only concerned with the fact that forcephoids emanate out into the realm producing a forcephoidal field environment of some intensity and considerable complexity.

Forcephoids emanating from a proton obey the inverse square law for EIL with distance. Thus forcephoids would still be monitored as nominal EIL peaks at considerable distances from the source. Forcephoids may traverse in the same general direction from different sources and as such may result in higher EIL peaks where phasing /

intersecting occurs. Forcephoids traverse at E-Flow velocity and as such may possess a variable velocity through the realm, dependant upon the linear momentum of the proton from which it originated (Fig 1.14 and Fig 4.29).

In the case of widely spaced protons the forcephoidal field would be extensive but with very low EIL peaks. A space traveller could venture through such a zone without noticing its effects. In the case of the compact mass body, the forcephoids originate from the same local area and phasing

would produce very high EIL peaks close to it but which diminish rapidly with distance. In our world today the effects that we tend to observe are only those caused by the very high EIL forcephoids. There is one other characteristic in forcephoidal fields from mass objects that we must resolve. Since the pattern of forcephoidal peaks does correspond somewhat to the general internal atomic structure of that mass then another mass object upon entering such a field could find its own structure induced to respond by creating its own internal free proton pattern. It must be apparent to the reader that forcephoids are indicated here as being the cause behind magnetic type effects. We shall be discussing that in the next chapter and it is the purpose here to simply lay the foundations for such a discussion.

Let us now consider the actual effect that forcephoids have upon mass structures. Consider the proton in Fig 6.1. The forcephoid source is on the left hand side of the proton. The forcephoids traverse towards side A of the proton

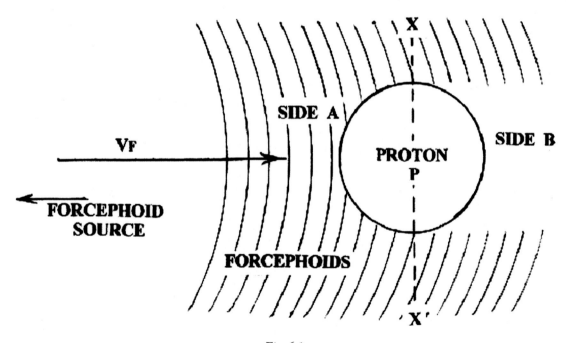

Fig 6.1.

and are absorbed in the E-M/Xs there. In chapter 2 Fig 2.16 we indicated that the proton received an impulse in a direction opposite to that of the forcephoidal traverse. At that point we simply mentioned the principle of the proton being induced with an impulse that was up an EIL gradient. Our representation of forcephoid and proton was not to any relative scale and was presented as a visual concept only. We shall now continue our presentation of that concept and assume the width of the forcephoid energy medium peak to be small in comparison to the diameter of the proton. In Fig 6.1 each forcephoidal peak is represented by a curved plane traversing from left to right with a realm velocity V_F. As each forcephoidal section phases with the proton surface EIL Grid (Fig 4.6) and then with the proton body, the Mps that sequence there have a high probability of an E-M/X taking place in a forcephoidal peak zone. These Mps will thus possess a cyclic rate that is marginally slower than if there were no forcephoids. As the forcephoid traverses into the body of the proton all of its peak energy medium gets progressively absorbed into E-M/Xs. A forcephoid is thus unable to traverse through a proton body. As such, in the diagram it is the side A of the proton that is influenced by the forcephoids. Side B experiences none of the forcephoids that influence side A and it is this fact that sets up a large EIL differential between their surface Mps. The situation is represented in Fig 4.8 from which we know that there is a higher frequency of Mp cyclic events on the side B of the proton. The centrally directed internal impulses are no longer evenly balanced and cause the proton to acquire a net impulse in the direction from side B to side A, i.e. towards the forcephoidal source. It is interesting to note that the behaviour of the proton in this instance is similar to that of gravitational impulse discussed in Chapter 4. There is however one important difference. In the realm grid datum the EIL differential on opposite sides of the proton is normally very small. The resultant effect produces a very small impulse in a direction up the EIL gradient. Of course, the greater the EIL gradient the greater is the gravitational impulse. However in the case of forcephoids the equivalent distance EIL differential is always the forcephoidal peak EIL. The resulting forcephoidal impulse within the proton is thus correspondingly large and is many times greater than its gravitational counterpart albeit in the same direction.

Let us now consider the effect of forcephoids upon an Mp-shell section AB as represented in the Fig 6.2. We have shown the Mp-shell section as an independent identity divorced from the PA structure simply to explain its independent reactions. We shall see (much later) that such a section does in fact occur for a limited duration under certain conditions. Since an Mp-shell, like a wave front, cannot remain stationary let us impart to it a specific

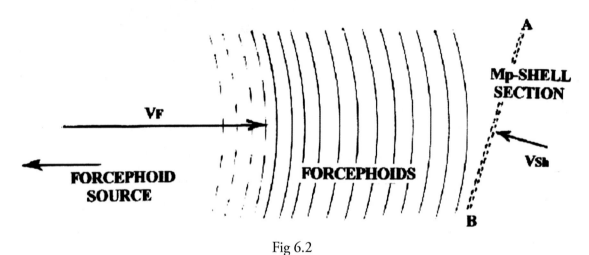

Fig 6.2

velocity Vsh as shown. Also consider that the Mp-shell achieves this velocity by a quantum step type action similar to the propagation of a wavefront. The Mps must continually reform at a uniformly advanced position in order that the Mp-shell energy medium is not dissipated. The propagation of the Mp-shell therefore implies that the Mp sequence is in operation all along the traverse path and as such the E-M/Xs absorb any energy medium quantities that are in its sphere of action. This means that linear momentum of that energy medium quantity is also absorbed.

In Fig 6.2 the forcephoids and Mp-shell section have velocity components that are in opposite directions and it is clear that they must phase with one another. As such linear momentum of the forcephoidal energy medium is absorbed in the E-M/Xs of those Mps in the shell section that sequence within a forcephoid peak. Since the forcephoids trail one another at regular intervals then as phasing continues more and more of the Mps in the shell must acquire linear momentum in the direction of VF . The resultant traverse path of the Mp-shell section is such that it has a continually changing direction that tends to follow the path of the forcephoids. The Mp-shell thus acquires an impulse in the same direction as the forcephoidal velocity VF. The volumetric shape of the Mp-shell section is also affected by the absorption of linear momentum and becomes orientated in a decided manner with respect to the forcephoids. We shall discuss this aspect in a later section when we shall also assume the Mp-shell section to be split up into even smaller sections or shell units abbreviated SU.

We may summarise the differing responses of the proton and the Mp-shell within a forcephoidal field by stating that a proton experiences an impulse towards the forcephoid source whereas the Mp-shell experiences an impulse in the opposite direction. This is shown up in Fig 6.3 for a proton and Mp-shell section traversing initially at right angles to the field direction VF.

In the diagram we have indicated a greater deviation in the shell path as compared to that of the proton. This is because of the difference in their masses. However there are also differences in their velocities, the proton velocity being much slower, which we have not considered here since the aim was to simply illustrate a theoretical concept. We also introduced the concept of the Forcephoidal Field direction. In essence this is normally the direction of the traversing forcephoids. In our diagram we have shown a simple forcephoidal field which arises from a single proton. In the realm however there are a countless number of protons giving out such a field. When these overlap one another a very complex forcephoidal field comes into existence with some very interesting effects. We shall consider these in a separate section.

Earlier we mentioned how we considered that the proton had a positive energy effect upon the realm energy medium quantity while the Mp-shell had a continuously absorbing or negative effect. Subsequently in Fig 6.3 we may consider the proton as acquiring a positive impulse within the forcephoidal field, while the shell section acquires a negative impulse.

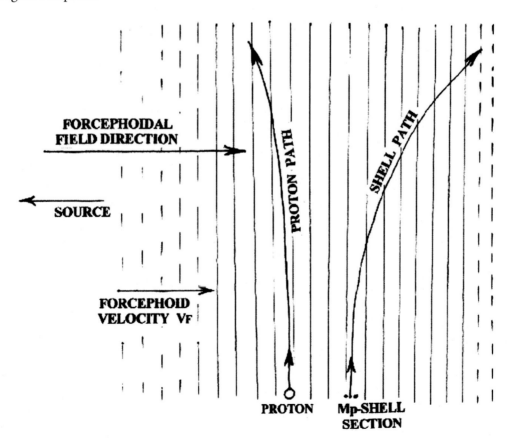

Fig 6.3

Let us now consider the interaction between the proton and its shell surround. It is of course most important that a clear mental picture of the full sequence of Mp-shell events be presented before pursuing such a discussion. Fig 6.4 helps to explain this. Here we represent successive forcephoids Fn, Fn+1, Fn+2, etc. traversing at the velocity VF outward from the proton position. Under certain conditions of realm EIL as explained in Fig 2.17 for proximity protons, a Mp-shell is formed around the proton at a set distance position. It is the aim here to highlight this sequence of events and to show precisely the nature of repetitive shell formation.

We shall assume the pre-condition that Mp-shells are already in sequential formation around the proton and that one such shell has just completed its overall event sequence. In Fig 6.4(a) we assume that the forcephoid Fn has just reached the location A at which it phases with the E-Flow from the M-E/X of the preceding shell. This energy medium phasing results in an EIL peak in excess of C_1 causing an E-M/X sequence to be initiated therein. When this E-M/X commences, the energy medium within the sphere of influence undergoes a subtle conversion into a mass medium. This new entity contains the resultant linear momentum of all the previous energy medium quantities absorbed. Since the E-Flow from the previous shell to the forcephoid Fn location A is opposite to the forcephoid velocity VF then it stands to reason that the resultant velocity of the mass medium entity must be somewhat less than VF. We can now assume that the forcephoidal energy medium is the second greatest contributor to the E-M/X quantity after the prevailing realm datum EIL, and so may consider that the resultant velocity of the mass medium is correspondingly influenced towards VF. Let this velocity be represented by Vs.

Fig 6.4(b) gives an enlargement of the shell sequence at (a) and commences with Fn at the plane position A. There are four event instants indicated to represent the start and end of each Mp phase.

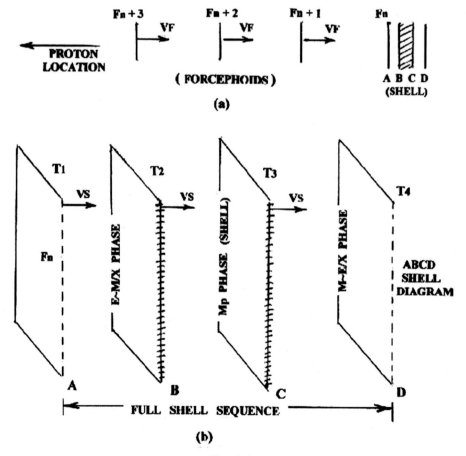

Fig 6.4

Each event instant is differentiated from the others by the elapsing MTUs of event duration as described for Proton Time in Chapter 5. Let the forcephoid Fn arrive at the plane position A at the event instant T1. At this instant the forcephoid energy medium quantity along with the phasing E-Flow section becomes an integrated mass medium quantity. As stated above this mass medium in the E-M/X phase of the Mp-sequence now has a resultant velocity Vs. The duration of the E-M/X event phase results in a traverse of this mass medium from the plane position A to plane position B. The completion of the E-M/X event phase means that all of the mass medium has sectionalized and imploded to form a Mp (Fig 4.2a) at each sphere of influence centre. All of these Mps sequence in perfect synchronism which thus sets up an instantaneous shell of Mps at the indicated plane position B at the event instant T2. From this we have the duration of the E-M/X as (T2 - T1). We shall observe later that depending upon the EIL conditions around the Primary Atom, the Mp sequence can alter around the shell such that shell position around the proton is not wholly uniform. As such an oval or distorted shell may occur. We can confirm that such localised differences in shell sequence are not uncommon and that the subsequent effects are testable.

The second event phase is the actual duration that the whole of the Mp entity remains in existence. As stated earlier (Fig 4.2) this is the period commencing when the mass medium in the implosive sequence attains optimum density. Although this implosive action ends when this optimum mass density is attained, nevertheless the linear momentum of the spheroid contraction must cause a brief but intense centrally directed impact action. This compresses the Mp entity to a condition well in excess of the optimum density mass and so sets the scene for an opposite reactive explosive action. This then causes the successive imaginary layers of mass to be expelled outwards progressively as an E-Flow. The second event phase is to be considered at an end with the commencement of this E-Flow. In Fig 6.4(b) this is represented at the event instant T3 at the indicated plane position C. We may thus assume that the life of the completed Mp is (T3 - T2) and that within this period the shell traverses the distance BC at the velocity Vs. As mentioned earlier the Mp life is considered to be 10^{-43} secs in Planck time.

The third or M-E/X event phase commences when the imaginary outermost material layer of the Mp is expelled as an outward E-Flow. This sequence of theoretical layer by layer emission continues until the entire Mp has reverted completely to the energy medium state. The main effect of this third event phase is the production of a uniform E-Flow in all directions from the location of the Mp* centre. The velocity of this E-Flow is constant relative to the Mp* and is the same as the forcephoid velocity VF relative to the proton. The duration of this M-E/X event phase is represented by (T4 - T3) on the diagram and we observe the diminishing mass of the shell traversing the distance CD.

The full shell sequence has a duration of one MTU (Mass Time Unit as defined in Chapter 5) and the mass quantities involved have traversed the physical distance AD. This diagrammatic view of the full shell sequence, Fig 6.4(b), is termed the 'ABCD shell diagram' and is an important concept in understanding shell behaviour and its response to external influence.

The ABCD shell diagram parameters are marginally variable with realm EIL. The greater the local EIL then the greater will be the overall duration T4 -T1 and subsequently the distance AD. The reason being that with more energy medium absorption all phases have a marginally greater duration. However there are other aspects that produce a reduction in the traverse distance AD albeit with the same overall duration which we shall discuss later. For now we are concerned with the formation of the next Mp-shell.

In the foregoing paragraphs we explained the sequence of events in forming an Mp-shell from the forcephoid Fn. It must be remembered that the forcephoidal peak is wholly absorbed into the E-M/X initiated and subsequently the forcephoid itself no longer exists. Now consider the forcephoid F(n + 1) following behind Fn and at the same traverse velocity VF. Let us relate in terms of notation each shell with its forcephoid progenitor. Thus we had the shell Sn forming from the forcephoid Fn. The next shell S(n + l) will involve forcephoid F(n + l), and so on and on. Now at the event instant T3 for shell Sn the third M~EX event phase commences. An E-Flow occurs in all directions from each Mp within the decaying shell Sn which continues upto the event instant T4. We have already dealt with the effects of the outward traversing E-Flow (Chapter 3) and so will currently concentrate our attention upon the E-Flow traversing inward towards the approaching forcephoid F(n + l). Since the shell Sn traverses with a velocity Vs then the E-Flow velocity from Sn is given by,

$$\text{VE-FLOW} = \text{VF} \pm \text{Vs} \quad \text{(Fig 4.29)}.$$

Subsequently the E-Flow emanating within the zone CD traverses back towards the plane A position at a velocity VF −Vs. Since Vs is the resultant of the energy medium linear momentum absorbed at T1 and since the forcephoid Fn supplied much of the energy medium absorbed, then Vs must be a proportion of VF. Hence VF −Vs is the velocity of the E-Flow towards the former plane A position.

In Fig 6.4(a) the distance between forcephoids is large in relation to the span of the ABCD shell diagram. At the event instant T3 consider that the forcephoid F(n+l) has not reached the former plane A position. Between the event instants T3 and T4 the E-Flow from shell Sn traverses towards the position A. Now consider that the arrival of the forcephoid F(n+l) at this position coincides with that of the early E-Flow from shell Sn such that once again the forcephoid EIL peak exceeds C_1. An E~M/X commences and the events of shell Sn are repeated for shell S(n+l). The same applies for the shell S(n+2), and so on and on indefinitely so long as the realm EIL remains high. Although we have become more aware of the formative aspects of the Mp-shell there is a great deal more to consider especially with regard to the inward E-Flow from the shell.

Let us now consider the nature of this E-Flow towards the proton centre. We know that this E-Flow emanates from each Mp-shell and is therefore discontinued at the end of each M~E/X event. We also know that it's relative velocity is less than the forcephoid velocity VF. Subsequently when the energy medium phases at the position A much of the E-Flow remains unabsorbed due to its late arrival at the sphere of influence. Thus if only a part of the entire E-Flow from Sn is absorbed into the next shell F(n+l) then the remaining E-Flow must continue its traverse towards the proton. If we assume that it is the leading portion of the E-Flow that is absorbed - since this is the first to phase with the forcephoid F(n+l) - then it is the later tail-end of the E-Flow that continues on. If we consider the entire E-Flow from a Mp* to be wholly uniform then it really does not matter where the energy medium absorption occurs. There is however a major difference between the E-Flow from a Mp-shell and that from a proton under normal conditions. The E-Flow velocity from a proton does not vary greatly and is given by (VF + Vp). The proton velocity Vp is usually quite small. The total E-Flow from each M~E/X therefore spans a distance that is the product of VF and the duration of the M~E/X. Consequently this energy medium quantity is spread over a very large volume. The outward traverse Vs velocity of the Mp-shell is large when compared to that of the proton. This means that the outward E-Flow velocity from the shell is (VF +Vs) and the inward E-Flow velocity is (VF - Vs). Now we know that (VF - Vs) is small compared to VF and in Fig 6.5 we make a comparison between the normal E-Flow emitted outwards from a proton surface Mp and that emitted inwards from a Mp* in the Mp-shell. It will be noticed that for each case we have indicated a representative E-Flow section for energy

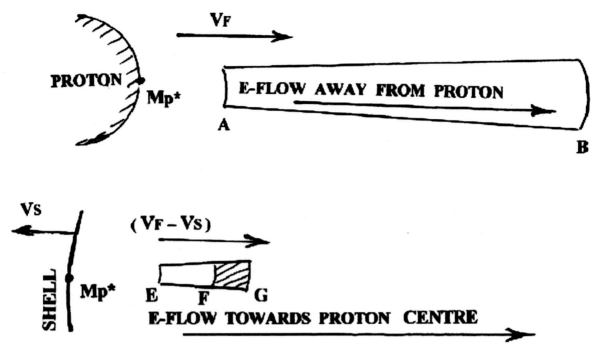

Fig 6.5.

medium emitted over an entire M-E/X duration. Therefore each section may be assumed to contain the same quantity of energy medium. If all velocities are assumed relative to the proton then the E-Flow from the proton is represented by the section AB traversing away from the proton at the velocity VF. The velocity of the inward E-Flow from the Mp-shell is only (VF - Vs), and the section of energy medium emitted is represented by EG. Clearly if each section represents the same energy medium quantity then the energy medium in EG will be at a much higher EIL than that in AB. Thus we have an E-Flow of a higher than normal EIL traversing between the Mp-shell and central proton at a velocity considerably less than VF. This high EIL section has its lead portion absorbed into the E-M/X of the next shell as stated earlier. The remaining section EF will continue its traverse and under certain very rare conditions may undergo a further depletion through absorption into a second E-M/X sequence with the next trailing forcephoid. Under these special conditions therefore it would be possible for more than a single shell to exist around a proton structured nucleus. However this is unusual and not the case for the PA with its single proton nucleus and therefore we shall leave such a discussion for now.

In chapter 4 on gravitation impulses we indicated how a proton reacts within a zone of EIL gradient i.e. it acquires a net impulse up that gradient (Fig 4.13). Now although this attraction impulse obeys a proportionate rule, it does so only under a certain set of EIL conditions. When the realm EIL in the zone of action approaches very high levels (albeit below C_l), then the proportionality rule alters for the Mp-cyclic rate to realm EIL as shown in Fig 4.21. Subsequently the impulse of attraction does not respond as much as before to new EIL gradient increases. To the contrary, Fig 4.22 shows a decided reduction in this impulse quantity. When the E-Flows contain such high EILs then the factors indicated in Fig 4.24 apply and a proton within the 'X' zone of that diagram must succumb to a net impulse in the direction of the E-Flow from the other source.

Let us consider how all of this relates to the primary atom. Under normal conditions of equilibrium the proton is at the geometrical centre of the Mp-shell sphere. Since the shell consists of Mps that are in complete cyclic synchronism then their inward E-Flow sections propagate towards the proton at equal velocities. This means that the E-Flow from every side of the PA reaches the proton surface at exactly the same instant.

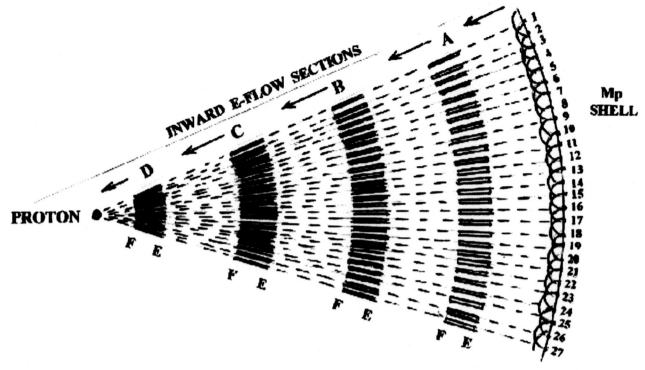

Fig 6.6

In Fig 6.6. we have highlighted the section EF of the E-Flow from the shell Mps (1,2,3,4,5,.....) that proceeds directly towards the proton centre. Because of their relatively slow traverse, E-Flow sections from earlier shells are also depicted en route along the same path to the proton. The reader may note how these sections get closer together on approach to their proton focus thus resulting in a considerable energy medium EIL buildup. The effect is the same as that depicted for protons in close proximity to one another (Fig 4.24). The concentration of E-Flow from each proton is such as to cause them to acquire a repulsion impulse away from each other - i.e. in the direction of the other's E-Flow. Subsequently, in the present case we may conclude that the conditions of E-Flow approaching the proton are similar and that the proton responds to this Mp-shell E-Flow by acquiring a net impulse in the direction of that E-Flow. Since the proton is at the centre point of the shell sphere then this impulse is equal in all directions and therefore the resultant impulse is zero. Should the proton be marginally out of centre then an imbalance in the acquired impulses causes a resultant impulse to push the proton towards a centralised position. A brief description of this effect was given in chapter 4 (Fig 4.32).

In Fig 6.6 it will be noticed that the inward E-Flow from the shell to proton centre occurs at a regular interval. Subsequently the proton acquires an impulse only when this E-Flow encounters the proton. There are thus intervals between these E-Flow encounters when there is no centralising effect and the proton is subject to the normal gravitational impulse for the prevailing realm EIL grid datum gradient (Fig 4.17).

Early in this chapter we stated that it would be nearly impossible to explain the aspects of any one phenomena without some interrelated effects upon others. It is thus that we must state here the relevance of the proton centralising effect in the quantification of the gravitational impulse of chapter 4. When the PA is within a realm grid datum with an EIL gradient then the shells conform to a varying ABCD diagram according to the realm EIL at which they occur. In Fig 6.7 we show the PA in an EIL gradient zone such that the EIL is marginally higher on the right hand side as indicated by the Realm Grid lines.

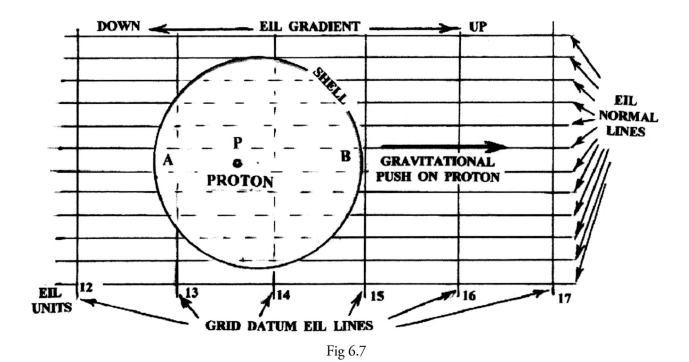

Fig 6.7

The Mp-shell will thus have an ABCD diagram with a marginally greater duration at site B as compared to site A. This will cause the shell at the B side to re-form consistently at a position further out from the proton than the shell at A. This means that the proton is in a non-central position with respect to its Mp-shell surround and as such will experience a centralising action pushing it away from A and towards B. This is in the same direction as the previously described (Fig 4.13) gravitational push up the EIL gradient.

Thus we see that the gravitational impulse that we observed earlier consists of an additional factor. It must however be remembered that while the centralising impulse is active, the proton is inside a very high EIL zone such that the grid datum gradient appears non-existent. It would appear that our earlier gravitational impulse action applies only in those intervals when the inward E-Flow sections EF are not impinging upon the proton. However since both of these effects complement one another, the proton apparently experiences a continuous impulse in the afore stated direction and there is no logical reason to alter the profile of the attraction curves of Fig 4.24.

Let us consider the transfer of linear momentum between colliding PAs. Under normal conditions of contact or collision, it is the Mp-shell that primarily undergoes physical distortion. At the instant of such a distortion of the Mp-shell we know that there are in fact several sections of inward E-Flow (EF) from earlier undistorted shells already en route to the proton. The effect of a collision is therefore not instantly conveyed to the proton but must wait until the E-Flow from a distorted shell gets there. This delay to the effectiveness of the centralising action contributes to the inertness of the PA which we shall discuss in a later section. For the present let us consider the manner of shell distortion when apparently impacted by another PA. We shall assume the velocity of relative movement of the PAs towards impact is small in relation to the rapidity of the ABCD diagram characteristics. The intention here is to indicate that impulse is transferred from one shell to the other not by the violent impact of Mp against Mp but by absorption of linear momentum from each other's E-Flows into their successive Mp-shells. In Fig 6.8 we illustrate the stages of such a transfer.

Consider the two primary atoms A and B such that the proton PA is at absolute rest within the realm while the proton PB has a velocity V directly towards PA. In the events sequence of Fig 6.8 we shall however be mainly considering the actions of their Mp-shells. Let us represent these as shell A and shell B. Alongside each shell is also given the corresponding ABCD shell diagram which indicates the variance in shell positioning when external influences are prevalent.

At Fig 6.8(a) the shells A and B are still a distance apart and we assume a negligible influence of one upon the other. Shell B traverses towards shell A. The ABCD shell diagrams for both shells may be taken to be wholly normal i.e. no external influence. At (b) the shells make theoretical positional contact. It is important to note that the Mp sequences of both shells are not required to be in the same event phase for the outward E-Flows of one to be absorbed into the next shell formation of the other. As such there is no real contact between the shells although the proximity effect from their E-Flows is equivalent to a full contact.

The E-Flow from A has the velocity VF towards B. Since B already has a velocity V towards A the relative E-Flow velocity that B experiences from A is (V + VF). Through similar reasoning the E-Flow velocity from B towards A is also (V + VF). This means that both shells exert the same external influences upon each other and therefore the ABCD shell diagrams for both are affected

by the same amounts at the XX contact location. The ABCD diagrams are forced to event with a reduced spacing and at a successively inward position for the start of each sequence i.e. position A in Fig 6.4 occurs marginally closer-in to the atom centre. By this means progressive distortion of each shell takes place until a limit is attained as indicated at Fig 6.8(c) when the distorted shell results in the proton being in an off-centre position and the centralising action just taking effect. The proton PA now acquires a net impulse in the direction away from B. Similarly proton PB also acquires a net impulse in the direction away from A. Each primary atom thus has the tendency to accelerate in a direction away from the other. So long as the absolute velocity of B towards A is greater than that of A away from B then the situation represented at (c) will continue. Eventually however the continued acceleration of proton PA and deceleration of proton PB must result in velocities that cause the distance between them to increase. This is represented at Fig 6.8(d). It will be noted that the shells are no longer in theoretical contact and subsequently there is no interrelated influence upon their shells. The ABCD shell diagrams are thus once again in their normal state. Since the primary atoms A and B are of equal masses we have indicated the end result as both atoms traversing in the same direction albeit at half the initial velocity of atom B.

Now we know that the shells A and B have separated from a contact situation. This indicates that at some stage the velocity of A must have been greater than the velocity of B for this to happen. There is apparently more to the collision aspect between primary atoms than that which we have discussed. When the shells are in contact action and cause distortion in one another, the centralising action upon the proton does not commence until the respective inward E-Flow sections reach the proton. This delay in impulse effect also applies when the shells cease to be in contact. The centralising action continues to be effective until all of the inward E-Flow sections from previously distorted shells complete their traverse to the proton. When a proton is under acceleration then each shell formation would be from forcephoids emanating from an earlier proton position. Thus at any instant the overall shell velocity tends to lag behind that of its accelerating proton. This automatically places the proton in a non-central location with respect to the shell at any instant and even more so by the time the inward E-Flows have traversed back. The proton subsequently experiences a centralising action opposite to the accelerating impulse. The greater the accelerating impulse then the greater will be the centralising action resistance. We shall include this phenomenon in our later discussion on the inertia of masses.

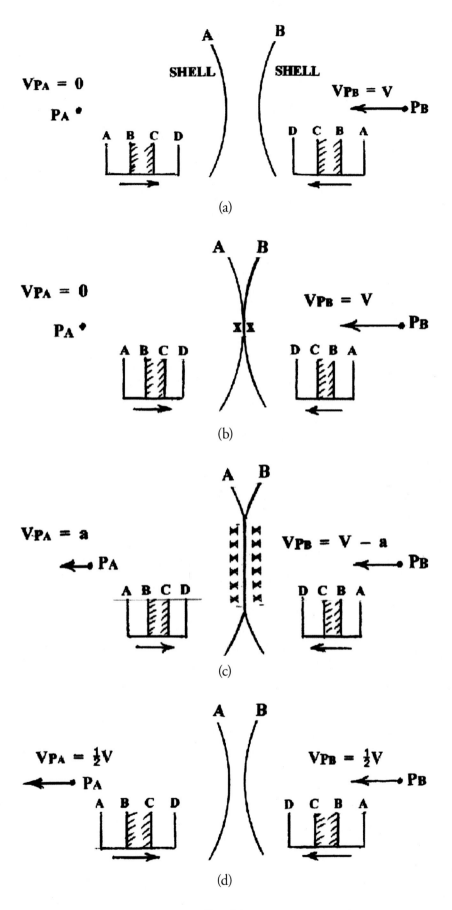

Fig 6.8

In our example of Fig 6.8 the transfer of linear momentum from B to A must continue until the shell A is free of any influence from shell B. Thus A will continue to receive impulse even after it has achieved the velocity ▱ V and shell B will similarly continue to decelerate to a velocity marginally less than ▱ V. Shells A and B will subsequently move apart. The contact zones of the shells now revert to a normal ABCD shell diagram configuration but with the proton still under the former acceleration. The opposing centralising action does not have a delayed effect since the inward E-Flow sections are already arriving at the protons altering location. This results in the rapid deceleration of proton PA back to the velocity ▱ V. In a similar manner Proton PB also reduces its loss of velocity and settles at ▱ V. We observe that the initial linear momentum of atom B has been redistributed equally between the colliding atoms.

This principle of linear momentum transfer between adjacent Mp-shells may be applied to grosser bodies with some additional allowances for the delays caused through progressive transfer of momentum from atom to atom with that structure. The greater the quantity of atoms within a body then the greater is the lag in achieving its acceleration. In cases where shell distortion cannot be achieved through the mass thickness then the body will not accelerate. We may conclude therefore that colliding bodies transfer their linear momentum through a process of progressive energy medium absorption in their shell structure and not from any direct mass to mass contact. This conclusion does not include free protons which in the absence of a shell surround are able to collide

physically with other free protons as represented in Fig 4.28. Also to be considered on a separate basis are conditions of very high velocity for both shell and proton which will be discussed fully in chapter 8.

Let us now consider the effect of the simple forcephoidal field (as represented for Fig 6.3) upon the traverse path of a primary atom. We have already indicated the responses of a proton and a Mp-shell section in such a field and we must now determine their response as an integrated atom structure. In Fig 6.9 let us assume that the primary atom is an ideal one with a perfectly spherical shell around a central proton. Let there be no external realm influence upon this atom structure. It is immaterial whether we consider it to be at rest or moving with a uniform velocity V. Now let us

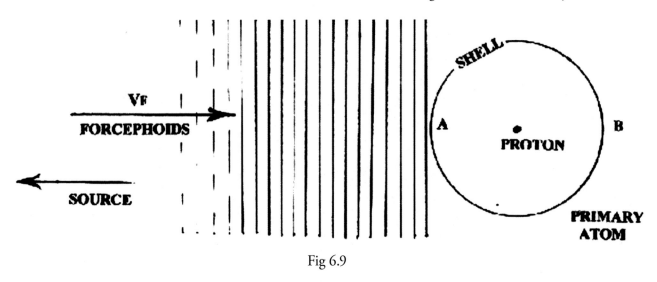

Fig 6.9

permit forcephoids from a single proton source to approach this atom from the left hand side at the usual forcephoid velocity VF. The atom shell will of course absorb energy medium from the forcephoidal peaks whenever these coincide with the E-M/X sequence. The net result will be a marginal shift of the entire Mp-shell away from the forcephoid source i.e. towards the right hand side. This places the proton in an off-centre location and as such the proton must receive an impulse in the direction from A to B due to the imbalance of the inward E-Flow sections. It must be remembered that forcephoids are not extensively absorbed by the Mp-shell and therefore they will propagate through the shell zone and on towards the central proton. The effect upon the proton is to cause an impulse in the direction from B to A whenever the inward E-Flow sections are not impinging upon it. These impulses appear to balance each other since the observed resultant impulse is zero. We now have a primary atom with an off-centre proton that is in perfect equilibrium within the realm. Let us now examine the actual behaviour of the shell in this situation in a bit more detail. The fact that a primary atom can remain in perfect equilibrium with an off-centre proton simply implies a greater complexity in the structuring and actions between shell and proton.

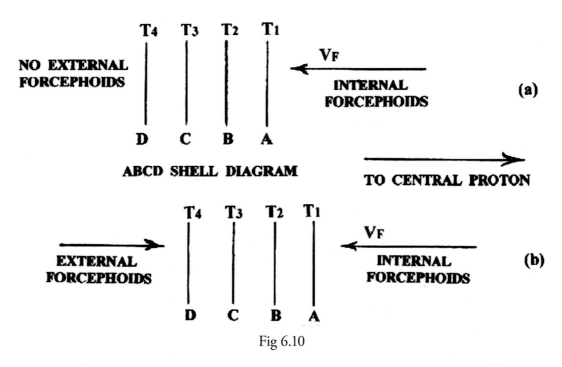

Fig 6.10

The ABCD shell diagram at A is shifted marginally towards the proton. Since T1 (Fig 6.4) commences at a location closer-in to the proton then the E-M/X also contains energy medium from a forcephoid at a marginally higher EIL. In Fig 6.10 we show the relative positions of shell before and after being acted upon by forcephoids. The situation at Fig 6.10(a) is before the action of external forcephoids. The ABCD shell diagram is considered normal with a shell velocity Vs in the outward direction. Let the proportion of forcephoidal energy medium in the shell make-up be X%. Now at Fig 6.10(b) the external forcephoid adds some of its energy medium to the E-M/X causing the ABCD shell diagram to shift closer to the central proton P. This in turn causes the E-M/X at Tl to commence in a region where the internal forcephoids are at a progressively higher EIL. A state of balance is very soon attained in which the energy medium constituents are in the same proportion as the previous Fig 6.10 (a) situation. This means that although the Mp-shell at Fig 6.10(b) contains a greater quantity of energy medium than at Fig 6.10(a) the proportion of internal forcephoidal energy medium remains unchanged at X% of the whole. This also means that the shell outward velocity remains unchanged at Vs. There is one major difference in the ABCD shell diagrams of the above two cases. Due to the greater energy medium quantity involved at (b) it stands to reason that the ABCD shell duration at (b) is greater than that at (a). Thus there would appear to be a lesser difference in the relative positions of the event instant T4 than that of Tl. This description of events may now also be applied to the collision aspects of the two atoms of Fig 6.8.

So far we have assumed that the internal forcephoids emanating from the central proton have a uniform EIL in all of the directions around the proton. This is only true if there are no external influences exerted on the proton to cause non-uniform response in its surface Mps. We know that within the realm there are not only forcephoidal fields to be contended with but also the basic realm datum EIL grid (Fig 4.12). As such no protons are free of these influences. In Fig 6.9 the proton is under the influence of the external forcephoids on its A side only. This means a reduced quantity of proton surface Mps contributing to the forcephoidal make up on side A as compared to that on side B. Thus the forcephoidal peak EIL will vary around the proton and in our example will be marginally less at A than at B. The action of the external forcephoids is to cause the shell at A to move closer-in towards the atom centre while the shell at B moves further out. It is because of the combination of all these factors that the proton in Fig 6.9 experiences a zero resultant impulse in a forcephoidal field. Although the atom does undergo a structural distortion resulting in an eccentric shell-proton configuration there is however, no centralising impulse. As such the atom does not deviate in its path through a forcephoidal field and is considered to have a neutral response therein.

Let us now consider the effect that a change in the realm EIL datum has upon the repetitive formation and positioning of the Mp-shell. In Fig 6.4 we indicated the formation sequence of each successive Mp-shell. In Fig 6.11 we have extended that diagrammatic view of the shell by comparing it with the same sequence of events but at a slightly higher realm EIL datum. At Fig 6.11(a) is shown the events as described earlier (Fig 6.4). Let us assume that as the forcephoid Fn reaches the location A at the event instant T1, the realm EIL datum increases to the higher level. The ABCD shell diagram will now sequence at the higher EIL and subsequently the E~M/X from T1-T2 will absorb a marginally greater energy medium quantity. This means that the entire Mp sequence or ABCD shell diagram at Fig 6.11(a) has a longer duration than that in Fig 6.4. Under previous conditions the next shell Sn+1 would have commenced its E-M/X sequence at the same physical location as shell Sn. However due to the extended ABCD sequence at the higher realm EIL, the commencement of the inward E-Flow is marginally delayed such that when it does eventually phase with the next forcephoid Fn+1, the location is somewhat further

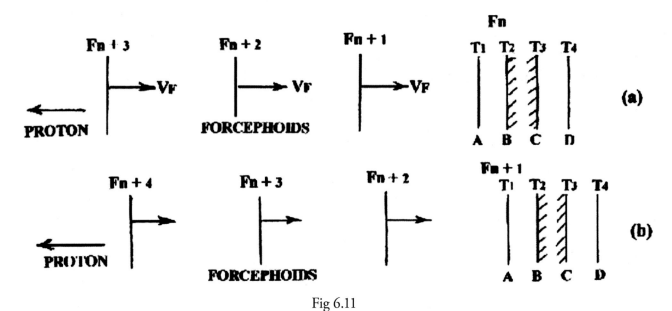

Fig 6.11

outward as indicated at Fig 6.11(b). At the instant of E~M/X commencement at Fig 6.11(b) the trailing forcephoids will have reached a more advanced position. Apart from their increased distance from the proton the relative positions of forcephoids and the ABCD shell diagram will remain similar to that of earlier and the sequence of new events may be described as for Fig 6.4. We may thus state that with the new shells repeating at a position farther out from the proton, the atom structure may be considered as having enlarged. This atomic enlargement is marginal but is in direct proportion to the realm EIL. This of course is subject to the limitations stated earlier specifying the two main reasons for absence of a shell around the proton.

The above phenomenon was due to the prevalence of a generally higher realm EIL environment causing the proton and shell to respond to the isotropic nature of the E-Flows making up the realm datum EIL grid. There is however a second factor that may produce a similar result without considering a change in realm datum level. In chapter 3 we described the energy medium content and stepping character of EIL oscillations. We shall now consider how the energy medium within a wavefront affects the ABCD shell diagram configuration of the atom. Each wavefront propagates forwards in very small quantum steps i.e. the minimum Planck length of 10^{-35} metres. At each quantum step position of the wavefront there is a flow of energy medium forwards, backwards and transverse to the direction of propagation. The propagation plate has a velocity Vp in the forward propagating direction (Figs 3.10 and 3.11) which ensures a greater E-Flow forwards than back. We have already specified the character of EIL oscillations in chapter 3 and a revision at this stage is advised.

Consider a stationary atom located in the path of EIL oscillations propagating at the velocity VC and frequency f and successive wavefront spacing l. By definition we know that the Mp-shell has a repetitive rate of one ABCD shell sequence per MTU (1 MTU = duration of a Mp event sequence). The wavefront propagates at approximately one quantum step per MTU. Successive wavefronts are however spaced apart by a very large number of intermediate quantum steps. This means that when a wavefront section has propagated through a Mp-shell zone, then a very large number of shell repetitions will have occurred before the arrival of the trailing wavefront at that location. The ABCD shell diagram will absorb energy medium from any external sources provided that the E-Flow is in the correct location i.e. within the sphere of influence when the E-M/X commences. When a wavefront steps into or near this zone then a part of the ensuing E-Flow from the wavefront centres will be absorbed into the shell sequence causing a marginal increase to ABCD duration. At each wavefront quantum step we know that a backward E-Flow emanates (Fig 3.14) so that the ABCD shell diagram continues to be influenced, albeit to a progressively lesser extent after the wavefront has propagated past. In reality EIL oscillations emanate from a large number of individual sources congregated together in a mass body. Subsequently, wavefronts from adjacent atoms on the surface of this mass may arrive at our stationary atom location either simultaneously or in closely trailing proximity of one another. The complexity and energy medium content of these emissions would of course depend upon the status of that mass as explained in Fig 3.36. This aspect puts considerable additional energy medium within the sphere of influence of the stationary atom (ABCD shell diagram) on a more frequent arrival basis with the proportionally higher frequency of absorption into the repetitive Mp-shells. Once again the situation as indicated in Fig 6.11(b) prevails with the Mp-shells repeating at a position farther out from the proton. Our conclusion here is that the Mp-shell is able to absorb energy medium from propagating EIL oscillations and thereby result in an expansion outwards. Since all EIL oscillations emanate from other similar structures we are able to observe an important realm phenomenon. This is the transfer of energy medium at wavefront propagation velocity between atom structures spaced quite apart in the realm.

In chapter 3 we defined the term 'Temperature' as simply a measure of the realm energy medium that is absorbed by the Mp-shell structure over and above its basic repetitive requirement and which also represented the corresponding amount of energy medium that could be given up by it. The minimum energy medium level for Mp-shell structure existence was correlated to the absolute zero of temperature with a marginal variation of this as the decay of the proton progressed and the shell structure contracted inwards. In view of the recent discussion perhaps the reader is now just a bit more aware of the role of our present temperature parameter and its relationship to the energy medium quantity, i.e. the relativity in energy medium content of different atomic structures is directly represented by the relativity of their indicated temperatures.

So far we have looked upon the atom as a single identity. However we do know that there are in existence countless numbers of atoms within the realm and that many of these are structured together in large cohesive bodies. We have already shown the effect of impact of one atom upon another (Fig 6.8), and it is our intention now to consider the case when two atoms become bonded together. Let the two primary atoms A and B approach one another at a very low velocity but not in a line for direct impact/interference of their shells as shown in Fig 6.12. A and B are depicted to be moving towards a glancing peripheral contact at the plane XX. If we assume the duration of this contact is large in comparison to the event duration of an inward E-Flow section (Fig 6.6) traversing between shell location and central proton, then the resulting influence of shell A upon shell B and vice versa is of prime importance. Let us consider the effect upon shell B of the EIL oscillations emanating towards it from shell A. In Fig 6.12 we have indicated a direct line between the proton centres PA and PB. Where this line passes through shell B is the point of greatest influence upon shell B by the EIL oscillations from shell A. The ABCD shell diagram at this point thus responds as described earlier causing this part to expand outwards. At locations successively farther away

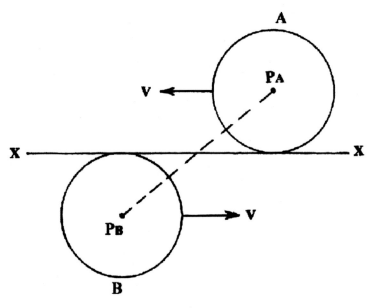

Fig 6.12

from this point position the enlargement of the shell is proportionately less. We are thus left with a shell profile around PB that has decidedly an egg shaped ovality. This causes the proton PB to be in an off-centre position and the subsequent centralising impulse pushes PB in the direction of the (marginally earlier) position of PA. In a likewise response, a centralising impulse is set up in atom A by the EIL oscillations emanating from shell B. This pushes proton PA in the direction of the position of PB. As the two atoms get closer together the centralising impulse gets greater and greater - due to the enhancing mutual effect upon each other's ABCD shell diagram configuration. When shell contact is made (in theoretical terms), this impulse towards the other atom can remain only if the shell at the theoretical contact does not alter from the above egg shaped ovality. We know from Fig 6.8 that such contact distortion cannot be avoided because the very mechanism of shell repetition does not permit Mp-shells to phase (physically) with one another. This distortion at contact face produces a change in the apparent location of the formerly non-central protons. In fact the change is opposite to that produced earlier through the action of EIL oscillations which thereby sets up a condition of equilibrium for the two atoms. Perhaps the above sequence of events are better illustrated in Fig 6.13.

Let the two atoms A and B each possess the velocity V as indicated. Each will be influenced by the other such that the peak of the shell ovality occurs on the side closest to the other atom. At Fig 6.13 (a) the two atoms are shown some distance apart but already there is a centralising impulse I upon their Protons PA and PB as indicated. Thus atom A has an impulse I towards atom B, while B has an impulse I towards A. Each atom will thus acquire a linear momentum towards the other. At (b), atom A just makes (theoretical) shell contact with atom B and at this stage we assume that as yet no shell distortion has occurred. Therefore the protons P_A and P_B are still under a large centralising impulse I (as indicated), and as such the atoms are not in equilibrium but rather have a maximum impulse in each other's direction. As P_A and P_B move towards each other we know that the shell interface at XX remains fixed. The ABCD shell diagram for each shell at the contact interface must now absorb greater quantities of higher EIL energy medium in a direction

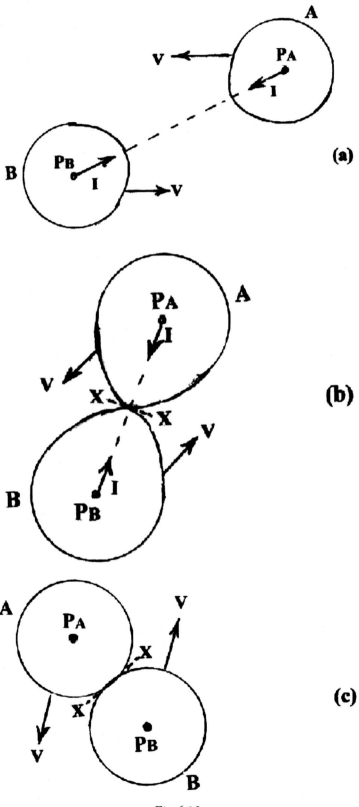

Fig 6.13

opposite to Vs (Fig 6.4). This results in a negative distortion or contraction of the Mp-shell at that location i.e. at XX. This influence is not restricted simply to the contact zone but is distributed in relation to the distances between the shells on either side of XX. This distortion thus results in a uniform adjustment of the former egg shaped oval shell to one of a more uniform concentricity around each proton. Just prior to the achievement of an equilibrium of the atoms, each proton receives an impulse in a direction opposite to I in order to allow it to 'settle' at a position of zero centralising impulse. We see the achievement of this at Fig 6.13(c) and must note the return of each atom to its nearly spherical shape. Because of the initial velocity V of each atom, we observe a change to an angular velocity after the atoms become joined. Since V was small initially then there will be insufficient distortion to cause atom A to expel atom B away from it. Atoms A and B thus remain in a state of equilibrium with regard to each other so long as no influence is exerted to pull them apart. It must be stated that if such an influence is exerted then the two atoms will revert to the situation represented at Fig 6.13(b) when A and B move towards each other. We note that there is a strong bonding action between A and B. This phenomenon is called a Primary Atom Bond (abbreviated PA Bond) and the bonded pair is referred to as a 'Molecule'.

The PA Bond, although of great cohesive strength, is not as strong a bond as that between two protons as shown in Figs 4.25 - 4.27. A comparison of these two types of bond action is recommended at this stage. In Fig 4.28 we indicated the relative positions of four protons in a common bond situation. The four protons can be held together in many configurations of which only two were then represented. In Fig 6.14 are indicated several configurations for bonding four spherical structures be they protons or the larger PAs. It is clearly observed that some of these

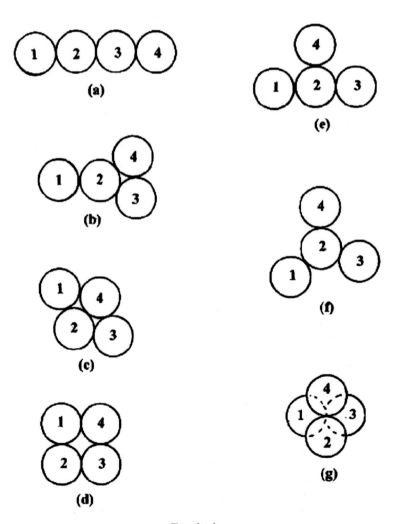

Fig 6.14

configurations are more compact than others. At (a) the combination is 'loose' while at (g) it is 'tight'. The strength of the compound bonding is directly proportional to the tightness of the configuration. Thus the bonding at (g) is stronger than the bonding at (a). Also the bond is strongest when only two spheroids are together. The strength of the bond however decreases with the addition of each new spheroid to the configuration. This is because the directional impulse action within the proton is divided in more than one direction and in many instances is neutralised by the oppositely placed spheroids. In Fig 6.14(a), (b) and (f) the spheroid number 2 is in a completely null-impulse location and is held in place by the combined action of the other three.

When a pair bond occurs the resultant velocity of the molecule is usually accompanied by a spin or angular momentum about a central axis. When an additional spheroid has to be accommodated onto this molecule it could not be rigidly held if it were in contact with only one of the original spheroids. Due to the angular momentum of the molecule the new spheroid gets 'rolled about' the molecular surface until it is in contact with both spheroids. This is a stable contact and rigidly positions all three spheroids with respect to one another. Each spheroid is now bound to two other spheroids and the whole will continue to spin at a new resultant angular momentum. Since each spheroid must now share its maximum directional impulse with two spheroids instead of one, the bond between any two of the above spheroids has only half the attractive strength of the basic molecule. The addition of a fourth spheroid to this group would similarly 'settle' into a position of maximum contact as has been shown in Fig 6.14(g). The bond between any two spheroids is further reduced for the same reason stated earlier.

Let us now consider a large group of PA molecules. We know that each molecule has an independent spin. When one molecule approaches another at some random velocity they are attracted only marginally by their internal centralising impulse mechanisms. Since each molecule spins about its own axis the impulse mechanism cannot remain orientated towards the other molecule and is therefore quite ineffective in comparison to the action set up between two PAs (Fig 6.13). What is important is the resulting effects of an accidental collision between molecules. This collision may be compared to the contact made when two spinning tops touch each other as represented in Fig 6.15. They may collide with a mild impact action for contacting surfaces that are moving in the same direction although perhaps at different velocities. This type of contact is represented at (a). Or they may collide with a violent impact action when their contacting surfaces move in opposing directions as represented at (b). There are of course other variations of contact between spinning molecules but in each case collision and transfer of angular and linear momentum is always between the PA of one molecule and a PA of the other. The spin of some molecules may be slowed by collision while others are speeded up. Transfer of linear momentum obeys the conditions set out in Fig 6.8 for resultant impulse and is eventually distributed over the structural unit of the molecule.

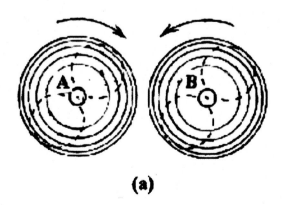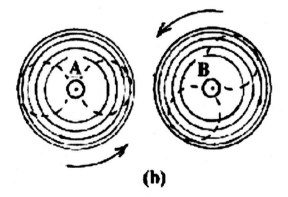

(a)   (b)

Fig 6.15

When a large number of molecules are grouped close together then collisions are frequent. Since most of the collisions result in the colliding PAs (of the molecules concerned) acquiring a net impulse away from one another, then a general tendency within the group would be for an outward dispersion to occupy a larger volume. For the same number of molecules in the larger volume the average spacing between molecules would be greater and the collisions less frequent. If however the molecular group is restricted to a limited volume then an outward impulse will be transmitted to the walls of the containment vessel whenever molecules impact against it. The group of free molecules is termed a 'gas' and under the above conditions is considered to express an outward impulse or 'pressure' upon the inside surface of the container.

In Fig 6.11 we showed how the ABCD shell diagram shifted to a more outward position for each successively higher Realm EIL datum resulting in the enlargement of the Mp-shell. The absorption of energy medium at the higher realm datum also resulted in a Mp-shell with a greater mass content. Since the realm datum is considered to be isotropic at any one location then the increase in mass of the Mp-shell would occur without a change in the equilibrium status of the primary atom as a whole. This means that the process of shell formation permits the overall mass and thereby linear momentum of the PA to increase without apparent effect upon the surrounding realm environment. The law of conservation of linear momentum is not violated since complete accountability remains with the forcephoids and E-Flows that emanate from the central proton of each shell. In the case of molecular spin an increase occurs in the angular momentum of the PA pair when the Mp-shell enlarges. The angular momentum of the molecule is thus dependant upon the volume mode of the Mp-shell.

When the shell is in the expanded or enlarged state due to realm EIL or 'temperature', the spin of the molecule results in a greater impact between shells whenever collisions occur. Thus it is that the molecules of the gas transmit greater impulses to one another under these conditions of realm datum and so convey a proportionately greater outward impulse to the walls of the enclosing vessel. Since the Mp-shell, by virtue of its forcephoid origin, contains a linear momentum (for each spherical section) in a direction away from its central proton, then there will always be this additional impulse to be considered when balancing the equation for linear momentum before and after the collision of molecules. It is the continued addition of this impulse quantity at every molecular collision that enables the outward impulse to be active upon the walls of the restricting container for an indefinite period without signs of appreciable diminishment. We will of course at this stage ignore the longer term effect caused by the overall decay in the mass of the proton. Since the molecules of the container are held together in a relatively rigid structural pattern the impacting gas molecules cannot effect a transfer of their linear momenta to the vessel. However if the container molecules were not too rigidly held then the container would allow its molecules to absorb linear momentum and so adjust to a slightly different structural hold upon adjacent molecules thereby causing the container to expand, as in the case of a rubber balloon.

In Fig 6.13(c) we showed the two atoms A and B in a molecular bond. Let us now look more closely into the character of this bonding. So far we have assumed an equilibrium in the status between the two PAs once the molecular bond had been achieved. We stated that any attempt to separate the bonded PAs would result in a centralising impulse pushing them back together. Similarly we also stated that an attempt to compress them closer together would result in a reactionary centralising impulse, through shell distortion, causing the two PAs to be pushed apart. From these two differing aspects we assumed that the two PA components of a molecule settled into a state of equilibrium with respect to one another. Apparently this is not quite the case.

We know that the PA has a repetitively forming Mp-shell. The formative nature of this shell surround is expressed by the ABCD shell diagram and if we could observe it we should see the PA as a pulsating spheroid. We should also see a continual emergence of a new shell on the inner side of the current (outward traversing) shell sphere. In

ABCD shell diagram terms (Fig 6.4) this means that a new shell emerges between the A and B phases after the previous shell had ceased somewhere between the C and D phases. Now when two PAs are brought together in a molecular bond there can never be a state of perfect equilibrium between them. The natural repetitive action inherent in each Mp-shell results in the molecular bond being a very active affair. It does not matter whether or not the shells are in synchronism with regard to their ABCD shell diagram configurations. Each shell will however be in a state of 'inner' (A - B) or 'outer' (C - D) configuration. Let us consider the 'inner' and 'outer' states of the PA as those which have the smallest and largest spheroid volumes respectively, ignoring realm EIL differences of course.

Let us now analyse the impulses that arise between two PAs constituting a molecule. We shall initially assume a set distance D between the protons of the molecular pair. We shall also represent each PA by two concentric circles denoting their maximum and minimum spheroid volumes. Let 'D' and 'd' denote the maximum and minimum PA diameters respectively (assuming identical PAs of course). For the moment let us ignore the delay in effects between the shell and proton centres in order to simplify our discussion. Once the concept of molecular impulses has been understood we shall then consider this delay factor as well. In Fig 6.16 we show a molecular pair with both shells in the (C-D) configuration with their protons at a distance D from one another. The theoretical contact of their shells is at the location indicated by the point c (or c_1). This also represents the PAs at their maximum ABCD volume. Let us assume that the shells are in the M-E/X phase and are on the point of non-existence. This means that a new shell is in the E-M/X phase at the location 'a' and 'a_1' and just about to form a complete Mp-shell in the Mp phase. From Fig 6.4 we know

that this shell traverses outward at a velocity Vs. However at this instant the shell diameter is 'd' while the protons are still a distance 'D' apart. The primary atoms A and B are not now in any theoretical contact and as such the situation is represented by that shown in Fig 6.13(a). As such a centralising impulse acts upon the protons A and B causing them to move towards each other thus reducing the distance between them to less than D. Each shell follows its ABCD shell diagram sequence and thus once again each shell makes theoretical contact with the other. However, because the PAs have moved closer together i.e. less than D, the shells have not as yet completed their Mp-phase. This results in the situation represented by Fig 6.8(c), i.e. shell distortion at the theoretical contact zone. The result of this is a centralising impulse acting upon the protons A and B causing them to move away from each other. Due to the existence of a generally higher EIL around the contact zone (from the previous M-E/X phase), the ABCD shell diagram configuration there tends to a greater event duration than the rest of the shell. As such the repulsive action between A and B is high and the protons move apart to a distance somewhat greater than D before this shell exchanges to energy medium and the next shell commences. We are now back to the situation of Fig 6.13(a) except that in this case the PAs are even further apart. The attractive centralising impulse

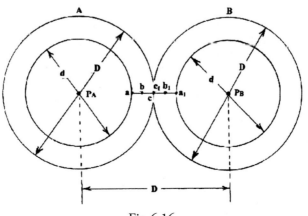

Fig 6.16

now causes the PAs to be pushed back towards one another and into the same theoretical shell contact. Once again shell distortion causes the PAs to be mutually repulsed. Under uniform realm EIL conditions this to and fro motion settles into a constant periodic rhythm and displacement amplitude. This phenomenon is termed

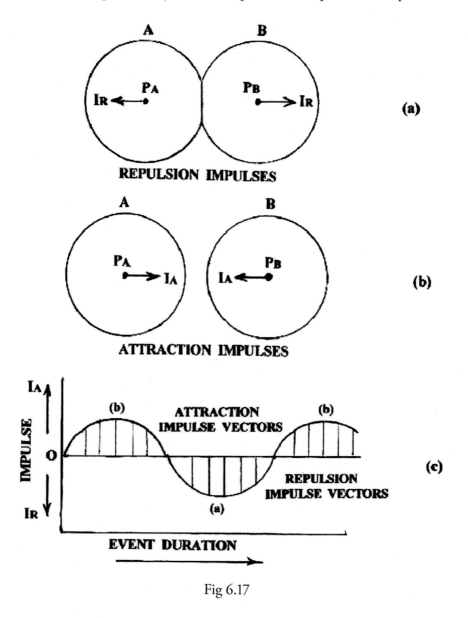

Fig 6.17

'molecular bounce' and clearly shows the impulse activity within the molecular bond. Fig 6.17 highlights this action. At Fig 6.17(a) the PAs of the molecule are in the closest theoretical contact and therefore the repulsion impulse IR is at its greatest. At Fig 6.17(b) the PAs are shown apart but with an attractive impulse in action. There is obviously some finite separation distance at which this attractive centralising impulse attains a maximum and we leave that to the readers imagination and to later developments. At Fig 6.17(c) is a graphical representation of the alternating aspect of the impulses that are responsible for the molecular bounce within each molecule. It should be noticed that each impulse (attractive or repulsive) builds and decays in a regular manner as indicated by the smooth continuity of the graphical curve represented.

Now we know that there is a delay factor in the responses by the central proton to any distortions or changed conditions at its Mp-shell (see Fig 6.6). If we assume the molecular bounce to have settled at a regular beat then we can show that the impulses causing this are simply out of phase by several cycles. However each impulse cycle has the same bounce frequency and hence its effects appear directly linked to the Mp-shell changes. This is explained more easily with the help of Fig 6.18.

Here we represent the events occurring at the shell relative to those occurring at the proton. If events at the shell always occur at the event sequence zero then because of the traverse duration of the inward E-Flow sections, these events are not sensed by the proton until perhaps event sequence 6 or 7. Since the current event (sequence zero) was repeated at event sequence 2, 4, 6, 8, etc., then providing that the molecular bounce is in phase with the relative proton response the action will continue. That is, each repulsion or attraction between the PAs will appear as an instantaneous response to proximity conditions as in Fig 6.17.

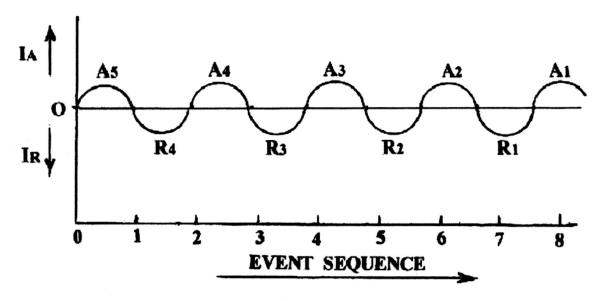

Fig 6.18

Let us suppose for the sake of argument that the molecular bounce is out of phase with the appropriate proton response. This would mean that for the proximity condition of Fig 6.17(a) the protons could be subjected to an attractive centralising impulse instead of that shown. Subsequently shell distortion at the contact zone would increase even further meaning that when the inward E-Flow sections from this zone reached the proton, the response would be a greater than previous centralising impulse (repulsion). This would cause the amplitude of the bounce to increase along with a change in its cyclic duration. At some stage the bounce and proton response would get into step with one another and irregular responses would soon be damped out. The same reasons apply for the condition of Fig 6.17(b) if accompanied by a repulsion impulse action. If the repulsion impulses come into play when the PAs are already apart then once again the amplitude and natural cyclic duration of the molecular bounce is increased causing it to alter gradually to an in-step situation. When molecules collide with one another then it is probable that they get 'knocked' out of normal in-phase responses. We know from the above that this state is only temporary and that the natural molecular bounce is soon regained. We may thus conclude that regardless of the shell to proton distance the proton impulses appear to respond instantly to the molecular bounce phases.

In Fig 6.16 we assume both PAs of the molecule to be in the same ABCD shell diagram phase in order to simplify the explanation of events. Let us now suppose each PA to be in a different ABCD phase with respect to the other. Referring to Fig 6.16 again, let atom A be the phase corresponding to a shell diameter marginally greater than 'd' while atom B is in the phase with shell diameter marginally less than 'D'. Since the protons are spaced D apart then clearly there is no theoretical contact between their shells. An attractive centralising impulse will be active and the protons will move towards each other. When shell A approaches the ABCD central phase position 'b' the shell B approaches the end of its ABCD phase. Now A and B have moved closer together and are on the point of a theoretical shell contact. This however will not occur as expected since the shell B is suddenly dissipated in a M-E/X while a new shell for atom B is achieved at minimum shell diameter d. However when the protons are at the closer-in distance of ½(D+d) then the two shells are bound to make theoretical contact. Since each shell has a continuous outward traverse velocity Vs then shell distortion at the contact zone Fig 6.8 (c) will result. This causes a repulsion impulse (centralising impulse signal) to be transmitted towards each proton resulting in their movement away from one another. This signal continues until the two PAs are no longer in theoretical contact due to their physical separation and/or renewal of Mp-shells. Once again an attractive impulse signal is sent to each proton. Thus we observe a continued bouncing action between the PAs of a molecule even though the ABCD shell diagrams were not in any particular synchronised aspect or phase sequence. In all cases the impulses generated within each proton of a molecule are equal in magnitude (on average) and in opposing directions. We use the term 'on average' because the shells in contact may be at differing ABCD diagram phases resulting in one shell being marginally smaller than the other. As such the smaller shell undergoes a greater distortion and thus sends a greater repulsion impulse signal to its own proton. As the ABCD shell diagram sequence progresses the shells are renewed and the size roles are reversed. Thus it is the turn of the other proton to be sent the greater repulsion impulse signal. On average therefore these impulses are equal and opposite. Apart from this side to side jiggling action the molecule remains in equilibrium. Perhaps the reader would like to evaluate the effect upon this jiggling and bouncing molecule of changes in the zonal realm EIL.

So far we have presented a rather simple working arrangement for the bouncing molecule. The repetitive formation of the Mp-shell is extremely rapid and it is unlikely that the mass of the proton is able to physically jiggle at the same frequency. The proton is more likely to position itself at a location at which the required response to all the centralising impulses (i.e. attraction and repulsion) averages to zero. The subsequent molecular bounce will then be wholly dependant upon interaction of shells due to the variation in their diameters i.e. from d to D. It must be borne in mind however, that the protons of the molecule are continually in receipt of either attractive or repulsive impulses even if these alternate too rapidly for effective response. If the protons are of different ages (see chapter 5) then the frequency of their ABCD diagrams may be different and the contact between their shells will vary between shells at minimum diameter (d) to those at maximum diameter (D). This would cause the protons and the PAs to physically jiggle/bounce at a frequency well below that of the shell ABCD diagram. Subsequently we perceive the molecule as a spinning, bouncing and jiggling pair of PAs held together not too closely and not too far apart - by alternate centralising impulse action upon its protons to produce a molecular bond. This bond is obviously of limited strength and it would not be too difficult to cause the PAs to separate again by external means. The reader may speculate upon this aspect.

By now a detailed mental image of molecular action should be perceivable and it is important that this be borne in mind whenever other topics are under discussion. Alternatives to molecular bonding may exist but for the present we shall adhere to our theory. Let us therefore now consider the behaviour of our molecule in an EIL gradient zone. For purposes of simplifying the discussion we shall temporarily ignore its angular momentum aspect. In Fig 6.7 we showed the response of a single PA within an EIL gradient situation. In the case of the molecule the effect upon each PA is simply a compounding of the impulses that arise due to the EIL differential across the proton and the centralising action from the bounce effect. In each case the resultant impulse is the vector sum of both as represented in Fig 6.19. We shall assume that the impulse due to the EIL gradient is generated as a continuous impulse and therefore concurrent with the centralising impulses of the molecular bounce. (Fig 6.6 will indicate the need for this assumption).

In Fig 6.19 (a) the molecular shells are in theoretical contact and the shell distortion causes a repulsive impulse IR to be active. Let the EIL gradient i.e. EIL normal lines, be in a direction perpendicular to the axis of the molecule - the imaginary straight line through its protons. Let the realm EIL be higher towards the top of the diagrams. Then IG will be the gravitational impulse pushing each proton up the realm grid gradient. Since we consider IR and IG to be active at the same event moment then they may be added vectorially to give the resultant impulse RR as shown. This means that not only will the two PAs be pushed apart as before (Fig 6.17), but in addition they will acquire linear momentum in the direction of IG. When the PAs have bounced apart the situation is represented at Fig 6.19(b). Here an attractive centralising impulse IA is active along with the gravitational impulse IG. Since these are active (more or less) simultaneously their vectorial addition gives the resultant RA as shown. This means that the PAs of the molecule not only move back together but continue to receive an impulse up the realm grid gradient. At the

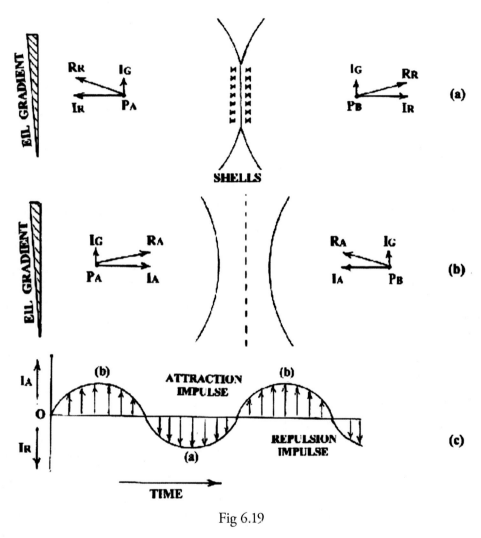

Fig 6.19

Fig 6.19(c) we show the progressive trend of the centralising impulses of the molecule and it should be noted that these are identical in magnitude and direction to those of Fig 6.17(c). This is because the resultant impulses RR and RA have for their molecular axis components the same centralising impulses IR and IA as before. From this we may infer that the molecular bounce remains unaffected within a realm EIL gradient zone and that the molecule as a whole acquires linear momentum towards the higher EILs. If we now consider the angular momentum aspect as well, we have simply to reckon with a relative change in the direction of the gravitational impulse IG with respect to the molecular axis. Since the rotational velocity of each proton is small in comparison with the forcephoid velocity VF we can assume the spin of the molecule to occur in a stepping fashion. As such we could analyse each step for resultant impulses upon the protons. The extreme cases would be represented when the molecular axis was in the same direction as the EIL normal lines (Fig 6.7). The resultant impulses would then be as under;

$$RR = IR \pm IG$$
$$RA = IA \pm IG$$

In all cases the molecular bounce remains unaffected while the whole accelerates up the grid gradient. The probable velocity of the two PAs about one another is small in comparison with the repetitive aspect of the Mp-shells and as such further analysis of any effects due to molecular spin may be ignored. However, if the reader undertakes to examine this further then perhaps it might also be worth considering proton spin and its influence - if any - upon the properties of the Mp-shell. (See Fig 3.15).

Let us now consider the effect of the simple forcephoidal field upon our molecule. In Fig 6.9 we showed that a single PA had a neutral response in such a field. We have also shown that the molecular bond is not a rigid affair but rather is the result of an over-riding attraction impulse between the two PAs whenever these are separated. The PAs of a molecule retain their individual character throughout the bounce phenomenon and as such may be assumed to respond to external influences on an individual basis. We observed this in the response to the realm grid gradient where the resultant response was the vector sum of the various impulses imposed on each PA. The same will apply to our molecule in a simple forcephoidal field. Since each PA has a neutral response to forcephoids, then the molecule - as a whole - will also appear neutral. The bounce, spin and jiggle of the molecule therefore remains unaffected by the forcephoidal environment.

Let us now define a complex forcephoidal field. Such a field arises when a large number of free protons are closely present within a limited realm volume. Each proton emits a continual series of forcephoids which traverse outwards to eventually phase with other forcephoids. When two or more forcephoids phase i.e. intersect at a point, the resulting EIL at that point is the sum of the individual forcephoidal EIL peaks. Let us initially consider the forcephoids from just two protons as represented in Fig 6.20.

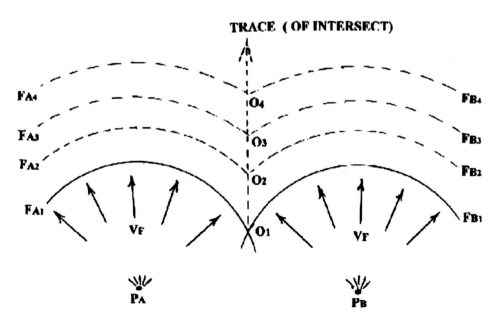

Fig 6.20

PA and PB are the free protons. FAl and FB1 are just two of the many forcephoids that traverse out from each proton at the normal velocity VF. At any instant let FA1 and FBl intersect one another at a point O1 as shown. At progressively later event moments these forcephoids are assumed to occupy the positions represented by FA2, FA3, FA4 and FB2, FB3, FB4 intersecting one another at the points O2, O3 and O4 respectively. The trace of these intersecting points (in the plane of the paper) is a straight line through O1, O2, O3, O4, etc., as shown.

In the more realistic 3-dimension aspect the forcephoids FAl and FBl are spherical and as such their intersection is in fact represented by a complete circle. The centre of this circle lies on the imaginary line joining the two protons while the circle itself is in a plane perpendicular to the line (and the paper) and operates through the points O1, O2, O3, O4, etc., in a progressive manner as mentioned above. Since forcephoids are emitted at uniformly spaced intervals then a succession of these intersecting circles seem to originate from the imaginary point X on the line between PA and PB as shown in Fig 6.21. The intersecting circles are like very fine rings (comparable to the expanding circle of ripples in a pond) with a peak EIL that is the sum of the EILs of FA and FB. We shall call these as forcephoidal rings since they are very nearly in a 2-dimension configuration. We assume of course that the forcephoid is a very narrow band of peak energy medium. The velocity of expansion of these rings is initially greater than VF at the indicated O1 and O2 positions, but in the limit this reduces to VF. This can be observed in Fig 6.20 by comparing the traverse distance from FA1 to FA2 with the traverse of the intersecting point from O1 to O2. If the relative velocity between the two protons is zero then the forcephoids from each will traverse at the same relative velocity VF. Also, since they are in the same local realm environment then the spacing between consecutive forcephoids from both protons will be similar. This means that each and every forcephoid from proton PA will in turn intersect with each and every forcephoid emanating from proton PB - and vice versa. These intersections will repeat at the same locations resulting in a parallel array of forcephoidal rings at selective

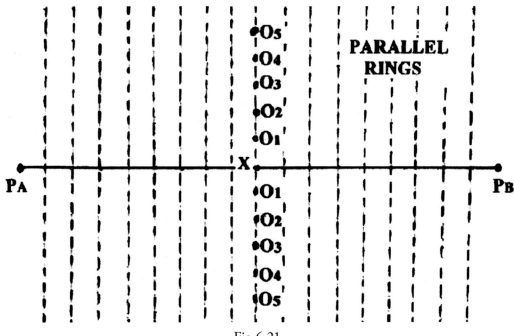

Fig 6.21

intervals along the imaginary line between the two protons. The selective intervals will of course relate to the original forcephoidal spacing and to closeness of either proton. These are represented in Fig 6.21 by the parallel lines on either side of the point X. It must be noted that the EIL peak of these rings is never very high and as such may be considered simply as a background effect. Perhaps it performs an obscure role in some phenomenon as yet unknown.

Let us now consider the much more complex forcephoidal field which is the result of forcephoids emanating from a very large quantity of proximity protons. We shall assume that all these protons come under similar realm influences so that their decay rates are identical. This means that the regularity of forcephoid emissions are also identical. Since each proton emits forcephoids that are spaced an equal distance apart then let us assume this inter-forcephoidal spacing to be approximately of the order of a few quantum steps only. See Fig 6.4 and compare this spacing assumption with the represented spread of the ABCD diagram.

As the forcephoids from each proton traverse outward they must intersect with forcephoids from other protons. The forcephoids from each proton must eventually intersect with all the forcephoids from every other proximity proton. With such a large number of intersecting events and with such a fine spacing between them it is highly probable that much more than just two forcephoids achieve intersection / phasing at the same point and at the same instant. It is highly probable that under closer proton positioning an even larger quantity of forcephoids intersect at such a point. Let us for the moment overlook the fact that as free protons get closer together they tend to set up Mp-shells around themselves (Fig 2.17). We shall assume (without justification for the Primary Atom), that the Mp-shells may occur as multiple shells in the case of the higher order atoms and that single proton type forcephoids are emitted beyond the shell zone. We shall discuss multiple shells and multi-proton nuclei a bit later on.

If we assume all the above protons to be stationary and fixed in positions relative to one another, then the multiple forcephoidal intersections will recur at the same relative points as successive forcephoids flow past. Fig 6.22 illustrates this as a simple intersection of a few forcephoids. The protons are shown at P_1, P_2, P_3, P_4, etc., with multiple intersections of their forcephoids occurring at the indicated point X. Obviously the EIL at the point X is the sum of all the intersecting EILs and is thus relatively high. If the protons happen to be situated at regular distances from one another - and

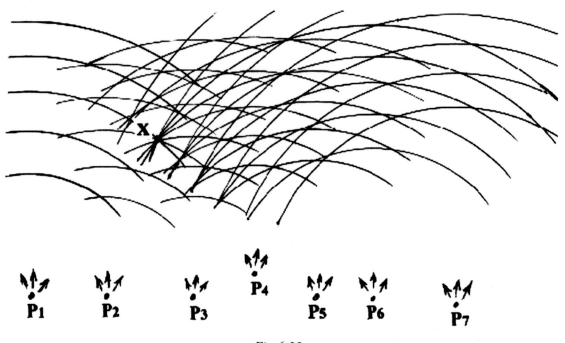

Fig 6.22

since the forcephoids are assumed to possess a regular spacing - then it is most probable that there will be many other locations where the point X type event occurs. The EILs at these other locations will of course depend not only upon the number of intersecting forcephoids but also inversely as the square of the distance from the proton group centre. Fig 6.22 only depicts one example of these multiple forcephoid intersecting points for the purpose of simply illustrating the topic. In reality there would be a very large number of such points occurring in a systematic pattern all around the proton grouping and relative to one another. We shall refer to these high EIL points simply as 'EIL Nodes' or just 'Nodes'.

Quite obviously when we talk of intersecting forcephoids the impression conveyed is rather vague since we do not know the quantity of forcephoids involved in each EIL Node. For the sake of argument let us assume that the EIL Nodes that interest us are those that comprise a quantity of forcephoids that represent between 20% and 40% of all forcephoid emitting protons in the group. It is estimated that above this range the EIL Nodes will become fewer than is necessary to present a close pattern of peaks, while below the range they will have a lower EIL and therefore not be worth considering. At the present time all of this is speculative conjecture and the reader is welcome to pursue an independent line of reasoning. Fig 6.23 is a graph showing the probable distribution of forcephoids within EIL Nodes.

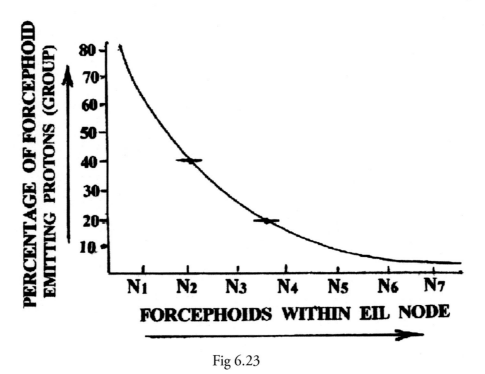

Fig 6.23

Let us now consider the nature of these EIL Nodes. When the protons from which they arise are fixed in location, then each EIL Node will recur as an independent energy medium peak at the same realm location. The duration of the EIL Node is extremely brief since it represents the phasing of forcephoids traversing at the velocity VF. The duration 'D' of the EIL Node may thus be given approximately by the formula,

$$D \text{ Node} = \Delta f / VF, \text{ where } \Delta f = \text{forcephoid width.}$$

However if the inter-forcephoidal spacing is f, then the EIL Node will repeat at intervals 'I' given by,

$$I \text{ Node} = f / VF$$

i.e. the frequency of occurrence of the Node is dependant upon the arrival rate of forcephoids at that particular realm location.

Now referring back to Fig 6.4 we observe that the repetition rate of the Mp-shell is also dependant upon the arrival of forcephoids at the 'A' plane of the ABCD shell diagram. Since we have already stated that the ABCD shell diagram has a duration of one MTU then by deductive reasoning we infer that forcephoids traversing at the set velocity VF must arrive at a set realm location at intervals of one MTU. If we ignore any proton velocities then the EIL Node defined above must recur at intervals of one MTU. The distribution of these recurring EIL Nodes must inevitably reflect in some manner the distribution pattern of the forcephoid emitting protons from which they arise. If the protons remain fixed in relative position then the pattern of Nodes will also remain constant.

Let us commence another thought experiment. Suppose we are able to re-position the emitting protons and that at the same instant we are able to (magically) view the pattern of EIL Nodes that result. Let us continue to re-position the protons either closer together or farther apart until suddenly we seem to get some sort of correlation between the positioning of the protons and that of the EIL Nodes. When this correlation is high we consider the group protons to be in a 'Node-synchronous' layout. At this stage we can assume that the pattern of positioning of the EIL Nodes has spacing similar to forcephoidal spacing and is therefore much more closely knit than the assembly of protons. A probable pattern of EIL Nodes from a synchronous layout is indicated in Fig 6.24.

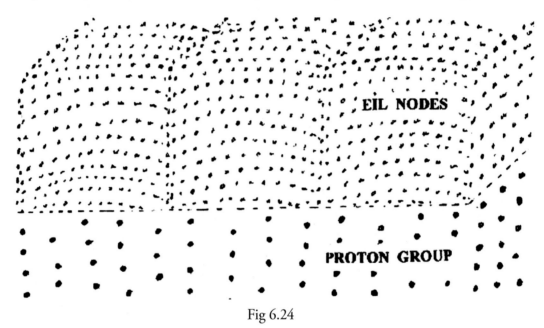

Fig 6.24

In the next chapter we shall deal with this concept of synchronous EIL Nodes in a bit more detail. We shall also indicate the interaction of one proton group with another when each lies in the other's 'field' of synchronous EIL Nodes. The interaction must of course be the result of the responses due either to the EIL gradient across a proton or to the change in centralising impulses from the Mp-shell or shells as the case may be. Marginal changes in the relative proton positioning within a group may also be induced by such a 'field' provided that the correlation in their respective layout patterns is high. We shall at that stage also indicate a directional aspect in the pattern of synchronous EIL Nodes.

Before we conclude this chapter there are a few features in the behaviour of protons and Mp-shells that require to be discussed. Although these features or events do not occur for the primary condition of the atom it is important that they be mentioned here since all atoms - no matter how complex - originate from the basic one proton one shell hydrogen atom structure. Protons are bound to one another within the nuclei of the more complex atoms and these bonds are observed to be relatively stable. In chapter 4 we showed the impulse action between two bonded protons (Fig 4.25) and the general effects when a separation was attempted (Fig 4.27). The bond is strongest when the distance between their centres is a minimum. As the contact area decreases so also does the bonding impulse action. Thus as the protons are pulled towards the point of separation a progressively lesser pulling-apart impulse is required. When separation is finally attained as shown in Fig 4.27(e), then the two protons experience a net repulsion impulse from one another as indicated by the impulse curve of Fig 4.24. This means that in order to maintain a condition of equilibrium, i.e. for the protons to be held close to each other, an impulse must be exerted causing each to be pushed in the direction of the other. A condition of natural equilibrium does however exist for protons that are placed at a distance 'X' apart as indicated by the impulse curve.

When large numbers of bonded protons occur as in the nucleus of a higher order atomic structure, then each proton will obviously exist in a very high EIL environment. Each proton has its own high EIL surface grid (Fig 4.6), and when protons bind together this energy medium grid is considerably enhanced. In our theory we have stated that response of protons alter in a C_1^+ EIL environment. There are very few clues to the type of behaviour that we can expect from protons in such an environment. However, the zone of the multi-proton nucleus is suspected of being at a very high EIL (approaching C_1), and so we shall assume this to be the case. Perhaps this is one of the reasons for the very slow rate of decay of these mass entities as indicated in chapter 5. We must nevertheless continue to assume that an E-Flow emission still occurs from each proton and also that the gravitation type impulse curves are still relevant but perhaps with a slight change in characteristics. In the zone of the nucleus we would expect the repulsion impulses to dominate events whenever possible.

As yet we are not certain of the disposition of protons in a nucleus grouping but it is considered that this depends upon the mode of manufacture. Since this is known to be done by the fusion of lower order nuclei at extremely high EILs, we would expect a uniformity in the combination patterns of the progressively more complicated nuclei. Observations indicate that by the addition or subtraction of a few proton masses in the nucleus the nature of the atom is altered. As it is actually possible to do this, we can only assume that the combination patterns are 'loose' rather than 'tight' as represented by Fig 6.14. At this stage in our awareness of atomic structure we are not able to describe the intricate responses of each proton within the nucleus but we are nevertheless able to speculate upon a possible interpretation based on the clues at hand. It is important that the reader realise this and as such is perfectly free to discard all or any part of the theory presented in favour of a more logical one thought out personally. We feel that it is important to initially present a theory because only then can an improved theory finally evolve.

Let us first of all consider the combination of atomic nuclei. Under normal realm datum conditions each atom nucleus is surrounded by at least one Mp-shell. In the case of the PA molecule the bonding action arises from the interaction of the Mp-shells causing a centralising action upon the protons. This action has the dual function of not only keeping the two atoms together but also of maintaining a set distance between their protons. Now if this shell were to become non-existent then the central protons would approach one another more closely and perhaps even collide to become bonded as represented in Fig 4.26. Early in this chapter we stated two main reasons for the absence of a shell structure around a proton. The first was if the protons were so widely spaced that there were insufficient EILs in the phasing energy medium of their E-Flows and forcephoids for a synchronised shell of Mps to occur. The second reason is opposite to the first with the atoms locating in an EIL zone in excess of C_1. This causes the Mps in the shell to loose synchronism with one another. If Mps reform in the shell ABCD diagram position without perfect synchronism then a continuous shell cannot occur and the PA must simply revert to a single proton entity (with forcephoids). In the case of the PA molecule the situation evolves to the condition of two proximity protons with impulses induced in each according to (revised) impulse curves of the type shown in Fig 4.24.

As already stated protons in the circle 5 phase behave in a different manner to those in the circle 6 phase of the realm energy medium curve. This is not because the laws of physics have altered but because the environmental conditions are different. In the circle 5 EIL conditions protons would still obey the same laws i.e. acquiring impulse up an EIL gradient. Unfortunately, this aspect does not necessarily coincide with the direction of the next nearest proton because in the circle 5 phase there is a profusion of Mp events occurring all around. The subsequent E-Flows from the M-E/Xs of these Mp events result in a realm datum with EIL gradients that are not focussed at proton mass locations. As such, protons (or for that matter any multi-proton mass entity) would appear to acquire linear momentum that does not conform to laws that would apply in the circle 6 phase. When there are large quantities of protons within a confined zone of C_1^+ EILs (circle 5) there is a high probability that frequent collisions occur between them. Because of the E-Flow emitted by a proton there is still the repulsion impulse action between protons that get very close to one another. When two protons are on a collision course the repulsion action reduces the violence of impact. In cases where the collision is gentle the two protons may become bonded (Fig, 4.25) and usually this bond continues indefinitely, even when the realm EIL reduces below C_1. Incidentally, atoms of any configuration cannot exist in the circle 5 phase simply because synchronicity of their shell Mps cannot be preserved. This is due to the abundance of external Mp-events. Thus when atoms are put into a C_1^+ EIL environment they lose their Mp-shell surround and become free nuclei which may collide and bond to other free nuclei. Such bonding of nuclei cannot easily occur in the circle 6 phase EILs because the Mp-shell usually acts as a contact barrier.

In Fig 4.27 we showed how two protons behaved in a thought experiment when they were forced to separate by some magical means. Just prior to the separation they were linked by a relatively narrow band of the proton Mp-congregation represented by the mass in zone 'C'. As stated earlier the mass of zone 'C' had too few Mps for actual repetitive E-M/Xs to continue as in the body framework of the proton. As such, zone C will rapidly convert to energy medium leaving the protons A and B as wholly separate entities. The zone C Mps had completely exchanged into energy medium over a relatively few exchange cycles and this was highlighted by a mixture of very high EIL forcephoids and E-Flows from this zone. Our conclusion here was that additional energy medium was only released when proton entities were forced to separate. There can be no release of additional energy medium from the simple process of two single protons (or two nuclei) becoming bonded together. However we cannot consider each in isolation since bonding and separation usually occur side by side in the fusion process as we shall now show. We shall assume that we have placed a large quantity of protons in a zone of C_1^+ EIL. There is a high probability that frequent collisions will occur between them. These collisions may be gentle enough to permit a permanent bonding, or may be of excessive violence in which case the bond or contact has a duration of a few MTUs only. There may also be instances when protons in a permanent bond are separated by the impact of a third high velocity proton. Since each separation results in the release of energy medium to the realm (as per Fig 4.27), the fusion process is able to build and subsequently maintain its own C_1^+ EIL environment. The number of proton or nuclei permanent bonds that result may be in the proportion of just 1% of the number of collisions that occur. This ratio is stated for example only and may in fact be very much greater (or lesser).

Let us now consider the relative positioning of the protons that have become bonded to one another and constitute the structure of the atom nucleus. We shall commence our discussion with a single proton and progressively allow additional protons to become bonded to the growing nucleus. We assume of course that the EIL environment is at C_1^+ and will permit this fusion process to occur. When two protons are bonded they can only occupy one relative position which is shown at Fig 6.25(a), and where proton A is bonded to proton B.

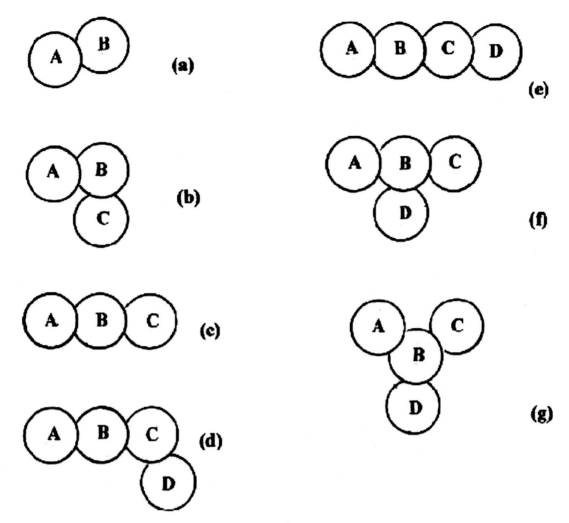

Fig 6.25

Let us now permit a third proton C to attach itself to the 2-proton nucleus. Due to the probable linear and angular momentum of the (A+ B) nucleus it is very unlikely that proton C will contact both A and B at the same instant. What is more probable is for C to become bonded to either A or to B. At (b) we show the contact bond to have occurred between B and C. Now A and C are both bonded to B but not to one another. If we just for an instant ignore the presence of proton B then we observe that protons A and C are in very close proximity to one another but are not in physical contact. The resultant impulse curve (Fig 4.24) indicates that both A and C will acquire impulses in a direction away from each other. Thus the situation at Fig 6.25(b) is simply that proton A has a bonding impulse towards proton B but a repulsion impulse from proton C. And proton C has a bonding impulse towards proton B but a repulsion impulse away from proton A. In the representation at (b) the impulses in A and C will not be in symmetrical balance and as such the bond configuration will undergo a change. The protons A and C will re-locate themselves (while still bonded to proton B) such that they are as far apart as possible. Thus they will move into position on opposite sides of proton B where their impulses are in perfect symmetrical balance as shown by the nucleus configuration at 6.25(c).

If we now go one step further and permit a fourth proton to join the above nucleus group then this proton D will either bond to one of the end protons as at (d) or to the central proton B as shown at (f). In either case the nucleus configuration must occupy positions where all protons are in perfect equilibrium (and symmetrical balance) and these are shown at (e) and (g) respectively. The reader must remember that the proton has internal impulses that are generated through the repetitive Mp-events within its congregation and that bonding is a strong though flexible hold that one proton has upon another. As such protons must never be considered as rigid entities that get glued to one another in a solid manner. Therefore it is conceivable that bonded protons may each possess an angular momentum of their own for a short duration after a collision with a high velocity free proton. Fig 6.26 illustrates this concept although it is considered that eventually a resultant angular momentum will prevail.

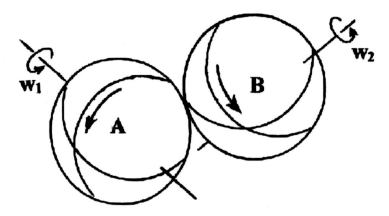

Fig 6.26

As the number of protons bonded indirectly to one another in a nucleus increases substantially, varying configurations are possible. One aspect is certain however, which is that bonding of protons will occur more often as a string of protons as represented at Fig 6.25 (c), (e), and (g). We shall refer to this pattern of proton bond as a nucleus string configuration. Since the impulses between proximity protons within the nucleus are repulsive, this string configuration can be maintained for a large quantity of protons grouped in the same nucleus. In the fusion process it is possible for two separate nuclei (both with protons in the string configuration) to collide and to subsequently become bonded at a single proton interface as shown in Fig 6.27. At Fig 6.27(a) we have the two configurations (A - M) and (P - V) on a collision course. We shall assume a gentle collision between the protons V and E such that they form a permanent bond and that the two configurations remain intact and so become a single new nucleus. We will now have a new and more complex string configuration nucleus which is a combination of the two previous nuclei and is represented at Fig 6.27(b). It will be noticed that the configuration adjusts its proton strings into positions that are symmetrically balanced with regard to the repulsion impulses that adjacent strings exert upon one another. Thus proton P will position itself in an equilibrium status with regard to the impulses from protons A and F. Similarly protons J and K are orientated so as to be in a balanced situation with regard to the impulses from protons B and M. We observe therefore that the nucleus configuration has all its component strings held in set positions by the inter-proton repulsion impulses.

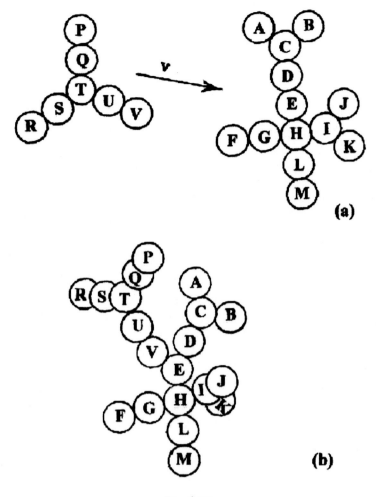

Fig 6.27

If we apply the theory of Fig 4.24 to the zone of the nucleus spaces, i.e. the spaces between the strings of a nucleus, we must conclude that as the number of protons increases so will the distances between protons on the ends of those strings. In Fig 6.27(b) protons R and F are several proton diameters apart. If this distance is greater than 'x' (of Fig 4.24) then R and F will experience an attractive impulse towards each other. Let us assume that as a result of this attractive impulse R re-positions itself closer to F. The amount of this re-positioning will depend upon the resultant impulse that each proton experiences from the nucleus as a whole. In our world today nuclei form part of the atom structure and as such experience the centralising impulses caused by the inward E-Flow from the Mp-shell (Fig 6.6). This means that there has to be some adjustment to the original string configuration after entering the lower EIL atomic state. It is considered that individual strings of protons must be limited in extent for purposes of stability of configuration. Very long strings would have a weaker bonding near its central protons and may have a tendency to oscillate in a whip-like manner. This becomes a probability in nuclei with very large configurations (in excess of 150 protons) and could result in smaller sections becoming separated when a particular string of protons within a nucleus configuration is excessively lengthy (consisting of say 8 or 10 protons in a continuous chain). We would term this as a floating proton chain because it 'floats' between the various impulses acting within the nucleus. Since these impulses are never 'steady impulses' we may assume that each nucleus configuration has an inherent vibration characteristic that is unique. Fig 6.28 will help to illustrate the floating proton chain and the probable break point when separation occurs.

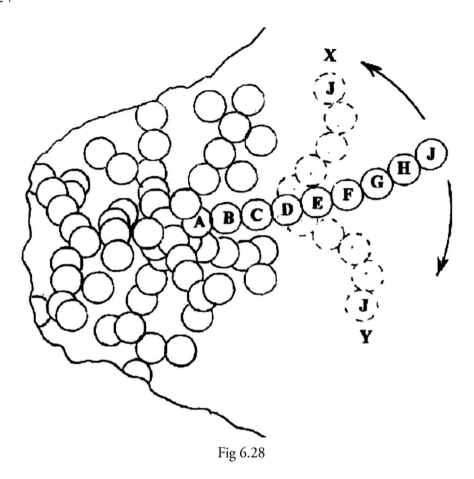

Fig 6.28

Here we show a portion of a large nucleus configuration and highlight the floating proton chain (A-J). In this chain the protons A, B, C and D are held in the indicated positions by virtue of all the repulsion impulses exerted upon them from other proximity protons and proton strings. Let us assume that proton D is at the limit of distance from other proton strings for it to be repulsed by them. The next proton E in the floating chain would thus be under an attraction impulse influence from the rest of the nucleus configuration. This would also apply to protons F, G, H and J. Thus the protons E - J of the floating chain would tend to position themselves closer to the bulk of the nucleus configuration. Let us assume that protons E - J shift to the 'X' position as shown and that they are then in a state of symmetrical balance with the other nearby protons strings. This situation would continue so long as the impulses within the nucleus remained constant. This however is not always the case since there are varying influences not only within the nucleus but also exerted upon it from without i.e. Mp-shells and other atoms. Subsequently, as indicated earlier the nucleus strings have an inherent oscillatory characteristic. In the case of most proton strings the oscillations are merely adjustments to the changes in symmetrical balance within the configuration. In the case of the floating chain there is usually more than one position for a chain to settle into. In Fig 6.28 the floating chain (A – J) can locate in either of the two positions 'X' or 'Y' for perfect symmetrical balance. However when adjustments to that balance are necessary it is possible that the impulse conditions are such as to cause the protons E – J to flex from the X-position all the way across to the Y-position. When this occurs it affects the entire nucleus configuration to a minor extent which then exerts a greater repulsion impulse upon the E - J proton string and may cause them to flex back towards the X-position. If the inherent oscillatory character of the nucleus has a frequency that is a whole number multiple of the period of oscillation of the protons E - J about the X and Y positions then these protons will oscillate indefinitely. Due to external influences occasional hiccups may occur in the oscillatory character of the nucleus and this swinging between the X and Y positions may either be dampened out or else given an additional impulse boost. If the latter is the case then it is possible for the repulsion action upon the proton string E - J to impart to it linear momentum in a direction away from the inner protons A - D of the chain. Obviously proton J will acquire a greater outward linear momentum than proton H which in turn is greater than that of proton G, and so on. In Fig 6.28 the protons F, G, H and J have a free movement while proton E is restricted to movement about the relatively rigidly positioned proton D. Now if the combined linear impulse of the protons F - J is less than the bonding impulse between protons E and F then they will simply flex across to the alternate X or Y floating chain position. If however the combined linear momentum of the protons F - J is greater than the bonding impulse then these protons will break away from the nucleus configuration and speed away at high velocity. The break away protons F–J continue as a single proton string identity and may eventually develop an Mp-shell of their own. The breakaway of such proton strings takes place as a random event within the more complex nuclei (150 plus protons) of an atomic grouping. Depending upon the origin and velocity of the break away string it may do one of two things. It may collide with another nucleus and be captured (temporarily) or it may escape outside the atomic grouping. The reader is free to pursue the former case with an own line of logical reasoning in order to come up with some other aspects of nucleus behaviour.

Let us now return to a scrutiny of the grouping of the bonded protons in a nucleus. It is apparent from Figs 6.27 and 6.28 that there are basically two types of physical set-up. A proton that has physical contact with only one other proton of the nucleus is defined as a mono-bonded proton. A proton that has physical contact with two or more protons of the nucleus is defined as a multi-bonded proton. In Fig 6.25(a) A and B are mono-bonded protons. At Fig 6.25(c) A and C are mono-bonded while B is a multi-bonded proton. At (e) A and D are the mono-bonded protons while B and C are both multi-bonded. At (g) A, C and D are mono-bonded while proton B is multi-bonded. Since multi-bonded protons are always in between other protons we shall refer to them as 'Intermediary Protons' (abbreviated I-Protons). Similarly because mono-bonded protons are always at the ends of proton strings we shall refer to them as 'End-Protons' or E-Protons. As such in Fig 6.27(a) the nucleus on the left has four I-Protons and three E-Protons, while the other nucleus has seven I-Protons and six E-Protons. At Fig 6.27(b) when these combine into a single nucleus there are twelve I-Protons and eight E-Protons. It would seem that proton V has changed its status from an E-Proton to an I-Proton after the amalgamation of the two nuclei.

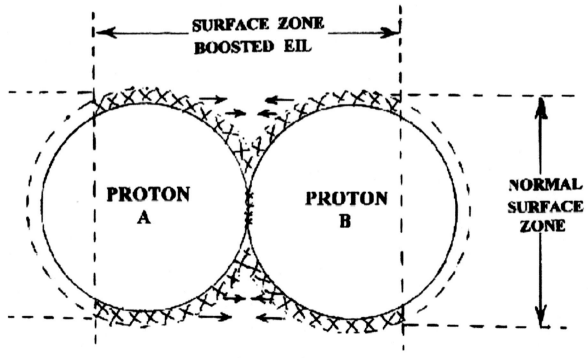

Fig 6.29

All protons whether 'End' or 'Intermediary" are of the same structure. The differences, if any, are entirely within the surface EIL Grid of the proton. Fig 4.6 shows in detail the surface zone of a single proton entity. When such a proton becomes bonded to another proton then the surface EIL Grid is supplied with a much greater quantity of energy medium in those areas directly in line with the other's E-Flow. This is highlighted in Fig 6.29 for the two bonded protons A and B.

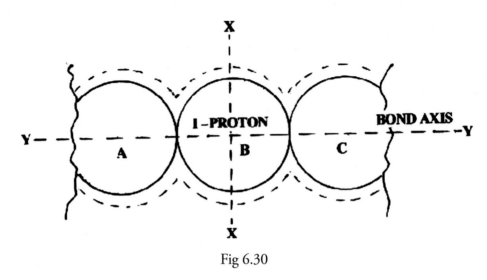

Fig 6.30

The E-Flows from proton A can only reach certain areas of the surface of proton B and vice versa. These areas are clearly shown by the hatched surface grid area which will subsequently possess a higher than normal EIL. Each proton has an area that is hidden from the other's E-Flow and is therefore at a normal EIL. This represents the situation for all E-Protons. I-Protons can have no such blind spot since its entire surface is always in full view of the protons on either side. Fig 6.30 clearly shows this to be the case. Proton B is bonded between the protons A and C and as such is

an I-Proton. The status of protons A and C is immaterial since their effect upon the surface EIL grid of proton B is the same whether they be I or E-Protons. Each face of the I-Proton B receives E-Flow from either proton A or from proton C. Since A and C are similar protons then the surface of B receives equal quantities and intensities of E-Flow. There is however a zone of E-Flow overlap which occurs near the indicated XX plane and it is considered

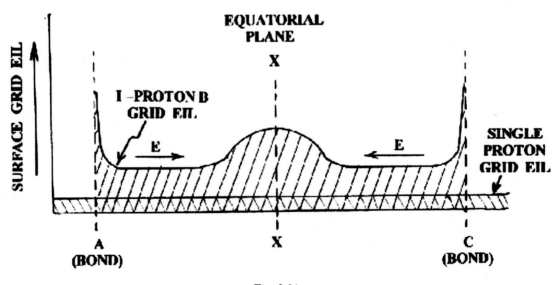

Fig 6.31

that a zone of very high EIL exists here. In theory we thus have a band of high EIL circumventing the I-Proton at a position that is equatorial to the bond axis YY. Fig 6.31 represents the EIL of the surface grid of Proton B along a trace between its contact points with the protons A and C on either side. For purposes of

comparison we have shown the normal surface grid EIL of a single proton in the lower EIL region of the graph. The distance AC represents the longitudinal trace on B joining its contact points with the two protons A and C. The reader will notice that not only is the general surface grid EIL much higher than normal but that an additional EIL peak occurs at the XX plane. This is the basic difference between I and E-protons and it is the effect of the higher surface EIL upon the proton's emissions that separates each in their observable effect upon the surrounding realm.

Essentially the I-Protons tend to be hidden from realm view not only because of their positioning within the nucleus configuration but also by the Mps that proliferate within their surface grid zones. It is presumed that conditions within the surface grid of I-protons are such as to prevent the emergence of the usual forcephoidal patterns. Perhaps the Mps that form within this high EIL grid conform to a set repetition pattern that distorts the setting up of the usual forcephoidal peaks. Since considerable energy medium enters the I-proton surface grid in the general direction of its equatorial plane XX, we would expect the surface grid energy medium to possess a distinctive directional movement. This would subsequently also result in all Mps that formed within the grid to acquire linear momentum in that same general direction i.e. towards the equatorial plane XX. It is not hard to imagine a build-up of energy medium in the vicinity of the equatorial plane XX and that the concentration of repetitive Mps in this area could also set up a sectional band of Mps as shown in Fig 6.32. This band of Mps forms a Mp-shell type belt around the I-proton which acts as a virtual shield to prevent the emission of normal

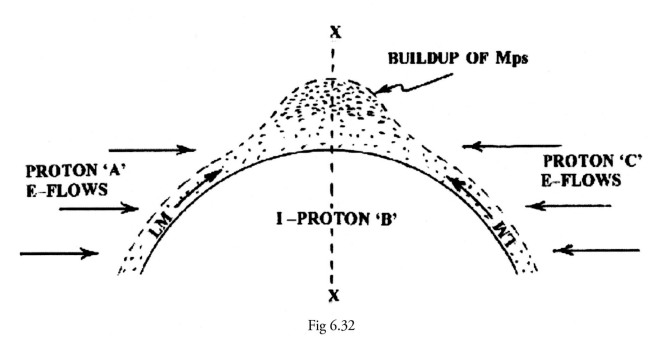

Fig 6.32

forcephoids. Between the protons A and C and this Mp-belt, the I-Proton B is more or less prevented from any forcephoid type action upon neighbouring atoms. An E-Flow does however emanate through the surface grid and as such the I-proton does contribute to the realm datum grid and thereby to gravitational impulses. It would appear that the I-proton with its Mp-belt and enclosing protons behaves somewhat in the fashion of a mini-atom so far as forcephoids are concerned.

Now the Mp-belt continues a buildup through the input of energy medium from the linear momentum aspect of the Mps in the surface grid. This causes the Mp-belt to enlarge. At some stage the Mp-belt can no longer be contained in the surface grid zone and its Mps must then absorb energy medium directly from the I-proton's outward E-Flow. The Mps in the belt acquire linear momentum in the outward direction away from the proton surface. This outward Mp-belt expulsion has only to take its first quantum step for it to burst forth in the manner of wavefront propagation. Because of the string configuration within the nucleus there will obviously be many obstructions in the propagating path of this Mp-belt and subsequently much of its energy medium quantities are re-absorbed by other protons in the same nucleus. There will however be sections of it that exit the nucleus structure and proceed towards and through the Mp-shell of the atom. By the orientation of an atom's nucleus we may ensure a directional preference in exit windows for these Mp-belt sections.

In the primary atom we have a nucleus that consists solely of one proton. The forcephoids from this proton are uniform in their extent and character and subsequently make up an Mp-shell that is a complete sphere. In the more complex nucleus configuration it is only the E-protons that emit forcephoids. These forcephoids are only emitted from one side of the E-proton that has a normal surface grid EIL. Thus the Mp-shell from these forcephoids extends over a limited area only.

However the layout and configuration of E-protons in the nucleus is such that successive shell sections form a 'patchwork' shell structure around the nucleus. There is always an overlap of these shell sections and each 'patchwork' pattern is unique to a particular nucleus configuration. As such the nucleus governs the 'texture' or character of the atom's apparent Mp-shell faces as represented in Fig 6.33 .

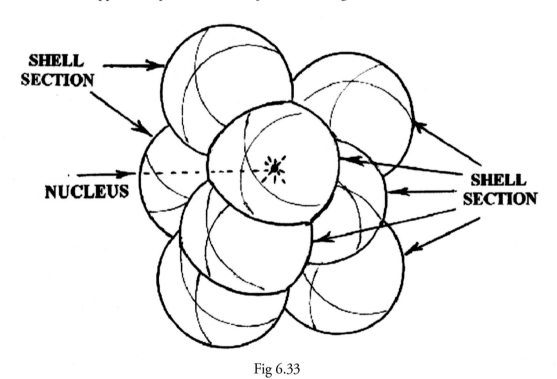

Fig 6.33

The shell sections shown are of course considerably exaggerated as is their symmetry of disposition. Perhaps a better mental view of the complete Mp-shell surround would be to compare its layout to that of fish scales. If we assume the Mp-shell formation to occur at a more or less constant distance from each E-proton then we may imagine the atom as being nearly constant in shell diameter for all nuclei. One may imagine the surface scale size to diminish as the quantity of E-protons in the nucleus increases.

It is not the intention here to make a detailed comparison between the theoretical properties of the Mp-shell, I-protons, E-protons, Mp emissions etc., and those of the current atomic model i.e. electrons, neutrons, protons, etc. There are considerable differences in concept even if the observational aspects appear to have a parallel. For example, how does the I-proton survive outside the nucleus environment? It doesn't, since it must become an E-proton if it is on its own. The neutron has a life of about 15 minutes. Is this the duration for the surface EIL grid of an expelled I-proton to revert to normal? It would be impossible to consider every aspect of the behaviour of Mps within and around a proton or for their effect upon subsequent E-Flow and forcephoidal type emissions. Each area would require an intensity of deductive interrelated reasoning that is collectively an excessive task. For the moment therefore, we hope that a general though slightly fuzzy concept of the atom structure is enough to enable readers to conduct more detailed investigations on their own. Perhaps some of these could point us towards a more correct theory than has been indicated so far. The proton string configuration of the nucleus may not be 100% correct but no alternative could be envisaged that would fit in with the rest of the theory. As such we see the Helium nucleus as a single proton string containing two E-protons and two I-protons. A listing of the progressive proton content of the known elements is given in the Fig 6.34 Table of Elements which shows the jump in I-protons for each additional E-proton in the Nucleus.

TABLE OF ELEMENTS

ELEMENT SYMBOLS	PROTONS E	PROTONS I	ELEMENT SYMBOLS	PROTONS E	PROTONS I	ELEMENT SYMBOLS	PROTONS E	PROTONS I
H	1	0	Se	34	45	Ho	67 9	8
He	2	2	Br	35	45	Er	68	99
Li	3	4	Kr	36	47	Tm	69	100
Be	4	5	Rb	37	48	Yb	70	103
B	5	6	Sr	38	49	Lu	71	104
C	6	6	Y	39	50	Hf	72	106
N	7	7	Zr	40	51	Ta	73	108
0	8	8	Nb	41	51	W	74	110
Fl	9	10	Mo	42	54	Re	75	111
Ne	10	10	Tc	43	56	0s	76	114
Na	11	12	Ru	44	57	Ir	77	115
Mg	12	12	Rh	45	58	Pt	78	117
Al	13	14	Pd	46	61	Au	79	118
Si	14	14	ag	47	61	Hg	80	120
P	15	16	Cd	48	64	Ti	81	123
S	16	16	In	49	66	Pb	82	125
Cl	17	18	Sn	50	68	Bi	83	126
A	18	20	Sb	51	71	Po	84	126
K	19	20	Te	52	75	At	85	125
Ca	20	20	1	53	74	Rn	86	136
Sc	21	24	Xe	54	77	Fr	87	136
Ti	22	26	Cs	55	78	Ra	88	138
V	23	28	Ba	56	80	Ac	89	138
Cr	24	28	La	57	81	Th	90	142
Mn	25	29	Ce	58	82	Pa	91	140
Fe	26	30	Pr	59	82	U	92	146
Co	27	32	Nd	60	84	Np	93	144
N	28	31	Pm	61	86	Pu	94	144
Cu	29	34	Sm	62	88	Am	95	146
Zn	30	35	Eu	63	89	Cm	96	149
Ga	31	38	Gd	64	93	Bk	97	149
Ge	32	40	Tb	65	94	Cf	98	150
As	33	42	Dy	66	96			

Fig 6.34

The quantity of E-Protons is also the stated Atomic Number of each element. It may be a useful exercise to attempt to obtain a pattern to the successive additions of E and I-protons to the elements in ascending Atomic Number order. Fig 6.35 is simply one analysis that attempts to relate the total additional protons in each successive element and the frequency of that occurrence throughout the listing.

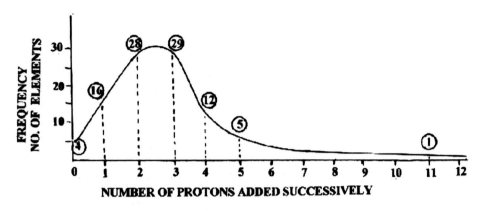

Fig 6.35

So far we have not been able to establish any special significance to this portrayal other than the indication that the most frequent occurrence is the addition of two and three proton groupings. It would be pure speculation to presume that at some state in the evolution toward the complex nuclei that two, three and four proton entities were made available in the high EIL fusion conditions. Our table of elements does not show some of the other nuclei combinations that perhaps existed for a short period as an unstable configuration. There are many isotopes to each stable element and perhaps these ought to be taken into account in the formulation of a graph similar to that in Fig 6.35. Finally, as we bring this chapter to a close we would like to speculate upon the proton layout of some of the more basic nuclei. These are shown in Fig 6.36.

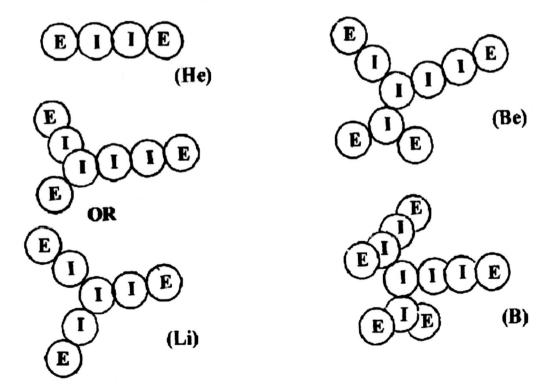

Fig 6.36

It would be worth considering the nature of the Mp-shells that are set-up from these nuclei, bearing in mind that only the outer face of the E-protons can generate forcephoids.

We have thus far only outlined the basic structure of atoms. Considerably more work needs to be done not only in extending the theory but in providing more detail within its assumptions. In the next chapter we shall he looking at the external influences that are exerted by large groups of atoms (mass bodies) upon each other and within the realm. Much of those effects will be based upon what has been discussed here and we shall of course refer back to this and other chapters quite frequently. There is much research that can be undertaken by the reader at various points within this text and we hope that such will be the case.

Chapter Summary

1. The mass of the proton in its present circle 6 phase can be directly related to the realm EIL at which it formed.

2. The proton is a supplier of energy medium to the realm and therefore has a positive function. The Mp-shell absorbs energy medium repetitively and as such plays a negative energy medium role within the realm. The PA (Primary Atom) as a whole does not allow forcephoids to be emitted beyond its Mp-shell and is therefore considered a neutral energy medium entity.

3. A realm forcephoidal environment or field is the result of free protons being in existence around or near that location.

4. The absence of a shell around a proton is for either of two reasons. Firstly the protons may be too widely spaced resulting in insufficient EILs for a Mp-shell to form. And secondly the protons may be too tightly positioned causing too high an EIL resulting in a loss of synchronism among the Mps of the shell.

5. Forcephoids obey the inverse square law for EIL with distance.

6. Protons entering a simple forcephoidal field receive an impulse towards the forcephoid source. An Mp-shell section in the same field receives an impulse in the opposite direction. A PA traversing through such a field remains unaffected.

7. The Mp-shell is a repetitive sphere of perfectly synchronized Mps that occur when successive forcephoids phase with the returning E-Flow from the M-E/X of a previous shell. This shell event sequence is expressed by the 'ABCD shell diagram'.

8. E-Flow sections traverse backwards from the Mp-shell position to the central proton. This results in a centralising impulse upon the proton.

9. Theoretical contact between two PAs results in shell distortion which in turn upsets the equilibrium of the centralising impulses causing the proton to re-locate in a new centralised position.

10. The overall diameter of the Mp-shell is enlarged when the EIL of the realm environment increases and vice versa.

11. Temperature is a measure of the realm energy medium contained within the Mp-shell structure over and above its basic subsistence level. It thus represents the quantity of energy medium that may be given up by the atom without altering its basic structure.

12. Molecular bonding is a direct result of shell distortion and subsequent centralising impulses when two PAs have theoretical shell contact.

13. The Molecule is not a static entity. It is a spinning and bouncing pair of PAs bonded together by an alternating centralising impulse action.

14. A complex forcephoidal field is a pattern of EIL Nodes that are set up within the realm by a group of free protons. The pattern of the EIL Nodes is indirectly related to the positions of the free protons of the group.

15. Proton bonding usually occurs as a string of protons. Although there is a strong bonding action between them, their configuration is the result of repulsion impulses that are exerted upon one another.

16. In the nucleus configuration there are two types of protons. These are the 'Intermediary protons' and the 'End protons' depending upon whether they are within a proton string or at one end of that string.

17. An I-proton develops a very high EIL surface grid near its equatorial axis with a build-up of Mps in a belt type surround. I-protons do not emit forcephoids and subsequently make no contribution to the shell structure.

18. The shell structure of an atom is made up of a large number of shell sections that equate to the quantity of E-protons in its nucleus. The shell sections are knitted together in an overlap fashion and determine the profile and texture of the atom's exterior. This gives to each atomic element a uniqueness of character. As such atoms have a property of orientation in a specific direction.

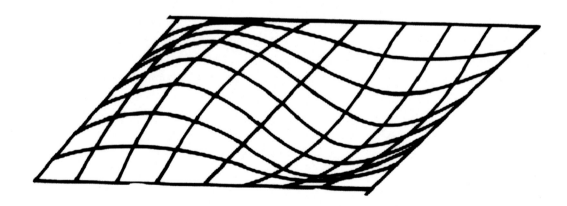

A realm grid section.

CHAPTER 7.

Forcephoidal Effects

In the early part of the previous chapter we introduced the concept of a simple forcephoidal field (Fig 6.1 to Fig 6.3). We showed this field to comprise of forcephoids traversing outwards from a single proton source. These forcephoids were assumed to traverse at the velocity VF which is also the normal velocity for E-Flows. However this was a simplified concept and its sole intention was to impress upon the reader a theoretical indication of how impulses can be generated within mass bodies as a result of their presence in such a zone. In reality forcephoidal fields are a much more complex overlap of forcephoids emanating continuously from large groups of atoms. The intersecting forcephoids give rise to EIL Nodes (Fig 6.22), and we assume that the profusion of these are laid out in some positional pattern. EIL Nodes are simply forcephoidal type peaks that occur repeatedly at relatively stationary point locations within a realm zone. We assume of course that the forcephoidal sources (the protons) are relatively fixed (in the short term) at their realm location. Fig 6.24 indicates the relativities of proton and nodal positions.

It was with intention that we introduced the more complex nucleus structures in the previous chapter. The primary atom has a complete Mp-shell structure surround and as such all proton emitted forcephoids are absorbed into its repetitive make-up. The next stable atom in the series is that of the element Helium containing two E-Protons and two I-Protons in a linear string (Fig 6.25e). Since it is only E-Protons that emit forcephoids, then the Mp-shell around the Helium nucleus will be made up in two separate semi-spherical sections as shown in Fig 7.1. At this stage

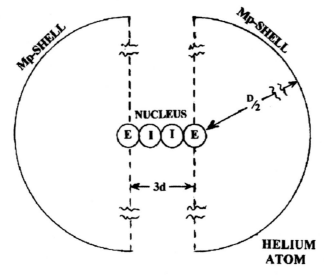

Fig 7.1

we shall assume that the E-protons emit forcephoids from the outer half of their surface area only. This may in fact be less if the influence of the E-Flow from the bonded I-Protons causes a greater 'boosted' EIL effect in the surface grid zone as represented in Fig 6.29. It will be noticed that the Helium atom in Fig 7.1 does not possess a complete Mp-shell surround. There are two semi-spherical Mp-shells, each with a radius of ⌐D and positioned symmetrically with regard to one another. They are separated at their closest by a distance approximately equal to three proton diameters i.e. 3d. Since the proton diameter is extremely small in comparison to its shell size we can assume '3d' to be practically negligible. As such we may consider the Helium shell to be very nearly complete and assume that under normal conditions forcephoids are not emitted beyond its boundary.

However as we proceed further down the table of Elements (Fig 6.34), these shells become more sectionalised and their arrangement pattern more complex. The physical character and 'texture' of an element is due to the configuration of its Mp-shell sections. The orientation of the proton strings within its nucleus may be altered when the atoms are part of a large molecular bonded group. In this state Mp synchronisation within the shell can be upset resulting in a directional emission of forcephoids outside the atom. We may speculate upon the mass body structure in which most of its nuclei have their configurations in a parallel orientation. Thus if forcephoids have a tendency to 'leak out' at a particular shell gap then all the orientated atoms of the mass body will 'leak' in the same general direction. This means that EIL Nodes can be set up from such a mass body structure in a zone outside itself and in a set direction. Depending upon the structural pattern of the atoms the EIL Nodes will also conform to some layout pattern.

The configuration of shell sections are also responsible for the manner in which atoms prefer to bond. We may assume therefore that atoms of the same element when grouped together as a cohesive body establish themselves consistently in the same atomic combination or body structure. As such different elements will have differing body structures. It is not the intention here to speculate upon the actual configuration of any particular element but rather to present the reader with a basic concept. This is a concept which upholds the principle that a body structure when located within a zone of EIL Nodes experiences a marginal displacement of its structural pattern. It is also considered that this displacement however marginal results in the atomic Mp-shell molecular bonds being distorted out of their normal 'close-in' positions. An impulse of the type represented in Fig 6.17 for molecular bonding becomes active and will continue until the upset is normalised. The result however is that so long as the various impulses remain out-of-balance, the entire body structure acquires an impulse in the direction of the original distortion.

Before we accept such a vague impulse concept we must establish that EIL Nodes under exceptional conditions of pattern can be made to influence a correspondingly unique structural combination of atoms. We do in fact presume that one group of atoms can set up a unique pattern of EIL Nodes that has an effect upon another group of atoms. Later we shall see that each group must have a similarity of structural layout and that each is automatically in the other's field of EIL Nodes. Also, that their effect upon one another is mutual and opposite with the same impulse magnitude. It must be remembered that an EIL Node is not a mass entity but simply a peak of energy medium that occurs due to the phasing of forcephoids at a set realm location. There is no resulting E-Flow from an EIL Node. A profusion of EIL Nodes occurring in a systematic though intricate layout pattern may cause particular nuclei configurations to be re-orientated in a preferred direction. It is this tendency for orientation of the nucleus into a new directional position along with the shell's centralising action which is the basis of action in the above impulse concept.

Consider the nucleus configuration represented in Fig 6.27(b). Assume that this provides a stable atom and that shell bonding characteristics are such that a strongly cohesive structure is possible. Let us in the first instance assume this structure or mass body to possess a random orientation of its atoms. This means that the preferred bonding between Mp-shell sections is not selective in any particular direction. Subsequently the forcephoids that are emitted from each nucleus would be in random directions and the layout spread of EIL Nodes would not conform to any regular pattern. If a similar mass body were then located within this inconsistent nodal field there could be no relationship between nucleus orientation and EIL Nodes. Therefore these EIL Nodes do not play any part in re-orientation of the atoms within this structure. As a second case for the Fig 6.27(b) nucleus let us assume that the mass structure does have a shell with a preferential characteristic for its molecular bonding. This means that although there may be a looseness of bonding connections, the orientation of shells and subsequently that of their nucleus centres will possess a rather high degree of directional correlation. Since the gaps between shell sections through which forcephoids may be emitted are also part and parcel of the same orientation plan, then the EIL Node pattern that is set up will also have a high degree of correlation with the mass body structural layout. The spacing between adjacent EIL Nodes is of course variable with distance from their mass origin but is assumed to be less than 'd' the proton diameter. It is this factor that presents a 'realm datum EIL grid' type of environment which plays an important role in nucleus orientation.

Before we can proceed further with this topic we must show how a nodal pattern can cause the nucleus string to be pushed into a preferred orientation. Consider the principle of the simple balance weighing machine. The weighing is done by balancing one mass quantity with another of known weight as shown in Fig 7.2. When the scale arm is level then the weights are considered equal. However a small difference in weight will cause the scale arm to acquire an alternative stable position which is not level. The gravitational impulses upon each weight are unequal but

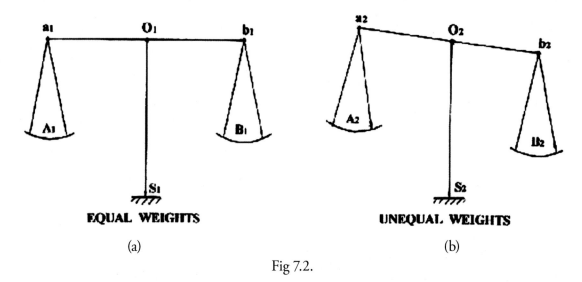

Fig 7.2.

are brought into balance by the re-orientation of the balance-arm as shown in Fig 7.2(b). The reason for this new balance position arises from the fact that although both arms of the balance are designed to be equal in length, the pivot edge is never a perfect point. Tilting of the balance arm as at (b) causes the pivot point to shift marginally towards the side of the heavier weight B2 such that a new position of equal torque is once again attained. This is given by,

$$A_2 \times a_2O_2 = B_2 \times O_2b_2 \text{ where } A_2 < B_2 \text{ but } a_2O_2 > O_2b_2$$

The position of balance is only attained by equal torque which relates a greater difference in the weights of A and B by a correspondingly greater difference in the effective lengths of a_2O_2 and O_2b_2 caused by an increased tilt of the balance arm.

Now if you were to apply a downward pressure at the a_2 position of the weighing machine to bring the balance arm to a level position and then to maintain that position the above equation will become,

$$A_2 + P = B_2,$$

since $a_2O_2 = O_2b_2$ in the level position and where P is the downward pressure exerted by your finger. If this system is in equilibrium then the pressure P of your finger upon a_2 is resisted and balanced by an equal pressure impulse in the opposite direction. In other words an impulse is being exerted continuously upwards against your stationary finger by the balance arm at a_2. Quite obviously all impulses in this example are due to gravity.

However let us now conduct a thought experiment using the balance arm but without the suspended weight pans or any weights. Also let the balance arm be totally weightless. By this assumption we have ruled out any impulses caused by the earth's gravity. As such we may now also dispense with the support stand O_1s_1 and O_2s_2 upon which the balance arm was pivoted. So now, as if by magic we have a balance arm a_2b_2 suspended motionless in mid air. Let us assume a large quantity of these magical balance arms grouped together and positioned at random angles of tilt but with a maximum tilt variance between ± 30° from the horizontal. We must also assume each arm to be flexibly connected to the other arms on either side as represented in Fig 7.3. Let us place this set-up within a field of EIL Nodes with nodes arranged in a horizontal pattern. Assume each balance-arm to be made up of a single string of protons.

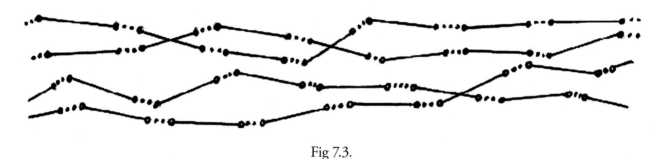

Fig 7.3.

Now although EIL Nodes are relatively stationary features, we must remember that they are simply the intersecting points of a number of regular traversing forcephoids. We must also remember that forcephoids cannot traverse through a proton (Fig 6.1). As such the set up of EIL Nodes on the downstream side of a proton have a marginally lesser energy medium content than if the proton had not been there in the first place. The very existence of a mass body within a nodal field causes a weakening of that field within and in the leeward side of its structure. We must also consider that any group of protons within a nodal field may achieve an orientation of its own structure and thereby set up a nodal field of its own. This may in turn affect the orientation of the structure setting up the initial nodal field. This may only occur if both proton structures are compatible not only with one another but with their nodal field patterns as well.

In Fig 7.3 we assume each balance-arm to be made up of a single string of protons. We shall initially only concern ourselves with the E-protons of the string. Since the orientation of each balance-arm is random within the limits specified (± 30° variance) then E-protons on the same string are quite likely to find themselves in nodal zones of different node EIL. The effect of this is best explained by the illustration of a basic Nodal field set-up as in Fig 7.4. Here we represent the balance-arm AB as being located within a simple Nodal field set-up. The EIL Nodes are arranged in a horizontal line configuration. The nodes in each line make up a nodal row which is denoted by a nodal row number that increases with distance from their origin.

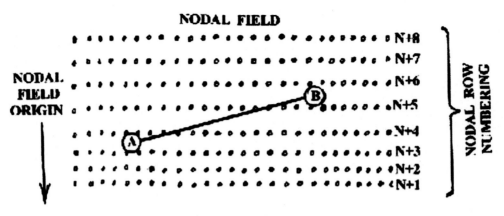

Fig 7.4

Nodes in the same Nodal Row are at the same peak EIL by virtue of having similar set-up conditions. The larger the nodal row number the further are its nodes from their proton body origin and the lower is their individual peak EIL. We shall refer to nodal rows by their Nodal Row Number or NR number. Another characteristic of the nodal field that we must assume is that of inter-nodal spacing. The smaller the NR number the closer are the positions of the nodes to one another within that row. Also the distance or spacing between nodal rows increases in line with the NR number.

In Fig 7.4 the E-proton A is shown to be in a nodal field zone with NR numbers ranging from (N+3) to (N+4), while proton B is in the field zone (N+5) to (N+6). This means that A lies in a higher EIL field zone than B. Now in Fig 6.3 of the previous chapter we indicated that a proton placed within a simple forcephoidal field received an impulse in a direction opposite to that of the traversing forcephoids. In the more complex forcephoidal field however we must also consider the effect of EIL Nodes. As stated earlier a node is a relatively stationary feature and as such should not affect the proton in the same manner as forcephoids. Since the nodal field has an extremely close-knit configuration of EIL nodes, many of these are bound to occur within the surface EIL grid of the proton. It is the contention here that such locations will trigger E-M/Xs to produce Mps in a regular and repetitive manner. Since Mps by virtue of their M-E/Xs produce an outward E-Flow then it is considered that as a result of the nodal field the proton will absorb this additional E-Flow directed upon its surface Mps. Because forcephoids cannot traverse through protons these nodes can only occur within the upstream face of the proton's surface EIL grid. Although nodes do occur in the areas behind the proton none can occur in the downstream surface EIL grid which lies in a sort of forcephoidal shadow. Behind each proton lies a zone that is devoid of field nodes, assuming of course that they have a single field origin to the front of the proton. Incidentally, this feature of forcephoidal absorption may be used to shield an area from a nodal field. Fig. 7.5 indicates the forcephoidal field shadow on the downstream side of the proton.

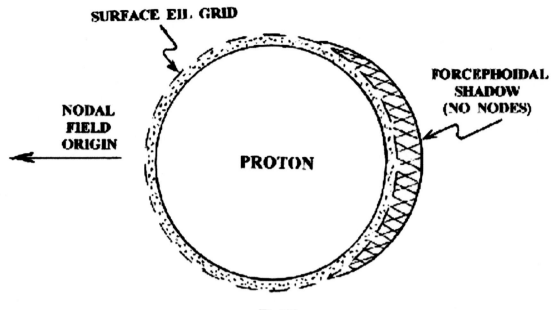

Fig 7.5

As a result of this additional E-Flow onto one face of the proton, additional linear momentum is absorbed here and transmitted towards the proton centre. As described in Chapter 4 this must cause an imbalance in the centrally directed impulses which causes a resultant impulse in the direction away from the nodal field origin. From the above we arrive at two conclusions on the impulse generated inside a proton by a Nodal field. These are similar in principle to the dual nature of gravitational impulses in that two separate impulse activities are always in existence - one towards and the other away from the origin zone. The resultant depends upon which impulse activity is the greater. At very low NR number locations the nodal spacing is small, indicating a larger quantity of high EIL nodes being active inside the proton's surface EIL grid. Consequently, the away impulse or repulsion of the proton from the nodal origin over-rides. At the higher NR number locations nodal spacing is larger and node EIL is lower and may even be too low to be effective within the proton surface EIL grid zone. Consequently the simple forcephoidal effect of Fig 6.1 over-rides and a resultant attractive impulse prevails.

We shall now return to our thought experiment and to the effect of the Nodal field upon the balance-arm setup of Fig 7.4. Let us put a constraint upon the arm AB in the form of a pivot point at its geometrical centre. The analogy here is with that of a complex nucleus string quite unsymmetrical in its configuration and constrained in its position by the centralising impulses from its Mp-shell. This pivot point in conjunction with our assumed flexible connection to other balance-arms will limit the over-all displacement of each arm within the structure of the mass body. The impulses generated within the E-protons A and B will be unequal, since each is within a different NR number zone of the Nodal field. Proton A will most certainly experience a greater repulsion impulse (due to nodal spacing and EIL) than proton B. Proton A also experiences a greater attractive impulse (due to simple forcephoids) than proton B. However the resultant impulse of one type against the other is considered in favour of the nodal impulse (repulsion) when the NR numbers are small. We now have a simple balance-arm situation in which the pull or push upon one side is greater than the other. Assuming our balance-arm structure to be rigid, we would thus observe it to rotate or orientate itself to a position where the impulses within the protons A and B become equal. This can only be achieved if both protons are in similar NR-number zones. Due to our assumed flexible connections between the arms, (analogy of intermolecular bonding), complete freedom of orientation does not exist. As such the balance-arm AB in Fig. 7.4 will attempt to align itself in a direction parallel to the nodal rows (N+4) and (N+5), and will be partially successful only. We shall return to this aspect a bit later.

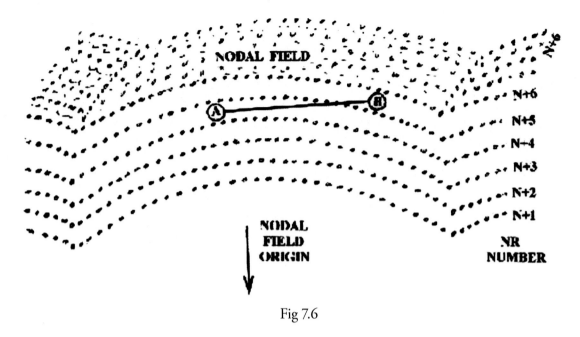

Fig 7.6

Nodal rows as depicted in Fig 7.4 and which occur in straight lines or flat parallel planes are considered quite improbable. It is much more probable that nodal rows occur in some form of a curved repetitive pattern that is symmetrical with regard to an imaginary axis through the field and its origin. The size of the repetitive nodal row pattern is as important as the pattern itself when considering nucleus orientation. Let us continue our thought experiment with the balance-arm AB in such a nodal environment.

Fig 7.6 shows that although the balance-arm AB is in the horizontal plane, A and B are not in the same NR number zone. Therefore the impulses generated within them will not be equal and orientation of AB will occur until both are in the same NR-number zone. In Fig 7.6 we have tried to represent the Nodal field as being made up of a large number of curved sections. As such each NR number actually applies to a curved plate that has a repetitive profile. The curvature and the repetitive sections determine the pattern of the Nodal field.

Let us now return to Fig 7.3. Here we have a large number of interconnected balance-arms positioned at a random orientation. The random orientation has a limited variance of ± 30° about a mean horizontal plane in a nodal field somewhat similar to that represented in Fig 7.6. Now each balance-arm will try to orientate itself so that each of its E-protons are in a zone with the same NR number. There will obviously be some shifting and orientation of the balance-arms but a complete re-orientation is not possible because of the inter-connecting constraints. As such a considerable proportion of imbalance will remain within the structure.

In Fig 7.6 we have imagined the Nodal rows to form an undulating curvilinear-sectional pattern and we would expect their effect upon the orientation of balance-arms to be one of pattern correlation. Since each nodal pattern is symmetrical across a 3-dimension cubic area, balance arms at the same NR number are not necessarily parallel to one another. As such we must appreciate that a nodal row is a section with x, y and z type co-ordinates. For simplicity of explanation however we shall continue with 2 dimension flat diagrams and corresponding descriptions. Consequently Fig 7.7 represents the re-orientation of the flexibly connected balance arms of Fig 7.3. We have assumed a high level of pattern correlation between Nodal field and mass body structure which is quite

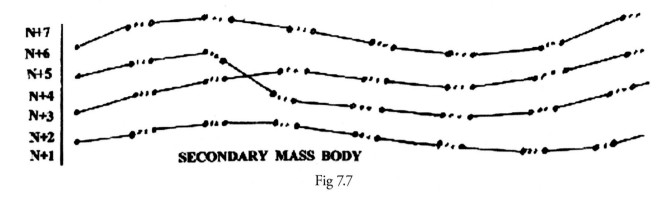

Fig 7.7

unusual but which is necessary for the purpose of our discussion at this stage. The NR numbers listed beside the balance-arm linkages are simply to indicate nodal field direction. Earlier we stated that for every grouping of atoms a particular pattern of EIL nodes is set up in the immediate area around it. As such the nodal field set up by the nuclei distribution of Fig 7.3 will be quite different from that set up by the Fig 7.7 configuration. Let us assume that the nodal pattern set up by the Fig 7.7 nuclei corresponds closely with its own structural layout. Let us also assume that the initial nodal field that caused this orientation came from a mass body with similar structural layout. This initial field is termed the Primary field. Although all atom groups generate nodal fields with some sort of pattern we shall only consider those Primary fields which are able to induce extensive orientation in another body structure termed as the 'Secondary mass body'. The nodal field that is then set up by the nuclei of the secondary mass body will be termed the 'Secondary field'. The mass body from which the Primary field originates is termed as the Primary mass body. Obviously the above relationships depend upon the primary and secondary mass bodies being of a compatible or similar structure. That is to say, of the same basic nucleus /atom /element design. In the entire range of known stable elements as listed in the table of Fig 6.34 there is only one element that possesses a close correlation between the pattern of its mass body structure and its nodal field. That element has a nucleus configuration containing 26 E-protons and 30 I-protons. Perhaps the molecular bonding making up the structure has a preferential bond direction that results in a naturally semi-orientated condition. The shell is made up from 26 shell sections in relative positions that correspond directly with those of the E-protons in its nucleus. The molecular bonding between the atoms of this element will of course be stronger in some directions than in others and it is this feature that we have considered as a flexible connector between adjacent balance-arms. The uniqueness of this structure to produce a similar nodal pattern stems from its own nucleus orientation and over-all shell size. Under conditions of excessive realm EIL the shell size enlarges (Fig 6.11). As such an increase in the atomic 'temperature' may result in a loss of this structural correlation. Perhaps due to the greater centralising impulses from the shell (Fig 6.6) the orientation of the nucleus alters sufficiently to cause both structure and nodal patterns to lose their correlation to one another. Also perhaps at some other higher or lower EIL condition, one of the other elements may develop a structure with nodal compatibility. All further references with regard to Primary and Secondary mass bodies are directed at the Iron (ferrite) element structure under current EIL conditions.

As a result of the secondary mass body of Fig.7.7 a secondary nodal field is set up that pervades the area of the Primary Mass Body. This must influence the orientation within the Primary mass body by an improvement to its nuclei orientation. In this case the curvature of the secondary field pattern is in the same directional phase as the Primary mass body structure. Any improvement in orientation is immediately reflected in a stronger Primary field which in turn further improves the orientation in the secondary mass body which then strengthens its secondary field and so on and on. However, limitations in nodal intensities and inherent structural resistances prevent any increase in orientation add infinitum. It must be stressed that changes in the structural configuration within a mass body are extremely small, and our diagrammatic representations of orientation movements are exaggerated for explanatory purposes only.

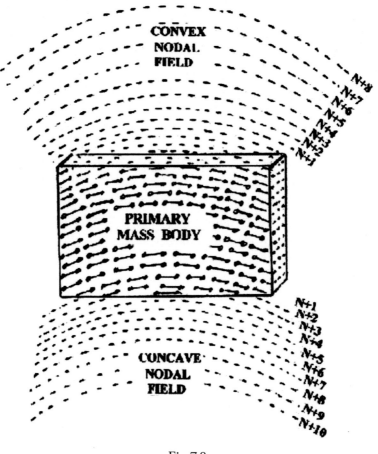

Fig 7.8

Consider the Primary mass body cube in Fig 7.8. Let us assume that the structural configuration of its balance-arm linkages is orientated mainly about the horizontal plane as shown. The subsequent Primary field adopts a nodal pattern that is in parallel with this structural configuration. As such, there appear to be two identical Primary fields on opposite ends of the Primary mass body. One field is at the top end and the other at the bottom end. It would also appear that the nodal pattern is an extension of the body structural pattern making it seem as though there was one continuous field throughout the diagram. By definition we consider a nodal field to commence at the surface of the Primary mass body and with NR numbers increasing with distance from that surface. As such, the 'Top' nodal field is very different from that at the 'Bottom'. Both have NR numbers that increase with distance from the structural origin. However, if we view each surface in turn the top nodal field may be considered as having a convex nodal pattern while the bottom field has a concave nodal pattern. The reader will notice that we have not indicated nodal fields originating from any of the vertical faces of the body cube. This is because the nodes that occur in these directions do not conform to a relevant pattern and so cannot have an orientation effect upon the nuclei of another 'Iron' body. As such we may disregard them in total.

Let us for a moment consider the Primary Mass Body. We have stated that its structure is made up purely from the physical element Iron. In order that this produces an effective nodal pattern its structural linkages must already be in an orientated state. We shall assume that at some previous Realm Date (Fig 5.10), an orientation of its structure was induced and that this re-orientated state was retained as a permanent structural feature. Let us now position this Primary Mass body A in the nodal field of another similar Primary Mass body B as represented in Fig 7.9.

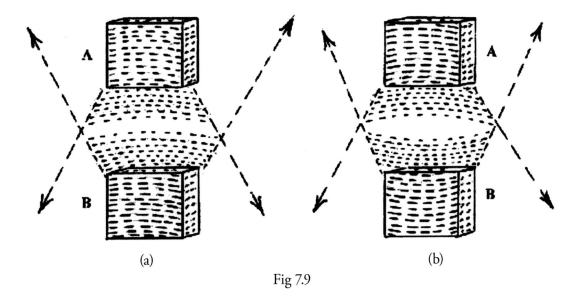

(a)   (b)

Fig 7.9

In the setup at (a) both A and B are positioned so that their structural orientations are in parallel. Each structure will be within the nodal field of the other and their patterns will also be in sympathy, i.e. with the same curvature direction. The orientation within each structure is already in the same direction as the nodal pattern and subsequently a strengthening of the existing orientation levels will be induced. In the set up at (b) however the Primary Mass bodies are positioned with their

structural orientation patterns in an opposite curvature direction. Thus the nodal field of A will not be in sympathy with the body structure of B and will attempt to induce the structural pattern of B to fall in line with its own nodal pattern. The same will be true of the nodal field of B with regard to the structural pattern of A. If we class both A and B as structures with a permanently fixed orientation, then the effort exerted by one nodal field upon the structure of the other would seem to have been quite ineffective. Perhaps no re-orientation was achieved but as we shall see later there appears to be quite an interesting side effect.

We know that a physical body made up solely from atoms of the Iron element has the property of relative softness and malleability. This means its structural molecular bonding is rather flexible and its nuclei can be re-orientated quite easily. Such a mass body could not hold to a permanent orientation pattern and can thus only be classed as a secondary mass body. However since no mass body has 100% purity there must be inclusions of other elements in its make up. There are certain other elements e.g. carbon, which if mixed in with the Iron in small proportions cause the entire property of the mass body to be altered to a tougher and harder material. It is considered that carbon interferes with the original Iron structural linkages and so reduces the former freedom of orientation movement. As such these impure structures are more likely to retain their orientated structural patterns over long periods of time. Such mass bodies may be considered as semi-permanent or even permanent primary mass bodies depending on how well and for how long their orientation pattern is retained. Permanent Primary Mass Bodies are usually referred to as natural magnetic bodies or 'magnets' and the end faces from which nodal fields are emitted are referred to as its 'magnetic poles'. In Fig 7.8 we shall refer to the 'Top' end of the Primary Mass Body from which a convex type of nodal field is emitted as its 'North' magnetic pole. The bottom end which emits a concave nodal pattern will be the 'South' magnetic pole. The line through the centre of the North and South pole faces is called the magnetic polar axis. It must be stressed that the structural pattern inside the Primary Mass Body is directionally consistent i.e. completely uniform along its polar axis. It is considered that the curvature of each nodal row within the nodal field increases with NR number or distance from the pole face. This applies to both concave and convex patterns. As such we find that the curvature of the nodal rows is always greater than that of the structural curvature of the mass body within its field. If we represent the structural pattern of a mass body by a solid curve and the nodal row of a nodal field by a dotted curve then Fig 7.10 shows their curvature relativity.

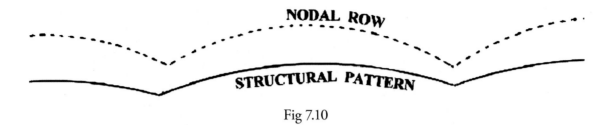

Fig 7.10

So far we have attempted to portray the mass body structure and its emitted nodal field in principle only. At this stage there is insufficient theoretical detail to permit a more detailed and precise concept with any degree of certainty. As such we cannot state precisely the manner in which linear momentum is induced within the Primary Mass Body by a nodal field. However we can speculate upon a reason for such an impulse based upon the forcephoidal principles developed earlier. It is considered that further orientation of the nuclei within a Primary Mass Body is resisted by the Mp-shell molecular bonding action and centralising impulses (Fig 6.6). Consequently the Mp-shells are also a factor in the structural pattern of the magnetic body. It is highly probable that the impulse generated within the mass body is the resultant of not only the nodal and forcephoidal effects upon the nucleus protons (Fig 7.5), but also of their effects upon the Mp-shell. The latter in turn affect the nucleus with a marginally altered centralising impulse. It is apparent that there are a number of separate impulse quantities that are active within the Primary Mass Body and that under normal conditions these are in balance with a zero resultant. However upon entering a nodal field this balance is upset by the orientation effect upon the overall structural layout. Consequently this impulse imbalance occurs throughout the mass body linkages such that the body itself cannot oppose the impulse. The body then acquires linear momentum in the polar axis direction. The reader will notice a similarity between this concept and that of the gravitational impulse induced within the proton body from an EIL gradient (Fig 4.8). In each case the balance of impulses within the structure was upset by external influences.

Although we have only generalised upon the nature of the impulse imbalance created in a mass body we are nevertheless able to indicate the resultant impulse direction under differing nodal field conditions as a defined set of rules. In Fig 7.11 the configuration of a Primary Mass Body structural pattern is represented by the solid curve AB. The mass body is located within the nodal field which has its format indicated by increasing NR numbers. The curvature of nodal rows and of the structural linkage pattern are in the same curvature mode (convex or concave). Since the curvature of the nodal field is always greater than that of the structural pattern, it is considered that the nodal rows will exert an orientation effect upon AB to increase its curvature. Since AB has structural

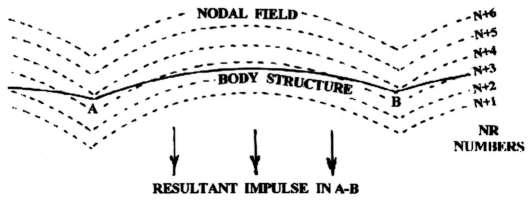

RESULTANT IMPULSE IN A-B

Fig 7.11

constraints that prevent this from being achieved then the impulses within the structure can no longer remain in perfect balance and the primary Mass body experiences a resultant impulse. The direction of this impulse is towards the nodal field origin.

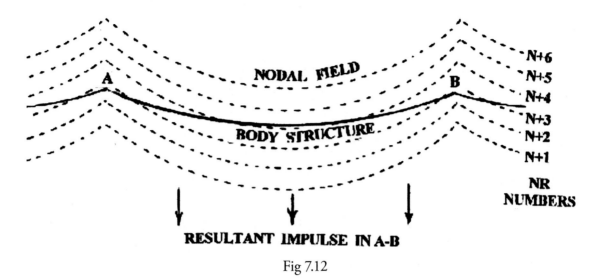

Fig 7.12

In the next case shown in Fig 7.12 we have simply reversed the NR numbers of the previous case and inverted the diagram. The curvature of the nodal rows and of the structure pattern of AB are still both in a similar curvature mode. As before there is an orientation effect upon AB to increase its curvature to that of the nodal rows. The structural stiffness resists this effect resulting in the imbalance of internal impulses. Once again the Primary Mass body AB experiences a resultant impulse in the direction of the Nodal field origin. In both of the above cases (Figs 7.11 and 7.12) we observe that so long as the curvature status of the pattern for Nodal field and primary mass body structure are in the same curvature mode then the resultant impulse in the mass body will always be directly towards the nodal field origin.

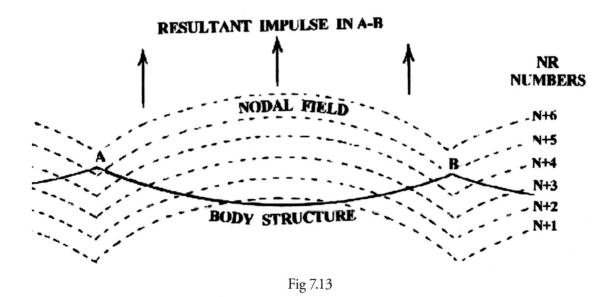

Fig 7.13

In Fig 7.13 we take up the case in which the curvature status of nodal field and Primary mass body structure are in opposite curvature modes. Since the curvature of the nodal rows are in complete opposition to the current orientation of the nuclei in AB, it is considered that the nodal field in an effort to orientate AB towards the field curvature, actually attempts to disorientate the structural pattern. This means that the nodal field must first cause the curvature of AB to become flatter and flatter in order for it to orientate towards the opposite curvature. Once again structural constraints resist this tendency and an imbalance in the internal impulses results. This causes AB to acquire an impulse in a direction away from the nodal field origin. Fig 7.14 is another similar case in which the curvature status of the nodal field and Primary body structure are in opposite curvature modes. The same arguments apply with similar results. The resultant impulse in AB is in a direction away from the nodal field origin.

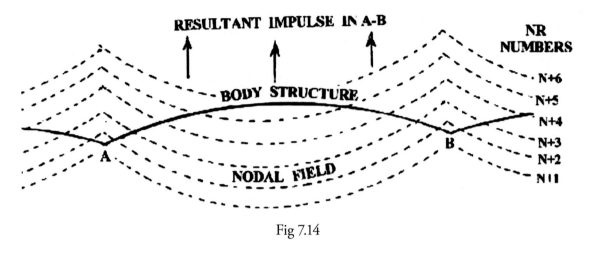

Fig 7.14

There is one other case to consider and that is of a Secondary Mass Body within a nodal field. Initially the mass body is considered as having no defined orientation pattern (Fig 7.3). When located within a nodal field the structural units within the mass body adopt a configuration as close to that of the nodal rows as is permissible (Fig 7.7). We now have the situation represented by the Figs 7.11 and 7.12. As such the Secondary Mass body always acquires an impulse in the direction of the Nodal field origin.

The amount of the impulses generated within two adjacent mass bodies is dependant upon the intensity of the nodal field within which each body is located. The field intensity or field strength is directly proportional to three factors. These are the node EIL, the population density of nodes within the nodal row and the closeness of nodal rows to one another. Each of these factors is greater the closer the nodal row is to the field origin. The field origin is the magnetic pole of the Primary mass body. As such we observe that the impulse is inversely proportional to the distance between the facing magnetic poles of two adjacent Primary Mass bodies. Without going into detail we shall assume these impulses to obey the usual inverse square law for distance.

In Fig 7.9(a) both of the mass bodies A and B have their structural patterns in a parallel mode. Consequently the nodal fields of each are also in the same curvature mode as the opposite body's structural pattern. In this case Fig 7.11 applies to A while Fig 7.12 applies to B. As such the impulse generated within A is in the direction towards B. Similarly the impulse in B is toward A. We observe that the magnetic south pole of A (concave pattern) faces towards the magnetic north pole of B (convex pattern). In Fig 7.9(b) the mass bodies A and B have their internal structural patterns in opposite modes. Each structure is in the other's nodal field such that the situation for A is represented by Fig 7.14 and for B by Fig 7.13. In this case both A and B experience impulses away from each other. We observe that the magnetic south pole of A (concave pattern) faces the magnetic south pole of B. From the above we may generally conclude that magnetic bodies receive impulses towards one another when positioned with unlike poles facing one another and receive impulses away from each other when positioned with similar poles facing one another. Or that unlike poles attract and like poles repel one another. Fig 7.15 shows the impulse direction for these magnetic pole combinations.

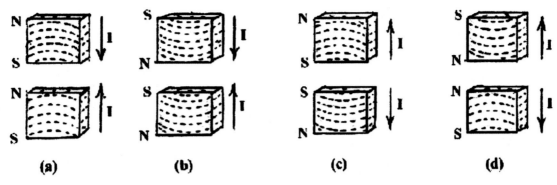

Fig 7.15

The magnetic poles are indicated by N for north and S for south, while the impulse direction is indicated by I. It is of the utmost importance that the reader does not interpret the above facts as meaning that magnetic bodies either attract or repel one another. They do not. Each magnetic body is only aware of the EIL nodes of the nodal field that pervades its structure. Its response is to that nodal field only, and if such a field could be produced by other means then the result would be the same i.e. impulse in a particular direction. We shall consider this aspect in a later chapter in our pursuit of a unidirectional impulse mechanism for a futuristic propulsion system.

Let us now consider the relationship between the intensities of the impulses generated within the magnetic body pairs of Fig 7.15. We have indicated that each is in the nodal field set up by the other, and that the intensity and concentration of these field nodes are dependant upon the structural orientation within each Primary mass body. Let us in the first instance assume that we have a Primary Mass Body A and a Secondary Mass Body B positioned near one another. The Secondary Mass Body B does not of its own accord emit the required pattern of EIL nodes until it comes under the influence of a Primary field. The structure of the Secondary Mass Body B will then be orientated in sympathy with the Primary field. However there are always restrictions to orientation within the body structure caused by the presence of other non-iron elements and so the level of orientation is limited. The strength of the Secondary field then emitted by the Secondary Mass Body B is at a correspondingly reduced level. The primary mass body A is now in the zone of this Secondary field and the response of its structure (Fig 7.11) will be directly proportional to the strength of that field. We observe therefore, that although the Primary Mass Body A may possess a very high level of permanent orientation, it comes under the influence of a relatively weak field. The Secondary Mass Body on the other hand is at a lower level of structural orientation but within a relatively stronger nodal field. The impulses in each structure are directly proportional to the product of their structural orientation level and the strength of the nodal field in which they are placed. The orientation level may be defined as the quantity of balance-arm units within a mass body that are correctly orientated with respect to the nodal field rows. We may consider this orientation level in terms of orientated units within a given structural volume or as the orientated Number per unit volume. The reader may consider this as the orientation density.

Thus for a given distance between A and B we have,

$$\text{Impulse in A} = K_1 \times \text{Field Strength of B} \times \text{Orientation Density of A}$$
$$\text{Impulse in B} = K_1 \times \text{Field Strength of A} \times \text{Orientation Density of B}$$

Where K_1 is a conversion constant and the impulse in A is opposite to that in B.
Now we also know that the intensity of EIL Nodes at a given distance is directly proportional to the orientation Density of the emitting mass body. Thus,

Field Strength of A = K_2 × Orientation Density of A

and,   Field Strength of B = K_2 × Orientation Density of B,

where K_2 is a conversion constant.

By combining the two sets of equations we have,

Impulse in A = $K_1 K_2$ × orientation Density of B × Orientation Density of A,

and,   Impulse in B = $K_1 K_2$ × orientation Density of A × Orientation Density of B

thus,   Impulse in A = Impulse in B

From this we may conclude that the magnitude of the impulses in each of the two pre-positioned magnetic bodies of Fig 7.15 are equal and opposite, and each is directly proportional to the product of their orientation Densities. Thus when a Secondary Mass Body is placed within the nodal field of a Primary Mass Body, both experience an impulse of equal magnitude in the direction of each other.

In the case of two Primary Mass Bodies being located as in Fig 7.15, the same arguments apply and with the same conclusions. In each paired configuration equal and opposite impulses are induced according to the rules of Figs 7.11 to 7.14. Let us illustrate this with an example. Consider Body A to have an orientation Density that is half that of the Body B. On this basis the nodal field of B (at a given distance) will be generated by twice the number or orientated nuclei than the field generated by Body A. As such the field strength of B is twice the field strength of A. Let the field strengths of A and B be denoted by F_A and F_B respectively and their orientation densities by N_A and N_B. Then we have,

$F_B = 2 F_A$, and $N_B = 2 N_A$ ....(a)   and

Impulse in A = $K_1 \times F_B \times N_A$ ....(b)

Impulse in B = $K_1 \times F_A \times N_B$ ....(c)

By transposing (a) in (b), we have

Impulse in A   = $K_1 \times 2 F_A \times$ 🗁 N_B

= $K_1 \times F_A \times N_B$

= Impulse in B.

Alternatively by transposing (a) in (c), we have

Impulse in B   = $K_1 \times$ 🗁 $F_B \times 2 N_A$

= $K_1 \times F_B \times N_A$

= Impulse in A.

The reader should remember that the impulses in A and B are of opposite sign when considering overall directional configurations. We must now leave this magnetism topic temporarily in order to continue our discussion into areas of other equally interesting forcephoidal effects.

Let us now consider the I-proton and its functional relationship to forcephoidal type effects. In the last chapter a basic description of the atom nucleus was presented. We subsequently observed the nucleus to be constructed from protons bonded together in a series of inter-connected strings. We showed that a strong repulsion impulse existed between the adjacent proton strings due mainly to the intensity of the E-Flows ensuing from them. The E-protons provide the atom with its sectionalised Mp-shell. The I-proton on the other hand is in a location sandwiched between protons and as such receives a much greater overall surface E-Flow within the nucleus than the corresponding E-proton. Fig 6.32 shows how a belt of Mps is set up around the I-proton as a result of this additional E-Flow. The probability is high that this zone of proton surface Mps increases in its Mp content. As is also the probability that with an increased Mp population a section of it may acquire sufficient outward linear momentum for it to break away from the I-proton. This event would be brought about chiefly because the E-Flow within the proton surface grid is essentially in the direction of the central equatorial belt. We know that the surface grid zone (Fig 4.6) is generated by the E-Flow(3) from the Mps on the surface of the proton as represented in Fig 1.13 and Fig 4.3. The secondary Mps within the surface EIL grid of a single proton are normally at zero relative linear momentum. However in the case of the more complex nucleus the surface EIL grid of the I-proton is quite another matter. The secondary Mps in this case acquire linear momentum towards the equatorial belt zone as indicated by the directional markers in Fig 6.32. Although these secondary Mps each have a cycle duration of only 1 MTU, progressive absorption from E-Flows can develop large momenta in those nearer the equatorial belt.

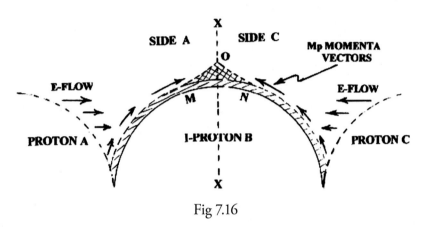

Fig 7.16

If we consider the I-proton as having two symmetrical sections on either side of the equatorial plane XX then the E-Flows and Mp momenta in its surface grid zones (on each side) will be equal and opposite as shown in Fig 7.16. The I-proton B is shown to be in a sandwich bond between the protons A and C. Also indicated are the Mp momenta vectors showing an increasing quantity on approach to the equatorial plane XX. At the equatorial plane the opposing momenta must intersect and the resulting Mp E-Flows will phase. This energy medium phasing occurs across a specific area near the equatorial plane and is represented by the triangular section MON. Secondary Mps occurring within this area must possess a resultant linear momentum that is in the outward-direction along xx. This triangular zone extends all round the I-proton along the equatorial plane and the secondary Mps within have already been referred to as the equatorial belt. This is diagrammatically represented in Fig 7.17. The reader may probably visualise the I-proton as a wallnut shaped entity.

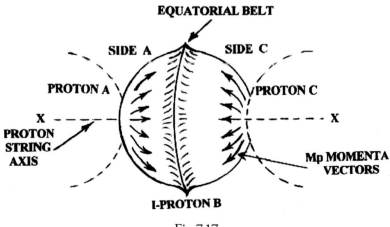

Fig 7.17

In the helium nucleus the I-protons possess an equatorial belt zone that is quite symmetrical in all respects (Fig 7.1). This is because there are no other adjacent proton strings to cause high EIL E-Flows to upset the Mp momenta vectors. In the more complex nucleus configuration the E-Flow influence of one proton string upon the I-proton surface EIL grid of another adjacent string is of considerable importance. In Fig 7.17 we assume the equatorial belt zone to be of even width and height all around the proton in the equatorial plane. This is so because of isotropic conditions of E-Flow upon it from sources other than the protons A and C. In the more complex nuclei however, such isotropic conditions do not exist although each proton string is in equilibrium with respect to its neighbours within the configuration. This implies that the inter-proton string repulsion impulses upon one side must be balanced evenly by an opposite impulse upon the opposite face in each string. Since the repulsion impulses are simply an absorption of linear momentum from high EIL E-Flows, then there must be some reduction in the outward momenta of secondary Mps within the equatorial belt in the direct path of such downward acting E-Flows. This is illustrated in Fig 7.18. Here we show three adjacent proton strings of the same nucleus. Consider only the effect of the E-Flows from the proton strings A and C upon the I-proton of string B. Let the shaded surround represent the equatorial belt of the I-proton B. The E-Flow from proton A is absorbed by side A of proton B resulting in an impulse away from A. Similarly the E-Flow from C results in a similar effect of an impulse in proton B but away from C. If these two impulses within B are equal and opposite, then B will be in equilibrium with regard to A and C. If the impulses are unequal then string B will acquire momentum enabling it to alter its position until an impulse balance is achieved. We shall assume that such a state of equilibrium exists for the conditions presented in Fig 7.18.

Now the plane of the equatorial belt is transverse to the proton string axis and as such is directly influenced by the nucleus inter-string E-Flows. In Fig 7.18 the equatorial belt is in the plane of the paper and is represented on a horizontal axis X_1X_2 and a vertical axis Y_1Y_2. The A, B and C string axes are shown perpendicular to the plane of the paper. The nucleus inter-string E-Flows are

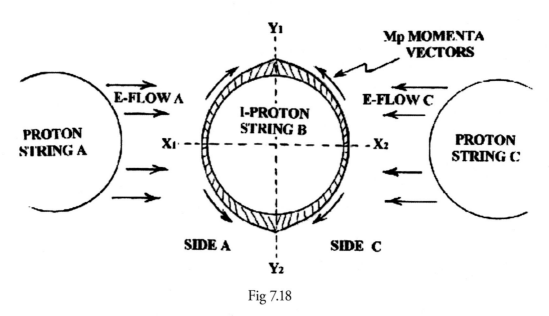

Fig 7.18

absorbed into the secondary Mps of the equatorial belt near the X_1 and X_2 zones thus resulting in a reduction of their outward momenta. This is represented by a flattening of the equatorial belt peak in this region. In the vertical axis zones represented at Y_1 and Y_2 the equatorial belt is influenced by the inter-string E-Flows in a manner somewhat similar to the Fig 7.16 conditions. As such the secondary Mps within the equatorial belt will not only possess momenta as represented in Fig 7.16, but also in the direction from the X_1 and X_2 zones towards the Y_1 and Y_2 zones. This is shown in Fig 7.19 where the vectors indicate the momenta of the secondary Mps within the equatorial belt.

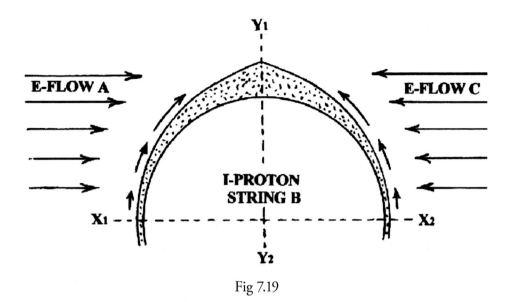

Fig 7.19

Since each secondary Mp has a life cycle of only one MTU, we cannot state that there is a migration of Mps towards the Y_1 or Y_2 zones. However since each Mp produces an E-Flow in its M-E/X phase, then those with linear momenta towards Y_1 or Y_2 must therefore also send out a greater proportion of their E-Flow in these directions. Successive Mps will absorb this E-Flow resulting in an increased linear momentum aspect towards the Y_1 or Y_2 zones. This aspect further increases with increasing distance from X_1 and X_2 in the direction of Y_1 and Y_2. As such it would seem that there is a migration of the Mp within the equatorial belt towards the Y_1 and Y_2 zones. Subsequently we may infer that a higher EIL exists at the Y_1 and Y_2 locations within the equatorial belt. Also indicated is the fact that these zones possess a higher population of Mps with a linear momentum biased in a preferred direction. The equatorial belt thus has a much flatter profile at the X_1 and X_2 zones than at Y_1 and Y_2. We also observe in Fig 7.19 that the Mps in the equatorial belt at the Y_1 location occur relatively further out from the protons surface. We must continue to assume that these equatorial belt aspects are still a facet of the surface EIL grid of the proton (Fig 4.6) but with a bit more detail. The Mps at the Y_1 location perform a key role in that they constitute a group or defined cluster of secondary Mps, the formation of which cannot be considered a random event. Later we shall show that a small section of these Mps may under certain conditions of external influence, be caused to separate away from the I-proton's surface grid.

As the nucleus becomes more complex, its proton strings tend to exist in closer proximity to one another. The repulsion impulse from the E-Flows of one string upon the protons of another adjacent string will have increased considerably. We would of course expect this to obey the inverse square law as indicated in Fig 4.23. Of prime consideration at this stage is the relative effect that this would have upon the profile and character of the equatorial belt zone. We shall assume that the configuration represented in Fig 7.19 is still applicable and that the effect from the E-Flows A and C can only be enhanced. As such, the profile of the equatorial belt would be thinner and flatter at the X_1 and X_2 zones. Whereas at the Y_1 and Y_2 zones the belt would acquire increased depth while its Mps acquired further outward biased linear momentum. This is the situation assumed for the atom nucleus with only a few proton strings. However in the complexity of juxta positioned proton strings of the heavier nuclei, the majority of I-protons receive a balanced E-Flow from more than just two proton strings, and at distances that could be relative to a few proton diameters (or less). This is represented in Fig 7.20 where we show the I-proton of string B to come under the influence of high EIL E-Flows from three separate directions. These E-Flows are from the adjacent strings A, C and D which act directly upon the profile of the equatorial belt. We do not show the E-Flow from the proton B as this would only complicate the setup and confuse the readers' view of the action upon B. Thus for now we shall ignore the effect of proton B upon A, C and D. The effect of the E-Flows upon the equatorial belt of B is to cause a flattening at the X_1, X_2

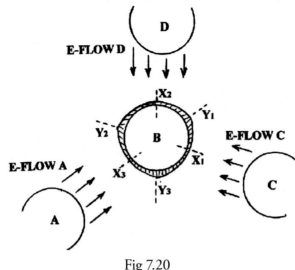

Fig 7.20

and X_3 zones as before. However on this occasion the profile of the equatorial belt at the Y_1, Y_2 and Y_3 zones is also partially suppressed due to the E-Flows directed downwards upon them. The cluster of Mps at Y_1, Y_2 and Y_3 thus possess a marginally lesser outward momentum than those in the same zones of Fig 7.19. All this is taking place inside the same atom nucleus.

The existence of an even greater all round E-Flow onslaught is to be found near the inner regions of complex nuclei from where the proton strings appear to originate. Complex nuclei may be considered to comprise two main regions. An Inner or Base region which lies near the physical centre of the nucleus and in which only I-protons are present. These I-protons are also considered

to be in a multiple bond i.e. bonded to at least three other protons. As such these I-protons would not possess a recognisable equatorial belt zone because of the multiple type of contact. The reader should visualise such a condition and consider the various contact configurations. It is not possible to discuss all such possible groupings but one configuration of a base zone is shown in Fig 7.21, but only as a probability. The radiating dotted lines are simply indicators showing the continued nucleus structure into its more outward regions. It will be noticed

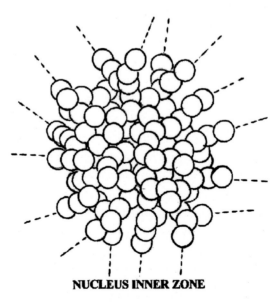

NUCLEUS INNER ZONE

Fig 7.21

that the protons have bonded together in many instances to form a closed chain loop. The progressive development of the more complex nuclei may have occurred by the fusion of two or more nuclei as shown in Fig 6.27. The very large number of proton strings in such close proximity must surely have resulted in contact between E-protons which would then have become bonded together thus creating a closed loop. The sudden increase in the I-proton number from 125 to 136 in the Table of Elements (Fig 6.34) when the E-proton content only increased by one may be due to such a loop formation.

The second of the two main nucleus regions is the 'outer' region. This comprises the proton strings containing much of the dual-bonded I-protons but essentially all of the E-protons. Because of the probable synthesised (or accidental) mode of nucleus manufacture, this outer region is not necessarily of a symmetrical nature, nor is it concentrically organised about the Inner or Base region. It is this characteristic of physical eccentricity between these two regions which permits the whole nucleus configuration to 'swing' into a preferred orientation when acted upon by external E-Flow conditions. In the simpler atoms the Inner nucleus zone may comprise only a few I-protons which usually equal the number of E-protons. However as the nucleus builds, the ratio of I-protons to the E-protons progressively increases with each added proton. In Fig 6.28 the I-proton A may be considered to be just within the Inner nucleus zone while proton B is not. As indicated in Fig 7.21 we may assume the protons of the inner zone to be bonded together quite rigidly while those in the outer region could have a greater freedom of movement when the E-Flow conditions permit. It is our contention that such relative movement does occur when external conditions cause sudden imbalance in the impulses between proton strings of this outer zone.

Let us now return to our consideration of the varying profile of the I-proton Equatorial belt. Since our interest is centred upon the changes that are induced in its profile from impulse imbalances or from relative proton string movement, we can temporarily ignore protons of the Inner nucleus zone. These protons are usually sheltered from centralising action E-Flows (Fig 6.6), and are bonded in a rather rigid manner which does not permit of any sudden relative movement. Their equatorial belt zones are also under considerable multi-bonded influence resulting in the secondary Mps within the 'Y' zones of the belt having momenta that are directed inwards upon the proton surface. In the nucleus' outer zone however the proton strings are usually distinctly separate proton string entities. Although these proton strings settle into positions of equilibrium they can be easily upset by the centralising impulse action from the Mp-shell. Initially in chapter 6 we simply applied this centralising action concept to the single proton nucleus of the Primary Atom. This same concept is however applicable to all nuclei, but with the additional effect of changes to its proton string equilibrium configuration. This implies that a complex nucleus under the influence of a sudden centralising impulse will initially respond by a relative movement of its outer zone proton strings.   As stated previously each proton string receives repulsion impulses from those strings adjacent to it and which together are in perfect balance for a condition of equilibrium. When impulses from an external source are generated within this zone, sudden alteration in its normal equilibrium configuration will occur. Although this will eventually result in the acceleration of the nucleus and the atom in a particular direction, it is the initial proton string response that is of special interest to us at this stage.

Let us for a moment consider Fig 7.18 again. Assume that this represents a part of the normal equilibrium proton string configuration of our nucleus. Let us now suppose that a distortion occurs at the Mp-shell which then sets up an altered centralising impulse action upon the nucleus. We shall assume that this causes the proton string C to be moved into the new though temporary

position shown in Fig 7.22. For the moment let us ignore all the other inter-related movements and just concern ourselves with the impulse action of strings A and C upon the equatorial belt profile of the I-protons in string B. It will be noticed that the high and low EIL grid profile points have altered their positions. The profile at the X_1

and X_2 zones is very nearly the same as before. At Y_2 the profile is much less prominent due to a larger proportion of the E-Flows A and C being generally downwards upon it. The profile at Y_1 however is seen to be considerably enhanced in the outward direction due to the E-Flows A and C combining such that the secondary Mps therein acquire linear momentum in the outward direction. It is our contention here that conditions may prevail that result in the outer regions of the Mp-cluster at Y_1 acquiring sufficient linear momentum which allows a group of Mps there to physically breakaway from the equatorial belt.

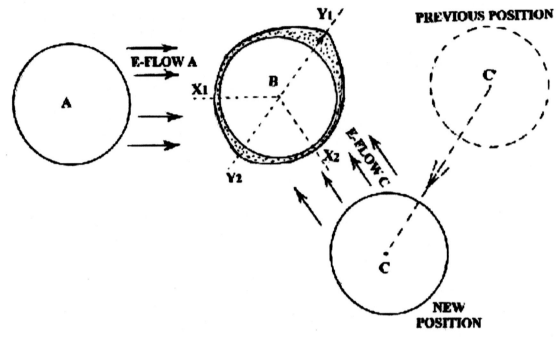

Fig 7.22

Let us now examine the probable mechanism of this Mp-cluster breakaway. In Fig 3.3 we showed how the EIL oscillation wavefront commenced its propagation outwards from the Mp-shell. In actual fact we presented this wavefront as simply a continuance of a synchronised cyclic repetition of the Mp event at a progressively forward quantum step location. In order that Mps could recur at the next quantum step position, sufficient energy medium must first have been transmitted there by the E-Flow process from a previous M-E/X event. This requires the wavefront Mps to possess a velocity Vp (Fig 3.35) in the direction of propagation. It is the contention here that the uppermost Mps of the equatorial belt at the location Y_1 in Fig 7.22 may acquire linear momenta equivalent to such a velocity Vp away from the proton. It is also considered probable that these uppermost Mps are arranged in a cluster having a patterned layout with a set spacing arrangement. The reasoning behind this is that the equatorial belt is a combination of the surface EIL grid E-Flows that have a unique phasing at the I-proton equatorial plane. Secondary Mps that would normally have occurred in random formations within the proton surface grid must now be influenced by a defined pattern of E-Flow on either side of the equatorial plane. We would therefore assume that the Mps within the equatorial belt must occur in some particular positional pattern. We cannot as yet speculate upon the aspect of such a pattern of Mps, but we would consider it to be of a repetitive character similar in principle to the event cycle of the Mp-shell. It is our contention once again that when these Mps acquire the necessary outward momentum their positional pattern alters towards a wavefront type parallel row configuration. We would assume that at the instant of physical breakaway this Mp-cluster or group has all its Mps arranged in separate though parallel planes spaced at a defined number of quantum steps apart. The physical nature of the breakaway may be compared to the commencement of the EIL oscillation wavefronts described in chapter 3. The backward E-Flows from each wavefront affects the Vp velocity of those trailing behind. As such the breakaway of one Mp-cluster reduces the chances of such an event occurring in rapid succession from the same equatorial belt location. Fig 7.23 shows the Mp-cluster at the instant before breakaway. The Mps in the hatched section at the highest point in the profile of the equatorial belt have acquired the velocity Vp, and E-Flows from the M-E/X of each Mp result in a wavefront type of Mp formation at the next forward quantum step position. Since the breakaway Mp-cluster functions like a series

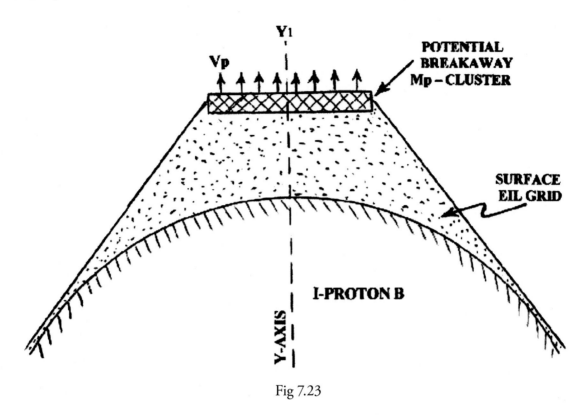

Fig 7.23

of parallel wavefront sections spaced a number of quantum steps apart, then like the wavefront an enlargement of the cluster will occur. This enlargement must of course diminish the energy medium density of the Mp-cluster over large propagation distances. Let us consider the mechanism of this propagation.

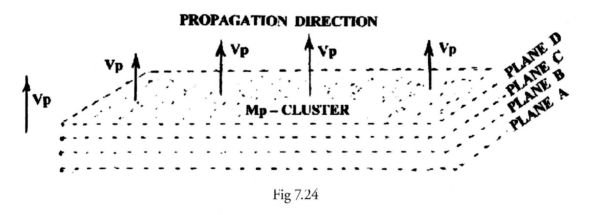

Fig 7.24

Let Fig 7.24 represent a fractional part of the Mp-cluster propagating in the direction of the Vp vectors. Let the Plane A section of this cluster arrangement be the trailing wavefront of this series. If we assume that the wavefront plane B is a single quantum step in advance of plane A, and that their Mp centres are also in synchronism, then we can expect that when the plane A wavefront propagates to the plane B location so will the wavefront B have propagated to the plane C location and so on and on. We see that the wavefronts at the planes A, B, C, D, etc, all propagate forward by one quantum step at the same time. As such the entire Mp-cluster propagates in the Vp direction at approximate wavefront velocity. Even if the wavefronts are separated by more than a single quantum step, the principle and the effect will be the same. Now Fig 3.31 defines a backward E-Flow emitted from each wavefront which gets absorbed by a closely trailing wavefront resulting in a reduction in the Vp factor of the wavefront velocity. Since the Mp-cluster has all of its active wavefronts at minimum or small quantum step spacing then the reduction in the Vp factor due to the backward E-Flows must be at a maximum. Therefore according to chapter 3 theory relating frequency or wavefront spacing to propagation velocity (Fig 3.31), we must conclude that the Mp-cluster will propagate at a velocity somewhat less than that of the visible light-wave spectrum. We must assume that the Mp-cluster, like the lower frequency wavefronts, traverses at a set velocity. However we feel that because of the limited quantity of propagating plates within the Mp-cluster that there will occur a marginal difference in the Vp factor between the leading wavefront and those trailing behind. Our conclusion is that this results in a gradual increase in the spacing between the wavefront planes A, B, C, D, etc., as propagation progresses. It would appear that the Mp-cluster increases in volume (though not in mass content) as it propagates. A sort of Mp-cluster Red shift with time.

At this stage it would be pure speculation on our part to estimate the extent and depth (in quantum steps) of the Mp-cluster although we would expect this to vary, dependant upon conditions at the moment of breakaway. However we can assume that during propagation it exists as a finite entity with finite mass and finite velocity. We suggest therefore, that this Mp-cluster entity be given similar importance as the other mass entities such as the proton and the Mp-shell. We would also suggest that this Mp-cluster be considered in parallel with the Electron entity concept as in conventional physics and as such be similarly termed.

This Mp-cluster or 'Electron' is therefore simply a group of Mps located by circumstance into parallel planes separated one or more quantum steps apart and all propagating in the same direction. Propagation is of course according to the principles laid out for wavefronts in chapter 3. This mass entity must therefore be in a continued state of directional propagation for it to continue its repetitive existence. If it were to be slowed to a level for Vp approaching zero then perhaps propagation would cease and the Mps in the cluster would simply dissipate in M-E/Xs. Perhaps under high EIL conditions an Mp-cluster or Electron could be induced to slow down as in the reflection or refraction phenomena of Fig 3.65 resulting in two separate clusters propagating in opposite directions. Since the Vp factor for the cluster planes is much less than for the single wavefront, it is considered that there is also scope for acceleration of the Mp-cluster as a whole by external influence. A nodal field may therefore play an important role in increasing the linear momentum of the Mps in the cluster by simply causing the Vp factor to increase. However a limit may be reached beyond which Vp cannot increase but simply results in its wavefront mass centres absorbing relatively large quantities of energy medium from the realm datum through which it traverses. We shall be considering these velocity type aspects in the next chapter and so will postpone discussion till then.

In Fig 7.23 the reader will notice that the electron as a breakaway Mp-cluster is a very small fraction of I-proton size and mass. In size between one tenth and one hundredth of the proton diameter. Since the equatorial belt is always in a plane transverse to the proton string axis (Fig 7.17), then the emission of all electrons will also be in that same plane. In general the string configuration in the nucleus outer zone is such that the proton strings are positioned with their axes in an approximately radial format. This means that electron emission, as we shall term the Mp-cluster breakaway phenomenon, is directed laterally towards adjacent proton strings of the same nucleus as represented in Fig 7.25. Thus although there may be a profusion of electron emissions within each nucleus, very few of these (if any) actually propagate all the way through the nucleus maze and out towards the Mp-shell. It would appear that electrons have no difficulty in propagating through the Mp-shell because of the repetitive character of their formation. Thus it would be a simple enough journey for an electron to propagate from the nucleus zone of one atom to that of another.

We know that the nucleus configuration consists essentially of an inner zone and that these are not necessarily symmetrically constructed about one another. Electrons are emitted from outer zone I-protons and these may be induced to occur at random intervals and directions. There are however, certain conditions that occur externally to the nucleus and which may cause its orientation in a preferred direction. These conditions or influences are also active upon the proton strings and this results in the emission of electrons being more frequent in a preferred direction. The configuration of the nucleus' proton strings then settle into a new equilibrium status resulting in additional paths for electrons to exit out towards the Mp-shell. These paths through which the electrons emerge are termed as 'nuclei windows' and usually occur in a preferred direction because of some external directionally exerted influence. The nucleus configuration of each element is different from any other but all may be induced to develop a preferred direction for the exit of electrons. This path is not a choice but depends upon the free pathways possible between the outer zone protons. There are considerable deflections caused by the intense E-Flows between the strings which result in the electron being diverted along circuitous routes that may or may not permit an exit from the nucleus.

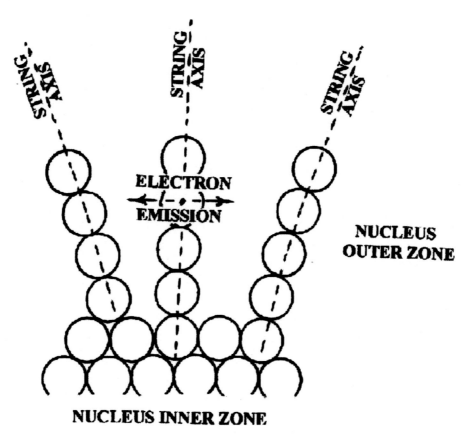

Fig 7.25

Let us for a moment consider the electron as a mass entity. Here we have an extensive group of Mps operating in near synchronism. We assume that the spacing of these Mps has a similarity with those on the proton surface that produce the overlapping E-Flows indicated in Figs 1.17 and 1.18 of Chapter 1. Since these are the origin of the proton's forcephoids we consider that something similar must also be emitted from the Mp-cluster that is the electron. The difference lies in the fact that the proton is spherical while the electron is more like a plate as shown in Fig 7.26. In the case of the proton its E-Flows occur from all around its surface producing a continuous forcephoid. In the case of electrons there are sufficient E-Flows only in the forward and backward propagation directions. The velocities of these E-Flows are indicated on the diagram. The forward E-Flows must therefore produce forcephoids as shown. We have not shown the backward acting forcephoids so as not to confuse the reader.

These forcephods have a different spacing and relative velocity as compared with those emitted by the proton. For a review of backward and forward E-Flow aspects see Fig 3.11. It is assumed that forcephoids conform to the curvature of the Mp layout from which they are emitted. As such the forcephoids emitted by the electron will be of a flat section and about the same area as the propagation face ABCD, which is very small indeed. This applies of course to both the forward and backward E-Flow emissions.

As stated earlier the electron is diverted along a circuitous route within the nucleus and then may or may not exit therefrom. When this path results in impact upon another internal proton then the electron is absorbed and ceases to exist as an independent entity. When it does however propagate outside the nucleus there are still very strong

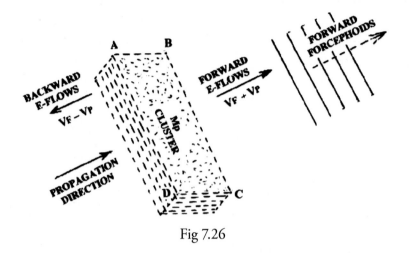

Fig 7.26

E-Flows and EIL gradients that cause it to deflect in some particular direction. These path deflections are not however imposed upon the forcephoids emitted by those electrons. As such the forward E-Flow forcephoids traverse ahead of the electron at a velocity which is rather greater than the electron propagation velocity. The ratio of the two velocities is given by,

$$(VF + Vp)/_{MTU} : 1 \text{ quantum step.}$$

where VF and Vp are the distances traversed by the E-Flow and Mp in the duration of a single time unit (MTU) as defined in Chapter 5. It would be pure speculation on our part to attempt an estimate of VF in comparison to VC the propagation velocity. In chapter 3 Fig 3.8 we indicated that VF was at least greater than VC by a factor that was represented by,

$$\frac{VF}{VC} = \frac{1}{Sin\ 60} = \frac{1}{0.866}$$

Or VF = 1.155 VC

If   VC = 186,000 miles per second, then,
       VF = 214,830 miles per second, (at least).

(A more detailed calculation is presented in Appendix 1 and shows VF to be far greater than this).

Now once these forcephoids race ahead of the electron they will continue along an absolute straight path until they are absorbed in an E-M/X. Since the electron (as a sort of complex wavefront system) is easily influenced into a deflected path, then the former forcephoids, emitted earlier, could not be easily traced back to their electron origin. We note therefore that these forcephoids must traverse a straight path unlike the electron from which they originate. Since forcephoids are emitted from each quantum step location then a succession of differing forcephoidal traverse paths occur as the electron propagates. This is represented in Fig 7.27.

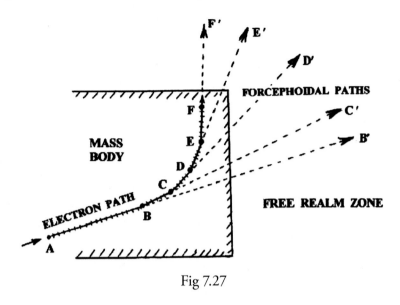

Fig 7.27

Consider the propagation path of an electron within a mass body. Let this path be along the route ABCDEF as shown. At F we assume that the electron ceases to propagate. This may be due to absorption by a proton or simply because Vp has reduced to zero resulting in a dissipation of its energy medium to the surrounding realm zone. Now at each quantum step forcephoids are emitted. For the moment we shall only consider the forward emitted forcephoids. If we assume that a continuous series of electrons follow the same electron path then we would also have a continuing forcephoidal emission along certain paths. In order to simplify matters we have shown the electron path to consist of five directional sections. Forcephoids emitted from the path section AB proceed along BB'. Those emitted from the section BC proceed along CC'. Similarly forcephoids emitted from the path sections CD, DE and EF traverse along DD', EE' and FF' respectively. Thus each section sends out forcephoids ahead of the electron and in the direction of its propagation. However in reality the electron path is deflected along a smooth curve between the path points A to F. This means that forcephoids would be emitted uniformly over the entire range of directions between the initial and final direction of the electron path. From this we note that although this type of forcephoid must originate at the electron face they must subsequently follow an independent straight line path. Obviously only a limited number of forcephoids get emitted from each quantum step position but this tends to be supplemented by a succession of electrons flowing along the same general paths. Later we shall be considering types of Nodal fields that may be set up when large quantities of electrons traverse in parallel paths and in continuous succession. We shall also then consider the effect that this type of field has upon the orientation and emissions from nuclei.

In our consideration of electron paths and nucleus windows we must state that there are many such paths that allow electrons to escape. However, it is important to understand that when a nucleus is under the influence of an external action from some particular direction, the escaping electrons tend to propagate along preferred directional paths. This means that the external influence causes deflections in the electron path within the nucleus such that escape windows are more easily achieved upon a particular side of the nucleus. The greater the external influence the greater is this bias for nucleus windows to occur on a particular side of that nucleus. Quite obviously the paths of these escaping electrons are not wholly parallel and therefore this stimulated directional emission will occur over an expanding conical area as shown in Fig 7.28. We would consider this to be within the limits of a conical angle of (say) approximately 25°. We would ask the reader to visualise the nucleus in Fig 6.27(b) being stimulated from the left hand side such that electron emissions occur within a conical zone towards its right hand side. This nucleus is then considered to be the one represented in Fig 7.28. Alongside the physical representation of the cone of emissions is a graph indicating the assumed population density of electrons within that cone. It is a reasonable assumption that the peak density occurs near the centre of the emission cone and this has been represented as such.

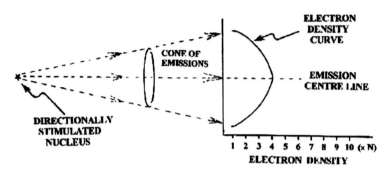

Fig 7.28

Now of all the electrons that escape from this complex nucleus of protons along a cone of emissions, there is a high probability that some of these will enter another atom nucleus when atoms are in close proximity as in a solid mass object. Let us assume that the atoms in such a mass object have atoms that also have their bonding in preferred directions. This will result in a mass body with a particular type of structural pattern and would be unique to each atomic element. Of course not all elements develop such structural patterns and as a basic concept we may segregate elements into two main classes of whether or not a such a structural pattern exists. For the moment let us only consider that class of element leading to mass objects with a particular structural pattern.

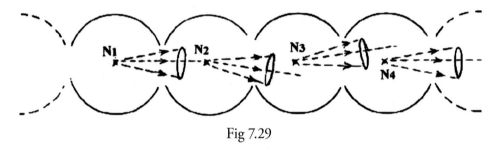

Fig 7.29

Subsequently when electrons from such a stimulated nucleus propagate across to an adjacent atom's nucleus, then because the orientation of that nucleus is similar, it will acquire stimulation from those electrons and produce an emission in a near parallel direction. This is termed the 'Domino Effect' for electron emissions and the principle is highlighted in Fig 7.29. Here we show four adjacent bonded atoms with their respective nuclei as N_1, N_2, N_3 and N_4. Consider the nucleus N_1

as having been stimulated into the emission of electrons. If we assume that the stimulation upon N_1 is such as to cause a succession of electrons to be emitted along a conical zone in the direction as shown towards the adjacent nucleus N_2, then some of these electrons must impact upon and get absorbed into the protons of N_2. If the stimulatory influence upon N_1 is reasonably consistent for a set duration then so will be the emission of electrons towards N_2. The consistent absorption of electrons by the outer zone I-protons of nucleus N_2 results in linear momenta inducements (impulses) upon its proton strings to reposition for equilibrium. This in turn causes an upset in some of the I-protons equatorial belt conditions resulting in the breakaway of Mp-clusters (electrons) along preferred paths with nucleus windows in the preferred locations. The subsequent emission of electrons from nucleus N_2 is shown in Fig 7.29 as being generally towards the right hand side and in the direction of the adjacent atom nucleus N_3. The reader may notice that we have shown that nucleus N_3 does not lie on the exact centre line path of the emission cone from N_2. This is because we do not consider that the orientations of all the nuclei within a mass object can be 100% directionally true. As such the emissions from the nucleus N_2 may not be wholly in parallel with those of N_1. Now since N_3 is not on the centre line of the emissions from N_2, then according to Fig 7.28 the density of the electrons received will be somewhat less than those received by N_2 from N_1 (N_2 being on the centre line of the emissions from N_1). Subsequently the stimulatory influence upon N_3 may be assumed to be somewhat lesser than that which was induced in N_2. Nevertheless nucleus N_3 would still be induced to produce stimulated emissions in the general direction of the nucleus N_4. We consider that electrons from the emission cone of N_1 may even propagate as far as N_3 and N_4. This would ensure their partial stimulation had N_2 been orientated in an excessively out of line direction. The relevance of these variances will become apparent as we progress with this topic. From the above we note therefore that there exists a trend within a mass body with a regular structural pattern for the progressive stimulation of much of its nuclei from a single stimulated plane.

In a large mass structure this 'domino effect' can and does produce a considerable directional flow of electrons. However the domino effect does not apply to the nuclei of atoms that occur at the surface of a mass body. This is because these atoms have a unidirectional molecular bond resulting in their nuclei being maintained in a set orientation. This orientation cannot be easily disturbed or altered and is such that when under external stimulation, results in the emission of electrons towards the mass body interior. When electron emissions from within the mass body impinge upon the nuclei of surface atoms, the corresponding emissions seem to obey the simple rules of wavefront reflection as represented by Fig 3.42. It must however be stated that the direction in which the electron emission occurs from these surface atom nuclei does not depend upon the direction from which they are stimulated but upon their own state of orientation. In the case of a simple molecular bond (Fig 6.13), the orientation of the nucleus is considered to remain fixed relative to the other atoms. Similarly, the nuclei of atoms at a mass body surface have an orientation that remains fixed relative to the adjacent atoms.

Now because the mass body is limited in its extent, these electrons can only propagate upto an end face. Here the domino effect is reversed and a new succession of emissions occurs across the mass body. With the initial stimulation still effective, we would have two sets of emissions propagating in opposite directions within the same mass body. When both of these electron emissions impinge upon a nucleus but from opposite directions, then the result is a neutral condition with no emissions in either of those directions. Thus we could state that in a limited mass body the domino effect gets neutralised by the action of its own emissions reflected as it were by the nuclei at the body surface. In the domino effect we have presented a basic principle of the progressive continuity in the stimulation of emissions from nuclei within a mass body.

So far we have only considered a mass body with a regular structural pattern. In the table of elements (Fig 6.34) there are a great number of intricate nuclei configurations and we consider that all are prone to stimulation by the impact of electrons. However it is also considered that many of these are unable to adhere to any constant nucleus orientation or a consistent structural pattern within a mass body. These then are the other main class of elements. In reality there is no clear cut demarcation of structural and orientation class since the variation can be graded through a scale from a near zero level to practically a maximum of a hundred percent. Those near the low orientation levels would have very little preferential bonding positions for their atoms. Nucleus stimulation at a Fig 7.29 N_1 location is therefore unlikely to produce a progressive domino type unidirectional emission of electrons in other nuclei positioned on its downstream side. Successive emissions are in random directions and are quite likely to be as shown in Fig 7.30. Here the stimulation of N_7 results in an indirect random route of stimulated emissions that loops back upon itself. In this case there is no unidirectional domino effect to produce an effective electron flow across the mass body dimension. No matter how much external stimulation is applied to N_7 the result will be the same. The opposite flowing electrons neutralise the stimulation effect upon a nucleus. We would conclude that this particular mass body is not conducive to the flow of electrons along a unidirectional path. As such it is termed a non-conductor of the electron wavefronts. There may of course be other reasons why a continual unidirectional domino effect cannot be achieved. Perhaps the nuclei of a particular element are not able to be stimulated sufficiently for the lift-off of Mp-clusters, or that there are as yet unknown obstructions to their free passage between internal proton strings, so reducing the effective number of nucleus windows (perhaps dramatically). We leave the reader to speculate at length upon this aspect.

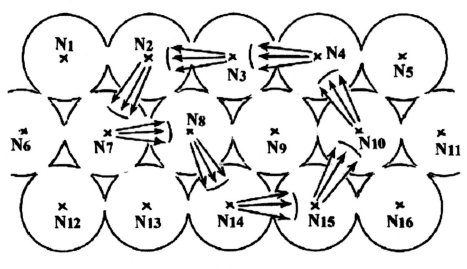

Fig 7.30

With structural orientation at a high level the nuclei stimulation and electron flow pattern will be as indicated in Fig 7.29. When the stimulation is removed the electron flow also ceases. However, there is a way around this in which the external stimulation of N_1 may be withdrawn without sacrificing the flow of electrons. To enable this to occur it would be necessary for a very long chain of the domino effect nuclei to be formed into a closed loop such that the nth nucleus has its cone of emissions directed at N_1 from the left hand side. A complete circuit of the domino effect is thus achieved meaning that indirectly N_1 is able to stimulate its own emissions. As such, the flow of electrons in this circuit should not find the need for external stimulation necessary. However, this is not the case because of structural irregularities and incorrect orientation of molecular bonded chains at shell EILs (temperatures) that are well above an absolute zero condition. This causes a progressive diminishment of the electron flow at each stage of the domino effect. By lowering the Mp-shell EILs within the mass body loop to a condition that corresponds to an absolute minimum (in a good conductor of electrons), then the orientation of nuclei is maintained more accurately. Subsequently the cone of emissions is directed towards the next nucleus with greater consistent centre line accuracy and a correspondingly improved electron density stimulation (Fig 7.28). Since the flow of electrons within the loop of atoms is of itself an energy medium source to the Mp-shell, we may assume that a high electron flow is destructive towards the correctness of nucleus orientation. As such a balance is struck for the stimulation of N_1 by the nth nucleus. This means when excessive stimulation produces a high electron flow in the loop, then the subsequent reduction in orientation levels causes a diminished domino effect. The overall electron flow therefore does not increase proportionately. The inverse is true when the stimulation is reduced. From this we note that a perpetual flow of electrons within a closed loop circuit is possible without the continued application of an external means of stimulation. In practice this has been achieved by cooling the loop to near absolute zero temperature and then removing the external stimulation. In chapter 5 (Fig 5.2) we defined the conditions for the absolute TN-line. As such the true absolute zero of an EIL condition (i.e. Mp-shell temperature) could only occur in a void region. In the current realm with its relatively high circle 6 realm datum we could never hope to even approach zero EIL.

Mass bodies in which orientation levels are high and which permit a unidirectional flow of electrons are termed as electron conductors. Under the current realm datum conditions we could not expect to find mass bodies that are perfect conductors. Thus all known conductors would cause some restriction to the normal flow of electrons. This restriction is termed as a resistance or impedance. Differing structural materials would possess different levels of impedance to the electron flow. The impedance is therefore simply a measure of the reduction in the total flow of electrons in a progressive domino effect situation. The conductivity of a mass body which is the inverse of its impedance status is thus directly proportional to some pattern of nucleus orientation within its atomic/molecular structure.

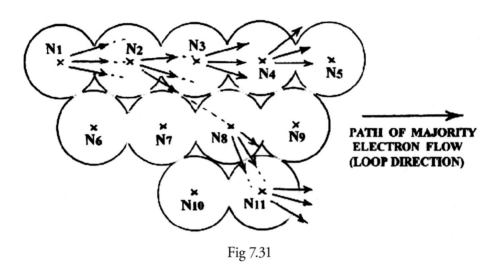

Fig 7.31

At this stage it would be appropriate to consider the conditions of electron flow in a bit more detail. The true situation is somewhere close to that represented in Fig 7.29 but with several stimulations in varied directions. Also perhaps a wider angle of conical emissions per nucleus, resulting in the stimulation of more than just one adjacent atom nucleus. Fig 7.31 shows how the cone of emissions from the nucleus N_2 is not quite symmetrical with regard to the location of nucleus N_3. This causes a diminished stimulation within N_3. However the N_2 emissions are also shown as being in the direction of the nucleus N_8. If we assume the orientation status of N_8 to be low then on being stimulated it would probably emit a cone of electrons in the direction of N_{11} as shown.

Let us for the moment assume that a cone of emissions also propagates from N_6 to N_7 and from N_7 towards N_8. We note therefore that N_8 receives stimulation from two different directions and which do not appear to be in opposition. One would normally expect the stimulatory effect upon N_8 to increase resulting in increased electron emission. Unfortunately this is not the case. This is because stimulation of the nucleus in a particular orientation status has nucleus windows in locations that depend firstly upon its orientation and secondly upon the direction from which electrons arrive. When stimulation is from a single direction, the lift off and propagation of electrons along paths leading to nucleus windows has a complication factor of (say) one. But when stimulation is from more than one direction then this complication factor for electrons propagating within the nucleus increases manifold resulting in far fewer nucleus windows. Subsequently a much lesser electron emission ensues from these nuclei. We note therefore that the flow of electrons in the loop direction of Fig 7.31 is somewhat less than optimum due to the incorrect orientation of a few of its structural atoms. If we consider a conductor string of infinite length with a stimulated emission at one end, then we could expect a gradual diminishment in the quantity of unidirectional electrons with distance. If the x-sectional area of this conductor has a finite dimension then the loss of lengthwise electron flow may be a corresponding gain in the transverse direction. However, electron flow in the transverse direction is quickly damped out by the reflection process at the conductor surfaces as explained earlier.

In a good conductor the impedance is very low and the decrease in electron flow in the progressive domino effect is minor. However, when the quantity of electrons propagating through a conductor increases dramatically we must also consider their effect upon the surrounding Mp-shells. In Fig 6.11 we discussed the effects of the realm EIL datum upon the energy medium content of the Mp-shell. We showed that the Mp-shell enlarged with an increase in the datum EIL. Subsequently, we also showed that the Mp-shell is able to absorb energy medium from EIL oscillations propagating through its zones. This resulted in an expansion or enlargement of its structure. Since increase in energy medium content of the Mp-shell is reflected as an increase in its 'temperature', we must assume that a high density of propagating electrons within a conductor must also cause a similar 'temperature' increase. In Fig 6.16 we explained the mechanism of molecular bonding. As the energy medium content of the Mp-shell increased so did the bounce in its molecular bonding. An increased bounce must surely have an adverse effect upon the nucleus orientation and thus cause a further loss of electron flow. This is usually represented as an increase in the conductor's impedance. Thus we note that an increased electron flow in our infinite conductor results in a temperature increase which in turn causes the impedance to rise. We shall return to this aspect a bit later on.

Now we know that an infinite length of conductor material is physically impossible to achieve. We are however able to produce an infinite path for electron flow by the simple means of creating a closed loop or endless circuit. If a conductor of finite length is manipulated such that each end meets the other then a closed loop is created. However there can be no flow of electrons in such a circuit until nuclei in some part of the circuit are stimulated into electron emissions. The extent and intensity of that stimulation could vary, which in turn would cause a proportionate flow of electrons. We know also that the impedance factor in the conductor or in any part of the circuit causes some of the electrons to be in an opposite or turned around direction. These electrons then oppose the external stimulation such that the overall flow of electrons in the circuit is reduced. The flow of electrons in all parts of the circuit however is the same although some parts of the loop may

have a lesser impedance factor than others. We can demonstrate this with the aid of Fig 7.32. Consider the circuit ABCDEFG which is externally stimulated at AA' causing a flow of electrons in the direction AB. Let us assume that the section BC has a high impedance and that the rest of the circuit is a very good conductor. Let us assume that the stimulation at AA' results in 100 electrons

flowing towards BB'. Thus initially there is a flow of electrons in the direction A to B only. The domino effect within the section AB is such that all 100 electrons arrive at the plane BB'. When these electrons propagate past BB' and into the zone of high impedance the domino effect is reduced resulting in much fewer electrons propagating past CC'. Let us assume that the impedance effect in BC is such that of the initial 100 electrons starting out from AA' only 70 electrons now propagate along CD. We have somehow lost 30 electrons between B and C. Actually many of these would have been reversed in their propagation direction (as at Fig 7.30) and would now proceed in an opposite or reversed direction past BB' and towards AA'. Let us assume that 20 of these electrons now propagate along the circuit from B to A. Since 10 electrons are unaccounted for we assume that they are lost in a transverse flow between the side wall faces of BC which is self neutralising but with energy medium absorption in the Mp-shells of its structural atoms as a temperature rise.

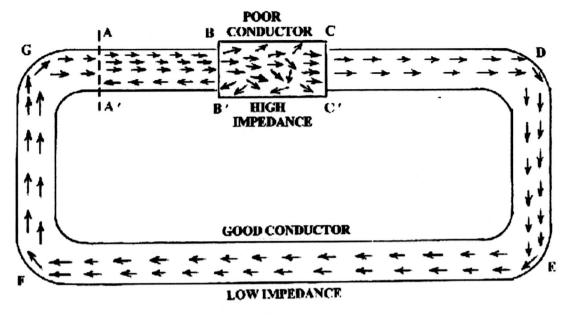

Fig 7.32

Let us consider these 20 electrons that have been reversed and are now propagating in an anti-clockwise direction toward AA'. These electrons act upon the nuclei in their path and alter their orientation sufficiently to produce a structural condition that only permits the clockwise flow of a limited number of electrons. This condition is not to be confused with impedance which is a natural structural state. It is more like the shutting down of some of the paths that electrons could flow along within the conductor. The number of paths that are shut down are directly proportional to the number of electrons that are reversed by a resistance element anywhere in the circuit loop. When the loop is opened then all the electrons are reversed by the reflection type process so shutting down all the flow paths. We see therefore that the electron flow in a stimulated conductor loop is inversely proportional to the impedance in any part of that circuit. Because of the effect that an impedance has upon the electron flow within the loop, we observe that the density of the electron flow remains uniform in all sections of the loop. Perhaps a hypothetical example will help to explain the principles behind this. Let the above 20 electrons propagate anti-clockwise by the domino effect. When this electron flow reaches the section AA' it will counteract and neutralise a proportion of the external stimulation effect. This then causes a reduced number of electrons to flow in the original clockwise direction. Anti-clockwise electrons flowing past AA' will propagate towards G. Here the reflection type process comes into play and the 20 electrons can only follow the paths towards F and then to E,D and C. At CC' about (say) 14 electrons propagate towards BB' while (say) about 4 electrons are reversed and join the original clockwise flow from C to D, (2 electrons being lost in transverse flows between the side wall faces of BC). So now we have 74 electrons propagating in the clockwise loop direction and 14 in the opposite direction. When these 14 electrons reach AA' their neutralising effect upon the external stimulation is not as much as it was on the previous occasion (with 20 electrons). This then causes a slight increase in the electron flow from AA' to BB'. This reduction and increase in the number of electrons flowing around the loop continues for several laps until a balanced flow is achieved in the original clockwise direction.

In Fig 7.32 there are three sections to the complete loop. These are the sections (firstly) AA' to BB', (secondly) the high impedance section BB' to CC', and (thirdly) the remainder of the loop from CC' through D E F and G and back to AA'. In order to explain the theory behind electron flow in a loop we can consider each section as an independent unit. Each unit is however dependent upon both of its neighbouring units for stimulation and feed back of electron flows. Consider the section AA' to BB'. Here we have the external stimulation at AA' causing a flow of electrons towards BB'. However at BB' we have a flow of electrons theoretically entering this section and propagating towards AA'. At AA' these electrons oppose the external stimulation effects which then results in a lesser flow of electrons towards BB'. However at AA' we also have electrons arriving from the D-E-F-G section which tends to boost the emission of electrons in the direction of BB'. The high impedance section BB' to CC' receives clockwise electrons at BB' and causes a theoretical electron scattering which results in one portion continuing through to CC' and another portion being turned around to flow back towards BB' in the anti-clockwise direction. So now we have a variable quantity of electrons flowing (theoretically) in both the clockwise and anti-clockwise directions in these two sections. The net effect being that each section has the same resultant electron flow in the previous stimulated direction. At CC' this resultant electron flow enters the section D-E-F-G and continues on towards AA'. The overall electron flow in the loop is therefore that which corresponds to the net effect of all the stimulatory influences at AA'. The poorer the conductor status of a particular section the greater will be the opposing action to the clockwise stimulation at AA'.

The reader will have noted that initially the electron flow from AA' to BB' was greater than that being emitted from CC' in the same direction due to the high impedance of the section BC. We consider that although in practice a uniform electron flow seems to exist in all parts of the loop, that in fact there is a loss of electron flow as the domino effect progresses. The difference is extremely marginal but each section of the loop in Fig 7.32 must be seen to possess an electron flow that is unique yet totally dependant upon the flows within the adjoining sections. In a unidirectional electron flowing system we must assume therefore that the upstream loop section becomes the stimulation factor for the downstream characteristics.

Since the impedance factor tends to set up a contra-flow of electrons we can conclude that the net electron flow through the conductor loop is (firstly), directly proportional to the extent of the external stimulation factor, and (secondly), inversely as the impedance within that loop. Let us now relate these factors to the terms used in conventional physics. The conductor loop with its external stimulation zone is referred to as an electron flow circuit or simply electrical circuit. The net flow of electrons are considered as an electron current or simply current denoted by the symbol 'I'. The impedance inhibits the unidirectional flow of current and is termed as the circuit resistance denoted by the symbol R. It usually indicates the poor quality of the conductor in any part of the electrical circuit. The overall resistance within the circuit is the sum total of all the individual resistances therein. Compound loops are not considered here. The much broader term of electrical impedance refers to the more complex mode of electron flow inhibition that may be induced by forcephoidal action from another adjacent electrical circuit. We shall consider this aspect in a later section.

Finally we must qualify the extent or intensity of the external stimulation at AA' that is the basic cause of the electron flow in the circuit. We know that external stimulation causes orientation of some nuclei which then have their proton strings re-adjusted to a new equilibrium format. This results in emission of electrons along defined paths towards new nucleus windows. The overall quantity of these exiting electrons depends not only upon the total number of nuclei that undergo orientation but also upon the degree of orientation that each nucleus achieves. Thus the intensity of electron emission from a nucleus depends really upon the positional aspect of its proton strings attained from the varying degrees of induced orientation.

Let us grade this theoretical degree of nucleus orientation from zero to 100%. Let us also assume that the emission of electrons increases in direct proportion to the degree of nucleus orientation and then represent this on a logarithmic scale. The total number of electrons emitted per nucleus per MTU is then given by the graphical line representation in Fig 7.33.

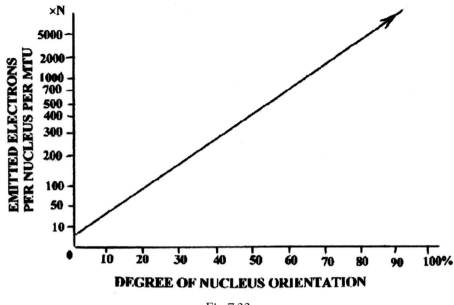

Fig 7.33

The rate of electron emission can be seen to increase rapidly in number per MTU as the orientation level approaches 100%. We must now also consider the overall quantity or percentage of the nuclei within the stimulated plane AA' that are induced into a partial orientated state in order to establish the total electron flow or electrical current 'I' within the loop. The reason for the electron flow is the directional orientated aspect of the nuclei at AA' and this is achieved by external stimulation. We now have a quantified concept of the stimulation factor for electron flow in an electrical circuit. The percentage orientation of electron emitting nuclei at AA' may thus be considered as a directional impulse type action that causes a quantity of electrons to flow in a particular direction. The overall effect is to drive the electrons around the loop and this is termed as the 'driving voltage'. This is denoted by the symbol 'E' and its effect is unidirectional at any instant. As indicated by Fig 7.33 the driving effect E can attain very high levels under certain conditions. We saw in the loop of Fig 7.32 that the flow of electrons around the circuit arrives at AA' from the direction of G and so adds to the effect of the external stimulation there. Now if there was no impedance in the loop then the stimulation effect at AA' would simply go on increasing with the continued arrival of increased levels of electron flow from G to AA'. After a few such laps a very heavy current would flow in the circuit which should in theory eventually reach infinite proportions. These conditions are referred to as 'short circuit' conditions and should be prevented from occurring in the practical field. Since no material is a perfect conductor we usually observe a large absorption of energy medium in the Mp-shells followed by a probable change in the character of the inter-molecular bonding. The driving voltage is thus the initial generator of electrons from where the domino effect commences. All other effects in the electrical circuit occur as a consequence of this initial setup.

In Fig 7.34 we show a simple electrical circuit with an in-built driving voltage E producing a resultant current I that flows through a circuit resistance R. We assume that the remainder of the circuit loop is a perfect conductor. The driving voltage E, the electron flow I and the resistance R are the three main components of an electrical circuit. The voltage E and Resistance R are independent components while the resultant electron flow or current I is a dependant component. This means that the current I has a quantity value that is directly proportional to the intensity or strength of the driving voltage E but inversely as the resistance R which hinders its flow. This is represented by the formula,

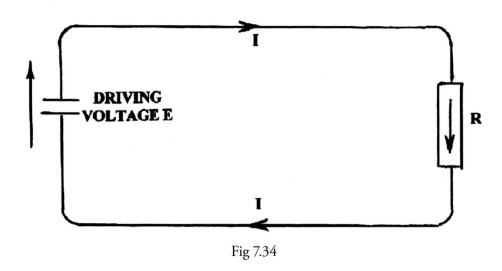

Fig 7.34

Current I $\propto$ $^{\text{Voltage E}}$/Resistance R

or simply as, I = E/R when using standard units of E in Volts, I in amperes and R in Ohms. If we know two of the above circuit components then the third can always be determined since their inter-relationship is consistent. However it must be remembered that of the three, E and R are the cause while I is the effect. As such I is the only true variable in the circuit.

So far we have described the 'domino effect' which shows how the electron flow is maintained across a conductor circuit. We have also indicated the reason for a mass body either being a partial conductor or a non-conductor. We have however, only assumed an external factor that stimulates nuclei and causes them to emit directional electrons. The initial driving voltage is usually produced by an electron flow generator that is present within the circuit. The electron generator (which we shall term as the voltage generator) is a system that causes the emission of electrons from within a nucleus group along flow paths that are orientated in approximately the same general direction. The intensity of this action determines the extent of the driving voltage. By itself, a voltage generator simply indicates a potential for an electron flow. Thus when an electrical circuit does not exist, the electrons that are emitted from a voltage generator simply 'reflect' within its body structure and cause nucleus disorientation with a similar electron flow in an opposing direction. The net electron flow being nil. Let us explain this aspect with the help of the theoretical voltage generator block represented in Fig 7.35.

Consider the generator as a material block consisting of the three zones A, B and C as shown. The active zone is represented at B and this is where the stimulation of permanently orientated nuclei produce electrons that are continuously emitted in the direction towards zone C. At this stage (in our presentation) zone A, which lies behind zone B, is to be considered as a dormant zone. This means that its nuclei are not fixed in their orientation and as such any emissions are in a random direction. Zone A is also considered to be at the back end or rear of the A-B-C generator block.

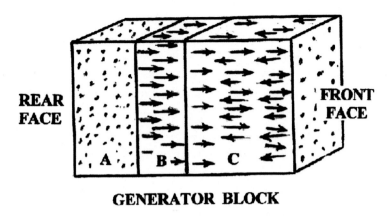

GENERATOR BLOCK

Fig 7.35

Zone C on the other hand has a very active function in that it is the recipient of a very large influx of electrons from zone B. The domino effect induces the nuclei within the zone C to emit electrons towards the front face of the generator block. Due to the phenomenon that the surface atoms of a conductor material have a very different nucleus orientation from similar internal atoms, we find that the domino effect does not easily propagate across a mass structure boundary into a realm datum zone. The principle of molecular bonding indicates that the surface atom nuclei possess an impulse in the general direction of the over-all mass body and as such do not respond to the impact of directional electrons as per Fig 7.29.

These surface nuclei are usually stimulated to emit electrons in the same general direction as the nucleus impulses. Since these impulses may alternate as for the bouncing molecule (Fig 6.17), then electrons should be emitted backwards and forwards in an alternate manner. Thus in zone C we shall find electrons being emitted back towards zone B with the result that nucleus orientation at the zone B - zone C boundary is upset. Subsequently the progression of the domino effect from zone B to zone C is also diminished. In the present case all of the domino effect from B to C is reflected back at the front face of the generator block. This results in a total neutralisation of the domino effect within zone C and hence no current I. It must be noted however that the activity level of the

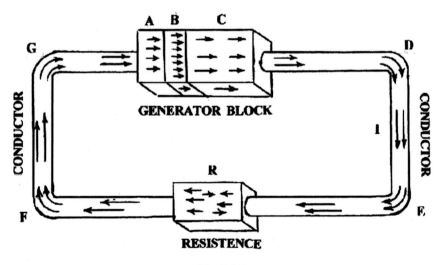

Fig 7.36

stimulation action within zone B remains virtually unchanged and continues its function regardless of what goes on in zone C. In this state of affairs an electrical circuit simply does not exist and as such electrons are unable to flow around an imaginary circuit to zone A.

295

Let us now consider the state of events within the generator block when it is integrated with a looped conductor and resistance element to form an electrical circuit. This means that there is now a path for electrons to flow from the front face of zone C to the rear face of zone A. As such the zone A becomes active in its response by permitting the domino effect to operate across it towards the already active zone B. The circuit is represented in Fig 7.36. Here the front face of the generator block A-B-C is connected to a resistance R through a conductor DE. The conductor FG completes the loop back to the rear face of the generator block. Let us now trace the path of the electrons emitted from the permanently active zone B. Electrons are directed into zone C which propagate towards its front face. Since zone C simply functions as a conductor then it must also possess the usual slightly imperfect structure of a few non-orientated nuclei, resulting in a small impedance to the otherwise perfect domino effect. This is termed as the internal impedance of the generator block and is usually the reason for a current limiting factor in short-circuited conditions as we shall discuss later.

In Fig 7.36 it is clearly observed that the front face of zone C has a much greater x-sectional area than the conductor DE. This means that electron reflections must still occur from the front face of zone C. The nature of these reflections however are such that nucleus orientation in the vicinity is nudged in the general direction of that part of the front face where no reflections occur, i.e. where electrons propagate from zone C to the conductor loop DE. This principle is illustrated in Fig 7.37 and can be favourably compared to the flow of water in a pipe system. Without dwelling too long on this property of the domino effect we can show that nucleus orientation and the ensuing electron flows are influenced by the internal electron reflection within a conductor, and that these subsequently cause the orientation of other adjacent nuclei and their emissions away from the reflection zone. It is by this means that the domino effect gets to be 'aimed' (so to speak) along a conductor path even when this contains apparent directional changes within its circuit. Fig 7.37 shows two cases in which the nucleus orientation or path of electron flow is directed (by this process of reflected influences) towards the only path of exit from the zone C to the circuit conductor. Extension of this directional theme is left to the reader.

Now consider Fig 7.36 where the electrons flow through the conductor BE and into the resistance R. We have already shown how the electron flow is restricted within R, and the subsequent current I within the electrical circuit. The limited current I then flows through the conductor FG and into the rear face of the generator block zone A. The originally dormant zone A must now accept this electron flow and by the phenomenon of internal reflections allows its nuclei to be orientated such that the domino effect occurs generally in the direction of zone B. Now we know that zone B is already a permanently active zone having its own stimulated production of electrons. The effect of the electrons flowing into zone B from zone A is to cause an additional stimulation upon the zone B nuclei, resulting in an additional emission of electrons into the zone C. However the internal impedance of zones A and C and the external resistance R prevent a runaway buildup of a unidirectional electron flow, and as indicated earlier (Fig 7.32) a balanced flow is achieved after several laps of the circuit. It also becomes apparent that in a circuit with little or no external resistance, a continued build-up of the stimulation within zone B of the generator block could lead to an exceedingly high level of electron flow in the circuit. This condition is generally referred to as a short-circuit situation and results in the Mp-shells within the material absorbing large quantities of energy medium. This causes a temperature rise and probable changes in physical property of the conductor. Perhaps it is worth a mention here that as electron flow in a conductor increases so also does the reflection process by the surface nuclei. This means that with a greater number of flowing electrons the greater will be the number that are affected by the reflection type process at the conductor surface. These electrons would then cause adverse orientation in other more internal nuclei, so resulting in an apparent increase to the resistance level of the conductor and a subsequently diminished current I. We leave the reader to contemplate further on this aspect.

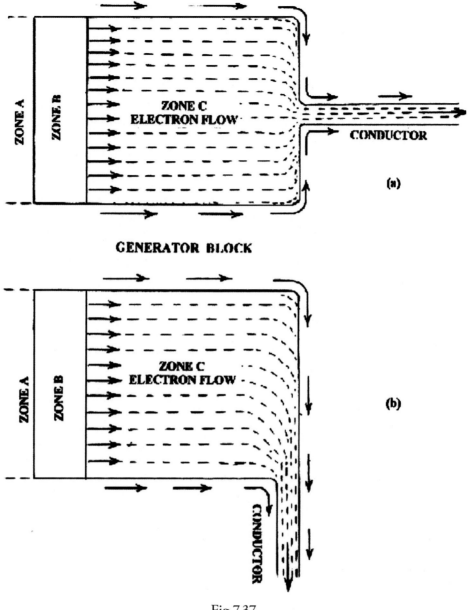

GENERATOR BLOCK

Fig 7.37

Let us now consider the nature of the action within zone B of the generator block. We assume that under conditions of normal nucleus equilibrium and orientation, electron emissions do not occur outside the nucleus. However we must consider that there is a high probability for electron emissions to be present inside the nucleus but that these get absorbed by adjacent proton strings. From this we assume that electrons do occur within the confines of the nucleus but that their propagation paths do not lead towards nucleus windows and hence no outside emission. In order to effect such an emission we would need to exert some kind of external influence upon the nucleus to direct those internal electrons along new paths that lead to nucleus windows and subsequent external emission. As such, the release of emission electrons in a preferred direction within the zone B of the generator block can be the result of chemical action by the alteration in the pattern of molecular bonding between groups of atoms, or due to a mechanical type action from the linear momentum in forcephoids and EIL Nodal fields.

In the case of the chemical type of action there occurs an alteration in the complexity or status of a molecular bond situation resulting in a sudden shift of the centralising impulse upon the nucleus. This in turn causes a temporary change to the proton string configuration within the nucleus, resulting in the emission of electrons along new paths and through nucleus windows. Continuous chemical bonding changes of this nature result in a continued emission of electrons usually in a singular direction due to a manufactured structural composition or layout. When the chemical action completes, i.e. when no further bond changes can occur, then the stimulation to the nucleus' proton strings also ceases and the emission of electrons gradually ends. As such the chemical voltage generator ceases to function. In some cases a flow of electrons impressed upon this exhausted voltage generator from an opposing direction causes the chemical bond changes to gradually revert to their previous status. Thus the potential exists for the re-charging of a chemical voltage generator.

In the case of mechanical action emissions external forcephoids in the form of EIL nodal fields are made to sweep across the paths of the electrons that propagate internally between proton strings inside the nucleus zone. These electrons would not under conditions of normal equilibrium configurations be able to exit from the nucleus zones. In order to cause a maximum shift or deflection in the electrons' propagation paths its Vp factor must be supplied with linear momentum from a direction transverse to that of this velocity Vp (see Fig 7.24). Consider the electron paths represented within the encircled nucleus boundary shown in Fig 7.38. Let us consider two electron

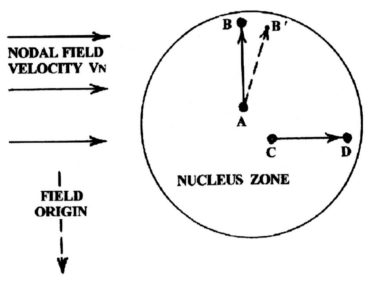

Fig 7.38

flow paths which are at right angles to one another. One path is assumed to commence at the I-proton A of one string and end at the I-proton B of another string. The other path is from the I-proton C of one string and end at the I-proton D of another string. In both these cases the electron has a limited propagation path that begins and ends inside the nucleus configuration. Now let the velocity of the EIL nodes that pass through this nucleus zone be in a direction parallel to the path CD as shown. Traversing EIL Nodes will be absorbed by the electron/Mp-clusters. In the case of the electrons following the path CD there may be a gain in the momentum of the Mp-cluster but without any change in direction. Thus the electron from I-proton C will still follow the same path and arrive at I-proton D and be absorbed therein. In the case of the electrons propagating along the path from A to B there is a gain in the linear momentum of the Mp-cluster in the direction of the Nodal field velocity. This causes the original flow path to alter from AB to AB' as shown. Since there is no I-proton string at the B' location it is probable that the electron path will continue beyond B' and through a nucleus window position. We can conclude that for all the internal electron paths within a nucleus, the greatest path deflections occur at right angles to the velocity of the traversing nodal field. This means that the resulting electron emissions are also at a maximum at right angles to that same field velocity.

In a length of conductor material large numbers of nuclei may thus be induced to emit electrons by this traversing nodal field method. The section of conductor so traversed by EIL Nodes thus becomes a sort of voltage generator block from which electrons are driven around a closed loop circuit. Now if the electrical circuit conductor is in the same direction as AB then the above traversing nodal field will cause emitted electrons to flow around the circuit. But if the conductor length is in the same direction as CD then no electrons will be emitted along CD and hence no current flows around the circuit. Electrons are of course emitted at right angles to CD, but flow in this direction is limited by their reflection off the conductor walls. This principle is illustrated in

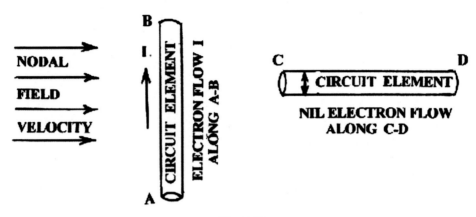

Fig 7.39

Fig 7.39 for both the above conditions. It would be interesting to consider the conditions of flow if a very wide section conductor were used in place of CD. Will the effect shown in Fig 7.37 prevail causing a flow of electrons around the circuit? Quite probable. However in practice such transverse flow action is considered as a wasted effect and so to make the system efficient, conductors are usually made up as a complex strand of a large number of wire elements bundled together. The natural skin effect of each wire element then limits any transverse flow action to a bare minimum.

Let us now briefly consider some of the factors upon which mechanical action emissions are based. A primary factor is the nature of the nodal field, i.e. whether it originates from a North magnetic pole of a primary mass body or from its South pole. In Fig 7.8 we referred to the top end of the depicted primary mass body from which a convex nodal pattern was emitted as its North magnetic pole. The other end which emitted a concave nodal pattern was termed the South pole of the primary mass body. In order to preserve a convention we shall ascribe to this magnetic field a symbolic directionality which correlates with that used in current day physics. As such, we shall consider the convex nodal pattern as directionally positive i.e. field direction will be in the direction of the increasing NR (nodal row) numbers from a North pole face. The concave nodal pattern will be considered as directionally negative i.e. field direction will be in the direction of decreasing NR numbers from a South pole face. Subsequently we may consider the primary mass body to possess a magnetic field that has a directionality from the face of the North pole to the face of the South pole as represented in Fig 7.40.

Another factor upon which mechanical action emissions are dependent is the relative velocity that exists between the conductor and traversing nodal field. Earlier in the chapter we indicated the general nature of the nodal distributions from the N and S pole faces of a primary mass body. We also stated that the fields from these bodies exerted an orientation effort upon other ferrite masses if these were located within the nodal field. The secondary mass body then developed a field of its own which then in turn exerted an orientation effect upon the primary mass body. Orientation of the internal body structure nuclei is possible in the case of ferrite material only because of the compatibility between field and body structural patterns. If however the above primary mass body is brought

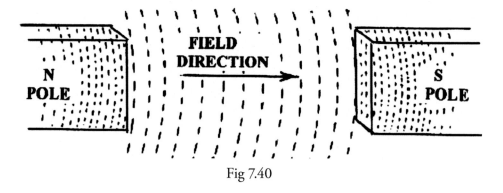

Fig 7.40

near a relatively stationary non-ferrite conductor body, a similar orientation effect does not occur. This is because the primary mass body field pattern is quite incompatible with the structural layout within the non-ferrite material, i.e. the nodal field is unable to produce a layout change in the non-ferrite as represented by the sequence in Fig 7.3 and Fig 7.7. However if the nodal field is allowed to traverse through this conductor then electron emissions should occur by the principle represented in Fig 7.38. There are several ways in which a nodal field can traverse relative to the length of the conductor and each of these has a varying effect upon the direction and quantity of electron emission.

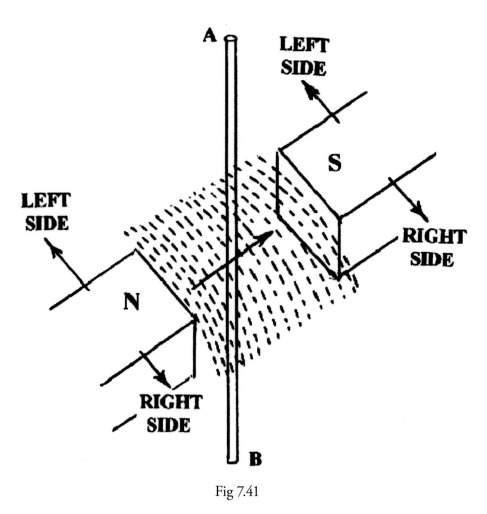

Fig 7.41

In Fig 7.41 we show a nodal field operating between a N and S pole in the horizontal plane. Perpendicular to this in the vertical plane and near the centre of the field is placed the EIL conductor element AB which we shall assume is part of a closed loop circuit. We know that the nodes of the field will pervade the structural body of

300

the conductor. So long as the conductor and nodal field have no relative velocity then there can be no resultant effect of the field upon the conductor, i.e. no mechanical emissions. If the conductor moves up or down its own length or in the direction of one of the poles, again there is no electron flow around the circuit. It is only when relative motion occurs between field poles and conductor either to the left or the right hand side as indicated in Fig 7.41, that electrons flow along AB (or BA). When the field pattern inside the conductor moves towards the right then nuclei orientation within the conductor atoms undergoes a series of jerks from the undulating nodal row patterns in such a manner as to cause emission of electrons through new nuclei windows in the direction from A to B, i.e. downwards in the conductor. When the field has relative motion to the left then a different orientation of nuclei occurs within the conductor resulting in the emission of electrons in the direction from B to A, i.e. upwards in the conductor. The operating rule is emphasized adequately by the well known 'right hand rule'. The quantity of flowing electrons depends upon the strength of the field and its traverse velocity through the conductor's nuclei structure.

In Fig 7.6 we assumed the nodal field pattern to possess nodal rows that were symmetrical in character. We have already shown that the nucleus has outer and inner zones that are not concentrically constructed relative to one another. This feature is the primary cause that makes the nucleus prone to an orientation from external influences. In the case of the traversing nodal field we consider that orientation of the conductor nuclei will have a differing aspect depending upon the field aspect and its direction of traverse. So far we have assumed that a ferrite nodal field has no effect upon non-ferrite material. The right hand rule disproves this concept and so we must assume that there is simply an overall neutral effect. By this we imply that the ferrite nodal field may act within the non-ferrite structure in a random manner producing localised north or south pole patterns. Thus the nodal field influence within a non-ferrite conductor does cause small zones of nucleus orientation, yet when considered overall these appear not to have any relative pattern. The structure does not conform as would a secondary mass body and so is considered non-magnetic.

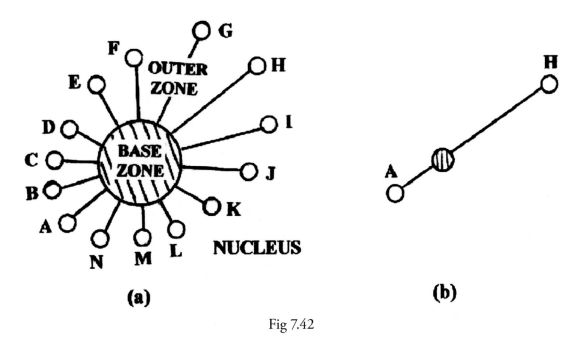

Fig 7.42

However when a nodal field is permitted to traverse through such a structure a ripple effect from the localised orientations runs through the non-ferrite structure. The nodal field aspect and the direction of traverse (which must be in a plane parallel to the nodal rows) result not only in a ripple of a preferred nuclei orientation, but also an emission of electrons through the Fig 7.38 type of action.

Let us consider these principles diagrammatically. Consider the nucleus as an eccentric structure as represented in Fig 7.42(a) . We observe the shaded inner base zone surrounded by the outer zone of radiating proton strings (see also Fig 7.21). It is this eccentric layout of the outer zone proton strings A, B, C, etc., that causes the nucleus to be orientated into varying directional positions. If we consider the 'centre of mass' for the nucleus to be located within the base zone then for a given influence the leverage by string H will be greater than that of string A (in the diagram). Hence orientation. Subsequently we may expect a shift in the directional aspect of the nucleus when its strings are acted upon by the drift of traversing nodal rows. As each nodal row whips through the nucleus, it sets up a flutter of orientation in the nucleus which is highlighted by the emission of electrons. Let us represent the eccentric nuclei by the unequal scale-arm AH shown in Fig 7.42(b) .

In Fig 7.6 we showed the effect of a static nodal field upon a ferrite (nucleus) scale-arm. In the present case for a non-ferrite conductor material we know that a ferrite nodal field can have no resultant orientation effect due to incompatibility between field and mass structure pattern.

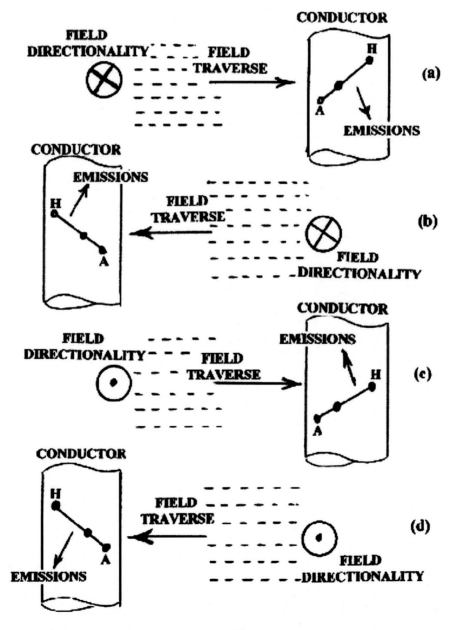

Fig 7.43

However, a repetitive orientation does occur as each localised zone of orientation traverses through the conductor structure. This is represented in Fig 7.43. The field directionality is represented by the arrows convention when perpendicular to the plane of the paper. The circle with the central dot ⊙ represents the tip of the arrow pointing out of the paper i.e. at the reader. The circle with the cross ⊗ represents its tail-end feathers and indicates the arrow pointing into the paper and away from the reader.

At (a), the nodal field has its directionality into the plane of the paper whilst also traversing through the conductor from left to right. The length of the conductor being in the plane of the paper and also at right angles to the velocity of traverse by the field. Now the nodal field traversing through the structure of the conductor results in a path of special orientation in some of the nuclei. This orientation is represented by the scale arm AH in the position as shown. Actually, the scale arm would probably occupy an orientated status somewhat similar to that shown but at an angle of approximately 45° to the plane of the paper with E-proton A being slightly nearer the reader. The direction of the nodal field has one effect on the orientation status while the traversing action results in linear momentum being imparted to the nucleus's internal electrons. This combination causes the emission of electrons through new nucleus windows in preferred directions as shown. The emitted electrons then produce the domino effect within the circuit such that this stimulated section of conductor functions as the generator block of the circuit. The rate of traverse of the field nodal rows through the conductor and the intensity of the field are both factors that increase the orientation aspect of AH such that emissions are more in line with the axis or length of the conductor. This results in a higher electron flow and hence also higher emission voltage (see Fig 7.33). In Fig 7.43 above we have only indicated the majority of the emissions, although we are certain that emissions must also occur in other directions as well. However, we can discount these since we are only concerned (for the moment) with the resultant domino effect within the circuit.

At Fig 7.43(b) we show the same conductor and nodal field as at (a) but with the field traverse in the opposite direction i.e. from right to left. The orientation aspect of the nucleus AH will alter and electron emissions occur as shown with the domino effect and electron flow in the opposite direction to that at (a). The case at (c) is simply a view of the case at (b) from the opposite side of the paper. The case at (d) is similarly paired with (a). In conclusion we observe that as a generator block the conductor element in Fig 7.43 is induced to emit a flow of electrons within an electrical circuit. The factors upon which these mechanical action emissions depend are,

1. the directionality of the magnetic field,
2. the direction of field traverse,
3. the velocity of field traverse,
4. the intensity of the nodal field,
5. the obliqueness between the conductor and the field directionality,

Of paramount importance in the above phenomenon is the conclusion that nucleus orientation inside a conductor is caused by a traversing nodal field, and that new nucleus windows do result in a biased direction. There is much theoretical detail that can be presented to support such a presumption and perhaps some readers will undertake such a justification case. Whatever the ultimate result, the observational facts will remain as the basic ruling guide.

In the setting up of a generator block we assumed nucleus stimulation to arise from either of the two main types of action - chemical and mechanical. We must now introduce a third form of nucleus stimulation termed as 'induction', which may be considered a facet of the mechanical type of action. Basically, induction means that a flow of electrons may be caused within a conductor if it is placed within the field of EIL nodes that exists around another current carrying conductor. Electron type fields that enter the structure of a conductor may cause the domino effect to be initiated within it (to a limited extent), resulting in a flow of electrons. In Fig 7.26 we represented a traversing electron as an Mp-cluster consisting of a synchronised grouping of wavefronts spaced only a few quantum steps apart. These were shown to be emitted by the nucleus in a directional cone of emissions (Fig 7.28), producing the 'domino effect' (Fig 7.29) which could not always be expected to follow a parallel path (Fig 7.31) with that of the circuit conductor. We already know that the atom nuclei at a conductor surface zone have a biased orientation towards the interior, and that the domino effect always reflects inwards. Fig 7.44 indicates the probable paths that the domino effect is likely to progress along within the circuit conductor. When the domino effect is due to the mechanical action generator of Fig 7.43, we observe that electron emissions are not quite parallel to either the conductor length or the plane of the paper. This would seem to indicate that the domino effect does not progress straight along the length of the conductor but rather continually reflects off the side walls in its propagation around the loop. The action of Fig 7.44 would seem to indicate a very gradual clockwise

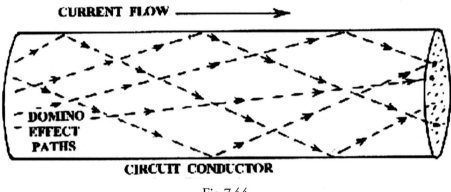

Fig 7.44

offset of the domino effect at each successive surface reflection. Fig 7.45 illustrates the principle of this rotation. At (a) we show the domino effect path along the conductor from left to right passed the internal reflection locations

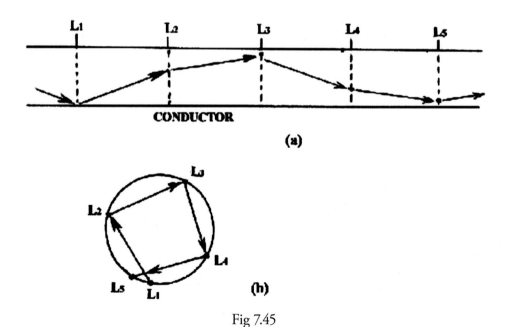

Fig 7.45

L_1, L_2, L_3, L_4, L_5, etc. At (b) we show an end view of the same conductor path as observed from the L_1 end. The relative positions of the reflection points are clearly indicated and a clockwise domino effect progression is quite apparent.

Now at each conductor reflection point there is a change in the domino effect direction. We assume that the domino effect progresses in an approximate straight line path. In Fig 7.27 we showed how a series of forward propagating forcephoids from the electron/Mp-cluster could be emitted into the realm zone external to the mass body. We must now assume such to be the case from each reflection point of Fig 7.45. Each electron flow path is the producer of forward forcephoids in the direction of propagation. At each reflection point therefore, these forward forcephoids will propagate outside the conductor in a line that is a continuation of the domino effect path prior to its reflection. Since the flow of electrons is usually at high density then reflection points must occur all along the conductor with an equally high density of occurrence. Subsequently, these electron forcephoids must be the normal emissions in all current carrying conductors. Their emission paths are indicated in Fig 7.46 by the two views at (a) and (b). We note that the forcephoid emissions occur at very acute angles relative to the lengthwise

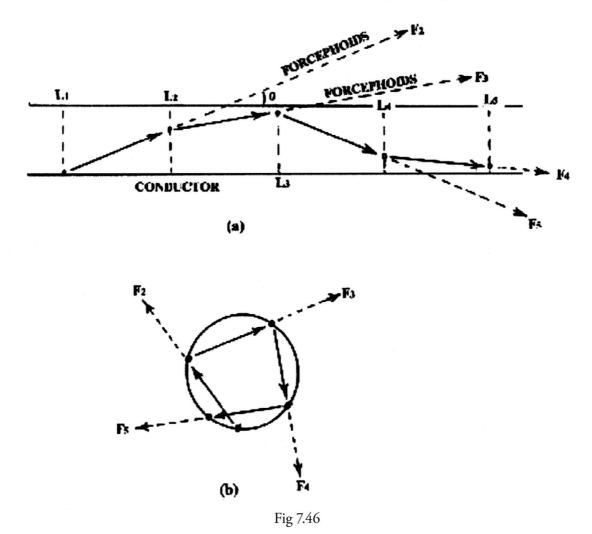

Fig 7.46

surface of the conductor. We also note at (b) that these emissions are not quite radial but are offset in the clockwise direction (the viewing being in the same direction as the electron flow). A large current in the conductor would subsequently produce a large quantity of forcephoids which could be represented by the two views of Fig 7.47.

Now not all of the domino effect paths actually reflect at the conductor surface. Let us consider two parallel paths within the conductor. Let path A be nearer to the conductor surface than path B. When path A reflects at the surface nuclei, path B will continue unaffected. Paths A and B are no longer parallel to one another and as such a theoretical intersect point becomes a probability. The cone of electron emissions from the nuclei in path A will now propagate in the direction of the path B nuclei. Fig 7.31 helps to visualise this effect. If the nuclei $N_1, N_2, N_3,$ $N_4,$ and N_5 are in path A, and if we consider a surface reflection to occur at N_2, then emissions from N_2 would alter inwards. Now if we consider the nuclei N_6, N_7, N_8 and N_9 to be the other domino effect path B, we note that altered emission from N_2 causes the emissions of N_8 also to be altered inwards along a new path through the nucleus N_{11}. The path B has thus been altered before it reached the conductor surface. This type of path change also tends to be a more gradual direction change than that which occurs at the surface. We may refer to this phenomenon as the inter-domino influence and we shall observe that this also results in the emission of forward

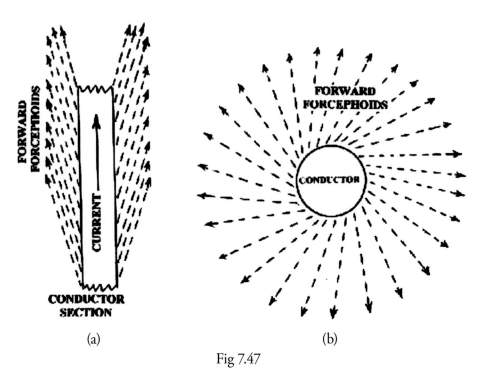

(a)   (b)

Fig 7.47

propagating forcephoids at varying angles to the conductor surface. Fig 7.48 illustrates this very well. Notice that whereas path A has a sudden change in its direction, the path B changes direction much more gradually. Now although the paths seem to become parallel again, the rotation aspect Fig 7.45(b) ensures that path A remains on the conductor surface side of path B. Also note that the forward propagating forcephoids from path B are more like those represented in Fig 7.27. The emission path angle (Fig 7.48) can be seen to vary from q_1 to Nil. Here we have also shown a more central domino effect path C. Perhaps the reader now has a better visualisation of the reflection process that we highlighted in Fig 7.37. Incidentally we can also conclude that the electron flow paths in a conductor are less indirect near the centre of the conductor. This means that although the domino effect may be considered as having a constant velocity its traverse along a conductor length may be deemed to be variable depending upon whether that traverse is nearer the centre or nearer the surface. Perhaps a laboratory experiment could test for this phenomenon.

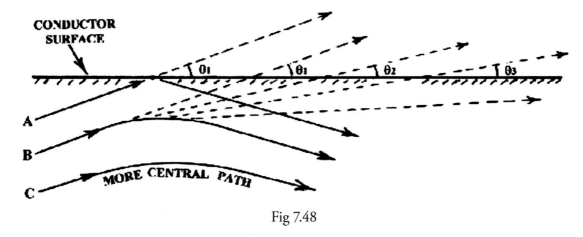

Fig 7.48

When a current flows through the structure of a conductor, the domino effect produces a very high proportion of nucleus orientation. The very fact that an electron flow occurs means that nucleus orientation has taken place. Each Mp-shell section of an atom is the product of a particular E-proton, and therefore the entire Mp-shell composite reflects the structural configuration of its nucleus. Re-orientation of the nucleus must therefore be followed by a similar orientation of its Mp-shell (Fig 6.33 and Fig 7.1). It must be remembered however, that the structural configuration of I-protons and E-protons within the nucleus as a whole remains unchanged. Thus current carrying conductors can have a high level of structural orientation. Previously this high level of nucleus and structural orientation was only observed in ferrous element mass bodies as discussed in Fig 7.7 and Fig 7.8. It was this exceptionally high level of structural orientation that resulted in the compatibility between its structure and nodal field patterns. We could argue at this stage that electron flow in a conductor also produces a similar relationship between field and structure. Unfortunately we cannot test for this in a direct manner since there is no end to the loop and so no top or bottom face as in Fig 7.8. However, we shall assume that such is the case as we shall prove later.

Now we have shown that the domino effect progresses down the conductor in a series of surface reflections that appear as a right hand screwed rotation. This will cause a corresponding structural pattern difference within the conductor (surface zone) such that the scale arm pattern is not quite symmetrical or parallel to the conductor axis. This subsequently results in the emitted nodal field also extending in a direction that is non-parallel to the conductor axis. This direction is the same as the direction of electron flow or domino effect, and also that indicated for the forward propagating forcephoids in Fig 7.47. If we assume the primary mass body in Fig 7.8 to be a surface section of a current carrying conductor, then the electron flow through it in the direction from bottom face to top face must produce a structural orientation configuration that is similar. This means that the forcephoids emitted from this structure will produce a compatible nodal field. The direction of this magnetic nodal field will be rather variable depending upon the 'depth' location of our selected section of conductor. However since we have chosen a surface section we shall assume that the nodal field has a directionality (Fig 7.40) parallel to the domino effect paths of Fig 7.45 and the forward forcephoid paths of Fig 7.47.

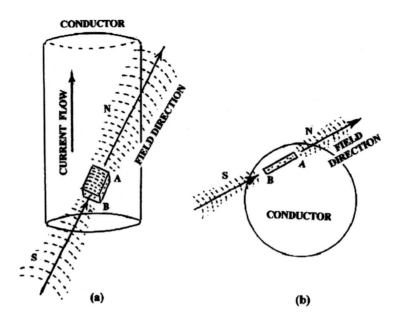

(a)   **(b)**

Fig 7.49 Consider a small conductor section AB located near the conductor surface as shown in Fig.7.49. If AB is similar in structural configuration to the primary mass body of Fig 7.8 then a nodal field will exist as shown. Subsequently, the end A will be considered as the north pole and end B as the south pole of the conductor section, but only so long as there is a flow of electrons through it from B to A. Also, the nodal rows initiated from each end will conform to the usual pattern of convex at N-pole and concave at S-pole. Since the electron flow through the conductor section is in parallel with the field directionality, then the forward forcephoids from the electrons will operate in the N-pole field while the backward flowing forcephoids (set up by the backward E-flows from the Mp-cluster Fig 7.6) will operate in the S-pole field. Also the structural nuclei within AB are fixed in location and therefore their field forcephoids must traverse at the constant E-Flow velocity VF. However this is not the case for the forcephoids emitted by moving electrons because forward forcephoids traverse marginally faster than backward forcephoids by the Vp factor. Since electron forcephoids and nucleus forcephoids operate in parallel then their intersection points (nodes) cannot be relatively stationary.

Consider the very simple case of forcephoid emissions from an electron source E and from a nucleus source N as shown in Fig 7.50. At a given instant let the two sets of forcephoids intersect at the points O_1, O_2, O_3, O_4, O_5,

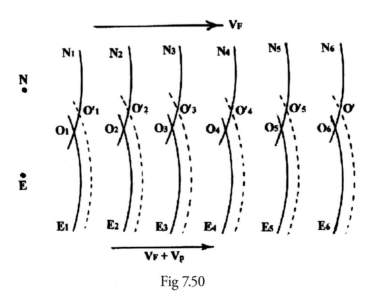

Fig 7.50

etc. If each set of forcephoids had the same velocity then the intersection points would simply traverse along the straight line joining O_1, O_2, O_3, O_4, O_5, and O_6 as depicted by Fig 6.20. However, since the forcephoids from E traverse with a greater velocity than those from N the earlier intersection points must progress to the new locations O_1', O_2', O_3', etc. We have shown the change in the relative positions of the N and E forcephoids by simply indicating the advanced position of the E forcephoids as a dotted curve. It will be noticed that the intersection point has altered not in the direction of VF but rather at approximately right angles to it. We could show that this trend will continue with continued traverse such that the intersection point progresses further along N_1, N_2, N_3, etc. Eventually the forcephoid E_1 will catch up to forcephoid N_2 and the entire shifting pattern will repeat. This principle will also apply when the setup involves the considerably more extensive system of forcephoids and EIL nodes. As such the result is that the nodal rows from this two velocity forcephoidal interaction setup are not stationary but will pulsate in the lateral direction. Now when another conductor loop is placed within its sphere of influence, mechanical action of the type indicated in Fig 7.41 becomes effective with the resulting directional nucleus stimulation and the emission of electrons. We observe therefore that the flow of electrons in one conductor can induce a flow of electrons in another conductor loop even when no relative structural movement occurs.

So far we have showed that the nodal field around a current carrying conductor consists of two distinct types. Firstly, there is the magnetic field of Fig 7.49 which is a relatively stationary field similar to those from the earlier primary and secondary ferrite mass bodies. And secondly, there is the electron generated field which operates in parallel with the magnetic field but with a continual and repetitive lateral shifting of its nodal rows. For any current carrying conductor both fields exist in conjunction and it would not be possible to isolate one from the other. Since these fields are operative in the realm zones surrounding the conductor and must extend some considerable distance from it, then we may assume that the conductor functions as a primary mass body (Fig 7.8). Let us for the moment ignore the induction effect just discussed and consider only the magnetic impulse conditions as represented by Fig 7.15.

We have shown how a nodal field can generate an impulse in a structurally compatible mass body (Fig 7.11 to Fig 7.14). In Fig 7.49 we showed that a structural element AB near the conductor surface could be considered as a primary mass body with a field of directionality along BA. It is our contention here that the domino effect between two consecutive reflection points as demonstrated in Fig 7.45 must result in such a magnetic element so as to produce a defined polarity and nodal field. $L_1 L_2$ would thus constitute a magnetic type element as would $L_2 L_3$, $L_3 L_4$, $L_4 L_5$, and so on. In the light of this let us consider the interaction between two parallel current carrying conductors that are positioned quite close to one another as shown in Fig 7.51. Here we have two parallel

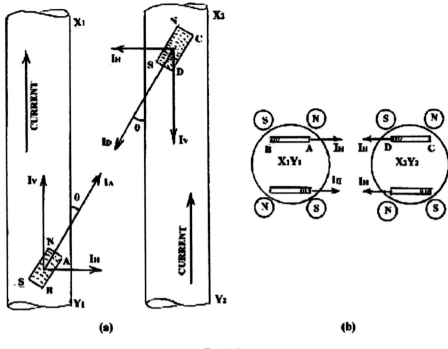

Fig 7.51

conductors X_1Y_1 and X_2Y_2 with the electron flow through each being in the same general direction. Let us consider the conductor sections AB and CD such that each lies within the nodal field generated by the other. The section AB has a magnetic type nodal field with directionality from B to A. As such the A side is the N-pole and the B side the S-pole. Similarly we can show that since the conductors have current flowing in the same general direction then the conductor section CD has a nodal field with the same directionality as for AB. As such the C side is the N-pole and the D side is the S-pole. If we now consider the elements AB and CD in isolation as primary mass bodies then the situation is similar to that represented in Fig 7.15(a). Subsequently each magnet acquires an impulse towards the other. The section AB is shown to acquire an impulse represented by IA in the direction towards CD, while the section CD acquires an equal and opposite impulse represented by ID towards AB. Let these impulses act along a line which makes an angle q with the length axis of the conductor. Due to the conditions represented in Fig 7.16 we know that q is small.

Now IA and ID each have a horizontal component IH and a vertical component IV as shown . We note that the horizontal component will always be in the direction of the other conductor.

$$IV = IA \cos \theta = ID \cos \theta, \text{ and } IH = IA \cos (90-\theta) = ID \cos (90-\theta)$$

Since θ is small then for each conductor section, IV is much greater than IH. However a mirror image of the Fig 7.51 representation will indicate to the reader that on the opposite side of the AB and CD sections IV acts in an opposite sense while IH is still in the direction of the other conductor. Thus we may ignore IV from any resultant action. On the other hand IH is cumulative and indicates that parallel conductors with electron current flowing in the same general direction acquire an impulse that may be termed as attractive towards one another. Assuming each conductor to be long, then they may be considered to contain a large quantity of elemental magnet sections and a subsequent resultant attractive impulse that is measurable.

In the other case for parallel conductors with electron current flowing in opposite directions we would have the two magnet elements represented by Fig 7.15(c) or (d). The same general arguments would apply except the

310

magnet elements would have impulses away from one another. As such the horizontal component IH would always act away from the other conductor, which means that the action causes the conductors to be repelled from one another. From the above two cases we may state as a general rule that long parallel conductors with similar direction currents are attracted to one another while those with opposite flowing currents repel one another. We leave the reader to speculate upon the magnitude of q and the subsequent minimum length of parallel conductor required. The contention that each surface reflection point of the domino effect (as discussed in Fig 7.45) represents the N or S pole of a magnet element, should also be examined

further to show that the subsequent nodal field is a reality for each such element e.g. L_1L_2, L_2L_3, L_3L_4,, and so on. A comparison between the structural orientation in ferrite primary mass bodies and that due to the domino effect in a current carrying conductor would be a worthwhile study. A prime zone of comparison would be the follow-on structural pattern within a U-shaped magnet and that of the conductor surface zone due to the reflecting clockwise spiral of the domino effect. Perhaps the reader would care to relate the emission of the forcephoids in Fig 7.47 to the corona effect that exists around conductors at very high stimulation voltages (Fig 7.33) and electron flows.

Let us now consider the effect of an EIL nodal field upon a current carrying conductor. We have shown that the domino effect (current flow) through a conductor material results in a structural orientation of its atom nuclei. The orientation pattern and molecular linkage configurations correspond to the direction of electron flow such that the structural pattern is relatively constant for a given flow direction. It is the domino effect in the electron flow that imparts a magnetic type of structural configuration to any conducting material. In Fig 7.7 we showed how a ferrite mass body could have its structural linkages influenced by the nodal field of a permanent magnet or primary mass body. We also indicated the principle behind the impulses generated within adjacent primary mass bodies due to their being in the zone of each other's nodal field. It was then explained that the nodal field of one body acted upon the other by attempting an alteration in its state of balanced structural equilibrium. Structural equilibrium means that for a nucleus orientated structure the bonded atoms have Mp-shells that exert a neutral centralising action upon their nuclei. When such a structure is placed within a directional nodal field then this balance can be upset. The nodal field will probably cause these nuclei to undergo a change in orientation. The Mp-shell bonding layout and centralising action pattern result in a resistance to such an orientation. Structural equilibrium is lost between the various impulse quantities and subsequently an impulse imbalance occurs throughout the mass body linkages. This imbalance causes a resultant impulse to become active within the body which then acquires linear momentum in a defined direction. This impulse is effective so long as the structure remains within the nodal field. Although we have only generalised upon the principle of an impulse imbalance being set up in primary mass bodies, we have nevertheless been able to indicate in Figs 7.11 to 7.14 the actual resultant impulse direction under differing nodal conditions.

Now if the body structure curve in Fig 7.11 is placed at right angles to the nodal field curve, then apparently there is no effect of one upon the other. This means that there is no upset in the body structural equilibrium by the field nodes and hence no resultant body impulse. This can be partly proven by the placement of two primary mass bodies (permanent magnets) in a T-configuration as shown in Fig 7.52 for the magnets A and B.

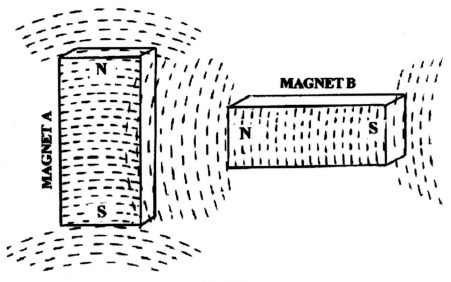

Fig 7.52

In a current carrying conductor the domino effect causes the structural configuration within the material to be maintained in a direction that is in line with the flow of its electrons. We have already shown the flow path of these electrons and it is our contention here that the structural configuration resulting from this electron flow is somewhat more complex than that in a ferrite magnet. However when a current carrying conductor is placed within a nodal field the structural equilibrium of its orientated configuration is upset in a similar manner. This means that a subsequent impulse imbalance occurs within the conductor body and a resultant impulse then causes the conductor to receive a push action in a defined direction. This direction is given by the well known 'left-hand rule' with the impulse effect being an optimum when the length of the conductor is at right angles to the field direction. This is represented in Fig 7.53 and shows a conductor AB placed at right angles to a nodal field's directionality. If the current in the conductor flows in the direction AB and the field direction is from N to S, then the conductor will experience an impulse in the direction XY as shown. If either the field direction or the current are reversed then so is the impulse. If both field and current directions are reversed together then

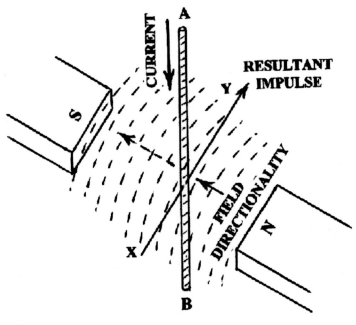

Fig 7.53

the impulse direction remains unchanged. If the conductor length is placed parallel to the field direction then no impulse is generated. For a given set of field and current parameters the impulse intensity is proportional to the sine of the angle between the conductor length and field direction.

Now in Fig 7.41 we described how the movement of a circuit conductor in a magnetic field resulted in a directional flow of electrons within that conductor according to the right hand rule. As soon as a current flows in the conductor we must consider the application of the left hand rule for directional impulses. In Fig 7.54 we show a circuit conductor AB positioned at right angles in a field of directionality NS. Let the conductor be given a movement

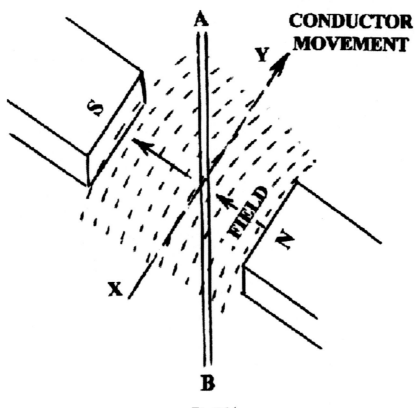

Fig 7.54

in the direction XY. Applying the right hand rule we find that a current is induced in the conductor in the direction BA (upwards). If we now apply the left hand rule to this current carrying conductor we find that an impulse is generated within the conductor in the direction YX, i.e. opposite to the initial movement of the conductor. It will be observed that no matter what the direction of the initial movement is, the induced current sets up an opposing impulse in the conductor. Thus in order to generate a continuous flow of current by this dynamic (dynamo) method a constant effort must be exerted upon the conductor to maintain its motion in a given direction.

The inverse situation is also true in the case of the impulse generated in a current carrying conductor that is in a nodal field. When the conductor is stationary there is only the conductor's own internal. impedance that inhibits the flow of electrons. Referring to Fig 7.53, a current in the AB direction causes the conductor to be pushed in the direction XY as shown. Now the instant that the conductor commences a movement in the field NS then the right hand rule will apply which indicates that a current flow is induced in the same conductor but in the direction BA i.e. in opposition to the prevailing current flow. This means that there is an additional opposition or inhibitor to the flow of current in the circuit. This is why electric motors have a very low start up impedance with the correspondingly very high start up current flows as compared to that when they are rotating at speed.

Let us now consider the nodal field generated by a current carrying solenoid coil. In Fig 7.44 we showed the domino effect paths within a conductor. This was a generalisation and was simply meant to convey a mental picture of the events within a circuit conductor when a current flowed through it. In subsequent diagrams (Fig 7.45 - Fig 7.47) we indicated more precisely the direction of the flow path, and also the pattern of forcephoids that were emitted to the realm zone outside the conductor. At that time we ignored the effects of any external influences upon the conductor. However when current carrying conductors are in close proximity and parallel to one another then not only do they have an impulse effect upon one another as described in Fig 7.51, but they must also have forcephoids from one conductor entering the structural zone of the other. The effect of forcephoids and EIL Nodes entering a conductor and altering the electron flow paths therein is termed as induction. The effects of induction are manifold but at this stage we would simply like to explain how this relates to the setting up of a magnetic type of nodal field when a current carrying conductor is shaped into a coil or solenoid.

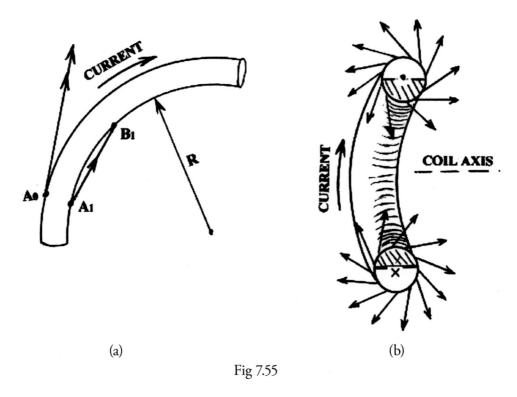

(a)   (b)

Fig 7.55

Consider a single straight section of conductor as represented in Fig 7.47(a). All of the forcephoids that are emitted from this conductor propagate continually away from its surface and as such can have no further effect upon that conductor. However if this conductor was to be bent into a circular arc or curve then the situation would alter as indicated by Fig 7.55(a). Here we represent a section of conductor in the shape of a circular arc with a curvature radius of R. Let A_1 and A_O be two points adjacent to one another but on opposite surface positions as shown. A_1 is on the inside of the conductor curvature while A_O is on the outer or convex face. Let forcephoids be emitted in the usual manner from both points A_1 and A_O when a current flows through the conductor at (a) in the clockwise direction. The forcephoid from A_O propagates out from the conductor surface and does not come anywhere near the conductor again. On the other hand the forcephoid emitted at A_1 is intercepted by the same conductor at the point B_1. We could in fact state that it is the conductor that bends round and intercepts the forcephoid, however the end result is the same. The relative distance of A_1B_1 will depend upon the curvature of the coil. Since there are a countless number of forcephoids being emitted continually at all points along the conductor then it is quite apparent that the inside surface of the conductor will be continuously under the influence of its own forcephoid emissions. At (b) we have shown a cross-section of the conductor and differentiated between its inner and outer regions. We have also shown the direction of forcephoid emissions prior to any self induction. It is our contention here that the effect of the impinging forcephoids i.e. A_1B_1 on the inside face of the conductor, results in a marginal shift in the domino effect paths therein. This in turn results in a change to the direction of emitted forcephoids from the conductor's inner region surfaces. In Fig 7.55(b) we have shown the rotation aspect of the forcephoids emitted at each of the end sections. It will be noticed that these produce forcephoids of opposite sense but with the same directionality along the axis of the coil. By applying a theoretical analysis of the normal forcephoids from a straight conductor (Fig 7.47) to the example of the coiled conductor, we would note that there is a defined directional angle of propagation for forcephoids that are emitted at A_1 and which impinge at B_1. This aspect is shown in Fig 7.56 where A_1B_1 is the impinging forcephoid on the inside surface of the conductor coil. Now we know that the domino effect within the conductor has a clockwise rotation in the current flow direction, as also

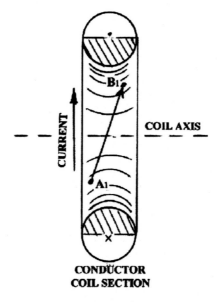

Fig 7.56

do the emitted forward forcephoids therefrom. With the coiled conductor a self induction situation is prevalent and we contend here that the domino effect rotation is enhanced in the inner zones of the coil. By this we mean that the self induction causes the magnet element AB in Fig 7.49 to have its alignment alter marginally towards the coil axis on the inner region surfaces only. Since the outer region of the conductor coil does not come under the influence of any such induction action then the domino effect and subsequently the alignment of the element AB will be unaffected in this region. Since each element AB functions as a primary mass body the net effect of the coiled conductor is to throw out a nodal field that possesses a directionality which is an average of all the AB elements. This has been found to be in the direction of the component of A_1B_1 (Fig 7.56) that lies along the coil axis as was expected. For a large number of coil turns the field directionality can be found using the right hand screw convention for current flowing in the coil. Fig 7.57 illustrates this.

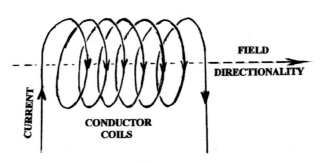

Fig 7.57

315

In the case of a very large number of coil turns that are wound alongside as well as on top of one another, then the induction effects become a bit more complex but with the same net result of field directionality. Since each turn contributes its AB element components to the overall nodal field, then the greater the total number of turns the greater will be the intensity of the resultant nodal field. Because the conductor section AB functions as a primary mass body then the overall conductor coil will also function as a single primary mass body so far as its nodal field is concerned. The reader must note that the field strength is the horizontal component of a magnet field that is aimed primarily in the vertical direction as represented in Fig 7.51, where the angle q is very small indeed. Imagine how much greater the solenoid field strength would be if this angle q could be increased considerably. We leave the reader to speculate not only upon this aspect but also upon the actual mechanics behind the self induction effect that results in the coil's inside conductor surface AB elements acquiring a different attitude/alignment.

Let us now consider a mass object that is placed within the axial zone of a current carrying coil. We can safely assume that the nodal field generated by the coil will also pervade the structural body of that mass object. If this mass object is in fact a secondary mass body, i.e. a randomly orientated ferrite structure as represented in the Fig 7.3 configuration, then the effect of the pervading nodal field would be to orientate its structural linkages as at Fig 7.7. So long as the current flows in the coil circuit this induced orientation in the ferrite body will remain. It will be noticed that the magnetic field of the coil and ferrite body together is much greater than that due to the coil on its own. This is because the orientation induced in the ferrite structure develops a better directionality than the AB sections of the coil, and which also causes the body to function as a magnetic unit (Fig 7.8). Subsequently an impulse is generated within this magnet unit which the reader may analyse in accordance with the principle of a primary mass body inside the nodal field of another (Fig 7.11). Since this magnet unit must also generate a nodal field of its own then the current carrying coil comes within the sphere of this field and so will acquire an impulse as represented by the conditions of Fig 7.53 in a direction defined by the left hand rule. The reader may well appear puzzled by this statement since we had specified earlier that the nodal field has a pattern that is parallel with the magnetic body's structural configuration, and that the side faces of that body did not emit a nodal field that conformed to any relevant pattern. In the case of the electric coil however, each of the AB sections' nodal fields emitted from the conductor's inside surfaces have an individual directionality that is very different from the overall resultant. This means that although the coil has an overall field directionality parallel to the coil axis, the ferrite body placed within it will come under the influence of a multi-directional nodal field from each AB section of the conductor. Subsequently, the structural orientation of the ferrite body will conform to a somewhat varied configuration. This will be similar to Fig 7.8 near the coil axis region but in a partly radial layout in the regions adjacent to the conductor. As such, we would assume that a related field is emitted accordingly. The field near the conductor is assumed to be effective only between the coil and the surface of the ferrite body as shown in Fig 7.58. This field has a directionality at a marginally radial angle to the ferrite body surface and subsequently causes an impulse to be induced in the conductor according to the left hand rule. The overall impulse induced in the coil will be in a direction parallel to the coil axis and opposite to that in the ferrite body.

In the case of a primary mass body traversing through the core space of a currentless coil circuit then the situation of Fig 7.54 can be recognised for relative movement between a conductor and nodal field. A current is induced in the coil conductor depending upon the traverse direction and directionality of nodal field.

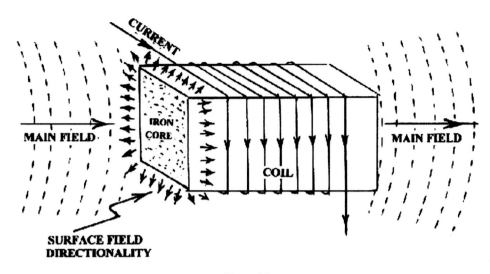

Fig 7.58

Let us now suppose the ferrite core of Fig 7.58 to be infinitely long or alternatively curved into a closed loop. The former conditions would be the same for the coil except that now there would be no exit face for the axial nodal field. The internal structural configuration of the ferrite core would still follow the axial line (indefinitely), and the surface radial field would exist as before. When a steady and continuous uni-directional current flows through the coil conductor the core body adopts the magnetic structural configuration and a nodal field is generated somewhat radially at the core surface. At the instant of switching on the current the rate of growth of this nodal field is extremely rapid yet nevertheless of a finite duration. Since the nodes must develop within the region of the coil conductor then this conductor must experience a shifting nodal row pattern during the core re-orientation and field growth period. The relative effect is equivalent to that of a circuit conductor moving in a nodal field. Subsequently the principle of Fig 7.41 for induced current in the conductor would apply. After the initial period of field development a nodal limit is attained and no further field growth occurs. In this steady state condition there is no relative movement of nodal row patterns and hence no induced current. When the current in the coil circuit is switched off the orientation of the ferrite core structure returns to its former random pattern (nearly). This change also takes place over a finite period which means that the nodal field from it must diminish at a corresponding rate. It is considered that the field decay characteristics are very nearly the opposite of the field growth conditions and as such the conductor will once again experience the equivalent of relative movement within a nodal field. Observation informs us that the field growth induces a current in the conductor that is opposite to that already flowing in the conductor but another current in the same direction when the field is in decay.

Let us now consider the case represented in Fig 7.59 in which two quite separate coils are wrapped around the same ferrite core. The primary circuit consists of a generator block V, switch S and coils A. The secondary circuit has the coils B connected in a closed loop to the galvanometer G. The closing of the switch S causes the needle of the galvanometer to kick once in one direction, while the opening of the switch S results in a needle kick in the opposite direction. Otherwise, with the switch either closed or open continuously the meter showed that no current flowed in the secondary circuit. The reason is obvious and shows that the structural orientation in the ferrite core is

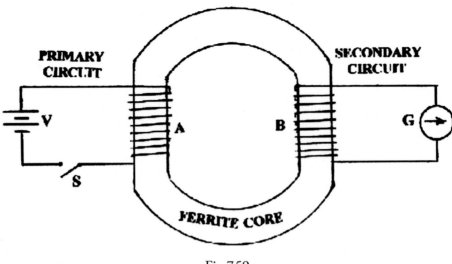

Fig 7.59

perpetuated around the whole of the core-loop and that nodal field growth and decay occurs across the whole core. This is why a directional current is induced in the secondary circuit only when the primary circuit current is in growth or in decay.

In the case of an alternating current in the primary circuit the reader will find it obvious that the current flow in A would act like a continuously current switching operation and that we could expect a corresponding frequency of current induction in the secondary circuit. At some optimum frequency of the alternating current in the primary circuit we would get its exact character duplication in the secondary circuit - albeit somewhat out of phase. Since the nodal field intensity is constant per core unit area then the amount of generator block induction for the secondary circuit will depend upon the length of the conductor in coil B. This really means number of coil turns in B. The greater the coil length i.e. B zone (Fig 7.35), the greater is the total voltage generated. Thus if there are Np coil turns in the primary circuit and Ns coil turns in the secondary circuit, and if Vp is the primary voltage then the induced secondary voltage Vs is given by,

$$Vs = Vp \times Ns/N\,p, \text{ or } Vs/Vp = Ns/N\,p$$

Now we know that a voltage can be generated in an open circuit loop when no measurable current is flowing. We have already explained the principle of internal reflection of the domino effect at conductor surfaces as at Fig 7.37. We also explained that when a conductor loop or circuit was open ended, that the domino effect electron flow emitted from a voltage generator simply reflected back from the circuit break point. This caused a backward flow of the reflected electrons in a direction opposite to that induced by the voltage generator. The net electron flow in any given direction was thus nil. Nevertheless a voltage or potential for current flow in a given direction was being continually generated. This principle applies to the secondary circuit of Fig 7.59. When this circuit loop is open no current can flow around it but the induced voltage is still Vs. Now we know that it is the current in the primary coils that causes the core structural configuration to be orientated and which subsequently results in the induction in the secondary coils. When the frequency of the alternating current is at an optimum then the alternation of the core's structural configuration and the subsequent growth and decay events of the nodal field lag behind the inducing current. This can be blamed upon the physical inertia in the core's atom nuclei orientating ability. The net result is that the field generated by the ferrite core lags sufficiently to cause a current to be induced in the primary coils that is nearly 180° out of phase with the original current. This means that an opposing current inhibits the flow of the original current and therefore the primary coil seems to develop an impedance of its own. Now when the secondary circuit is closed and a current is permitted to flow through it (by means of a variable impedance) then this current will also be nearly opposite in phase to that in the primary coil. We now observe two separate current flow conditions causing structural orientation in the same ferrite core. The current flowing in the secondary coil will now induce a structural configuration pattern change within the ferrite core that is more in phase with its own alternating current pattern. However once again due to the inertia of the ferrite core's atom nuclei orientation rate, the field generated by the core lags sufficiently behind to generate a separate current out of phase to that already induced in the secondary coils. Theoretically we should have the same self impedance situation in the secondary coil that we had in the primary setup. However this is not a similar case since we now have a more complex situation which functions as outlined below:-

1. The primary coil induces a voltage in the secondary coil and also an inverse voltage in its own coils. This self induction then acts as an impedance to the primary current flow.
2. The secondary coil functions in a similar manner when its circuit is closed and a current flows in it.
3. The secondary coil current now attempts to generate a voltage in the primary coil.. This voltage will be nearly 360° out of phase with the original primary voltage. However 360° and 0° are the same phases and so the primary voltage receives a boost. This also counteracts the inverse voltage in the primary circuit and so reduces the self impedance of the primary coil. A heavier current will subsequently flow in the primary circuit.
4. The generation of an inverse voltage applies to both primary and secondary circuits and its intensity is proportional to the number of coils in each. However the current in each circuit helps to reduce the inverse voltage of the other on a coil for coil basis. If the secondary circuit has more coils than the primary circuit then the reduction to the inverse voltage of the primary coil is greater. Subsequently the self impedance of the primary coil is lesser and its current flow greater than in the secondary coil. The current flow in the two circuits is therefore inversely proportional to the total number of coil turns in each.

From this outline of actions we may conclude the following relationship:

$$Ip/Is = Ns/N\,p$$

where Ip and Is are the current flows in the primary and secondary circuits respectively.

We have already expressed a functional relationship between number of coil turns and generated voltages. If we combine these then we have;

$$Vs/Vp = Ns/N\,p = Ip/Is$$
$$\text{or, } Vs\,.\,Is = Vp.\,Ip$$

This last relationship is the well known operating principle of the AC Transformer. We shall not pursue its operational characteristics here as we leave the reader to follow up on these and other magnetic type features such as resonance and the hysteresis loop using the general principles already presented.

As the reader must by now be quite familiar with the basic principles of electron current flow, perhaps it is also the appropriate moment for an extension of that theory to other related phenomena. If a rod of ebonite is rubbed with fur, it gains the power to cause light bodies such as pieces of paper or tinfoil to receive an impulse in its direction, i.e. to be attracted towards it. The discovery that a body could be made attractive by friction is attributed to Thales (640 - 548 BC). He seems to have been led to it through the Greeks' practice of spinning silk with an amber spindle; the rubbing of the spindle in its bearings caused the silk to adhere to it. The Greek word for amber is 'elektron' and a body made attractive by rubbing is said to be electrified or charged. The study of this type of phenomena is called electrostatics. The reader must be familiar with this topic as it is quite conventional and fairly well understood today. However there are certain basic principles of action that have been simply assumed to exist, such as; like charges repel and unlike charges attract one another. There is no magical wand and as such we should be able to explain this and other actions quite simply.

Consider a generator block. A flow of electrons is stimulated to flow in a particular direction. If a closed loop exists then an electric current builds up from the domino effect and is observed to flow through the circuit. However if there is no loop then we note that an electron flow is still generated within the generator block but this is reflected back and forth inside the conductor section. The reflections are always at the material surfaces and are the result of the inward orientation of the surface atom nuclei. The continual presence of a finite voltage within the block is indication enough of this internal activity.

Let us perform a thought experiment with such a generator block as represented in Fig 7.35. Here we show its three basic zones as A, B and C. Since the block has no circuit outlet for its stimulated electrons we assume that these are in a continual state of reflection from the front end face of zone C. Because of the small amount of internal resistance within the material of this zone the domino effect will deflect as represented in Fig 7.31. This will subsequently result in a proportion of electrons also reflecting off the side faces of zone C. The continued stimulation of the nuclei in the zone B ensures that a driving voltage exists across the block, the voltage simply being the potential for sending a domino effect flow of electrons around a theoretical conductor loop. Since this stimulation is permanent, the potential for current flow is also ever present. As part of our thought experiment let us assume the generator block to be completely isolated from all other conductor materials and supported upon a perfect insulator. As such we assume that there can be no outflow of electrons from any part of it. Now the level of reflected electrons reaches a stage where these interact with those being emitted afresh from the zone B. Subsequently the electrons simply reflect around in zone C. Since the quantity of electrons emitted by zone B is very nearly the same as those entering it from zone C, we may subsequently assume zone B to be in a state of emission equilibrium. As such let us magically expel the zone B function from the generator block. We are now left with zones A and C. As there is no stimulated zone to demarcate the boundary between zones A and C we may now consider the entire block to comprise a single conductor zone. Although we have eliminated the source of voltage generation we must still consider those electrons that were emitted earlier and which are still being reflected between the conductor surfaces. If these continue in a back and forth reflection mode then this internal domino effect activity could be considered a long term effect. As such we assume that the same driving voltage exists within the block as when zone B was in existence. However if a conductor loop were connected to this block the electrons would gradually escape down the line and less and less would be left to reflect within the block. The driving voltage would decrease. Eventually there would not be any electrons left in the block and as such its driving voltage would have reduced to zero. It would thus appear that we are able to store up a driving voltage within a conductor body (suitably isolated) and then to use it to produce an electron current flow simply by connecting it to another conductor circuit. Unfortunately, any such current flow is limited in duration by the depletion of its driving voltage.

In our original generator block the driving voltage had a defined directionality i.e. in the direction from zone A to zone C. However once we have removed zone B altogether, the imperfect structural quality of the block material causes the domino effect to send electrons along other multi-directional paths (Fig 7.31) such that eventually these electron reflections are distributed evenly over the entire block surface. Theoretically the driving voltage is more or less equal in all directions. Sharp corners however cause a higher concentration of the internal reflections in the block with the subsequently higher localised voltages. Fig 7.60 depicts these internal reflections in the stages just described. We note at (c) that the block has reflecting electrons that are more or less equally distributed over its entire surface. In a good conductor material the domino effect paths are mostly straight line paths and as such the frequency of reflections at its surface will be an optimum. If such a block is connected to a conductor wire then. its reflecting electrons will be drawn down the wire and the driving voltage dissipated very quickly. In a poor conductor however the domino effect follows a deviating non-linear path as explained in Fig 7.30. As such the lengths of the domino effect paths between successive surface reflections can be quite extended. Subsequently, the frequency of the reflections would be comparatively less and it would take relatively longer to draw out all the electrons in a flow through a connected conductor wire.

Any material body that is in the condition shown at Fig 7.60(c) has the potential to send an electron flow through a suitable conductor path. This driving voltage is not perpetual or self generating and can only be used once to produce a current flow. As such we cannot consider the body as a voltage generator block but rather as a reservoir that holds a fixed quantity of electron flow paths. This is quantifiable and determines the 'potential' or 'charge' within that reservoir. There is however an interesting experiment that may be performed to produce a perpetual current flow in a circuit containing such a charged body. Consider a material body with a shape as shown in Fig 7.61.

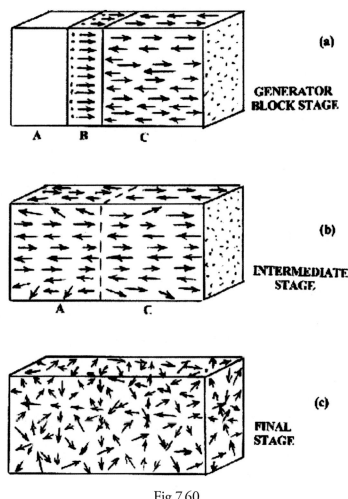

Fig 7.60

Connect a very thin conductor wire to this body at the point B and around to the point A. Place a compass needle at a suitable position as shown but not touching the wire. Isolate the entire apparatus in the perfect vacuum of 'space' and then let a generator block give the body a large 'charge. The reader must be quite familiar with this method of charging. It will be noticed that an electron current will flow along the conductor wire in the direction B-C-D-A as shown for a limitless period. The reason is that a greater frequency of electron path reflections occur in the confined material volume near point B than at the location of point A. As such the driving voltage at B is greater than that at A. Subsequently, the quantity of electrons flowing into the wire at B is greater than those flowing into the wire at A. The difference gives the net flow of current in the clockwise direction. Since (theoretically) all electrons flowing into the conductor wire at both A and B simply trace a path back to the charged body, then the current flow can be considered perpetual. However the domino effect paths inside the material body do undergo certain losses from disorientated nuclei etc., and so a natural depletion of driving voltage or 'charge' is inevitable.

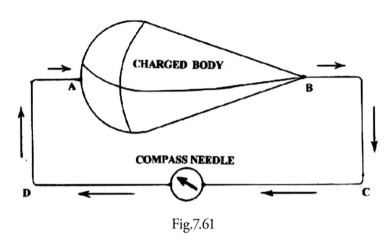

Fig.7.61

Let us now consider the characteristics of the reflections of the electron paths inside such a charged body. Of all the atoms bound in a mass body structure, those at the surface will have the greatest degree of orientation freedom. The effects of this are threefold. Firstly, the orientation state of the mass body skin will be easily influenced by regular forcephoidal fields emitted by other bodies. Secondly, re-orientation of the surface atoms causes them to emit electrons inwards into the body structure. This is not a continuous occurrence and lasts only for the duration of the actual orientation process (See Fig 7.22). As already explained this results in the body becoming electrostatically charged. Thirdly, the skin atoms, after having been induced into a particular orientation pattern, can cause the electron internal reflection paths to be given a definite directional quality. By this we mean that no matter what the angles of incidence of the domino effect paths at the skin are, the angle of the reflected path is more or less the same. It is logical to assume this if we consider that the stimulation of a pre-orientated atom will result in the emission of electrons in a defined direction and does not depend absolutely upon the approximate direction from which the stimulation occurs (Fig 7.30). One may compare this with the example of the loaded cannon which is primed and ready to fire. The powder can be ignited by the gunner standing in any of several positions and yet the shot will always be expelled in the direction that the cannon muzzle points. We must also consider the alternative case whereby the shot of one cannon strikes the stock of another and causes it to fire in turn. The sudden impact causes the aim of the second cannon to be displaced somewhat just before it goes off. As such the firing stimulation has also caused a re-orientation of cannon aim. We observed this in the principle of the domino effect Fig 7.29.

Now as the surfaces of all mass bodies are simply a combination of atoms in an adjacent to one another manner, then we must conclude that the configuration of atomic or molecular bonding at this surface is similar for all solid state elements. As such all mass body surface atoms should be equally prone to external orientation influences such as the nodal field emitted by the changed body in Fig 7.60(c). Let us establish the manner in which a nodal field is set up around such a charged body.

Compare the electron path reflections in this body with those in the current carrying conductor of Fig 7.44. The reflections therein produced a field around the conductor as shown in Fig.7.47 and with a field directionality as in Fig 7.49. There is absolutely no reason why the reflections of the domino effect should not produce a similar type of nodal field around the Fig 7.60(c) body. Fig 7.62 highlights the basic principles behind this. AO is the incident electron path that reflects at the point O on the body surface and then follows the path OC. Now we know that

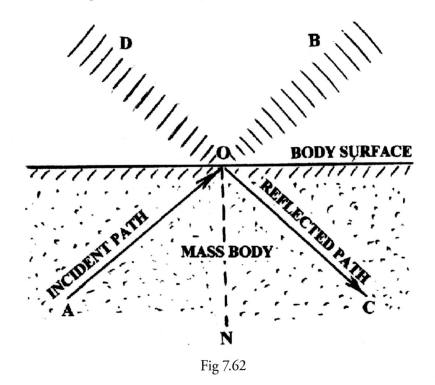

Fig 7.62

electrons propagating along the path AO will emit forward forcephoids in the direction OB. Similarly after reflection the electron path is along OC and backward forcephoids are emitted in the direction OD. For the moment we can ignore the backward forcephoids of path AO and the forward forcephoids of path OC since these propagate inside the mass body. With large numbers of reflections occurring in the localised area of point O, significantly large quantities of forward and backward forcephoids must be present outside the mass body. Under certain conditions these must produce nodal patterns that conform to the definition of a regular field. However, because of the differing velocities between forward and backward forcephoids their combinations are unable to produce consistent EIL nodes. As such only the same type of forcephoids can produce a relatively stationary nodal pattern. (See Fig 6.22 for node origins).

In the status represented in Fig 7.62 we have shown the incident and reflected electron paths to be quite symmetrical about the normal ON. Subsequently, the forward forcephoids will set up a nodal pattern as will also the backward forcephoids. Since both types will be equally present near the body surface no distinguishing nodal pattern will emerge. The pattern of one will be inter-laced with the pattern of the other. This however is a rare occurrence if we consider the skin atoms to possess a particular orientation. Depending upon the nature of this skin orientation it is considered that the reflection paths will be restricted to a narrow angular band width at a more or less consistent angle to the normal ON. Now we know that the incident path angles can vary from the extremely oblique to the perpendicular. This then provides a wide range of the forward forcephoid emissions outside the mass body as shown in Fig 7.63. Obviously the forcephoid EIL diminishes with distance as per the inverse square law. In the diagram we show the electron incident paths A, B, C and D at various angles to the normal line. If we select the outward forcephoid positions of a, b, c and d respectively such that the forcephoid EIL at each is the same, then their distances from the surface point O will also be equal. However we are not interested in this aspect but rather with the fact that positions 'c' and 'd' are much farther from the body skin than positions 'a' and 'b'.

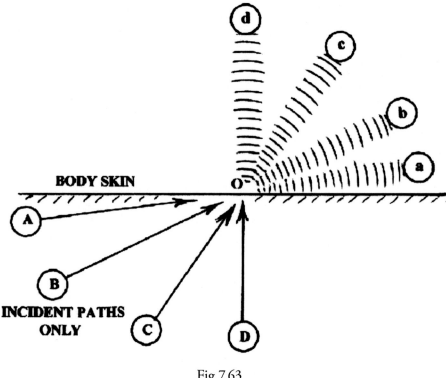

Fig 7.63

This means that the near-to-normal incident paths C and D produce higher EIL nodes than those from the oblique incident paths A and B at a given distance outwards from the mass body. We can apply the same reasoning for the emission of backward forcephoids from the various angles of the reflected paths.

Let us assume that the orientation of atoms in the body skin can adopt either of two basic modes. One of these causes all incident paths to be reflected along paths that are close to the normal line and the other mode causes all incident paths to reflect obliquely close to the surface plane as represented in Fig 7.64 . At (a) we show the skin orientation pattern to be of one mode, say type-x, causing only oblique reflections and at (b) we represent the other mode, say type-y. In each case we must also consider the effect of a path interference between incident and reflected electrons. At (a) for the skin type-x, all of the reflected electron paths are oblique. Incident electron paths that are also oblique will thus tend to meet these reflected electrons 'head on'. The net result is considered to be one in which the incident path deviates either towards the skin or away from it. The deviation towards the skin causes a shift towards the normal line while the deviation away can be ignored as the path is no longer incident. The reader will note that the term 'reflecting path' is only used locally since such a path must eventually become the incident path when it traverses across to the other side of the mass body. By the above process the majority of incident paths tend to approach the skin atoms from a near normal direction.

Subsequently, the main forcephoidal field that dominates the area around the mass body is that from the incident paths, i.e. from the forward forcephoids. The backward forcephoids are emitted at too oblique an angle to be of any consequence. In Fig 7.49 we showed the field setup in front of the flowing electrons as having a positive field directionality away from the source. We must therefore also consider the field setup around the above body as having a similar directionality. We may therefore consider the body with skin type-x orientation as emitting a positive field.

In Fig 7.64(b) we show the skin orientation pattern to be of the other mode type-y which causes reflection paths close to the normal line. Using the previous argument for electron path interference we can conclude that any incident paths that approach from the near normal directions are deflected

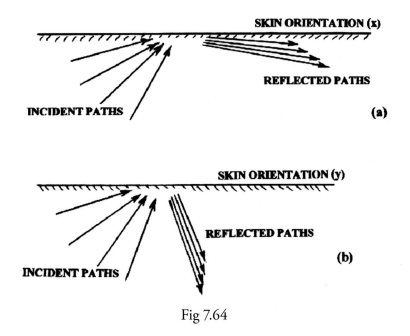

Fig 7.64

to a more oblique path. The result is that the majority of incident paths are in an oblique direction while the reflected paths are nearer the normal line. We note therefore that the main forcephoidal field set up around this body is from the emitted backward forcephoids. Also this field is considered to possess a directionality that is towards the surface of the body. Fig.7.65 highlights the main features of the above two modes.

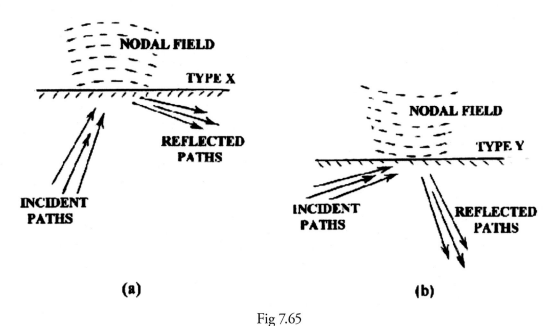

Fig 7.65

Now driving voltage is defined only as the ability of a charged mass body to send a flow of electrons along a conductor. This driving voltage is derived only from the incident electron paths within a charged body. Also, it is logical to assume that it is easier for electrons to flow out of the body if their incident paths are along the normal line or very close to it. Thus we may consider that a greater driving voltage is achieved from incident normal line paths than from oblique or shallow angle incident paths. A charged body with an orientation mode type-x will possess a greater driving

voltage than a similarly charged body with an orientation mode type-y. If a conductor wire is connected up between two such opposite mode charged bodies then a current of electrons will flow from the type-x orientated body to the type-y orientated body. It is considered that the type-y body will need to contain a charge many thousands of times larger than that in the type-x body to produce a balance in driving voltages. It is for this reason that convention declares that the type-x body be considered as positively charged relative to the other. Since the type-y body also contains a definite charge and nodal field, convention declares that this body be considered as negatively charged relative to the other. The nodal fields of each are different as indicated in Fig 7.65 and as such it is possible to ascertain type of charge from a distance.

Let us for a moment speculate upon events when two equal and oppositely charged bodies are connected to each other through a conductor wire. If body A has the positive charge and body B has the negative charge, then a current will flow down the conductor (for a short duration) from A to B as represented in Fig 7.66. Since this electron flow must initially contact the surface of B, its initial effect is upon the orientation of those surface atoms.

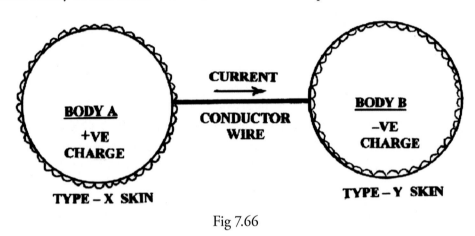

Fig 7.66

The effect is for the electrons to flow through the surface atoms and to cause an orientation away from the type-y and towards the type-x. This orientation can only occur in a progressive fashion, i.e. from type-y state to a non-orientated state and then to the type-x state. Since we have assumed a defined quantity of electron paths in A, then the current flow to B is only sufficient to alter the orientation in B from type-y to a neutral non-orientated state. We speculate here regarding the electron paths within B. Because these had not been drained away, a situation will prevail as indicated by Fig 7.62 whereby no defined nodal field dominates. However there should be some indication of a driving voltage within such a neutralised body. Perhaps an experiment such as at Fig 7.61 could be conducted to either prove or disprove this. The flow of electrons from body A results in their depletion within A such that its driving voltage ultimately reduces to nil. Subsequently no forcephoids can be emitted due to the non-existence of any electron paths within A, and as such it would appear that the charge of each body is simultaneously reduced to nil. The skin orientation state of A can only be maintained by the continuous internal bombardment of its surface atoms by directionally aimed electrons. The lack of these results in a surface de-orientation.

Early in this chapter we showed how the structural configuration of a ferrite body could be influenced by the nodal field of another and made to adopt a regular structural pattern (Figs 7.3, 7.6 and 7.7). The same principle applies when a secondary neutral charge body has its surface atoms placed in the nodal field emitted by a charged body. Assume that the charged body has a positive charge. Then its nodal field will be made up from the emitted forward force phoids. The pattern of this nodal field is such as to cause the surface atoms of an adjacent body to adopt the type-y orientation. The orientation process also causes the surface nuclei to emit electrons. The amount of orientation so induced is dependant upon the strength of the inducing nodal field. The emission of electrons is in turn dependant upon the extent of the induced orientation. Since the strength of the nodal field is in turn dependant upon the quantity of the positive charge then it can be shown that the induction effect is to produce a nearly equal and opposite type charge in an adjacent mass body. We use the term 'nearly equal' because of diminishing nodal field strength with distance. The same applies in the case of induction by a negatively charged body. As its nodal field is produced by the intersections of backward forcephoids, then its pattern of nodes is such as to induce a type-x orientation in the surface atoms of the adjacent body.

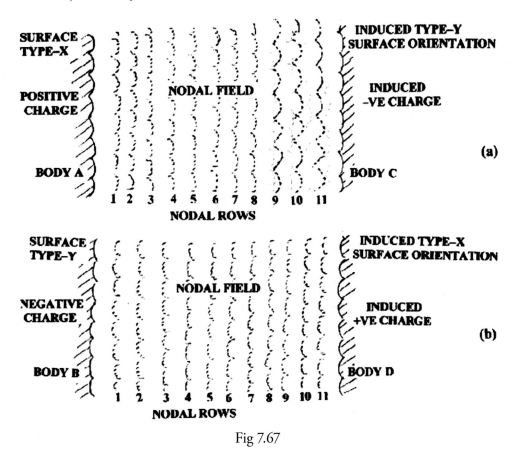

Fig 7.67

Let us illustrate this principle by imagined patterns for the above nodal fields and body surface orientations as in Fig 7.67. At (a) we present a theoretical pattern for the surface orientation of a positively charged body A. To the right side of this are the pattern of the nodes in the nodal field emitted by it. The reader will note that we have assumed the body surface pattern to consist of a regular series of outward facing convex peaks. Furthermore we have also assumed the pattern of each nodal row to be a facsimile of the body surface. Now from the extreme right hand side of the diagram we move a non-charged body C into this nodal field. The surface atoms of body C are then rapidly orientated into positions that are compatible with this field. This means that the surface pattern adopted by body C must be in parallel with the nodal row pattern. This also means that the surface patterns of C and A are in parallel. However since the body masses are on opposite sides of this parallel pattern, a pattern peak in body A is in actual fact a pattern trough in body C. The surface orientation patterns are thus similar but opposite in direction. We assume that the induced pattern in C corresponds to the opposite mode or type-y orientation and hence a negative charge profile.

In Fig 7.67(b) however, we commence with a negatively charged body B and show the induction of a type-x surface orientation and hence positive charge in the body D on its right hand side. The surface orientations and nodal row patterns that have been assumed in both of the above cases were for illustrative and conceptual purposes only to help explain a basic principle and should not be considered as actual. We consider that the surface pattern simply represents the status adopted by the nucleus and perhaps a marginal adjustment in the inter-shell bonding there. Experiment indicates that once a particular level of charge has been induced as above no further induction occurs to increase that charge. This means that electrons are emitted by the nuclei only when an orientation change occurs (Fig 7.25 and Fig 7.42). As already stated these electrons then cause a marginal increase in that orientation by their incident path pattern and so help maintain that status when the inductive influence has been removed.

Before we progress further it is important that the reader is able to differentiate between the magnetic field set up by a ferrite body as represented in Fig 7.8, and that set up by the electrostatic charge bodies just described. The difference lies essentially in the origins of those nodal fields. The magnetic field derives its pattern from the natural emission of forcephoids from a uniquely structured atomic mass body. The orientation pattern occurs throughout that body and can be maintained on an independent and permanent basis. In conductors carrying an electric current such an orientation pattern is maintained by the domino effect, and therefore this magnetic field is a temporary one. On the other hand the electrostatic body nodal fields are the result of the intersections of either forward or backward type forcephoids which result from electron flow paths within a mass body. The electron flow paths being the result of a nucleus orientation pattern that is only caused in the surface atoms of the mass body. There can be very little comparison between magnetic and electrostatic nodal fields since each has a unique node pattern. Each produces a nodal field that is only compatible with the body structure of its origin and therefore electrostatic and magnetic bodies do not interact. Magnetic body fields only induce magnetism in ferrite bodies but an electrostatic condition can be induced in the surface atoms of bodies of all the elements.

Let us now return to Fig 7.67 and consider the effects of charged bodies upon one another. Early in this chapter we described the effects between primary mass bodies of the element Iron (Fig 7.9). Now when a charge is induced electrostatically, the surface orientation becomes a permanent feature of that body so long as it contains its internal bombardment of electrons. In the setup at Fig 7.67 the surfaces of the adjacent bodies have patterns that are in parallel. Each body will be within the nodal field of the other and their node patterns will also be in parallel, i.e. with the same peaked curvature direction. The orientation within each body surface is already in the same direction as the nodal field it is in and so a strengthening of the existing orientation levels will be induced. The relativity of pattern between nodal row and surface structural orientation can be assumed similar to that described in Fig 7.10 for ferrite bodies. The arguments for the impulses generated in the surface atoms of the above charged bodies are also similar to the principles we discussed for magnetic bodies in Figs 7.11 to 7.14. It was then considered that further orientation of the pattern of nuclei orientation in the structure (or surface) was resisted by the Mp-shell molecular bonding action and relevant centralising impulses. It was apparent that there were a number of separate impulse quantities within the orientation plane and that under normal conditions these were in a state of balance. However upon entering a nodal field a certain amount of structural displacement was inevitable and a subsequent upset in that balance of impulses. A resultant impulse is thus a finite quantity. There was a similarity with the upsetting of the centrally directed impulses inside a proton body by the existence of an EIL gradient, and resulting in a gravitational type impulse. In each case the balance of impulses within the structure was upset by external influences.

In the case of Fig 7.67 where all the surface orientation patterns and prevailing nodal patterns have the same curvature or pattern direction, the resultant impulses generated are therefore in the decreasing nodal row number direction. This means that each surface structure acquires an impulse in the direction of the adjacent body. The impulses are therefore attractive. When similarly charged bodies are placed near one another as in Fig 7.68, then the curvature or pattern status of surface orientation and prevailing nodal field will be in opposite modes. Here we show two positively charged bodies adjacent to one another. Body A is in the prevailing nodal field emitted by body B, and vice versa. The situation is as for Fig 7.13 of earlier and the result is the same. The surface atom

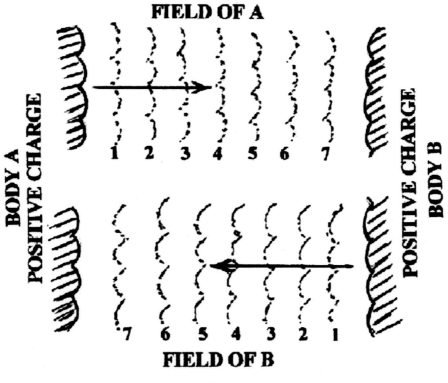

Fig 7.68

structure acquires an impulse in the direction of the increasing nodal row numbers, i.e. away from the origin of that field. Since each body surface undergoes the same type of nodal field influence then each acquires an impulse away from the other. The same applies if the two bodies have negative charges. Thus, similarly charged bodies acquire repulsion impulses away from each other.

The amount of the impulses generated is dependant upon the intensity of the nodal field within which the surface atoms of each body are located. The field intensity or field strength is directly proportional to three factors. These are, the node EIL, the population density of nodes within the prevailing nodal rows, and the closeness of nodal rows to one another. Each of these factors diminishes with distance from the field origin. Therefore, we observe that the impulse is inversely proportional to the distance between the surfaces of the adjacent bodies. Without going into detail we shall assume these impulses to obey the usual inverse square law for distance.

Consider the relationship between the impulses generated in two adjacent bodies of unequal charge. Let A have a charge twice that of B. Then B will come under the influence of a strong nodal field. Now the impulse is directly proportional to the product of their surface orientation level and the strength of the nodal field in which they are placed. Therefore for a given distance between A and B,

$$\text{Impulse in B} = K_1 \times \text{Field strength of A} \times \text{Orientation Level of B}$$
$$\text{Impulse in A} = K_1 \times \text{Field strength of B} \times \text{Orientation Level of A}$$

where, K_1 is a conversion constant and the impulse in A is opposite in direction to that in B.

Now we also know that field strength is directly proportional to the orientation level of field origin. Therefore for the same distance between A and B,

Field strength of A = K_2 × Orientation level of A,
and, Field strength of B = K_2 × Orientation level of B

where, K_2 is another conversion constant.

By combining the two sets of equations we have,
Impulse in A = $K_1 K_2$ × Orientation level of B × Orientation level of A,
and, Impulse in B = $K_1 K_2$ × Orientation level of A × Orientation level of B,
or, Impulse in A = Impulse in B.

From this we note that whatever the difference in charges between two adjacent bodies the impulse generated in each are equal and opposite. This is similar to the case for magnetic bodies as in Fig 7.15. It will be noticed that the principle for impulses is the same.

With these general principles in mind let us consider some electrostatic phenomena. When a cellulose acetate rod is rubbed by a duster it obtains a positive charge, while a polythene rod similarly treated obtains a negative or opposite charge. The reason is quite simply in the nature of each body's surface atom orientations. The contact in the rubbing action causes Mp-shell distortions which then send centralising impulses to the nuclei. Then according to the configuration of each nucleus structure, these impulses cause a re-orientation therein and also the emission of electrons therefrom. Thus a charge is obtained in the body being rubbed. The type of charge is then dependant upon the more permanent orientation that the surface atoms adopt as a result of both centralising impulses and internal electron bombardment.

In 1843 Faraday performed his famous ice pail experiment in which he suspended a positively charged metal ball on a long silk thread and lowered it into a pail without letting it touch the sides or bottom. He discovered that a positive charge was induced on the outside of the pail. This is explained by the induction process of Fig 7.67(a). The body A, in this case the charged metal ball, sets up a nodal field which causes the near surface of body C (the pail) to acquire an induced negative charge. However because the ice pail is made up of a relatively thin body the same nodal field will also induce orientation of the nuclei in the outside surface. Thus the outer surface adopts a pattern in parallel with the nodal field origin. We observe that the inner and outer surfaces of the ice pail are induced with opposite charges from the same positive charge source. Faraday repeated his experiment with a nest of hollow cans insulated from one another and showed that by simply inserting a positively charged ball as before into the central can, that equal and opposite charges were induced on the inner and outer walls of each. As before the inner can has a negative charge induced on its inside surface and a positive charge on its outside. This positive outer surface then also emits a nodal field which induces a negative charge on the inside surface of the next can and a positive charge on its outside and so on and on.. The negatively charged inner surfaces will also emit nodal fields which reinforce the induction effect of each can upon the other. Once a condition of 'charge' equilibrium had been achieved, and if one of the cans was to be suddenly discharged then the induction from the nodal fields of those on either side would result in its charge condition being rapidly recovered. The above also explains the means whereby the electrophorous seems to obtain by induction a practically unlimited supply of charge from its polythene base (which had been suitably rubbed).

When a voltage generator is connected to a body, electrons flow into that body giving it a positive charge. When this body is designed as two metal plates separated by an insulator then the device is able to store that charge in a directional manner. The device is called a 'Capacitor'. Let us carry out an experiment to show how the capacitor stores its charge. Fig 7.69(a) shows a circuit which may be used to study the action of a capacitor. C is a large capacitor and R a large resister. A is a galvanometer and K a two-way switch. D is a voltage generator (supplying

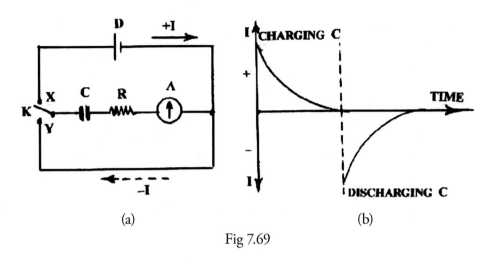

(a)   (b)

Fig 7.69

6 volts dc). When K makes contact at X the current in the galvanometer is observed to be initially quite high (approx. 60mA), and then as shown at (b) it gradually decreases to zero. Thus current flows to C for a short time even though the capacitor plates are separated by an insulator. But once the capacitor is fully charged no more current can flow through it. Why? Also, when K is disconnected from X and connected to Y so that a circuit loop is made, we observe a current of approximately 60µA to initially flow in the opposite direction through R and A and gradually to decrease to zero as shown at (b). The capacitor when charged has functioned like our voltage generator of Fig 7.35 for a short while. The reason for this behaviour is simple and is explained with the aid of Fig 7.70.

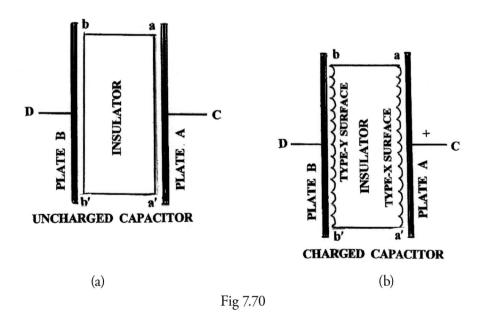

(a)   (b)

Fig 7.70

At (a) we have an uncharged capacitor. The central insulator is between the two metal plates A and B which are each connected to a circuit loop through the conductors C and D. Let a current flow in the circuit from the direction of C to the plate A and on towards the face aa' of the insulator. The electrons that flow into the insulator body set up a complex configuration of random domino effect paths not too unlike that shown in Fig 7.60(c). This results in the insulator body acquiring a voltage generator status. The flow of current from plate A into the insulator opposes the outflow of electrons from the surface aa'. It is considered that the flow of current to plate A also induces a type-x orientation at the surface aa'. The domino effect paths that are set up from this orientation must also eventually traverse to the insulator's opposite surface bb'. Reflections from surface aa' are oblique, Fig 7.65(a), and therefore any traverse towards surface bb' is not direct. Initially the electrons that bombard bb' can escape to plate B, and hence a current will appear to flow through the capacitor in the direction from C to D. However with continued flow of electrons driven by the external source, the type-x orientation level of the surface aa' will increase until it attains a voltage generator intensity that equals the external driving voltage. At his stage a state of voltage balance will be in effect and no current will flow in the circuit. Similarly at the surface bb' the continued bombardment of its internal surface by electrons results in a progressive type-y orientation of its surface atoms. As the internal bombardment continues so the type-y orientation level increases, until ultimately there are no further emissions of electrons to plate B. No further current can then flow in the circuit. As shown at Fig 7.70(b) we now have an insulator with a very high positive charge or orientation pattern on one face and a negative orientation pattern on the opposite face. Subsequently when the external voltage source is removed, and if the circuit is maintained, then a current will flow for a short duration from the positively charged face aa' through the circuit conductor to the negatively charged insulator face bb'.

This aspect was explained in Fig 7.66 for a positively charged body connected by a conductor wire to a negatively charged body. An experiment in which the charged insulator is removed and placed between another set of similar plates gives the same results, i.e. current flow. It would be a worthwhile study for the reader to examine the relative charge holding properties of insulators and conductors especially with regard to their internal electron path patterns.

We have now shown that all mass bodies can be given a charge. This means that electrons traverse inside the body and reflect internally at a surface. A body is considered neutral or uncharged if there are no such internal electron paths. However it would be illogical to assume that the atoms of a given body do not emit electrons now and again, and as and when the realm environment dictates. With the current high level of interactive influences that prevail in our realm zone we are bound to assume instead that all mass bodies must possess a defined frequency of electron emission from the nuclei within their structure. As such all bodies have a small but natural in-built voltage generator action. Since each material element has a unique nucleus configuration and subsequent molecular bonding, then each will respond differently to local realm environmental influences. Hence some materials will be affected more than others and so achieve a relatively greater voltage generator action. Experiments show that this is so and delicate measurements have shown a graded level of such emissions in metals and which has provided us with a 'galvanic action' series. Antimony, Iron, Zinc, Lead, Copper, Platinum and Bismuth are in a sequence of diminishing internal electron emissions. This grading of metals showed that a very small voltage differential exists between them and that a natural flow of current will occur between any two metals in the series. The frequency of electron emissions and voltage differentials are both extremely small, and connecting one to the other reduces the difference to nil by the principle explained in Fig 7.66.

In chapter 3 we defined the term 'temperature' as simply a measure of the realm energy medium that was absorbed into the Mp-shell structure over and above its basic repetitive requirement. Absolute zero of temperature could thus mean the Mp-shell at its most basic EIL. In chapter 6 (Fig 6.11) we discussed this aspect in detail and indicated that the increase in Mp-shell EIL also resulted in an increased amount of the centralising impulse action on its nucleus. We now maintain that an increase in temperature therefore also results in a higher natural emission of electrons from within the elements of the galvanic series due to the increased centralising impulses.

Let us now consider a length of iron wire AB . Let the temperature of end A be at 100° C, and that of end B be at 0°C. The internal electron emission at A will thus be at a higher level than at B.

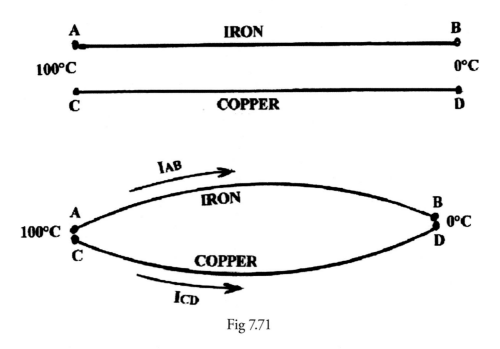

Fig 7.71

Subsequently, an initial current IAB must flow from A to B. As shown in Fig 7.71(a) consider a similar length of copper conductor CD with end C at 100° C and end D at 0° C. Once again end C will have a greater voltage generator action than end D and so an initial current ICD will flow form C to D. Since no circuit loop exists the internal electrons in each case get to be uniformly distributed and subsequently no current flows. However we know that Iron is higher up in the galvanic series than copper, and so A will be at a higher voltage than C. Also, B will be at a higher voltage level than D. We assume that the voltage differentials caused by temperature conditions are greater for elements higher up in the galvanic series and as such the initial current IAB will have been greater than corresponding current ICD.

Now let the 100° C temperature condition be temporarily removed and then let the end A of the Iron conductor be joined to the end C of the copper conductor, as also are the ends B and D as shown in Fig 7.71(b). Now re-introduce the 100° C temperature to the A-C end. A current IAB will flow along the iron conductor in the direction of A to B. Similarly a current ICD will flow in the copper conductor in the direction C to D. Since IAB is greater than ICD then the resultant current IR flowing in the loop will be given by,

$$IR = IAB - ICD$$

and will flow in the same direction as IAB. This phenomenon is called the thermo-electric effect and occurs only over a limited temperature range.

Photo-electricity should be an easy enough subject for the reader to explain, considering how easily some nucleus proton strings are influenced into the emission of Mp-clusters (Fig 7.22). There must also be many other related topics that the reader would like to have discussed here. However we must leave these explanations to the ingenuity of individual readers since the main object of this chapter was primarily to present a basic theory upon which explanations of magnetic and electrical behaviour could be based. Another chapter and a different topic awaits.

Chapter Summary

1. When the more complex atoms form part of a larger molecular group, the orientation of their nucleus proton strings may alter and Mp-shell sections may shift marginally permitting the exit of forcephoids. This results in the set up of EIL Nodes outside the group.

2. A nodal field comprises EIL nodes arranged in a pattern of rows. Each row is represented by a nodal row number increasing with distance from field source. Nodal rows occur in parallel planes and are represented in curved sections of a repetitive pattern.

3. Spacing between nodal row planes and individual nodes increases with distance from their source. Node EILs obey the inverse square law.

4. Mass bodies have their atoms bonded in a regular repetitive pattern. When there is a parity between this mass structure and its emitted nodal field pattern then nuclei orientation can be induced by that field in a similar mass structure.

5. Ferrite mass bodies possess an atomic structural layout that is compatible in pattern to their own emitted nodal field. Thus ferrite bodies can induce structural orientation in each other.

6. Structural orientation and nodal field patterns are both directional. Thus ferrite bodies are considered to possess a directional polarity. A North pole on one side and a South pole on the opposite side.

7. Linear momentum is induced in ferrite mass bodies by these nodal fields. This is the resultant impulse when the normal balance between molecular bonding impulses, centralising impulses and Mp-shell orientation is upset by the nodal field effect.

8. When mass body structure and nodal field pattern have the same curvature direction then the impulse generated in the mass structure is always towards the source of the nodal field. When the nodal field pattern has an opposing curvature then the impulse effect is the opposite.

9. The intensity of the impulses generated in magnet body pairs is always equal and opposite.

10. The surface grid zone of an I-proton has a much higher overall EIL than that of the E-proton. Mps in this surface zone of the I-proton are induced with a linear momentum towards a symmetrical equatorial plane.

11. Mps in the equatorial belt of the I-proton surface grid acquire a linear momentum away from the proton surface.

12. When nucleus strings inside the atom are suddenly displaced then a cluster of Mps may be expelled away from the I-proton. These Mp-clusters are called   electrons.

13. An Mp-cluster is a series of parallel wavefront sections spaced at intervals of a very small number of quantum steps.

14. Although electrons are frequently expelled from the individual I-protons of a nucleus string, very few actually emerge from the nucleus zone. Such external emissions only occur through nucleus windows.

15. An electron or Mp-cluster must propagate in a given direction as it cannot exist if stationary. In its propagation it emits forcephoids in both forward and backward directions.

16. Nuclei in a mass body can be orientated by nodal fields and induced to expel electrons through nucleus windows that are unidirectional.

17. The impact of a beam of electrons upon a nucleus can result in its re-orientation and the subsequent emission of electrons in a similar ongoing direction. This is termed as the Domino effect of electron emissions.

18. A succession of the domino effect in a conductor loop constitutes a flow of electric current.

19. Mass bodies in which nucleus orientation levels are directionally high are considered to be good conductors of electric currents.

20. There are three factors that operate and control the flow of electrons in a loop. The first is the percentage overall orientation of electron emitting nuclei. This is the directional pressure that drives the electrons through the loop and is referred to as its driving voltage. The second is the total number of directional electrons flowing through the loop. This is the current. And the third is the resistance to current flow. This is the impedance or resistance to the domino effect.

21. The driving voltage E and resistance R are independent components, while the current I is a dependant component. I is directly proportional to E and inversely proportional to R.

22. When the electron flow meets the surface atoms of a conductor, the domino effect flow path is reflected back into the conductor body.

23. When a nodal field traverses across a nucleus, then nucleus windows are created at approximately right angles to the traverse direction.

24. The electron or domino effect follows a path that rotates clockwise when propagating along the volume of a conductor length.

25. Whenever the electron path reflects at a conductor surface, forward and backward forcephoids are emitted outside the structure of that conductor. This sets up a directional nodal field around the conductor.

26. Electrons that propagate entirely within the confines of a mass body (by reflecting back and forth between opposite surfaces) impart an electrostatic charge condition to that body.

27. Surface atoms have the greatest degree of orientation freedom as compared to the other atoms in the mass body. As such it is the orientations acquired by these surface atoms that are chiefly responsible for all electrostatic effects.

28. Mass bodies with a particular surface orientation resulting in the setup of a nodal field from forward forcephoids are considered to possess a positive driving voltage. Mass bodies with a differing surface orientation and which emit backward forcephoids are considered to be negative in driving voltage terms. Electrons can only flow from a positive to a negative type mass body.

29. Forward and Backward forcephoids produce nodal fields of characteristic patterns. Forward forcephoidal fields induce a negative type surface orientation while backward forcephoidal fields induce a positive type surface orientation. Thus charged bodies induce opposite type charges in adjacent neutral mass bodies.

30. The emission of electrons is a natural event in many elements because of their existence in an active realm environment. This leads to a natural voltage generator capacity and the build-up of a positive electrostatic charge.

CHAPTER 8.

Velocity Effects

When a mass body is deemed to possess a velocity within the energy medium realm this is usually defined in a relative manner. We assume one object to be stationary and then determine another's motion relative to it. However, since all known astronomical objects have some sort of movement within the realm, the concept of an absolute velocity condition for any given object becomes an exceedingly difficult proposition. Due to the vastness of the realm it would be quite impossible to select and maintain a finite point within it that could be considered as a stationary bench mark, and relative to which all other motions could be defined. Since we have also assumed that our realm as a whole is in motion through the cosmic void at some unknown and un-measurable velocity relative to the Mother Realm (see Fig 2.3), then the concept of an absolute velocity status becomes even more remote. It is however essential that we establish a format for determining the velocity status of mass objects within the realm. We must therefore consider absolute velocities from a totally different viewpoint, and yet at the same time base this upon a system of realm medium relativity.

We could not accept the principle where all objects were considered as possessing a motion relative to a stationary observer. Instead we consider that each object within the realm environment should be capable of determining its own velocity status relative to the realm grid.

Consider the example of an airplane on a night flight through dense cloud. If we assume that the pilot has no radio or radar contacts then he would simply gauge his velocity from his compass heading and air speed indicator. Since the airspeed is also an indication of the air velocity over the wings, then the pilot can ensure that the airplane supports itself in flight and is not in danger of stalling. The airplane can thus gauge its own need to relate visibly to any other objects in order to know its air speed velocity through the clouds and atmosphere. Just as the pilot in the plane did not need to relate visibly to any other objects to know his air speed, so do we consider that each object within the open realm environment must be able to determine its own velocity status without the need to relate visibly to other objects around it. Since all objects exist within the realm energy medium grid, then it should be the flow of this energy medium grid past each object that is the dominant factor in evaluating and quantifying its velocity status. It follows therefore that velocity is a condition of relative motion through the realm grid. Unlike the atmosphere around the air plane however the realm grid is a complexity of nearly isotropic E-Flows. This makes the evaluation of an object's realm velocity somewhat complicated.

In chapter 1 we defined the realm datum as the prevailing EIL at the various spatial grid locations. The EIL was not considered to be uniform but compared somewhat to a 3-dimensional survey map where high and low levels are represented by contour type lines. Gradients can subsequently be determined, which has proved quite essential to our theory of gravitation. The realm datum EIL is the sum of all the E-Flows arriving and phasing at each grid location. Since all E-Flows arise from mass objects, we may consider the realm datum as the EIL status resulting from the combined phasing of all local mass decay E-Flows. Over the very large distances of the realm we consider mass distribution to be fairly uniform and constant (at least in the short term). As such, the decay E-Flows at most points in the realm grid will be very nearly isotropic. A test proton at rest within this grid would absorb E-Flows from all directions as shown in Fig 8.1. Here we represent the velocity of each E-Flow by the vector VF. The proton is considered to be stationary because the velocity vectors of E-Flows from all directions are nearly equal. Now because the structuring of the realm datum is based upon the E-Flows from mass objects, then if a local group of mass objects have a combined uniform motion in a particular direction, then the realm grid set up by these masses will also possess a similar motion. This means that the realm datum within such a mass group has a shifting EIL grid that follows the overall motion of the group. A test proton within this EIL grid would be considered stationary, i.e. zero realm speed, if its position was maintained constant relative to the mass group and the EIL grid.

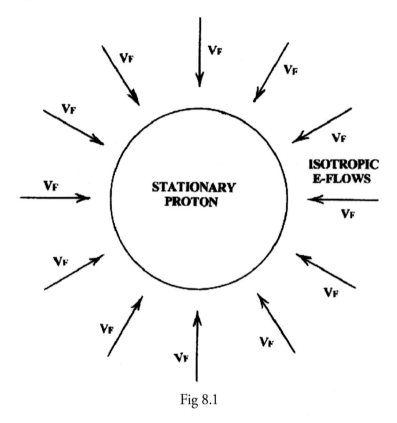

Fig 8.1

Because of the vastness of the energy medium realm some mass body groupings may have a visible relative movement to each other but without affecting the stationary status of test protons in each group. Thus test protons in different parts of the same galaxy may each be considered at absolute rest if the E-Flow conditions are isotropic. The same applies to test protons in different galaxies that are visibly in relative motion.

Since the realm grid datum at any location is defined by the sum of all the E-Flow EILs arriving there, and since E-M/Xs do not result from this energy medium phasing, then we can assume that the velocity of each E-Flow is inconsequential to the grid EIL. We know that E-Flows tend to possess the same velocity VF from their mass object sources. Since relative velocities between masses of a local group are usually very small then variation in E-Flow velocity VF at each grid location will be minimal. However, as mass locations alter then so will the realm grid datum also change. Consider a set location X in the realm grid of a local group. If all the realm masses surrounding this location X have zero relative motion then the E-Flows arriving at X will all possess the same velocity VF. If however one of these masses has a receding velocity from X of say, 10% of VF, then the E-Flow from this mass will traverse towards location X at a velocity that is only 0.9VF. From this we can conclude that a test proton at any stated location would experience a variation of E-Flow velocities from distant mass bodies that are in relative motion. By comparing these variations in E-Flow velocities with the standard of VF, it should be theoretically possible to evaluate the receding or approaching velocity of distant masses. It is not the purpose here to establish the velocities of distant masses but rather to obtain a condition of absolute velocity for our test proton.

Unfortunately at present we do not possess the technology nor the practical instrumentation to conduct a measurement of any such E-Flow velocities. Let us imagine that for the moment such an instrument exists and that it can accurately measure the velocity of an E-Flow arriving from a particular direction. There is obviously much to be considered in the makeup of such a device but we shall not speculate upon that aspect at the present time. Let us simply refer to such an instrument as a VF-meter. We could subsequently establish an absolute standard for the E-Flow velocity VF relative to source. We can then use this to establish the absolute velocity of a test proton in relation to the realm grid. To illustrate, let us consider T in Fig 8.2 to be a small space craft and the mass objects A, B, C etc., to be proton giants at large distances. Since T is unable to observe these masses it must

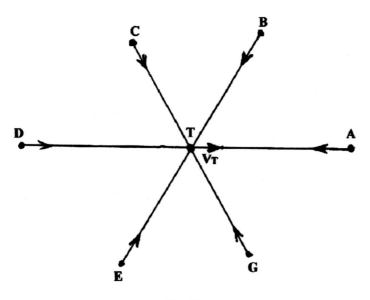

Fig 8.2

rely upon its VF-meter to evaluate its realm speed. We assume that T has a finite velocity VT which we must now determine. By simply positioning a series of VF-meters in the X, Y and Z planes, it would be possible to compute the direction and value of the greatest E-Flow velocity. The difference between this and the absolute standard VF will give the true velocity of T through the realm grid. In the diagram the greatest E-Flow velocity detected would be VF + VT from the direction of mass A. The velocity of T would then be established as;

$$\text{Velocity of T} = (VF + VT) - VF$$
$$= VT \text{ (in the direction of A).}$$

The reader will have noticed that we have ignored the EIL factor in the E-Flows. This is because we are not concerned with EIL gradients which have little bearing upon our definition of realm velocity. EIL gradients were discussed in relation to the impulses generated in the phenomena of gravitation. Realm EILs will play an important role in determining the effects upon masses that progress through zones of varying EIL (to be discussed later). But first we must be able to simply define for each mass a true velocity of traverse through the realm energy medium grid. The VF-meter permits us to do this.

Now by its very existence in the realm the proton is subject to a variety of incoming E-Flows. These E-Flows are absorbed into the E~M/Xs that occur at or near the proton's surface. In chapter 2 Fig 2.7 we indicated that the energy medium that phased into the zone of action of an E~M/X would be absorbed as part of the resulting Mp. The quantity of energy medium from each E-Flow that gets absorbed depends upon its relative velocity and EIL. We know that the E~M/X has an extremely short event sequence duration and that the E-Flow velocity would have to be proportionately high for additional energy medium absorption to be of any consequence. It is considered that the standard E-Flow velocity VF is not high enough for a significant quantity of energy medium to be absorbed into an E~M/X that is in progress. By this we mean that when the mass medium spheroid is in its contraction or imploding stage, the E-Flow velocity VF would not have phased into this zone to any great extent. We have assumed that the implosion to Mp is extremely rapid (the entire Mp cycle lasts 10^{-43} secs in Planck time). Let us now assume that the rate at which the mass medium spheroid contracts inwards is comparable to VF. This assumption is derived from the logic that the E-Flow velocity in the first instance is a result of the explosion - layer by layer - caused by an impacting of the mass medium beyond optimum density (Fig 4.2).

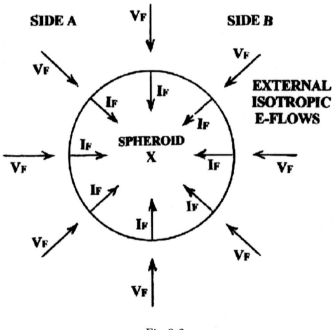

Fig 8.3

The greater the imploding impact, then the greater will be the exploding effect. It is considered that the two features are directly related. In Fig 8.3 we show a mass medium spheroid X in a state of contraction with the surface contraction velocity represented by IF. We consider that the rate of contraction accelerates from zero at the start of the E~M/X to nearly VF at Mp formation stage. We consider that the acceleration is greatest at the commencement of the E~M/X and that this reduces as the energy medium density increases. In Fig 8.4 we represent the contraction velocity in relation to the minimum energy radius R (see Fig 1.10). It can be clearly observed that the rate of contraction increases very rapidly to nearly VF within ▥R and then remains approximately constant though continually approaching closer to the velocity VF. This characteristic of IF means that only a very small proportion of additional energy medium gets absorbed into an E~M/X that is already in progress. In Fig 8.3 we assume the spheroid X to possess a stationary status relative to the energy medium realm grid. This means that there are E-Flows directed upon the spheroid X equally from all directions. At any instant of the E~M/X event let

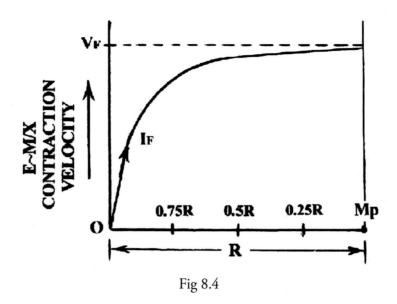

Fig 8.4

us represent the quantity of additional energy medium absorbed as EA, then,

$$EA = KA \times EIL \times 4 \pi R_1^2$$

where KA is the progressive difference between VF and IF. EIL is the prevailing energy medium realm grid datum and R_1 is the momentary radius of the contracting spheroid X. We note that as the E-M/X progresses, the value of KA gets less as does the surface area $4 \pi R_1^2$. When IF = VF, then KA becomes zero and thence EA is nil. We note from Fig 8.4 that IF is nearly equal to VF for a major proportion of the E-M/X duration. Therefore EA is negligible during this period. We conclude from this that the bulk of any additional energy medium absorption takes place during the early part of the E-M/X only, i.e. for a spheroid contraction from start at minimum energy radius R to about 0.75 R. Beyond this contraction stage we can ignore any further EA quantity.

Let us determine the volume of E-Flow that phases with the contracting spheroid X from start of contraction to 0.75 H. The duration T1 of this stage is given by,

>   T1   = 0.25R ÷ average IF
>   = 0.5R/VF (assuming IF averages 🗀 VF).

Now the distance d1 that the E-Flow traverses in this duration is given by,

>   d1   = VF × T1 = 0.5R

But since the spheroid surface has also contracted by a distance of 0.25R then the amount by which the E-Flow has phased with the spheroid will be given by,

>   E-Flow phasing length = d1 - 0.25R = 0.25R

subsequently the volume of E-Flow that is absorbed into a stationary Mp is given by,

>   E-Flow phasing volume V1 = ☺ πR^3 - J p $(0.75R)^3$

The energy medium quantity EA therefore is given by;

$$EA = V1 \times EIL$$

This volume is represented diagrammatically in Fig 8.5 by the hatched area. The inner and outer boundaries of V1 are concentric.

It can be clearly observed that the amount of additional energy medium entering the spheroid X is still a small proportion of the overall Mp that is formed. It must be remembered that the EIL of the energy medium within the sphere of influence just prior to the E~M/X would have been in excess of C_1. But the E-Flow that gets absorbed as EA is considerably below C_1. This means that the percentage of EA is actually less than that represented volumetrically in Fig 8.5.

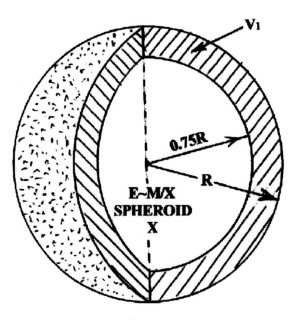

Fig 8.5

Let us for a moment step back and review the properties of the mass medium in the E~M/X. We have assumed that when the EIL of a quantity of energy medium exceeds C_1, it converts into the mass medium status. We also assumed that this was mass in an extremely stretch strained state. This caused the mass medium to implode towards a state of optimum density mass i.e. the mass point. While in the stretched format the imploding mass medium will absorb into its structure any additional energy medium from E-Flows that arrive therein. One could compare its behaviour with that of a piece of blotting paper which has the capacity to absorb water until it becomes saturated. It is considered that if the mass medium were to attain optimum density prior to its contraction to Mp size, then its absorbent quality would cease. When a realm volume attains an EIL above C_1, then that entire volume must convert to mass medium. We assume here that the stretched condition of the volume block results in internal stresses which causes simultaneous contractions in localised internal zones. Subsequently, the mass medium block splits up into defined size volumes containing set quantities of energy medium as explained in Fig 4.1. To illustrate the principle of this action consider the section of honeycomb in Fig 8.6. As contraction commences the entire block splits up into the individual cells A, B, C, D, etc. Each cell then implodes down to a Mp. As stated in chapter 4 each cell does not have to be of the same volume or shape - only approximately so. The hexagonal shapes represented are for illustrative purposes only. Also all E~M/X sequences will be in perfect synchronism with one another and so E-Flows will not easily reach the more centrally located cells/spheroids.

In the case of the proton we know that the E-M/X and M-E/X events occur in a somewhat random manner where the E-Flows from surrounding M-E/Xs cause an EIL build up and subsequent E-M/Xs. We could consider that Mp formation within such a congregation does not necessarily have to be perfectly in step. Indeed if it was, then it would be a very short lived entity indeed. Subsequently the E-Flows shown in Fig 8.1 are able to reach into

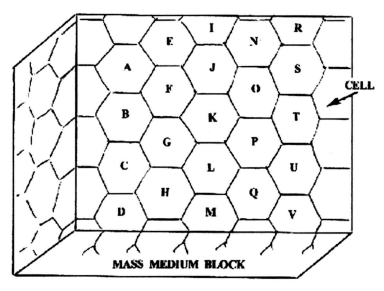

Fig 8.6

the more central regions of the proton before being totally absorbed into Mp formations. The reader is already familiar with the centralised impulses within the proton (Fig 4.4), and the part this plays in maintaining the proton Mps in a compacted state. We also know the effect that an imbalance of these impulses has upon the proton.

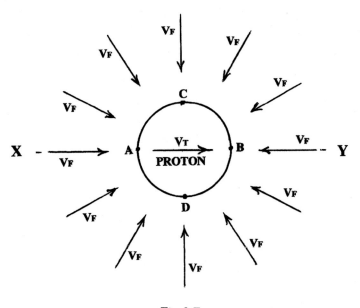

Fig 8.7

Let us now consider a proton that has a set velocity VT through the realm grid. Let the direction of traverse be as shown in Fig 8.7. Since the proton is traversing towards the right hand side, then all the E-Flows arriving at the proton surface from the realm location Y will possess a relative velocity VF+VT. Similarly, E-Flows arriving at the proton from the realm location X will possess a relative velocity VF-VT. Let us consider the E-M/X event at the location B on the upstream face of the proton. Let VT = 0.1 VF. This is shown in Fig 8.8. The E-Flow arriving at the E~M/X 'B' from outside the proton has the relative velocity VF+VT or 1.1 VF. Since the moving proton is

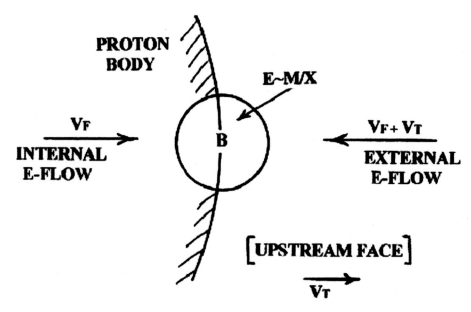

Fig 8.8

simply an Mp-congregation with all of its Mps in a grouped traverse, then the E-Flow arriving at the E~M/X 'B' from inside the proton will have a constant velocity VF. As in the case of the mass medium block, E-Flows cannot traverse very far into the proton before being totally absorbed into successive E~M/Xs. Thus we can discount any E-Flows from the location X of Fig 8.7 arriving at B. The volume of the E-Flow that phases with the imploding mass medium spheroid B will differ for the external and internal facing surfaces. As previously shown the E-Flow phasing length is given by,

$$\text{Phasing length} = d1 - 0.25R$$
$$\text{where } d1 = \text{E-Flow relative velocity} \times T1$$
$$\text{and } T1 = 0.5R/VF$$

Thus the phasing length on the external face of B is given by,

$$\begin{aligned}\text{Phasing Length BEXT} &= (VF + VT) \times T1 - 0.25R \\ &= 1.1\, VF \times 0.5R/VF - 0.25R \\ &= 0.3R\end{aligned}$$

Whereas the phasing length on the internal face of B is as calculated for a stationary contracting spheroid with relative E-Flow velocity VF.

$$\text{Phasing Length BINT} = 0.25R$$

The main conclusion from this is that because of the proton velocity there is an increased amount of the energy medium quantity EA absorbed by the E-M/Xs on the upstream face of the proton. The volume V1 in Fig 8.5 will increase marginally as represented by the x-section in the diagram of Fig 8.9. It will be noticed that the inner and outer boundary of V1 are no longer represented by concentric circles. The increase to V1 is however only upon the outer face of spheroid B. As stated earlier the Mp sequences inside a proton are not in step and so the external E-Flow is able to reach marginally into the proton structure before being totally absorbed. If we assume that the cyclic rate of Mp formation is reasonably constant then the greater the external E-Flow relative velocity the further will be the penetration of that E-Flow into the proton. As such we must note that the effect of proton velocity is not only to increase EA per upstream E~M/X as at Fig 8.9, but also to increase the total number of E~M/Xs so involved. Let d denote the diameter of the proton and let us assume that when the proton is stationary the external

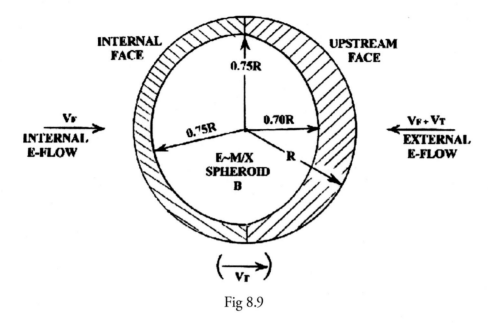

Fig 8.9

E-Flows at velocity VF (which constitute the realm datum setup) penetrate the proton to a depth of 0.1d as shown in Fig 8.10(a). The E-Flow penetration is therefore to a uniform distance all around the proton. However at (b) we represent the same proton which has been accelerated to a velocity VT. The relative external E-Flow velocity phasing with the upstream face of the proton is (VF + VT). Subsequently this E-Flow will penetrate the proton to a distance that must be marginally greater than 0.1d. As such a greater number of E~M/Xs in the upstream

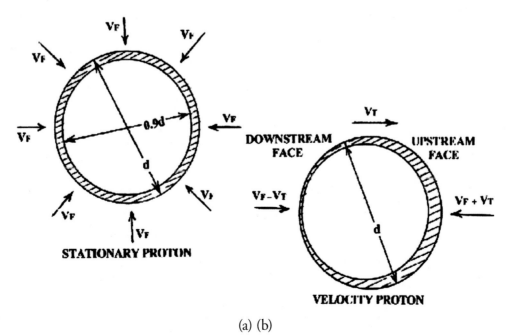

(a) (b)

Fig 8.10

345

half of the proton will have additional energy medium absorption as compared to the status at (a). The relative external E-Flow velocity phasing with the down-stream face of the proton is VF-VT. Subsequently this E-Flow penetrates the proton to a distance marginally less than 0.1d. A lesser number of E~M/Xs in this half of the proton will be involved in the additional energy medium absorption EA. It would seem that the additional volume of the EA involvement on the upstream half of the proton is balanced by the decrease within the down-stream half. This would seem to imply that for small values of VT the total EA (additional energy medium) absorbed into the proton is the same as for the stationary proton. This is not the case. In Fig 8.9 we showed that EA was marginally increased for E~M/Xs within the upstream part of the proton. We can show that EA is somewhat reduced for an E~M/X on the downstream face of the proton in Fig 8.11.

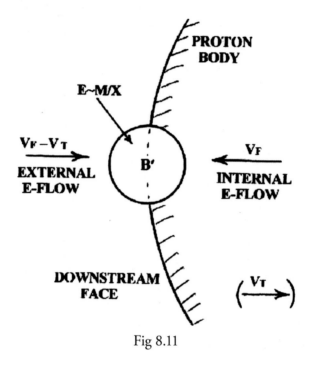

Fig 8.11

As before let the proton velocity VT = 0.1VF. The external E-Flow velocity is VF-VT, while the internal E-Flow velocity remains constant at VF. Then,

Phasing length BEXT   = (VF-VT) × T1 - 0.25R
                      = 0.9 VF × 0.5R/VF - 0.25R
                      = 0.15R

But the phasing length on the internal face of the E~M/X is 0.25R. As such the volume V1 of Fig 8.5 will reduce as presented in Fig 8.12.

By comparing this with Fig 8.9 the differences are quite apparent. The overall conclusion from this is that for any proton with a given velocity an additional energy medium quantity is absorbed by that proton as a result of that velocity. This extra energy medium absorption takes place in the upstream half of the proton body. The down stream half has a reduced EA factor which subsequently reduces the overall gain of energy medium by the speeding proton.

In the above illustrated examples the overall gain of energy medium by the proton is small. As the velocity of the proton increases towards VF then the gains become a more significant part of the proton's mass. If the overall decay rate of the proton is reduced then this will alter the TA Line profile (Fig 5.3) and so extend the life of the

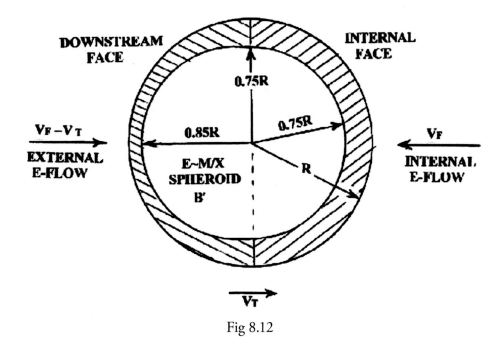

Fig 8.12

proton as though it were in a higher EIL zone. It is possible that at some very high proton velocity the overall gain in energy medium becomes equal to the loss through proton decay. Quite obviously the velocity that would be required to achieve this status would be lesser in a higher realm EIL zone and greater in a lower realm EIL zone. This being based on the assumption that the required quantity of energy medium 'scooped up' by the speeding proton must be the same in each case, ignoring any variations in proton age of course.

Let us for a moment consider the law of conservation of Linear Momentum. According to our energy medium theory all mass in our current circle 6 phase is in a state of perpetual decay. This means that the proton mass is decreasing continually. If the mass traverses through the realm at a constant velocity VT then its linear momentum is the product of the two, i.e.,

$$\text{Linear Momentum} = \text{Mass} \times \text{Velocity}$$

Since the mass is a decreasing quantity then it follows that the linear momentum must also decrease in Time. However the change in mass is so small in a human lifetime that we are unable to notice any differences. As such we assume all mass bodies to possess a constant mass quantity. Perhaps when our energy medium theory is more advanced we may be able to apply a factor to the linear momentum to allow for proton decay.

When a proton speeds through the realm it absorbs additional energy medium. This energy medium has a velocity of VF+VT relative to the proton. When this energy medium is absorbed into the Mps of the proton it must impart its linear momentum to the proton. This means that as EA is absorbed and accelerated the velocity of the proton must decrease. From our observations upon astronomical bodies in orbital paths we note that no such slowing down of any significance appears to take place. This is because of the principle of centralised impulses within the proton as represented in Fig 4.6. When one side of a proton is in a higher realm datum EIL than the other side then its centralised impulses become unbalanced (Fig 4.8) and an overall impulse accelerates the proton in the direction of the higher realm EIL (Fig 4.13). In the case of our speeding proton the additional energy medium quantity absorbed into the E-M/Xs on the upstream half of the proton results in a higher EIL status from the subsequent M-E/Xs. The effect is the same as if there was an EIL gradient across the proton - the higher EIL being on the upstream side. The internal centrally directed impulses are once again in imbalance and a gravitation type

347

impulse pushes the proton in the direction of its velocity. For the present we can only assume that the tendency for a slowing down in the proton velocity is exactly offset by the gravitational type impulse to accelerate it. The greater the realm EIL or realm velocity, then the greater is the quantity of E-Flow medium absorbed. This subsequently results in a higher EIL differential across the proton and the consequent directional impulse. We would expect a direct relationship between these two factors to prevail for velocities that are a fractional part of VF. At the very much higher velocities approaching VF we would expect the balance in this relationship to alter. The theory behind this is that explained in Fig 4.21 in which we considered that the variance in the Mp cyclic rate progressively diminishes at the very high grid EILs. The impulses generated inside the proton subsequently behave in the manner shown by the curve of Fig 4.22. Subsequently we consider that a proton could not maintain itself at the very high realm velocities approaching VF. The absorption of energy medium (along with its linear momentum) into the mass of the proton at these high velocities would require a considerable amount of net impulse within the proton to maintain such a velocity. The imbalance in the two quantities is then such that the proton undergoes a continuous retardation. Newton's First Law of Motion does not hold for velocities approaching VF.

So far we have only considered the effect of grid velocity upon a test proton. However, mass objects are made up of atoms and so we must also consider the Mp-shell structure in our deliberations. We know that the Mp-shell exerts a centralising action upon its proton nucleus (Fig 4.31 and Fig 6.6). We also know that the effects of a collision between two atoms are not instantly conveyed to their respective nuclei but must wait until the inward E-Flows from each distorted Mp-shell section reaches the centre (Fig 6.8). It is our intention here to evaluate whether the velocity of the atom has any effect upon the delayed aspect of this centralising action.

In Fig 6.4 we assumed the Mp-shell to possess a velocity VS away from its central nucleus. When the atom is at rest the E-Flows and their velocities are isotropic and therefore their effect upon the E-M/Xs in the Mp-shell are a uniformity of energy medium absorption as represented in Fig 8.5. If the atom is now given a velocity VT, then quite obviously any E-M/X within the Mp-shell should undergo a biased absorption of energy medium from different directions as shown for the E-M/X in Fig 8.9. Let us assume the Mp-shell velocity is to be shifted opposite to the direction of the velocity VT. In Fig 8.13 we represent an atom with a velocity VT as shown. The outward velocity of the Mp-shell at the position B should be marginally less than VS due to the additional quantity of energy medium absorbed from the E-Flows coming in from the direction Y. This E-Flow would have a relative velocity VF + VT. Now if the E-Flows and forcephoids from the proton P to the locations A and B were maintained

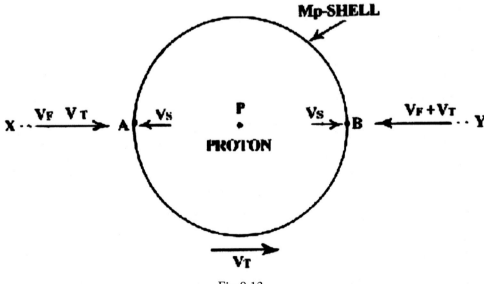

Fig 8.13

at a constant EIL regardless of the atom's realm velocity, then we could indeed claim that the outward traverse velocity of the Mp-shell at B was less than VS. Conversely we could also claim that the velocity of the Mp-shell at A was somewhat greater than VS. Logic would then dictate that an Mp-shell with a slower outward traverse would have an average location closer to the proton centre than the Mp-shell with a greater VS. One could then argue that this results in the nucleus being off-centre with regard to its shell and that subsequently a centralising impulse must prevail in a direction that opposes the velocity VT. We must however take into account the effect of relative realm velocity upon the proton as described earlier.

We know that the upstream half of the proton absorbs a much greater proportion of EA, the additional energy medium from E-Flow velocities. This means therefore, that the subsequent M~E/Xs will operate at marginally higher EILs. Thus in Fig 8.13, forcephoids that traverse out from the proton P towards the Mp-shell at B will have a higher peak EIL than similar forcephoids traversing from P towards A. Subsequently, the combination of energy medium momenta at the B location will result in a shell section that has a resultant outward velocity that is nearly constant at VS. By the same principle the outward velocity of the shell section at A can also be assumed to be nearly constant. We assume these conditions to exist for values of VT that are a fraction of VF. However, when VT becomes large and approaches the value of VF, then we consider that the outward velocity of the ABCB shell diagram at the B location must become less than VS. This means that the average position of the Mp-shell at B will be closer in to the proton P. Similarly we consider that the outward velocity of the shell section at A becomes greater than VS and thus farther out from P. Subsequently the proton P will be in an off-centre position with respect to its own Mp-shell and therefore a centralising impulse must act upon P in the direction of B to A. This means that the atom cannot maintain a velocity close to VF. Just as in the case of the test proton described earlier, our atom will also undergo a retardation. Once again we observe that Newton's First Law of Motion does not hold for velocities approaching VF. Experiments in which test particles are accelerated to very high velocities have shown this to be the case and the inertial mass of a particle has been re-defined in relation to its velocity.

The mass of a stationary object is termed as its rest mass M0. As the object is accelerated to a velocity VT, then its mass must increase by the value of EA. Subsequently the formula for acceleration a_1 of a stationary mass by a force F1 is given by,

$$F1 = M0 \times a_1 \ldots\ldots\ldots \ldots\ldots\ldots\ldots (1)$$

This must alter to,

$$F2 = (M0 + EA) \times a_2 \ldots\ldots\ldots\ldots\ldots\ldots(2)$$

when the body has attained the velocity VT. Since EA is extremely small when VT << VF, then it does not matter if we ignore EA. However, it is obvious by comparing (1) and (2) above that the force F2 will have to be greater than F1 if the acceleration values a_1 and a_2 are to be equal. Today, physicists accept that the inertia mass MV of a particle or body is obtained by dividing its rest mass M0 by the contraction coefficient.

$$MV = {}^{M0}\!/\!\sqrt{(1 - VT^2/_C2)}$$

where C is the velocity of propagation of the EIL oscillation wavefront and VT is the velocity of the mass object.

Let us now consider the effect of all this upon the delayed aspect of the centralising action. When the outward traverse velocity of the Mp-shell actually increases or decreases from the normal value VS then the inward flow velocity must also change. In Fig 8.13 let us assume that the atom velocity VT is very high and that this causes the outward traverse velocity of the Mp-shell section at B to decrease by an amount V1. This means that the outward velocity of the shell section at A will also increase by a corresponding amount. Thus we have,

$$VB = VS - V1$$
$$\text{and} \quad VA = VS + V1$$

where VA and VB are the Mp-shell outward traverse velocities at A and B respectively. Now in Fig 6.5 we showed that the E-Flow from the shell M-E/X towards its proton centre was given by,

$$VE\text{-Flow} = VF - VS$$

subsequently the inward E-Flow velocity from B is given by,

$$VE\text{-Flow B} = VF - (VS - V1)$$
$$= VF + V1 - VS$$

i.e. the inward E-Flow velocity from B has increased by V1.
Similarly the inward E-Flow velocity from A is given by,

$$VE\text{-Flow A} = VF - (VS + V1)$$
$$= VF - VS - V1$$

i.e. the inward E-Flow velocity from A has decreased by V1.

This means that the effects of a shell distortion at B will be transmitted to its nucleus more rapidly than before. Thus retardation of a speeding atom can be achieved with a lessened delay to its centralising action. However, a shell distortion at A will be transmitted much more slowly with a subsequent increased delay in commencement of the centralising action. From this we can conclude that it would require a shorter duration of applied effort to retard a speeding object but a much greater duration of applied effort to accelerate it in the direction of its existing velocity. Perhaps the reader can relate this to the principles of motion in the field of applied mechanics when objects are in collision with one another.

Let us now consider the velocity C and a famous experiment that was conducted by Michelson in 1881 and later with Morley in 1887. In chapter 3 we showed that the velocity of propagation of an EIL oscillation wavefront was not dependant upon the velocity of the atom from which it emanated (Fig 3.32), but rather upon the quantum step principle of Fig 3.20. We showed that;

1. The velocity of wavefront propagation within the energy medium realm was directly proportional to the quantum step size and inversely proportional to the realm EIL datum.
2. The propagating velocity was made up of two components: the quantum step component and the propagating plate (p/p) forward motion component Vp.
3. The p/p forward motion component Vp is directly proportional to the oscillation wavelength or inversely proportional to the oscillation frequency.
4. The quantum step size increases marginally with the distance propagated.
5. The direction of wavefront propagation tilts towards the higher realm EIL datum.
6. The velocity of wavefront propagation Vc is always less than the E-Flow velocity VF.
7. Changes in frequency are real only if they are the result of changes in motion of the Mp-shell source.

8. The wavefront cannot propagate in a realm above C_1 nor if the realm EIL is below a certain minimum level.

Although it may seem that there is quite a variable aspect to Vc, there is in fact a considerable uniformity of effect due to the extensive isotropic nature of the energy medium realm around us.

In Fig 3.33 we showed how two observers X and Y could each observe the same wavefront propagating along the path AB yet independently arrive at a different frequency measurement. We related this condition to the doppler effect. However, for a set realm datum status the velocity of the wavefront must be a constant velocity quantity (relative to that realm datum). Nevertheless, we showed in Fig 3.35 that the Vp component of wavefront propagation can be influenced by the input of external E-flows that are not isotropic. These E-Flows can cause an increase or a decrease in Vp. It was shown that subsequently Vc also increases or decreases accordingly. Under these conditions therefore, we may assume that Vp and subsequently Vc is greater in the direction of a diminishing EIL gradient i.e. in the direction of the major E-Flow, than in the opposite direction up the EIL gradient.

Scientists of the nineteenth century considered that if a simple 'ether' pervaded all of the universe or realm, then lightwaves must propagate through it at a constant velocity. They then considered that since the earth possessed an orbital velocity around the sun of approximately 30 Km per sec, that it should be possible to detect a flow of the ether medium past the earth at some stage in its orbit round the sun. The experiment undertaken was similar in principle to that of our two observers in Fig 3.33 and became known as the famous Michelson - Morley experiment with its equally famous negative result. More recently this experiment was repeated using a laser beam. The difference of frequencies for the two laser beam paths when the entire apparatus was rotated in the horizontal plane through 90° was still less than one thousandth of the change expected from the ether wind hypothesis. Why did the experiment yield such a negative result? The answer must surely lie in our theory consideration that the energy medium does not simply flow past the earth in a uni-directional manner, but originates from every decaying proton in the bulk of the earth and is emitted in a radially outward direction. This E-Flow is the main constituent of the realm datum at the earth surface. Other E-Flows that reach the surface of the earth from external realm sources are very nearly isotropic and at a much lesser EIL. Subsequently the realm datum near the earth's surface has an EIL gradient with the EIL normal lines in a radial format from the earth's centre as shown in Fig 8.14.

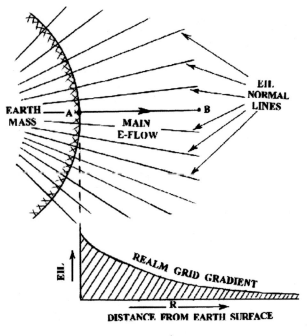

Fig 8.14

We observe that the main E-Flow from the earth's mass is along the radial outward direction as represented at AB. As such the Vp component of a light wave will be greater when it propagates in the direction A to B than from B to A. This means that the velocity of a lightwave is greater in a direction vertically upwards from the earth's surface than vertically downwards. The difference in these two velocities will be greatest at the earth's surface and will progressively diminish with distance from the earth. At a far enough distance the E-Flow EIL conditions become nearly isotropic and the velocity variance becomes negligible. However we must remember that there are other realm masses that could influence the realm datum and cause a directional variance in Vp. The E-Flow from our sun, nearby stars and even the giant proton mass of our galactic core could influence Vp. As such we suggest that the Michelson - Morley experiment be repeated but with the apparatus located and rotated in the vertical plane.

In our current realm energy medium theory we have progressed a considerable way from the simple ether concept. The realm energy medium entity comprises of a complexity of phasing E-Flows that combine to create our current EIL grid structural environment. All masses contribute energy medium to this structural environment in a continuous manner because of their proton decay characteristic. E-Flows traverse in absolute straight line paths at velocities much greater than that of a wavefront. Wavefronts can be refracted but E-Flows cannot. We consider that the principle within the Michelson - Morley experiment has an important place as a measurement device of the future for determining realm grid gradients in the vast structure of free realm space.

Let us propose a series of interference fringe displacement experiments to be conducted on the earth's surface in both the horizontal and vertical modes to determine the effect of other mass bodies upon the directional velocity of lightwaves.

Experiment 1. To be conducted in the horizontal plane but with a mountain range rising on one side and a geographical plain on the other. The experiment to be conducted down on the plain near the base of the mountain. Successive sites to be at progressive distances from the mountain base. The experiment to be performed with the sun vertically overhead to negate the effect of its E-Flow, and a predicted full moon condition since this means that it would be directly opposite the sun. Results should show a marginal increase in velocity C in the horizontal direction away from the mountain mass.

Experiment 2. To be conducted in the horizontal plane near the Equator and at either sunrise or at sunset under predicted new moon conditions. This means that the E-flows from the Sun and Moon are effectively combined and arrive at the experiment site from an approximately similar direction. The experiment should be performed at a site that is remote from any mountain range mass. The results should indicate a marginally greater lightwave velocity in the direction from Sun to Earth site than from Earth site to sun.

Experiment 3. To be conducted exactly as our previous Experiment 2 but under full moon conditions. This means that the E-Flows from the sun and the moon arrive at the experiment site from opposite directions. Results should show that the effect of the sun's E-Flow upon Vp is either greater or lesser than that from the moon's E-Flow.

Experiment 4. To be conducted in the vertical plane near the equator at noon under predicted new moon conditions. This means that the E-Flows from the sun and moon are effectively combined but are in opposition to the E-Flow from the earth. The results should show that the effect of the earth's E-Flow upon Vp is greater than that from the combined E-Flows of the sun and moon.

Experiment 5. To be conducted as the previous Experiment 4 but at midnight. This means that the E-Flows from the earth, sun and moon are all in the same direction at the experiment site. The results should show a maximum variance for lightwaves in the upward and downward directions. The maximum velocity will of course be in the vertically upward direction in the direction of the earth's E-Flow.

Experiment 6. To repeat experiments 4 and 5 but under predicted full moon conditions i.e. in which sun and moon are on opposite sides of the earthbound experiment site.

Other experiments could be devised by the reader to detect the effect of single masses on the velocity C. One such experiment could be to evaluate the mass of our galaxy's proton giant central core. Such a computation would be rather complex but we would initially need to establish the relative effects of the nearby planets and stars.

We consider the Michelson interferometer to be an important instrument for the detection of high EIL E-Flows, especially in the case of a proton giant (the proverbial black hole). Since a space traveller could not observe one of these then he or she should at least be able to detect it by its considerable outward E-Flow. Incidentally our realm traveller would be in no danger of being sucked onto the surface of the proton giant since according to Fig 4.24 there is a considerable repulsion impulse at distances closer than X (where X represents several proton giant diameters). We leave the reader to speculate further upon the sphere of mass objects that are held in equilibrium around such a proton giant and at a distance X from its surface and as already discussed in chapter 2 on galaxy formation (page 61).

Let us now consider the laws of ordinary gravitational attraction as discussed in chapter 4 and the effect of a realm velocity condition upon the impulse curve of Fig 4.24. We showed then that the resultant impulse IG generated within the atoms of each of the mass bodies A and B was given by;

$$IG = G \cdot {}^{(MA \times MB)}/R^2$$

where G is the gravitational constant as defined earlier, MA and MB the masses of A and B respectively, and R the distance between A and B. We assumed of course that the mass bodies were in a stationary condition with respect to one another and with respect of the realm datum grid. Now we know that the impulse generated within the proton is due to an imbalance in the centrally directed impulses that maintain the Mp-congregation as a compacted identity, the imbalance being caused by an EIL differential in the external realm E-Flows that arrive at its surface (Fig 4.8). The net impulse being in the direction up the EIL gradient, which was also towards another E-Flow source or mass body. However we have just shown that the velocity aspect of a proton results in a somewhat similar energy medium differential status due to the additional energy medium EA absorbed in its upstream face (Fig.8.10). Since EA needs to be accelerated to the velocity of the proton, we assumed (for realm velocities that were small) that the impulse to accelerate EA was obtained from the upset of the proton's centralised impulses resulting from that velocity. Let us term the impulse resulting solely from the realm velocity condition of the proton as the velocity impulse and denoted as IV. It is our intention to show the variance in net gravitation impulse that occurs in cases where two large mass objects possess relative velocities that are either towards or away from one another.

Let us first consider the case in which two mass bodies under each other's gravitational influence are in traverse towards each other. We assume that this traverse velocity is a small fraction of VF. When the two bodies are quite a large distance apart the net gravitation impulse IG in each mass is small yet sufficient to result in an acceleration of one towards the other. The EA quantity that is absorbed by each mass sets up a retarding influence IR upon its progress, but is overcome by the additional velocity impulse IV. Since IG and IV both act in the same direction then;

Resultant Impulse = IG + IV -IR

since VT is small then IV and IR are considered to be equal in magnitude and so cancel out the effect of one another. Therefore,

Resultant Impulse = IG = G - $^{(MA \times MB)}$/R^2

As such the gravitational impulse in the inward traversing mass objects can be considered the same as for stationary masses. However as the velocity VT of each mass towards the other increases then so does the retarding influence IR. As the two mass bodies draw closer together, the EIL of their E-Flows must also increase considerably resulting in a further increase in the EA quantity and subsequently in IR again. Now the combination of the realm grid EIL and velocity conditions could be such as to cause the resultant attractive impulse to have increased in a non-linear format similar to that shown for the high EIL condition represented in Fig 4.21. In that case IR will be greater than IV and therefore we shall find that the resultant impulse is less than IG. The resultant impulse in this case does not obey the curve of Fig 4.24 and the variance is represented here in Fig 8.15 for a given velocity status by the dotted curve. We consider that the greater the velocity status towards the other mass the lesser will be the attractive impulse between them. It should be possible to present a series of dotted curves (each one inside the previous one) for progressive levels of increasing velocity status. At the very high relative velocity of VF we consider that EA becomes large and hence IR dominates the impulse setup. As such there is no attractive impulse between the two masses. On the contrary a resultant repulsion impulse prevails causing a reduction in the velocity of

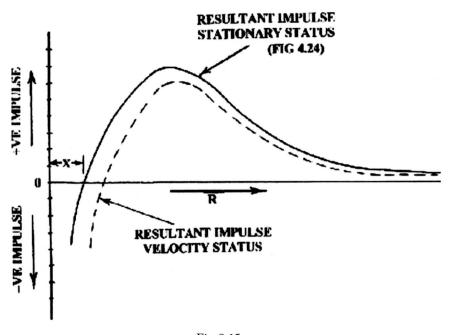

Fig 8.15

each mass towards the other. All of the above is applicable even if one of the objects is stationary and the other possesses all of the velocity. Since each object is under the influence of the other's E-Flow, then the E-Flow velocity reaching each mass is simply dependant upon the relative velocity of one mass with respect to the other.

Let us now consider our second velocity case in which two large mass objects possess a relative velocity in a direction away from one another.

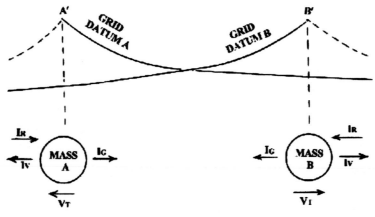

Fig 8.16

In Fig 8.16 we show the mass objects A and B moving away from one another and each with a realm grid velocity VT. If we consider all other masses to be very distant and in an isometric layout, then mass A will essentially be within an EIL gradient from the E-Flow of mass B. And B will correspondingly be in the EIL gradient from A. If A and B were stationary then each would acquire an impulse IG as shown. See also Figs 4.10 and 4.11. However, since A and B each possess a realm velocity VT away from one another, then IG for each will get progressively less and less. This is because each mass traverses into a lower region of the other's grid datum at the rate of the relative velocity between them i.e. 2VT. For velocities that are a very small fraction of VF we can assume that the velocity impulse IV and the retarding impulse IR are equal and opposite. Once again we have,

$$\text{Resultant Impulse} = IG = G - {}^{(MA \times MB)}/R^2$$

which is similar to IG for stationary masses at the distance R.

However, as the realm velocity VT of each mass increases, the quantity of energy medium absorbed in the upstream face also increases rapidly. At the same time the rate of E-flow arrival from one mass to the other diminishes by twice VT as given by,

$$\text{E-Flow Velocity (A to B)} = VF - 2VT$$

Subsequently, it is considered that IG goes into a state of rapid decline with increasing velocity status and that the centralising impulses in the protons of each mass must at some stage reach a situation of perfect balance. This state will be given by the equation,

$$IG - IV + IR = 0$$

In this situation the mass body protons will find themselves in a zone of zero grid gradient. That means the energy medium quantities absorbed by the opposite faces of each proton will be equal. This also means that the repulsion impulse component within IG (as represented in Fig 4.2), must exactly balance with IR the retarding impulse due to EA. Subsequently we may conclude that gravitation impulses cease to exist between two masses that are in recession from one another at a particular finite relative velocity. This velocity will of course vary according to the distance between the masses, the required velocity being less at the greater separation distances.

Earlier in this chapter we assumed that for small velocities we had IV balanced by IR such that a mass body continued in a state of uniform velocity through the realm if conditions permitted. In the above case however, there is an additional repulsion impulse component (albeit small) which acts in the same direction as IV. Subsequently, just after having achieved the above balance in the proton's centralising impulses, the continuing traverse takes the mass bodies an even further distance apart. If the velocity is unchanged then IV and IR are also unchanged. We also know that the gravitational repulsion impulse component is nearly constant for large values of R (the separation distance between mass objects) as shown in Fig 4.23. It is the attraction impulse component of IG that decreases more significantly with distance (Fig 4.22), and therefore there must be a resultant impulse in the mass body protons which acts in the same direction as the velocity. This means that the proton centralising impulses are sufficiently unbalanced to cause the mass bodies to acquire a gravitational type impulse away from one another. However it is considered that as the masses traverse further and further apart the variance between IR and the gravitational repulsion impulse component increases. As the gravitational effect (i.e. grid datum gradient) decreases with distance so the retarding impulse IR acquires a more prominent aspect. Subsequently IV and IR will relate simply as for isolated masses in which each possesses an independent realm velocity. At the lower velocities, IV and IR are equal and opposite in action and so neutralise one another. At the higher VF fractional velocities, IR becomes greater than IV and causes the grid velocity of the mass to reduce.

We have indicated above that the relative velocity between two masses results in impulses generated within those masses that may wholly or partially off-set the effects of gravitation. In order to ascertain whether these conclusions are correct we propose that a simple experiment be conducted at a site on the surface of the earth. The experiment involves a flywheel or disc fitted with a central frictionless bearing and mounted at one end of a rigid shaft as shown in Fig 8.17(a).

Let the flywheel disc be about 10 inches in diameter and an inch in width. We assume the flywheel to be in perfect dynamic balance i.e. vibration free at about 50,000 rpm. At Fig 8.17(b) we show a frontal view of the disc relative to the earth's surface. The E-Flow from the earth can be compared to the E-Flow (2) in Fig 1.15. This means

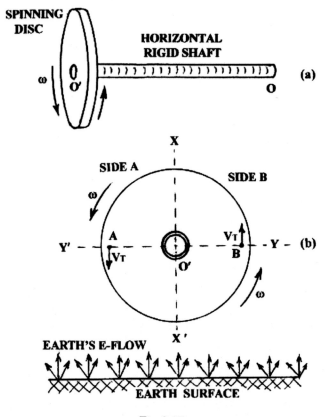

Fig 8.17

that the E-Flow is multi-directional near the surface, but since the bulk of the E-flow comes up from inside the earth, the resultant E-Flow can be assumed to be directly upwards. If we draw vertical and horizontal axes XX' and YY' respectively through the disc centre, then the points A and B can be symmetrically located within the disc as shown. Let the disc be given an angular rotational velocity of 'w' such that the velocity of atoms at positions A and B is VT in the directions as shown. As such, we may consider the mass at position A to possess a relative velocity VT directly towards the mass of the earth (our first case Fig 8.15). Similarly, we may consider the mass at position B to possess a relative velocity VT away from the earth (our second case Fig 8.16). According to our theory, the gravitational pull of the earth upon the masses at A and B will be somewhat less than that for a non-rotating disc. Subsequently the over-all weight of the rotating disc and shaft assembly should get less and less as 'w' increases.

Let us re-start the experiment. We ask the reader to lift up the non-rotating wheel and shaft and to hold this in the position shown at Fig 8.17(a). A considerable weight will need to be supported and the reader will require both hands to grip the shaft firmly and to maintain the assembly in the level position. It is doubtful whether the reader could maintain this effort for very long. Let a motorised device now be held against the disc causing it to acquire a very high rotational velocity. As the rpm increases the reader will notice that the assembly becomes easier to support. Eventually the reader should not only be able to support the rod in the horizontal plane using just one hand but that it would not be necessary to grip the free shaft end at all. It would simply require the reader to support the end O upon the palm of his hand as represented at Fig 8.18.

The explanation for this result is that the side-A of the spinning disc (Fig 8.17(b)) has an over-all velocity downwards towards the earth and at the necessary velocity the repulsion impulse component becomes large enough to

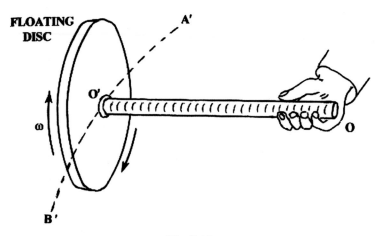

Fig 8.18

overcome the attraction impulse component. Subsequently the side-A mass acquires a net impulse in the upward direction. Similarly the side-B of the spinning disc has an over-all velocity upwards away from the earth. At the indicated rotational velocity the mass in the side-B comes under the influence of a shift in the protons centralised impulses, causing it to subsequently acquire a net impulse in the direction of its velocity i.e. upwards. Since the shaft does not rotate its gravitational attraction towards the earth remains constant. Therefore, for the Fig 8.18 condition of equilibrium we consider that the upward impulses generated within the disc is just sufficient to negate the effect of the earth's gravitation and support approximately half the weight of the shaft as well. The other half weight being supported by the readers hand at O. It would seem that our spinning disc is unaffected by the gravitational pull of the earth. In fact its state of balance implies that it is being repelled from the earth's mass. It would be interesting to consider the application of Newton's Third Law of equal and opposite reactions to the above conditions of our floating disc. His law of gravitation has already been re-defined by the impulse curves of Fig 4.24 and Fig 8.15. Results from our experiments should provide a range of disc speeds for the above status to prevail. We consider that this speed range will alter in direct relationship to the EIL of the earth's E-Flow i.e. distance from the earth. Higher rotation speeds being necessary at the higher EIL gradients.

During our floating disc experiment an interesting side effect was observed. If an effort is made to move the shaft in Fig 8.18 in an anti-clockwise direction about the hand position O i.e. towards the indicated position B', then we observe that the disc dips down towards the earth. However, if an effort is made to swing the shaft clockwise towards the position A', then the disc moves upwards away from the earth. The rate of upward rise or downward dip is directly proportional to the rate at which lateral effort is applied to the shaft. It is considered that after each such manoeuvre there may be a decrease in the rpm of the disc which may not be attributable entirely to the friction in the bearing. There are two possible explanations for this. In the first we must consider the oblique E-Flows from the earth that impinge upon each side of the spinning disc as shown in Fig 8.19. Here we represent

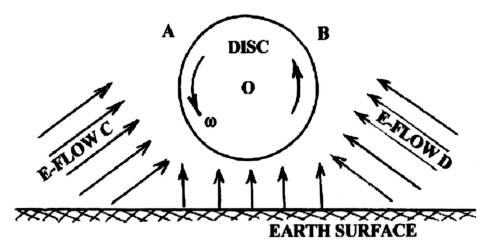

Fig 8.19

the oblique E-Flows on either side of the spinning disc. The reader should realise that the E-Flows from the earth are in a multitude of upward directions, but we have only represented those above to aid our explanation. When the shaft is stationary then the E-Flow C and E-Flow D have the same relative velocity with regard to the protons in the disc. Since they also have the same EIL, then each is absorbed equally by the sides A and B of the spinning disc. However if the disc moves horizontal to the earth in the direction of B to A, then the relative velocity of E-Flow C is increased and the retarding impulse IR in the side A protons subsequently increases. At the same time the relative velocity of E-Flow D with regard to the side B is decreased. This means that the grid velocity of the side B protons (which are in a direction away from the E-Flow B) would effectively appear as an increased velocity away from the earth. This could result in an increase in IV the velocity impulse for side B. Thus side A and side B have both achieved an increased impulse away from the earth causing the disc to rise upwards.

When the spinning disc is moved in the opposite direction from A to B then the E-Flow C velocity relative to the side A protons will be less and the retarding impulse IG will be reduced. Similarly the effective velocity of the side B protons with respect to the E-Flow D will also be less resulting in a reduced IV velocity impulse. Side A and side B now both have a reduction in their away from earth impulse causing the floating disc to dip towards the earth.

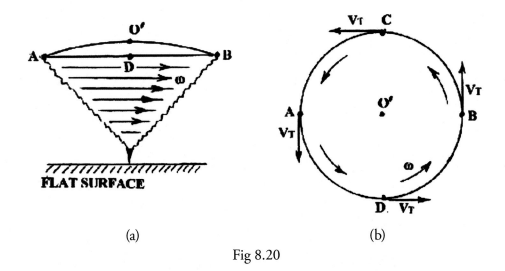

(a)   (b)

Fig 8.20

In our second explanation for the above phenomena we must resort to the example of a simple child's toy, i.e. a 'Top' spinning on a flat horizontal surface as shown in Fig 8. 20(a). Let the 'Top' spin at an angular velocity 'w' in the direction shown. At (b) we show a view of the 'Top' as seen from above. A, B, C and D are the grid co-ordinate points on the edge of the 'Top' such that they are uniformly spaced about the spin axis OO'. Let the velocity at each point be VT which will be in the direction of the surface velocity as shown. If we now take a rigid rod and just touch the spinning 'Top' at the co-ordinate position B then obviously the point B will respond as though it were in a collision. This means that the velocity VT at B must reduce with some of its linear momentum being transferred to the rigid rod. Let us assume that the VT at B reduces by 'n' such that the velocity at B is then given by;

$$\text{Velocity B} = VT - n.$$

At this instant the velocities at the co-ordinate positions A, C and D are still unchanged at VT. Since the 'Top' is a rigid body and all points on its surface must have the same angular velocity, the entire mass of the 'Top' responds by moving on the flat surface in the direction opposite to the velocity at B. That is, with a direction and velocity which is the resultant of the velocities at co-ordinates A and B. This is given by;

$$
\begin{aligned}
\text{Horizontal Velocity of Top} \quad &= \text{Velocity A - Velocity B} \\
&= VT - (VT - n) \\
&= n \text{ (in the direction C-D)}
\end{aligned}
$$

We would have achieved a similar directional result if we had caused a small object or piece of sticky putty to be fired into the 'Top' at the co-ordinate point B. The same experiment at point A would cause the Top to move in the direction D-C. Impact at C would move the Top in the direction A-B, while for impact at D the movement would be in the opposite direction B-A. From this we come to the conclusion that a spinning object will move in a direction opposite to the velocity direction of the point at which impact takes place.

Now in our floating disc experiment we can assume the same conditions as at Fig 8.20(b), except that the points A and B are re-positioned at the hub bearing locations. This is represented in Fig 8.21 by A''and B''.

359

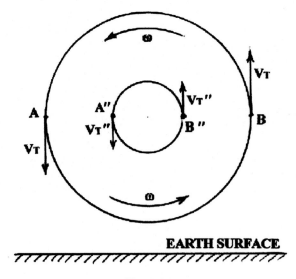

Fig 8.21

A'' and B'' are points on the outer-race of the shaft hub bearing. Results are similar to those achieved in our spinning 'Top' experiment if we use A'' or B'' as the reference points instead of A and B. Movement of the shaft in Fig 8.18 towards A' causes the friction of the Hub bearing to become concentrated at the position A'' (Fig 8.21). Subsequently, the disc responds to this "touch" by moving in the upward direction. Movement of the shaft in the direction of B' (Fig 8.18) will subsequently result in the disc moving downwards. We shall consider the practical uses for this phenomenon in the next chapter.

So far we have indicated the effect of realm velocity upon mass bodies in general. We should now consider the effect of such a velocity upon the aspect of Proton-Time as discussed in chapter 5. We had then showed that the decay rate of the proton was inversely proportional to the EIL status of the local realm grid. A proton at the Earth's surface is at a relatively higher realm grid datum and a subsequent lesser decay rate than that of a proton further away in free realm space. In Fig 5.5 we showed how we could evaluate the actual decay rate (TA-Line) by considering this in relation to the absolute decay TN-Line. Subsequently a rough estimate was made of the TA-Line for a proton at the surface of the earth as below,

$$TA\text{-}Line = (TN+ 0.616)\text{-}Line$$

In Fig 5.6 this was represented by one of the TA-Lines that fitted somewhere between the absolute TN-Line and the nil decay T0-Line.

The consistent decrease in the overall energy medium / mass content of the proton became our yardstick for duration measurements which we defined as Proton-Time. We also established that under varying realm EIL conditions the rate at which this proton mass decay occurred was variable. Subsequently we concluded that our yardstick of Proton-Time was also variable in character. We showed that it varied inversely as the EIL of the local realm datum because of the influence of the EIL upon the Mp-cyclic rate.

When a proton mass traverses through the realm grid at a defined velocity we have shown that additional energy medium EA is absorbed by its upstream face. We also showed how this additional quantity was absorbed into the upstream half of the E~M/X mass medium sphere (Fig 8.9) . The quantity absorbed increased with increasing grid velocity. Now consider a stationary proton within a set EIL grid datum. The rate at which its mass depletes is governed by the local EIL of the realm datum. Let us suppose that its decay rate is a mass loss of N energy medium units per MTU. Let this proton mass be induced with a realm grid velocity in the same grid EIL conditions. Since the proton will acquire an additional quantity of energy medium EA from this velocity status, then its actual decay rate is a mass loss of (N - EA) energy medium units per MTU. As this is a slower decay rate than that of the former stationary proton and applies for all velocity conditions, we can conclude that the rate of decay of stationary protons will always be greater than for those with a grid velocity in similar datum conditions. Similarly, it must also be concluded that under similar EIL conditions, the decay rate is inversely proportional to realm velocity. Since decay rate is synonymous with Proton-Time we conclude that Proton-Time slows down as the realm grid velocity of the mass body increases. We would expect the TA-Line for a mass traversing at a grid velocity approaching VF to be quite close to the T0-Line of Fig 5.6.

Before we conclude this chapter topic let us briefly consider the aspect of observed light wave frequencies in a space craft travelling through the realm at a high grid velocity. Now we know that wavefronts propagate at a constant realm velocity under uniform EIL conditions. In Fig 3.33 we represented an emission of wavefronts as traversing in a set direction AB at a constant velocity Vc. An observer X travelling in an opposite direction will intercept these wavefronts and measure a higher than normal frequency. However, an observer Y travelling in the same direction will measure a lower than normal frequency. Similarly our spacecraft traveller will observe light waves from stars in front of it to be blue shifted while those to the rear will appear to be red shifted. If we assume the interior of the space craft cabin to be painted white then at a high grid velocity the traveller would observe the front and rear walls in slightly different colours. The front wall should appear to possess a blue tint while the rear wall a pink tint. This is because the space craft walls would not be impervious to the realm datum E-Flows and so light waves inside the craft would be propagated at a velocity constant to the realm grid only. Therefore the status of Fig 3.33 would also apply inside the spacecraft.

However, it must be remembered that the space traveller is also made up of atoms and molecules which are equally affected by the high velocity and EIL conditions. This means that proton decay or Proton-Time for the space traveller is also slowed and subsequently his concept of duration will also diminish as represented by Fig 5.8. To the traveller the concept of duration or Event-Time will always appear constant and hence he may not actually perceive that the wavefront frequencies have altered with grid velocity. Proof that frequency changes have occurred will have to be extracted from the spectral absorption line data of starlight received from the front, rear and side directions. We leave the reader to consider this aspect more fully.

In this chapter we have attempted to isolate our explanation of the effects of a grid velocity upon the behaviour of a proton mass. However the reader is reminded to be simultaneously aware of all the other realm phenomena and conditions that are operative within the realm but which are not brought into focus when discussing individual aspects. Realm EILs and grid gradients also play an important role in the upset of the protons centralised impulses and in its decay rate or Proton-Time status. Subsequently, the effects of grid velocity upon the proton's net impulses and decay rate must be superimposed upon the corresponding effects from the realm EIL grid. Furthermore, the large number of I-Protons and E-Protons that cluster together in the nucleus of the more complex atoms also have their interactive effects, which must in turn also be superimposed upon the grid velocity and gravitation phenomena and vice-versa.

There can be no doubt that one day mankind will be able to bring his machines to approach the velocities we have just discussed. However there will be other problems to be resolved such as temperature conditions caused by the absorption of additional energy medium into the Mp-shells of the travelling atoms. In the next chapter we discuss some of these aspects along with the possible practical applications of theories presented in earlier chapters.

Chapter Summary.

1. The velocity status of a mass object is a condition of relative movement through the realm grid datum and is to be considered as an absolute realm velocity.

2. The realm grid is made up from a complexity of E-flows arriving from all directions at the various grid locations.

3. A mass object within the realm grid would be considered stationary i.e. nil realm velocity, when the E-Flow velocities from all directions are equal.

4. Due to the velocity of these E-Flows an additional quantity of energy medium is absorbed into each E-M/X event during the mass medium implosion stage. The amount of this additional energy medium absorption increases with increasing realm velocity of the mass object.

5. At grid velocities that are a small fraction of VF, the retarding impulses and acceleration velocity impulses set up in the protons of a mass object appear to be in perfect balance. As such, mass objects traversing the realm at these velocities do not slow down.

6. At grid velocities that are a larger fraction of VF, the excessive retarding impulses caused by the large absorption of energy medium will result in a slowing down of the mass object.

7. The Mp-shell is unaffected by grid velocity when this is small. At the higher grid velocities however the Mp-shell is displaced into a marginally downstream position with respect to its nucleus and the subsequent centralising impulse causes the atom to slow down. As such, the atom cannot maintain a velocity close to VF.

8. The velocity of the EIL oscillation wavefront near to the surface of the earth is greatest in a vertically upward direction and least in a vertically downward direction.

9. The Michelson-Morley Interferometer has the potential of an important instrument for the futuristic directional detection of high EIL E-Flows.

10. When one mass is within the gravitational field of another mass and has a relative velocity towards that mass, then the net attractive impulses induced in each are somewhat less than if the two masses had been stationary with regard to one another. At very high relative velocities (towards each other) a net repulsion impulse will exist between the two masses.

11. When two masses possess a relative velocity away from one another then the net attractive impulse between them is reduced. At some defined away velocity the net gravitation impulse reduces to nil.

12. A rotating disc near the earth's surface experiences the conditions of both items 10 and 11 above. Subsequently this disc can be freed from the gravitational pull of the earth by rotating it at a defined angular velocity in the vertical plane .

13. Proton-Time slows down as the grid velocity of a mass body increases.

CHAPTER 9.

Practical Implications

In the foregoing chapters we presented the reader with an overall view of the make-up of our realm Universe. One could assume that all of the activity that we observe around us today is simply a consequence of the continuing mass to energy medium decay process. In chapter 2 we showed by means of our 7 circle theory what we considered to be the past, present and future modes of our universe realm structure. Alas, the future presented was rather bleak for the mass entity. In the very much later stages of the circle 6 energy medium curve we consider that nearly all of the free mass structures (excluding the proton giants) would have reverted into the energy medium state. Ultimately of course there would be no mass entity left within the realm. From then on (in the realm cycle) EILs would progressively decrease and would be represented by the circle 7 phase of the energy medium curve (Fig 2.29). Much of this energy medium would have been in the mass state at some stage in the realm's life cycle and would have therefore gone through the mass to energy medium explosion (E-M/X) phenomena. Subsequently, this energy medium would exist in the form of E-Flows. These E-Flows must eventually exit from the former realm volume and would seem to cause its enlargement. However, these E-flows would all finally revert to our hypothetical E-Lines of Fig 1.12 once their EILs had reached C_0. The realm as such would gradually deplete of these E-Flows and its EILs would decrease further. Eventually these EILs would become so low that EIL oscillations would cease to propagate. Our former realm would then be left devoid of most of its original energy medium quantity and also of any character or activity. Its initial velocity within the cosmic void and relative to the Mother Realm entity (Fig 2.3) would however be unchanged. Is the fate of the realm to also be the ultimate fate for mankind? We cannot speculate upon the answer because there is nearly an eternity of scientific advancement yet to be achieved, and who is to say what new phenomena and laws may yet be discovered. Man's conquest of the realm Universe can only be achieved when he has fully understood its make-up and character. Only then will he be able to utilise its behaviour laws in a practical and advantageous manner. It is the intention here to show up some of these theoretical aspects and to highlight their potential for a futuristic beneficial practical application.

Before we delve into our pot of theoretical laws let us consider our concept of the velocity C at which lightwaves i.e. EIL oscillations propagate within the realm. Experiments have measured this at C = 186,235 miles per second in the EIL environment near the earth. The number of MTUs in one second is therefore at least equal to the number of quantum steps taken to cover 186,235 miles. Thus the MTU, which is the overall duration of the complete Mp event cycle, is an extremely small quantity. Conversely, the propagation of lightwaves has a very high frequency of stop-start operations (as represented in Fig 3.20), and we may therefore consider its rate of propagation as relatively slow in comparison to the E-Flow velocity. Let us demonstrate this slowness by a practical example. In Fig 9.1 we represent the relative positions of the Sun and Earth by A and B respectively. We know that the distance between them is approximately 92 million miles and that sunlight takes nearly 8 minutes to reach earth. For our experiment we would like the reader to draw a line from A to B at a rate that simulates the velocity C of lightwaves. In order to conform to a true simulation the reader must complete the line in not less that 8 minutes, a task that surely requires a considerable degree of patience and control. Alternatively, the reader may choose instead the length of his back garden to represent the above distance and to walk it at simulated light speed.

Fig 9.1

Fifty or a hundred yards distance would still require a slow walk on the part of the reader. The object here is to project a concept of the relative slowness of lightwave propagation. In contrast we have no measured value for the E-Flow velocity VF, but we know that it has a continuous traverse without any stop-start operations. As such we consider VF to be very much greater than C. For the present we shall assume that VF is at least twice the value of C. Subsequently the velocity of light can no longer be considered as a barrier or obstacle to higher velocity achievement. More on this in Appendix 1.

When a distant star explodes and then collapses into a neutron object or even a proton giant (black hole) there must occur a considerable surge in the E-Flow quantity emitted for that brief instant. Here on Earth we would experience that E-Flow surge well before the lightwaves that visibly depict the explosion. On February 23rd 1987 the explosion of a hot blue star, catalogued as Sanduleak 69°202 with 20 times the mass of our sun, was witnessed here on earth. The event was a relatively close 170,000 light years distant and came from within the Large Megallanic Cloud, the galaxy closest to our Milky Way. The explosion actually occurred 170,000 years ago since it took that long for lightwaves from the explosion to reach Earth. The E-Flow surge however would probably have taken only half this time to reach Earth i.e. nearly 85,000 years ago. Unfortunately this can be of no practical use to us today. However if we could now commence a record of all future E-Flow and forcephoidal surges received on Earth then perhaps in the centuries ahead we may even be able to connect these to another Supernova's visible lightwaves. It must be remembered that E-Flows travel in an absolute straight-line path while lightwaves follow a path governed by the realm grid gradients. Subsequently lightwaves and related E-Flow surges may seem to come from different directional positions within the realm. Allowances will have to be made for this.

In order to determine E-Flow velocity we could conduct an experiment in the form of a controlled fusion explosion at a distance of about 10 light hours from Earth. We would need to know the precise timing of the explosion triggering mechanism. Laboratories on or near earth could monitor all conditions as they arrived and so evaluate a true velocity for both VF and C. Magnetic effects could not reach Earth at such a distance but the traversing forecephoids would. These traverse at E-Flow velocity and should reach Earth at the same instant as the E-Flow surge. Perhaps measuring devices of the future would permit such an experiment to be conducted. It will also be noted that the lower frequency EIL oscillations would arrive prior to the higher frequencies as discussed in Fig 3.31.

Gravitation is the effect that a realm grid gradient has upon the centrally directed impulses of the proton (Fig 4.6). The realm grid datum is the result of all the proton decay E-Flows from local mass objects. The greater the number of protons congregated within the mass object then so much the greater are the E-Flow quantities from that object. The E-Flow EIL obeys the inverse square law for distance but is directly proportional to the mass quantity of an object. Mass objects that are at some distance from one another but which are within each other's E-Flows will experience a net attractive impulse towards each other. However, as the proton ages its mass content will reduce (Fig 5.2). From this we assume that as the proton decays its E-Flow emissions will also reduce. We consider however that the centrally directed impulses within the proton will remain in proportion to the mass of the proton and its rate of decay. As such we can assume that the compactness of the Mp-congregation (i.e. the proton) will be virtually unchanged. Subsequently the effect of grid gradients in causing impulse imbalances within the proton continues in a consistent and fairly constant manner. The ageing of the proton however, results in a decline of EILs within the realm grid which implies also of a general reduction in realm grid EIL gradients. The end result of this is that the net gravitational impulses generated within the protons of adjacent mass objects will progressively decrease with the elapsing of time.

The implications of this are far reaching. The orbits of the moon around the earth, of the earth around the sun and of the sun and other stars around the galactic core should therefore enlarge with time. However, we know that free protons exist throughout the expanse of the realm (having been formed in the circle 5 phase), and that these will now acquire linear impulses in a direction up the grid gradients. These free protons will thus gravitate towards other large mass objects. The decay losses of most mass objects may thus be offset by the addition of new atoms built from these free floating protons. In a mass congregation like the sun, the fusion of primary atoms results in the Helium atom which has a nucleus consisting of a string of four protons (Fig 6.36). The total E-Flow output from this nucleus is somewhat less than that from the former 4 separate individual protons. This is because the E-Flow from the E-protons and I-protons are partly absorbed by one another before leaving the nucleus. Incidentally, the greater the total number of protons within a nucleus the higher is the nucleus EIL environment and the slower is the ageing process and decay rate of its protons. Nevertheless, the greater population of protons within a nucleus must result in relatively higher initial E-Flow EILs being emitted, which must also have the accompanying characteristic of steeper EIL gradients. Subsequently gravitation effects at a given distance should be correspondingly greater.

In order to highlight this aspect let us conduct a theoretical experiment to compare the gravitational impulse induced in a test proton by a large mass grouping of primary atoms in the first instance, and then by another mass grouping of iron atoms in the second instance. Both mass groupings are to contain the same number of protons. In Fig 6.34 we showed the element iron to contain 26 E-protons and 30 I-protons. Since all Mp-shell sections tend to occur at the same relative distances from their E-proton source, then we can assume that all atoms are approximately of equal size (give or take a few proton diameters). Also, since each element has a slightly different configuration of protons in its nucleus then each element will have its Mp-shell sections arranged in a unique and corresponding pattern (Fig 6.33), giving it its equally unique physical and chemical properties. Consequently our iron group or planet will contain 56 times fewer atoms than the corresponding hydrogen group or planet and so will be 56 times smaller but 56 times more dense. We assume all protons to be the same age at the commencement of our experiment. If the hydrogen planet is 56,000 miles in diameter then the iron planet will be approximately only 1,000 miles across. What is the gravitational impulse induced in a test proton at a distance of 500,000 miles from the centre point of each planet? Also what will be the corresponding gravitational impulse induced when the protons of the hydrogen planet have aged by 50% (see Fig 5.2)? Assume the planets to be of a uniform homogenous structure and test protons to be stationary.

We can only speculate upon the answers. We must first consider how much the EIL of the realm grid datum is increased at the centre line position of each planet as shown in Fig 4.9. We consider that this EIL will be proportional to the density and proton quantity of the planet. Since the proton quantities are equal in each case then we consider that the iron planet will produce the higher realm grid datum EIL peak at its centre line. Also since E-Flow EIL diminishes according to the inverse square law, then the E-Flow from the iron planet should reach the stationary test proton at a higher EIL than that from the hydrogen planet. In an otherwise isotropic realm datum this means that the grid gradient at the test proton location is greater in the case of the iron planet than it is for the hydrogen planet. We consider therefore that the test proton has a greater gravitational attraction towards the iron planet. We are however unable to speculate upon the impulse differential for the two cases. Nevertheless, we can state that this differential must increase with the elapse of proton time. This is because, in general, the protons of the hydrogen planet will age at a faster rate than those of the iron planet due to their TA-Lines being different (Fig 5.6).

Perhaps the reader would prefer to contemplate a similar though simpler problem. In Fig 9.2 we present two isosceles triangles Al-Bl-Cl and A2-B2-C2 such that the height of A above the base BC is the same in both cases but the base of one being twice that of the other. If a test proton is positioned at A and an iron atom is located at B and another at C, then what is the gravitational impulse induced in the test proton in each case? We would of course, ignore all outside influences.

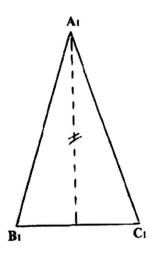 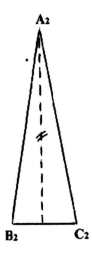

Fig 9.2

Now all mass objects are E-Flow sources and the impulse quantities generated in neighbouring masses tend to be equal and opposite. One may conclude therefore, that the resultant of all the impulse quantities so generated is zero. However, our understanding of the mechanics of gravitation leads us to speculate on whether there is any means by which we may alter this balanced state of affairs and cause a unilateral gravitation impulse to be generated.

By this we mean to investigate the possibility of a mass object A being attracted towards an E-Flow source B, but without the E-Flow source B being attracted in turn towards the mass object A. As represented in Fig 4.13 the E-Flow source B sets up an EIL gradient for the protons of the mass object A which then acquire an impulse in the direction up the EIL gradient. In order for the E-Flow source B not to be similarly affected we must assume that it is not a mass object and as such contains no protons. Subsequently we would define B as an artificial E-Flow source, while the mass object A would be considered a natural E-Flow source. Basically, an E-Flow is generated during the third phase or M-E/X of the Mp-event sequence as shown in Fig 4.2. Now we know that Mps were formed in great profusion during the circle 4 phase of the realm energy medium cycle resulting in an intense surge of E-Flows in every direction. If we could somehow duplicate such a C_1^+ EIL environment over a small volume B of say 0.1 cubic inch, then perhaps the resulting E-Flows from within this volume would set up a significant

EIL gradient within the local realm grid. Let us assume that at a distance of about 10 inches from this volume B we have an EIL grid gradient that is equivalent to the gradient set up by the earth's E-Flows at the distance of the moon's orbit. As such if a mass object A is positioned 10 inches from our hypothetical E-Flow source B then it should experience the same amount of attraction towards B that the moon experiences towards the earth. Yet in this case the E-Flow source B would not be affected by any grid gradient that was set up by the mass A. This is because B does not contain any proton structure masses.

In order to effect such a condition we require the EIL within the volume B to be increased to well above C_1. Mps would then form within this volume in profusion and at random. The subsequent M~E/Xs would then produce the outward E-Flows that we desire. In theory we now possess a massless E-Flow source. Unfortunately, the formation of Mps within B and the resultant outward E-Flow would result in a depletion of its energy medium content and a considerable drop in its EIL. When the EIL drops below C_1 then Mp formation would cease. Consequently for a continuous E-Flow to be emitted we must somehow ensure that an adequate quantity of energy medium is fed back into the volume B. If we could develop this principle in practical terms then the potential uses to which it could be put are simply enormous.

Consider a spacecraft with such a device mounted at its front and forming an integral part of its superstructure. As such, nearly all the atoms of the craft would then lie in an EIL grid datum having an upward gradient towards the device. The spacecraft would then acquire a continuous resultant impulse towards its front. In other words the spacecraft would be in a state of continual acceleration. When a condition of artificial gravity is required for the occupants within the spacecraft then similar devices will have to be operated at both front and rear locations. Objects (and occupants) closer to the rear device will gravitate toward the rear of the craft, while objects closer to the front will gravitate towards the front. This is shown by the floor arrangements within the spacecraft of Fig 9.3. As the occupants moved around and across the centre line of the spacecraft, the E-Flow sources B1 and B2 would be required to adjust their outputs in order to maintain perfect equilibrium. Gravity effects will of course increase as one progresses away from the centre-line and towards either B1 or B2. Subsequently, a gravitation environment similar to that on earth could be maintained within the spacecraft. Duration spent in space therefore need not have any adverse effects upon the physique of an astronaut.

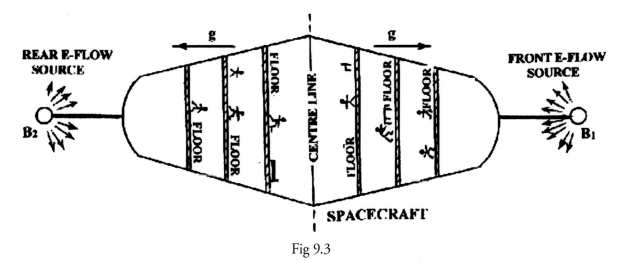

Fig 9.3

As these E-Flow sources are developed into more proficient and powerful devices there is no reason to suppose that they could not be used to gravitate much larger mass objects such as other planets in our solar system into new orbits around the sun. It may even be possible (and perhaps necessary) at some future date to move our sun and hence the entire solar system to a new position within the galaxy to avoid a potential collision with another star. Alternatively we may simply desire a more advantageous position for our sun (and earth) near the fringe of the galaxy as an aid to the long range exploration of inter-galactic space. Also, when the nuclear fires of our sun finally commence to diminish, then we should most certainly need to relocate the earth into an orbit around another much younger sun. There are of course other much less grandiose uses to which E-Flow sources may be put. We leave these for the reader to contemplate further.

Let us now consider whether it is possible to shield an object from another's E-Flow. We know that an E-Flow traverses in an absolute straight line and the only means of stopping it is for it to be absorbed by an E-M/X. This however will only absorb a small part of any E-Flow. Thus E-Flows can get past an Mp-shell as only a small part gets absorbed therein. However a proton is an Mp-congregation of such a density and profusion of Mps that any quantity of E-Flow gets completely absorbed therein. An E-Flow cannot phase through a proton. Unfortunately a proton is very small in comparison to the overall volume of the atom and so most mass objects offer a negligible shielding effect to any E-flow. In the more complex atoms we have large numbers of protons that are bonded to one another in groupings and strings within each nucleus (Fig 6.28 and Fig 7.21). Obviously such nuclei would absorb a bit more of the E-Flow but unfortunately this is still a negligible proportion of the overall E-Flow passing through the mass object. Towards the end of chapter 4 on gravitation, we speculated upon the formation of a Neutron Star and compared the proton grouping to that of marbles in a pouch bag. We noted that although the marbles were in very close contact with one another there were still interstitial air spaces between them. This means that the protons of the Neutron Star are individual entities bonded together but which could nevertheless be prised apart if we had the means. Now the compact structure of the Neutron Star offers a complete shielding effect against any E-Flows over its entire diametric volume. It is our contention that we do not need the entire bulk of the Neutron Star to absorb E-Flows but simply a single layer of protons arranged in a sheet configuration. Therefore, if we could obtain some Neutron Star material we could reposition the protons into a sheet-like structure with a width of one or two protons. We would consider this to be an effective E-Flow shield which would absorb well over 99% of all the E-Flows reaching its surface. Fig 9.4 depicts such a sheet of protons.

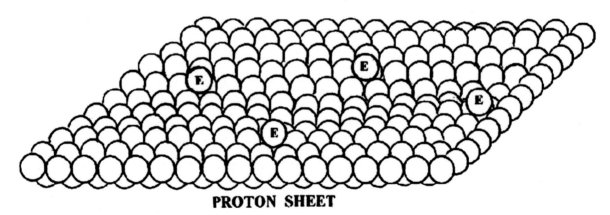

PROTON SHEET

Fig 9.4

We cannot however know the reaction between normal atoms and this blanket of protons. Perhaps random E-Protons positioned as shown on either side of the proton sheet could set up Mp-shell sections that would convert the sheet of protons into a custom-built overly large single nucleus of an artificial atom. Perhaps when we acquire a greater understanding of the character and structure of natural atoms it may be possible to synthesize new atoms from the composite material of neutron stars.

If such a proton shield sheet were to be placed flat upon an area on the surface of the earth and was large enough to perhaps cover a 10 feet square area, then any object positioned over its centre would be shielded from the normal gravitational pull of the earth. The object would however come under the gravitational pull of the proton sheet which could be quite large though perhaps somewhat less than that from the bulk of the earth. As such the mass object positioned over the shield should weigh less than normal. Incidentally, the air molecules over this area would also weigh less and a considerable updraft could result, perhaps even causing an upset in the local weather conditions. Perhaps such a device could be used to advantage to bring rain clouds to desert areas of the world. We would suggest however that all such shielded zones be kept as small as possible and be enclosed within a specially made airtight structure. Consideration will also have to be made for the fact that the above small area of earth will be shielded from the gravitation pull of the sun possibly affecting the current orbit of the earth. Therefore such shielding will have to be done on a very limited scale and that too as far north or south of the equator as possible. Ideally the plane of the proton shield should be parallel to the plane of the earth's orbit around the sun during the day periods. At night the shielding would not obstruct the E-Flow from the sun to the mass of the earth and it would be immaterial as to how much and where such shielding was positioned. Although consideration would have to be made with respect to the earth's gravitation pull upon the moon and the possibility of an upset to its current orbit. Any displacement could be corrected by our hypothetical artificial E-Flow source devices.

Once again the potential uses to which such a shield device could be put are enormous. Of apparent consideration must be the gravity wheel. The gravity wheel can be compared in principle to the simple water wheel in which the weight of water pouring into the wheel buckets results in one half of the wheel becoming heavier than the other thereby causing the wheel to rotate. In a gravity wheel the same result can be produced by making one half of the wheel lighter than the other. This is done by positioning a proton shield under one half of the wheel as shown in Fig 9.5. The side B will be under normal gravity while side A will be part shielded from the earth's gravitation

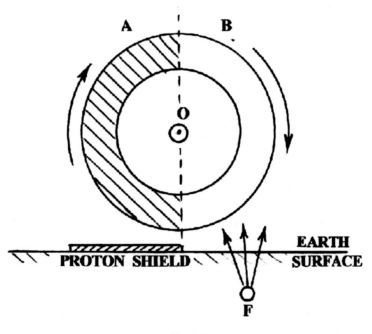

Fig 9.5

and so will be lighter than normal. Thus the wheel will rotate about its well-oiled axle O in the direction as shown. A means of enhancing the rotational power of this wheel would be to locate an E-Flow

source underneath side B as shown at F. Subsequently the gravity-wheel could become an inexhaustible power device to drive voltage generators for the production of electricity on a vast and very cheap scale. It must be noted that the gravity-wheel is not a perpetual motion machine although it would appear to be. The decay of motion will however be in line with the realm mass to energy medium decay life cycle and as far as we are concerned this has still got a very long duration to go in its current circle 6 phase. The proton shield simply manipulates the local realm grid gradient and it is the centralised impulses within each proton and the upset thereof that produces the driving power of the gravity-wheel.

In chapter 8 we indicated the effects of realm velocities upon mass objects. In order to traverse the vast distances of the realm in as short a duration as possible, spacecraft will need to be designed for realm velocities that exceed the E-Flow velocity VF. Consider the following scale of velocities required in order that we may traverse our own galaxy. We assume our milky-way galaxy to be approximately 150,000 light years in disc diameter. Thus if we were to travel at light speed C, it would take us 150,000 years to cross from one side of the galaxy disc to the other. But this is no good to the explorer since it is much too slow and time consuming. Faster traverse velocities are required as indicated by the following scale of hypothetical 'light year' (denoted LY) velocities.

1. At $12 \times C$ or 1 LY/month velocity the traverse across the galaxy would take 12,500 years.
2. At $365 \times C$ or 1 LY/day velocity the traverse across the galaxy would take 411 years.
3. At $8760 \times C$ or 1 LY/hour velocity the traverse across the galaxy would take 17 years.
4. At $525,600 \times C$ or 1 LY/minute velocity the traverse across the galaxy would take 104 days.
5. At $31.5 \times 10^6 \times C$ or 1 LY/second the traverse would take just 42 hours.

If we required to travel to our nearest galaxy Andromeda which is 2.5 million light years distant, then at a velocity of 1 LY/second it would take 289 days. At a velocity of 100 LY/second it would take only 70 hours.

Unfortunately we consider that the effects of these high realm velocities would be disastrous to the atomic structures of both spacecraft and occupants. The EA quantities absorbed by both proton and Mp-shell during these high velocity traverses would result in the equivalent of a progressive temperature rise. Eventually molecular bonds and material structuring must adjust to the new higher energy medium configuration. In other words the spacecraft and its occupants would just burn up. To prevent this occurrence during such high velocity traverses the spacecraft and its occupants must be effectively shielded from absorbing any abnormally high velocity E-Flows. A proton shield located over the entire upstream face of the spacecraft would effectively prevent E-Flows entering the craft from the forward direction.

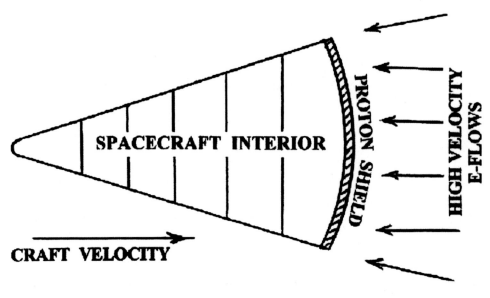

Fig 9.6

We would consider that such a spacecraft would have to be designed according to the principles represented in Fig 9.6. Here it conforms to a conical shape with the proton shield being positioned over the wide base section which is also the lead face of the craft. At the very high velocities required of inter-galactic travel, all E-Flows encountered would approach from a direction that was very nearly parallel and opposite to craft travel velocity. The conical shape of the craft therefore ensures that its entire internal structure is always in the E-Flow shadow of the proton shield. Quite obviously design and layout improvements would be made to such craft as more was learned about the intricacies of space travelling at high velocity.

Apart from being able to travel at high velocity, an added benefit of the proton shielding effect would be that the EIL environment within the spacecraft could be maintained approximately the same as it is here on earth. Subsequently, Proton-Time rates would also be about the same and astronauts could then traverse vast distances at extremely high velocity and not age in a disproportionate manner to their planet bound families. Consider the example of a spacecraft leaving earth, travelling at 100 LY per second to the Andromeda galaxy on a navigational charting expedition lasting 60 days and then returning to earth. The total elapsed time would be about 66 days both on earth and for the occupants within the spacecraft. There are obvious advantages to be had from maintaining a uniform proton-time rate within such conditions because this means that earth-time can not only be transported to remote galaxies but also maintained there as such. Future colonisation across the realm would then be able to relate directly to events not only on the earth but in other parts of the realm as well. A uniform system of realm dates could also be set up for the entire realm population.

There would of course be some very real dangers to the spacecraft's proton shield from its very high realm veolcity. Free protons that exist abundantly within the realm's open spaces between galaxies are bound to strike the forward shield at the spacecraft's relative velocity. These protons could either be absorbed into the shield or else their bombardment of it could displace one or more of the shield's protons, effectively causing the shield to be 'holed' or damaged. Sufficient numbers of these holes could impair the shielding effect and the interior of the spacecraft could then experience a temperature rise and change in Proton-Time rate. It is considered that a change in the profile of the shield could be made so that any such bombardment only resulted in a glancing impact upon the shield.. Such a profile would be similar to that used in streamlining the forward sections of present day atmospheric craft. An ideal profile would be similar to that of a twin cone or diamond as shown in Fig 9.7.

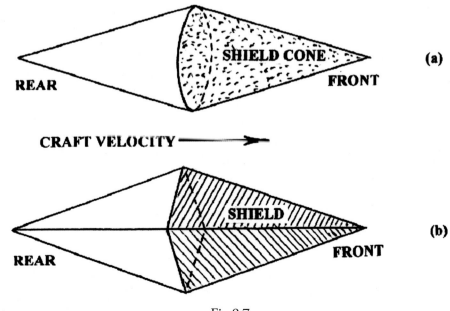

Fig 9.7

Just as an EIL grid gradient causes an upset in the centralised impulses inside the proton resulting in the gravitation effect, so too can a nodal field cause an upset in the equilibrium of all the inter-molecular and structural bonding impulses that go into the make-up of a large mass object. The resultant impulse is simply the forcephoidal effect of magnetism in a ferrite body described in chapter 7. We cannot state precisely the manner in which a net linear momentum or impulse is induced within the ferrite body by the nodal field but we can indicate a reason for this resultant impulse from the forcephoidal principles developed earlier. There are a number of separate and distinct impulse quantities that are active within the ferrite mass but which are all in a state of balance or equilibrium under normal EIL conditions. There are the centralising impulses that maintain the inertia and orientation of the nucleus (Fig 6.6). The Mp-shells in turn can undergo distortion and thereby impart impulses to the atom as a whole resulting in the phenomenon of molecular bonding (Fig 6.13). In the ferrite structure the molecules bond in preferred directions in the main and this is exemplified by our linkage diagrams of Fig 7.3 and Fig 7.7. These bonding preferences are a direct result of the 'shape' of the atom (Fig 6.33), which in turn is a result of the E-Proton configuration in the nucleus (Fig 6.27). When this structure comes under the influence of a nodal field (Figs 7.11 to 7.14), a new orientation of the atom nucleus (Fig 7.6) is partially effected. Now a shifting of the nucleus E-Protons will also result in a repositioning of the corresponding Mp-shell sections. As such the entire atom will become marginally re-orientated. However, we stated that the unique 'shape' of the atom resulted in its molecular bonding being stronger in certain preferred directions. As such the bonding will tend to continue in the same relative Mp-shell positions but with a certain amount of distortion to the shell. This in turn exerts an altered centralising impulse on the nucleus and which can upset its state of equilibrium. Subsequently, an impulse imbalance can occur throughout the mass body linkages and a resultant impulse can effectively push the whole of the mass configuration in a singular direction.

The amount of the impulses generated within two adjacent ferrite bodies is dependant upon the intensity of the nodal field within which each body is located. Ferrite compatible nodal fields, as far as we know, can only be generated by other ferrite masses or by electric currents in coils (Fig 7.55). We showed in chapter 7 that the magnetic type impulses so induced in two adjacent bodies and/or coils are always equal and opposite. One may conclude therefore that the overall resultant impulse is always zero. Earlier in this chapter we speculated upon the aspect of a unilateral gravitation impulse being generated and we subsequently evolved our new hypothetical non-proton E-Flow source (Fig 9.3). We should now like to consider whether the same principle may be applied to the setting up of a ferrite type nodal field to produce unilateral magnetic impulses. By this we mean that for two adjacent magnetic masses A and B only one of these is influenced at any particular moment. That is, if B is repelled or attracted towards A, then A remains unaffected by B. As represented in Fig 7.9 the nodal field generated by the mass A pervades the structure of mass B and causes a resultant impulse to be active. In a similar manner the nodal field of B will pervade A. In order for A not to be similarly affected we must assume that either A is not a mass object (or a coil) thus containing no atoms or structural pattern (Fig 7.7), or that the structure of A is not compatible with the nodal field of B. Subsequently we would define A as an artificial nodal field source. Similar consideration may be applied to the nodal fields from an electrostatic source and the impulses generated in the surface atoms of another mass object within this field (Fig 7.67 and Fig 7.68). It is our intention to speculate upon the practical design of a system for producing such a unilateral magnetic or electrostatic impulse. If the principles upon which we have based magnetic impulses are correct then it should not be too difficult to theorise upon such a unilateral impulse condition and then to eventually evolve a practical system.

In Fig 6.22 we showed how a number of forcephoids could intersect repetitively at the same point location to produce a relatively stationary high EIL node. Obviously, forcephoid EIL peaks obey the inverse square law with distance from proton source and therefore the distances at which high EIL nodes can occur is limited. As such all magnetic effects are at a relatively close range. Our speculation thus leads us to consider whether a device A, constructed from a non-ferrite material, could produce a ferrite type nodal field in the zone around an adjacent ferrite mass object B. Thus B would be influenced by the field of A, but A would not be influenced by the resulting field from B. The resultant impulse of the combination of A and B is thus in the direction from B to A. Our primary concern is therefore the practical design of device A. There are several alternatives to be considered. One of these is the focussing of non-ferrite nodal fields from synthetically engineered materials which combine at a set distance to create a limited zone ferrite type nodal field. The principle is indicated in Fig 9.8. The device A is represented as a bowl shaped object composed of a non-ferrite material, though perhaps constructed from a

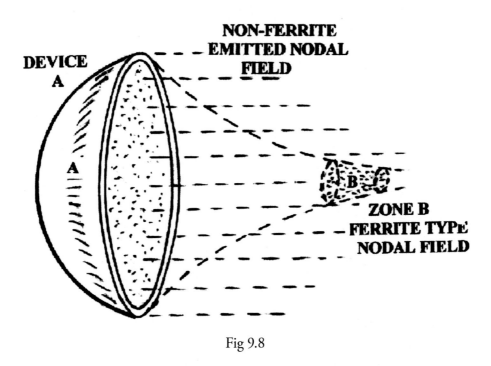

Fig 9.8

mixture of elements close to the iron element configuration of the Fig 6.34 table of elements. The nodal field set up from the forcephoids of this object will of course be incompatible with the ferrite mass structure. However in a limited zone represented at B the forcephoidal emissions may combine to create a nodal field that is compatible with the ferrite structure. As such, if a suitable soft iron object is positioned inside this zone B then it should come under a magnetic type influence (Fig 7.7) and acquire an impulse towards A (Fig 7.11). The nodal field emitted by this ferrite secondary mass body (Fig 7.8) within the zone B would not have any effect upon the structure of Device A since the two would be incompatible in pattern and structural configuration. The potential uses for such a device are enormous. The profile and the material from which Device A is to be constructed would of course need to be developed from considerable amounts of experimentation on a trial and error basis. The generation of a unilateral resultant impulse for a virtually indefinite period of the circle 6 phase of the realm energy medium cycle and from an apparently inexhaustible energy medium source is now not an unthinkable or an unreasonable proposition. The combination of Device A to the ferrite object in zone B could thus be a source of a seemingly perpetual impulse power. As in the case of gravitation, Newton's third law in which every action must have an equal and opposite reaction does not seem to hold true here. This is because Newton never considered that there were inherent internal impulses within the atom structure that could be harnessed under favourable conditions to produce a unilateral resultant.

Perhaps another practical method for achieving a similar result would be to introduce an oscillating nodal field that alternated rapidly between the magnetic North and South Pole configurations. In Fig 9.9 the mass object B is made of soft iron i.e. it is very easily magnetised and is considered as the ideal secondary mass body. No matter what the initial polarity of A, we know that an opposite polarity is induced in B causing B to acquire an impulse towards A. Now when the structural configuration within B acquires this induced mode, then B will create an

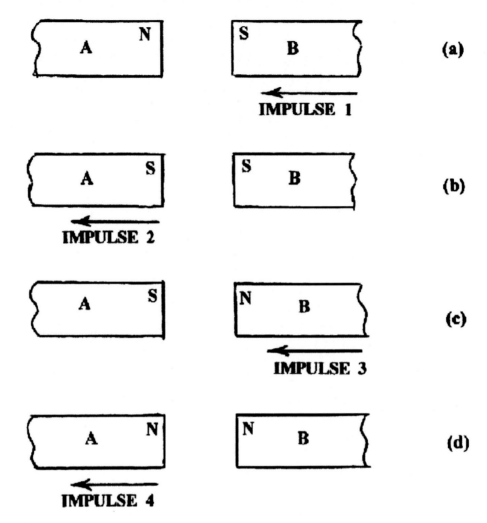

Fig 9.9

opposite type of nodal field around A which should cause A to acquire a net impulse that is towards B (Fig 7.9a). However, we propose to interfere in this system by causing the polarity of A to change just as the nodal field from B is created. This means that the effect of B's nodal field upon the structure of A will be like that of Fig 7.13 and will result in A being repulsed away from B as in Fig 9.9 (b). As such, both A and B have acquired an interim impulse in the same direction. The next sequence (c) is that A with its sudden polarity change sets up a new nodal field which induces a new structural configuration in B. Since B is a soft iron structure it will readily adapt and once again it will acquire an impulse towards A. B will then initiate a nodal field around A. Once again we interfere in this system and cause A to undergo a polarity change just as the new nodal field from B is created at (d). Thus A is again repulsed away from B. We note that with A as the induction agent, the net impulse acquired by both objects was always in the B to A direction. If we can cause the polarity of A to alternate sufficiently rapidly to meet the above conditions precisely then we may have another means of apparent unidirectional (reactionless) impulse power.

Another consideration could be the setting up of a ferrite type nodal field using a flow of electrons in an open realm environment and by deflecting them towards various predetermined points. We know that electrons emit both forward and backward forcephoids (Fig 7.26) and it is our contention here that electrons emitted in a uniform manner from a relatively expansive surface area are able to produce EIL nodes that could (within a limited focal zone) conform to a ferrite type nodal field configuration. Perhaps the combination of a large number of individual 'cathode ray' emitters aimed at a particular location could achieve the same result as Device A in Fig 9.8. As already stated this will need to be the subject of considerable initial research and experimentation. It may be that we are unable to produce a ferrite type nodal field from the above but instead are able to create an electrostatic type nodal field as indicated at Fig 7.65. The impulse quantities between two electrostatically charged bodies would then come into play and we should again be able to achieve a unilateral impulse condition for one of the bodies.

Finally, on the subject of unilateral impulses, we would like to speculate upon the feasibility of building an extremely complex network of current paths and electrostatic charges. The setup would be similar to that of a microchip except that it would be in a 3-dimension cubic format. The object is the creation of a particular type of nodal field in a unique direction and at a defined distance. The resultant nodal field configuration would probably depend upon a unique combination of current paths, current flow and qualitative distribution pattern of electrostatic charges within the elements of the network. A considerable variation in the pattern of the nodal field could probably be realised resulting in an electrostatic or magnetic type of attraction or repulsion in material objects at a distance. We know that within the animal kingdom all creatures have some form of a centralised control for their nerve network. This central control is the brain which operates by means of a system of micro electric current pulses through a complex neural network. We consider that the flow of any such electric currents, no matter how small, must emit forcephoids to the free realm zone. These emissions will thus be 'broadcast' to the surrounding area in a correlated pattern and therefore it should be possible for these emissions to be picked up by another intelligent brain and interpreted correctly. This could be the basis for telepathic communication between intelligent beings. In the case of the human brain it may also be possible to achieve a voluntary control of the flow paths for these electric pulses and their frequencies such that a unique nodal field could be set up at a defined distance. Perhaps mankind's brain has the capability of an electrostatic type induction at a distance. Or even has the capability to produce a sympathetic nodal pattern to suit a particular material object and thereby cause magnetic-type effects within that object. We consider that with proper control or training the human brain is capable of inducing a unilateral impulse within a specified material object at a distance. Telekinesis has been experimentally performed and witnessed but as yet remains unexplained. Perhaps we have now advanced a step closer to an understanding of this phenomenon. We shall not pursue this aspect any further at this stage since to do so would simply be a form of an even wilder speculation. However we leave this to the reader's imaginative reasoning and hope that some form of experimentation can be evolved to either prove or disprove this theoretical aspect.

Let us now pursue a more practical anti-gravity aspect. In chapter 8 on velocity effects we demonstrated the floating rotating disc phenomenon (Fig 8.17). We also showed that an additional upward impulse could be generated by giving the disc a linear impulse in the horizontal plane (Fig 8.18 and Fig 8.19) . We consider that there is considerable scope for incorporating this principle in a mechanical gyroscopic device which is then able to impart a vertical upward impulse to a capsule or mass object to which it is attached. One such mechanism is shown in fig 9.10. There are four rotating discs C mounted symmetrically in the horizontal plane. Each disc is driven by an electric

motor B which is attached to a rigid cruciform structure with its centre point at O. This structure is then in turn rigidly connected to a shaft which can be rotated in either direction by the large electric motor A. The bed-plate is an integral part of this electric motor and contains bolt holes for fixing to any mass object. The electric supply to each disc motor B is provided through the slip rings X and Y which are insulated from the shaft D. To operate the device we must first supply electric current to the motors B causing them to rotate each disc C at a very high rpm. All discs must spin at the same rpm and in the same relative direction in order that the device may impart a balanced and uniform upward impulse. In Fig 9.10 we have shown the discs to rotate in the anti-clockwise

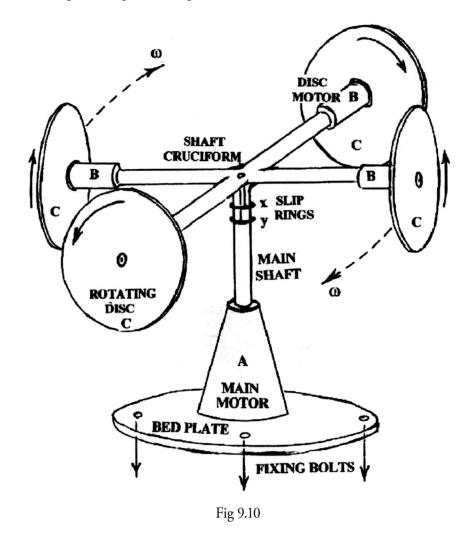

Fig 9.10

direction when viewed from the outside. With the shaft assembly in the stationary status we suggest that the rpm of the discs be increased to the optimum for upward impulse. This may or may not be sufficient to counteract the effect of the earth's gravitation upon the whole device. If not, then perhaps a larger diameter disc C should be used. However, an additional upward impulse can be generated by causing the shaft cruciform assembly to be given a rapid rotational acceleration in the clockwise direction (when viewed from above). This is done by operating the main electric motor A in a controlled manner. Since the earth's E-flow is mainly in the upward direction then the device proposed above can only operate in the vertical plane. Also, we consider that the upward impulse capability of our Fig 9.10 device is of a limited nature and hence in order to lift a much greater capsule mass it may be necessary to attach several such devices to the capsule. Once the capsule has been lifted to a certain height above the earth then horizontal impulses will have to be obtained from an alternative source. Rapid changes in height location can be achieved by the shaft D being given an angular acceleration in either direction. Clockwise for up or anti-clockwise for down. Quite obviously much experimentation will be required before such a device can become a reality.

The idea of using gyroscopes for propulsion is not a new one. Its origins go back many decades. Academics in Britain such as Professor Laithwaite of Imperial college and private researcher Sandy Kidd produced some very interesting work, which involved novel combinations of hinged gyroscopes. As a result of their work and that of others, gyroscopic propulsion did receive some coverage in the national press in the 1970s and 1980s, but it had little long-term effect on the investment into this research. In the late 1980s Hideo Hayasaka and Sakae Takeuchi of Tohoku University in Japan conducted an experiment that simply compared the weights of a gyroscope at rest and when spinning. Their results showed that for certain orientations of the rotor, an increase in rotor speed produced a small but measurable decrease in the rotors weight. This effect was only apparent for one direction of rotation. A number of independent tests have since been carried out with mixed results. A minority showed weight loss and the rest did not. The resulting consensus in the scientific community was a vague feeling that there was some flaw in the original tests. Gyroscopic propulsion has been investigated by some of the biggest names in the aerospace industry. For example, during 1990-91 a team from British Aerospace (Bae) tested a Sandy Kidd Inertial Thrust Machine. The drive mechanism was based on a pair of force-precessed gyroscopes revolving around a vertical axis. In their first experiments they detected nothing out of the ordinary. However, in their second set of tests there was one trial, which did appear to detect a small change in weight, which they could not explain. Unfortunately, the effect could not be repeated and no further trials were conducted.

Let us now once again return to the aspect of travel through the realm. We have already shown that the distances are vast and the required speeds of traverse are extremely high. However, it would be quite useless for us to travel about within the realm without the aid of maps or navigation charts. Visual 3D layouts of all the stars and galaxies within the realm would be most useful initially. As we developed spacecraft of the type shown in Fig 9.3 and Fig 9.7 and began to achieve faster traverse velocities of the order of one light year per hour (per minute or per second), then we would most certainly miss our intended target by a wide margin if we simply aimed at what we could observe. In a crowded system this would be a most inaccurate and unsatisfactory method since light waves are known to bend or refract as they propagate through the realm grid. Perhaps it may be because of this that we observe more stars in the realm than actually exist. That is, light waves from a single source can reach the same observer along two or more separate paths. When Einstein stated that space was curved he did not intend this to be taken literally in a physical sense. He meant that each zone of space had the property to cause all light wave propagation through that zone to bend or refract in the same general direction. From our theory we know that this is really the effect that the realm grid gradient has upon light waves i.e. to cause them to bend towards the higher realm EILs (Figs 3.51 and 3.52). The realm grid and its gradients are the result of the phasing E-Flows from all of the realm masses. We know that E-Flows travel in absolute straight line paths and so the realm grid datum can provide directionally accurate information about relatively close E-Flow sources. We consider that realm grid EIL charts that also show grid gradients could be a far more accurate means of indicating the true location of other realm masses (see Fig 4.9). Visual charts would not show the positions of proton giants and this omission could have disastrous consequences for the unwary space traveller. A 3-dimensional realm grid chart could be considered similar in principle to a 2-dimensional topographic terrain map. It would therefore be possible for a spacecraft computer to evaluate its realm location by recognition of particular grid zone features. One would assume of course, that the instruments of this future time would be capable of such an undertaking. Spacecraft travelling through the realm grid would need to take account of all grid gradients when plotting a course between distant locations. No matter what the craft velocity, its realm path will always tend towards the 'curvature' of the space EIL grid. The calculations would be complex since we would also need to account for the velocity effects listed in chapter 8.

The progressive 3-D charting of the realm grid - the science of which may be termed as Realm Gridology - would require a continuous updating program to allow for the relative motion of masses within the realm. Just as the planets of the solar system orbit around the sun causing local realm grid changes in a repetitive and predictable fashion, so also do the stars of our galaxy orbit its central core about once every 200 million years. Galaxies in turn belong to larger clusters and these may possess a defined path or orbit around the realm centre point as represented by the gradient

configuration of Fig 4.12. However we consider that all such mass motions are relatively slow and grid adjustments in the span of a human lifetime would be minor. We do not subscribe to the Big Bang theory assumption that all mass evolved in a single instant from a point of infinite density. Nor that the effects of that initial explosion are still apparent in the recession of all galaxies from each other as a conclusion drawn from the observed Red Shifts in the visible spectra of progressively distant galaxies. The origins of the mass within our realm are to be found in the circles 3, 4 and 5 representations of the realm's seven circle cycle (chapter 2). Admittedly there was no set time scale for this portion of the realm energy medium curve and hence the actual duration of the event sequence for mass formation may have been relatively short. However mass formation was a widespread feature and therefore there was no centralised location for the origin of mass formation. We have already explained the reason for the galactic red shift phenomenon in Fig 3.68 which is essentially based upon the diminishing character of the circle 6 energy medium curve. As such, we can conclude that the very distant galaxies are not rushing away from us at the phenomenal velocities as theorised by the Big Bang theory. Instead these would possess relative velocities that are comparable to those that are within our own galactic group. In fact the motions of all masses would be determined solely according to local realm grid gradients. If there is a reasonably uniform distribution of mass objects throughout the realm volume (see page 63), then in general the realm grid datum will be as shown in Fig 4.12 with an overall upward EIL gradient towards its more central regions. Thus, in general, all masses will tend to gravitate towards this central region or else be in a very large orbit around it.   Subsequently there should be no great difficulty in the continual update of the realm grid charts that would be stored in all computer memory banks. Perhaps these charts of the future can contain an orientation aspect to indicate the direction of the realm central region.

Let us now consider some practical aspects to realm navigation. During a high velocity traverse each spacecraft would be protected from all external E-Flows by a proton shield (Fig 9.6), and as such measuring instruments would not be able to sample the realm grid EILs to establish a realm location. However since each journey would have a known start point and a planned end point then it should be a simple matter to verify ones' grid position at the destination location. A true test of the grid navigation chart would be when a space drive malfunction occurred and the craft ended up at an unplanned destination. It would then become a matter of grid identification by the onboard computer to ascertain the craft's true realm grid position. It would then be a simple matter for the computer to indicate the direction in which one must travel to arrive at the originally planned destination point. We would of course need to have special instruments for the measurement of grid EILs and realm E-Flows. We do not propose to speculate upon the design of these devices since it is very often possible to arrive at a conclusive result by the interpretation of the responses observed in associated phenomena. Nevertheless we consider that scientific advances in most fields of study tend to correlate and so keep pace with one another. As such we assume that as realm travel progresses, so also will the design and development of energy medium measuring devices used for realm navigation and realm gridology.

Spacecraft that accidentally left the realm volume to venture into the cosmic void region would accelerate under their E-Flow source type drive conditions(Fig 9.3) to near infinite velocities. This is because the retarding impulses that existed within the realm are no longer present in the cosmic void. No matter what the velocity of the space craft the EA quantity will be nil (Fig 8.5). Since the cosmic void is devoid of all character or medium, then velocity as a relative quantity becomes meaningless. Light waves cannot propagate within the cosmic void and so visual sightings could not be made. Also, without a realm grid we should be unable to use EIL gradient charts. Perhaps detection of the E-lines of Fig 1.12 may be possible so that a reversed traverse could be effected to enable a return to our realm. Alternatively it may be possible to also detect E-lines from another nearby realm and thereby for mankind to undertake inter-realm travel. Perhaps when our own realm nears the end of its circle 6 mass cycle, mankind can undertake an exodus to a realm that is in an early circle 6 phase. Perhaps we could hop from realm to realm and eventually approach the Mother Realm (Fig 2.3). We leave it to the reader to speculate upon the conditions to be found there.

We must also consider another essential aspect when the realm undergoes a widespread colonisation. And that aspect is communication. Quite obviously light wave velocities are much too slow for effective communication to take place. Imagine an EIL oscillation message taking 200 million years to reach its destination. In realm size terms this is not a very great distance but the time taken for the signal is just not practical. A much faster method of message transmission is needed and this could be done by using an interruptible E-Flow source to send Morse code type signals between realm locations. This would be several times quicker than lightwave messages but over the vast expanse of the realm may still prove too time consuming. An alternative is for an accelerated form of E-Flow to be emitted. This could be achieved from a vibrating E-Flow source that would throw out E-Flow blocks at perhaps several thousand times normal velocity. Perhaps such high velocity communication blocks of data could only be achieved by using extremely high velocity robotic spacecraft. Whatever the ultimate method employed we consider that communications based upon E-Flows and in-built forcephoidal peaks would be a considerable improvement on EM transmissions over large distances. The E-Flow blocks would need to be emitted as a directional beam. A parallel block of energy medium would not diminish in EIL by very much over the vast realm distances. Perhaps we could allow a doubling of the E-Flow block volume over a distance of about 10 to 20 million light years. A beam of data could be directed at relaying stations that are located at suitable positions within the realm and especially near to other inhabited planets. Final transmissions to the planet could be in the conventional EIL oscillation format. We could have a destination code included in the data on the E-Flow block so that the relaying stations could decipher this and then retransmit the data block in the correct direction. As such, data may need to go through a series of relay stations before reaching the intended destination. We could compare such a system to our current telephone network (or e-mail system) but on a much grander scale. Imagine being able to send a message to someone who lives in another galaxy system about 100 million light years distant and getting a reply in a matter of a few weeks. In the very long term with continual improvements in communication technology, this interval might even be reduced to only a few hours.

In chapter 3 (Fig 3.68) we briefly discussed the meaning of the Red shift phenomena in relation to the circle 6 phase of the realm energy medium cycle. It has always been our intention to speculate more fully upon our current position on this circle 6 portion of the energy medium curve and we consider the observed wavelength shifts (always towards the lower frequencies) as a fundamental clue. Initially, when Hubble made his observations the red shifts were small and of the order of 0.004. As viewing techniques improved through the latest advances in telescopic photography, observations were possible of galaxies that were even further away. These provided spectra with ever increasing red shifts. In Fig 4.21 we expressed a relationship between the Grid Datum EIL and Mp event cyclic rate. Since the emission of wavefronts depends upon the repetition frequency of the Mp-shell, then we may conclude that the emitted wavelengths are inversely proportional to the prevailing realm EIL. We assume of course that EIL oscillations can only propagate in an EIL environment below C_1 and emanate only from the Mp-shell of the atom. This means that all EIL oscillations were commenced within the circle 6 phase only. Our observations of objects with increasing red shifted spectra thus take us closer in time to the start of our circle 6 phase. Lightwaves reaching us now from a galaxy some 20 million light years away would have been emitted 20 million years in the past. By analysing these wavefronts we are thus looking at atomic conditions as they prevailed then. We assume that in the past the realm datum EIL was higher than it is now (as represented by the droop characteristic of the circle 6 curve), and subsequently the Mp event cyclic rate must have been at a somewhat lower frequency. At the relatively lower EILs we consider the relationship between Mp-cyclic rate and EIL as approximately linear (as represented by the 'ab' portion of the graph in Fig 4.25). There is however a lower limit to the Mp-cyclic rate. Therefore as the EIL increases towards C_1, we consider the amount by which the Mp event cyclic rate alters gets progressively less. Ultimately the Mp event cyclic rate will be an absolute minimum. We consider that the frequency of all the currently emitted wavelengths would have been at a minimum when the realm datum was originally just below C_1. At this stage we consider that perhaps only protons and primary atoms existed, and that subsequently the earliest emissions would have been solely those from the hydrogen atom. Since these would initially be uniformly distributed over the whole volume of the realm, then their emissions would be received at any realm location in an isotropic manner. A location within the realm would be in continuous receipt of these emissions from 'early atoms' that were progressively further away. Due to the vastness of our realm, and despite the evolution of its mass into proton giants, stars and galaxies, all locations should still be receiving these early hydrogen emissions. It must however be remembered that these early emissions would have a much lower frequency than that given out by the hydrogen atom of today. We consider the frequency difference to be quite large and represented by the principle of Fig 4.21. This is now shown in Fig 9.11.

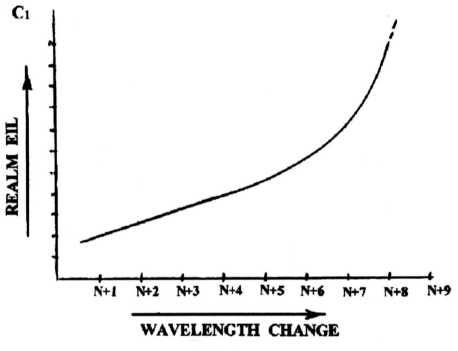

Fig 9.11

It is clearly observed that there is a limit to wavelength changes with increasing EIL. At the other end of the reducing EIL scale one would expect a similar limitation to apply, with the above curve taking on a considerable downward droop. We leave the reader to consider that aspect. Now if we are able to correlate the current hydrogen frequencies to those arriving from the realm, then perhaps we could speculate upon the propagation distance of those emissions. To do so would be to determine the age of the circle 6 phase and also our approximate EIL status on the Fig 9.11 curve. The next task would be to correlate Fig 9.11 to the circle 6 energy medium curve. It is very unlikely that we could meet with any success using current technology. However the future is another matter once we have defined and measured the EIL parameter. The essential factor is that we are aware today of our requirements for the future. Since these isotropic emissions are from progressively increasing distances, then our observed results must show this as a continuing change (by day, week, month and year). Perhaps these emissions have a connection with the cosmic microwave background radiation discovered by Penzias and Wilson in 1964. Since these are frequencies well below the visible range, it may be extremely difficult to obtain a conclusive wavelength shift relationship with today's hydrogen emitted frequencies. Earlier in chapter 3 we stated that the quantum-step size (10^{-35} metres) increased marginally with the distance propagated. However since this would result in a corresponding increase in both wavelength and propagation velocity, the observed frequency would remain constant. We may thus ignore the effect of propagated distances upon the wavelength and frequency.

Let us now consider whether the vast amounts of data received from radio galaxies and also from what have been termed as QSOs (quasi stellar objects) can provide us with some indication of the present age of our realm's circle 6 phase. All of the red shifted light spectra that we observe today must have been emitted by large galaxies from very great distances away. We do not know how long it took for all the observable galaxies to form, but we must speculate that this must have occurred some time after the end of the circle 5 phase and well into the current circle 6 phase as described on page 61. Light from many of these galaxies are now received by us though it is probable that the light from some of the more distant galaxies (in the further areas of the realm) has yet to reach our grid location. Others may be too faint to be observed by today's instruments. Nebulae with very large red shifts became known as QSOs and are considered to be very far away in cosmic terms, and are supposedly receding from us at tremendous velocities (upto 80% of light velocity). This was based upon Hubble's Law and upon the Doppler Effect being imposed as an explanation of the red shifts. Thus if the red shift of QSO 3C-273B at 0.158 obeyed Hubble's Law, then it should be at a distance of about 500 megaparsecs from Earth and receding at nearly 16% of light velocity (1 parsec being approximately 3.5 light years). Also since this object is visible to us with an apparent brightness of the thirteenth magnitude, then its absolute brightness must be tremendously high at about 100 times brighter than that of the brightest known galaxy. QSO 3C 48 has a red shifted spectra of 0.367 and so must be even further away and receding even faster. Over 1400 QSOs have been observed and with red shifts that go as high as 3.53. Subsequently instead of the classical Doppler-shift relation as in Hubble's observations given by,

$$Z = V/_C$$

it was now considered more appropriate to regard the red shift as a special relativistic Doppler effect given by,

$$1 + Z = \text{✩ } (C + V)/(C - V)$$

The red shift thus tends towards infinity as the recession velocity approaches C. For a red shift of 2.012 for QSO 3C 9, the recession velocity V works out to be about 80 percent of C.

In chapter 8 on velocity effects we showed that mass objects with very high realm velocities must absorb large amounts of energy medium in the quantity EA. This must then result in a large retardation impulse which will cause the high velocity object to slow down. As such we cannot accept the receding velocity interpretation for the observed red shifts. We also find it difficult to accept the estimated distances for QSOs for two reasons. Firstly, there must be a limit to the absolute brightness of any cosmic body and hence the distance at which it can be observed. And secondly on the principles of refraction. When the location of a QS0 is identified by its radio frequencies and later by a visible object at the same astral location, we consider that if the object was actually as distant as Hubble's Law presumes then the radio and optical frequencies should refract by differing amounts (as they propagated through the varying realm grid gradients). It would thus be extremely rare for the radio and visible images of such a distant object to appear to come from exactly the same sky location. If we accept that wavefront propagation in space follows a curved path (Fig 3.51 and Fig 3.52), then different frequencies must refract by differing amounts. The greater the distance propagated the greater is the difference. Therefore we contend that where radio and visual images coincide, the object cannot be as distant as was presumed by Hubble's Law for large red shifts. However, in order to explain such large wavelength shifting we consider that the QSO in question must be located within a very high EIL zone. The effect upon emitted wavelengths would then be according to the graph of Fig 9.11. One of the clues to be had from the observed red shifts is the probable range of the variation in the Mp event cyclic rate. In chapter 5 on proton time we then assumed a 'reasonable' range of variation in the Mp event cyclic rate with EIL (Fig 5.8). It may be that the Mp event cyclic rate varies by considerably more than we had at first presumed. This would mean that Proton Time, propagation velocities and gravitation impulses would also have a much greater variance than had been considered initially.

If high EIL zones do exist within the realm, then one of the probabilities is that these could be due to the presence of proton giants (black holes). These were defined earlier as having formed prolifically across the expanse of the realm as massive Mp-congregations (approximately to the size of our solar system) during the circle 4 phase when the realm EIL approached C_2 levels. Later in the circle 6 phase a proton giant could also form from the gravitational collapse of a dying massive star (10 times or more larger than our sun). These are considered to possess EILs of C_1^+ near their surfaces. The distance at which C_1 initially occurs is called the event horizon - because no events can be observed therein. Wavefronts cannot propagate in EILs of C_1 or above as described earlier. In the vastness of the realm, proton giants are to be considered no different in general properties than the ordinary 10^{-14} metre sized proton with its high EIL surface grid (Fig 4.6). As discussed in chapter 4, protons tend to repel one another if they approach to within the distance X as represented in Fig 4.24. By the same principle the I-Proton strings within the nucleus are repulsed by one another (Fig 6.27) and thus maintain the configuration of the nucleus and hence of the atomic shell. Therefore a grouping of proton giants could exist near one another without the tendency of being drawn into a single large entity. Obviously the larger the proton giant the further out from its surface will be its event horizon.

It is possible that some parts of the realm entered into the circle 6 phase before others, and so may be considered older (in realm date terms). The stars and galaxies in this older sector would thus have had much longer to evolve and many stars would have collapsed into proton giant masses. The age of a realm locality could thus be reflected in its population of dead stars and proton giants. With such an extensive proton giant grouping, the general realm EIL of the vicinity would be relatively high. Galaxies with their super-massive proton giant cores that drifted into such a zone would have Mp-shell oscillations at a slower rate and emitted lightwaves of a lower or red shifted frequency. We on Earth would eventually observe this light and correlate it with our own currently emitted frequencies but with the red shifted factor as the difference.

In chapter 1 we presented the concept of high EILs being produced by the multiple phasing of several realm entities (Fig 1.5). Perhaps the EILs within our realm are still maintained in part by such a continued action. If such be the case, then there are bound to be some zones in existence that are at extraordinary high EILs. One of the observed peculiarities among QSOs is that a red shift of 1.95 appears far more often than would be expected by chance, and as though it were a standard value for some existing phenomenon. In our theory we would contend that this is simply an indication of the prevalence of a defined realm EIL much higher than that which we experience in our locale, and which must occur over a rather large realm region. Earlier we had assumed that the observed red shift phenomena had a special significance which would have eventually assisted us in establishing the true age of our realm. However, we must now conclude that red shifts cannot provide us with the necessary means for gauging the age - through time distance calculations - of remote star systems. We do however see considerable potential in the correlation of the realm's background radiation (microwave or otherwise) with the hydrogen frequencies as they occurred near the start of this circle 6 phase. If we could predict the emission behaviour of the hydrogen atom near C_1 EILs, then perhaps our correlation data would indicate the differences in emission pattern with today's atom. We could then speculate upon the drop in EIL between then and now and so perhaps obtain an estimate of our current circle 6 status.

Another indicator that our current position on the circle 6 energy medium curve has progressed quite considerably is given by the pattern we observe in the configuration of all the galaxies visible to us today. A galaxy would have formed when stars were drawn towards its proton giant core in the standard 'cartwheel' rotational effect. Stars are basically composed of the hydrogen element fusing into helium atoms. Since the Mp-shell of the atom can only exist in environments that are below C_1 EILs, then galaxies could not have formed prior to the circle 6 phase of the realm energy medium cycle. However, as the mass of the proton core has continued to decay, the attractive impulses have reduced and the outer stars have gradually drifted to wider orbits. As such they are unable to keep up with the 'cartwheel' velocity and so begin a trailing spiral position. We observe most galaxies to be of this spiral variety which thus indicates to us the elapsed time in the circle 6 energy medium cycle of events. The current configuration status of a galaxy could be an indicator not only of its own age but also that of the elapsed time of the current circle 6 energy medium phase.

In chapter 5 (Fig 5.2) we defined the age of a proton by the quantity of energy medium that it had lost since it was at TMax. Since the decay rate is a variable, then protons under different influences would age differently. It is therefore highly probable that protons of varying ages co-exist near one another. We have to progress a long way before we become technologically capable of measuring the energy medium content of individual protons. However, when we do acquire this skill, then perhaps we could build up data on the ages of single atoms (primary) found in remote areas of the realm. We could assume that these hydrogen atoms or ions had never been molecularised and had been under no influences other than the progressively diminishing circle 6 phase EILs. We could then estimate a particular TN-line for their decay rate and subsequently evaluate an age from their measured energy medium content. From such data we could then obtain a fair estimate for the age of our circle 6 phase realm cycle.

During the course of this text we have attempted to convey to the reader the complexity of effects within the realm energy medium entity. Yet all these events are compounded from a simple origin that focuses around the Mp event cycle and its resultant E-Flow characteristic. As such we consider our 7-Circle Theory as a universal unified theory applicable to all activity within our realm, not only as it is today, but also as it was in the past and will be in the remote and distant future. This theory therefore allows us to predict the conditions that will occur within the realm over the entire period of its energy medium life cycle. Although we treated each phenomenon aspect on an individual basis, we nevertheless did highlight their correlation and inter-dependence. There are of course many questions that remain unresolved. In a vast and complex configuration of realms (as must surely exist) we could never hope to achieve the completeness of knowledge that we so desire. In the preceding chapters our topic discussions have been focussed and essentially introductory in nature. This was done to convey the overall concepts of the theory with a minimum of diverted description. Our theories are based upon logical considerations and cannot at this time be either proved or disproved. However, we do not expect to be absolutely correct in all of our energy medium hypotheses, but we do nevertheless consider it to be an exciting alternative to both the Big Bang and Steady State Theories. Exciting enough to start a whole line of work especially when we consider its relevance for progress towards the future of the human race. We can think of no better way to conclude this chapter than to present in outline some of our expectations from the 21st century and beyond.

If our theories are found to hold true then the potential for developing and applying its principles are enormous. We would initially expect a re-definition of the extent and volume of the Universe. This would include all other realms within the cosmic void including the hypothetical Mother Realm, thus making our new universe concept very much greater than we had at first supposed. Although we cannot as yet confirm the locations of these other realms, we must nevertheless accept that their existence has a certain level of probability.

As man ventures into different EIL regions of the realm his consciousness and understanding of Proton-Time would be enhanced. We already know that event duration -Time - only became a measurable quantity at the start of the circle 6 phase of the realm's energy medium cycle. As such any reference to the period 'before Time began' simply refers to the energy medium cycle prior to the circle 6 phase. Similarly the reference 'till the end of time' refers to the end of proton decay as we know it. In the future we would expect a more exact definition regarding the beginning and end of Time.

The physics of today does not as yet recognise the existence of a subtle ether type energy medium. When it finally does then there is bound to follow a surge of ideas and new concepts on such a scale that will quite revolutionise the machines we employ. What an early leap into the future man would take if only a few of the concepts presented in this text were proved to be correct. We would then find that we had at hand a near limitless source of power just waiting to be tapped. The inventor who perfects the E-Flow generator or the unilateral magnetic impulse system (Fig 9.8) will have achieved a tremendous breakthrough for man's future. Humankind would have acquired a new 'wheel' that permits him to leap out at the stars and distant galaxies with an ease that would appear frightening today. Interplanetary travel could then be quite a normal daily occurrence. Just like commuting to work in a nearby town. Fossil fuels and the machines that depended upon their use would be retired to the museum. The same would be true of nuclear power units as these are simply used as an indirect heat source. Nuclear energy would thus be redundant along with its lethal waste and disposal problems. Coal and oil would simply be considered as a mineral and useful ingredient for the chemical and synthetics industry.

The gravity wheel principle of Fig 9.5, when developed as a prime-mover would ensure the very cheap production of electricity. The entire world could then operate on the basis of a welfare state with electricity supplied at a nominal price. There would probably be a dual form of supply. A very high frequency supply for heating purposes and another more suitable for the appliances of the home - washing machines, refrigerators, mixers, etc., etc. Both forms of supply would perhaps enter the home through the same cable. Eventually stimulation of the nuclei of certain materials would cause electrons to flow into conductor wires (Fig 7.29) without the need of rotating magnetic fields. Each home could then possess its own independent and seemingly everlasting electric power supply. This may need to go through a process of re-configuration in order to suit the exact requirement of the utilities in each home.

Public transport and vehicular systems would be wholly revolutionised away from the conventional wheel. These would float through the air above ground and without any crude blast of thruster jets or apparent reaction effect (Figs 9.8 to 9.10). We should literally be able to build and maintain our 'castles in the air'. Large city islands could be floated to different locations at will and for reasons that are climatic or even political. Imagine an entire city of several square miles being built upon a framework for mobility. A very large number of E-Flow sources would have to be distributed uniformly over its upper surface in order to provide the necessary EIL gradient in the upward direction for liftoff. However these E-Flow sources would have to be used restrictively as they could cause minor changes in the Earth's orbit, especially if used over extended periods of time. There would of course exist a world governing body to monitor this and cause the appropriate adjustments of Earth orbit to be made as and whenever necessary.

Roads and motorways as we know them today would be obsolete. A very different highway code would prevail to maintain order in the skies for commuting craft. Streets in towns and cities would become pedestrian ways and simply a means for address identity. Parachute type devices would not be necessary since collars or belts containing unilateral impulse units could be worn that would allow the wearer to float gently down to earth. More elaborate flying suits could of course be made to enable the wearer to float upwards as well under some degree of control.

Travel to and from the earth would be easily achieved, and man could be made to feel quite at home in space. Living in space would of course require an earth type gravity to be ever present for the well being of its occupants. Craft would need to be in a state of constant acceleration and so would need to operate in a manner that can be compared to our current atmospheric flying machines i.e. aeroplanes. With unilateral magnetic impulse devices this could be achieved quite easily to provide a constant one G environment. Astronauts would of course be trained to 'fly in space' to best effect for their own and their passengers' comfort. A space academy institution would be essential to achieving this skill.

Since the E-Flow source acts equally upon both craft and occupant, then acceleration levels do not limit manoeuvrability and so can be exceedingly high. Navigation between nearby star systems would be by a built-in directional computer program. We assume the computer to be of the 'thinking type' and with a near infinite memory capacity. Journey times would be a matter of choice, though proton shielding for the very high velocities would probably not be available to the smaller private craft. However for very long distance travel it may be possible to 'hire' suitable shielding. This would have to be fitted by professionals and could only be carried out in the weightlessness of free realm space.

To some degree man of the future should find it possible to also 'play' with Time - the pace of event duration as discussed in chapter 5. We should be able to utilise the principles of proton time to slow the pace of life at selected locations on earth. By the use of symmetrically located E-Flow sources around a person it should be possible to setup a relatively high EIL zone and so cause a slow-down of the Mp event cyclic exchanges in their atomic structure. By this means some humans could extend their living into a more distant future - for whatever reason. There could of course be no reversal of events. Man could never re-visit the past. Proton Time can be slowed or speeded up but not reversed.

It is probable that the most valuable material of the future would be Neutron Star material. One of mans' chief concerns would be that of its manufacture under laboratory conditions. Even with the advanced technology of the future, man would not find this easy to achieve. Consider machine tools and drill bits having an edge made of this material. They would cut through ordinary metals as easily as a knife through butter. Industrial practices would be revolutionised. When man finally achieved the ability to manufacture this material, he could then use it to build the necessary proton shielding for his spacecraft. This would then permit man to attain hyper light speeds without endangering either craft or its occupants (Fig 9.6). Mankind could then truly migrate to the far reaches of the realm. Incidentally, the construction of these spacecraft would require large building yards on earth somewhat similar to the ship-building facilities of today. Any size and shape so constructed could easily be floated up into space using multiple E-Flow sources or unilateral magnetic impulse systems. The proton shielding would have to be attached later when the craft was in a gravity free environment. Large realm traveller trains, many miles in length, could be assembled in space and require only a single proton shield.

Technology associated with the manufacture of the proton shield as represented in Fig 9.4 would also enable man to design and manufacture new elements with nuclei configurations that offered unique properties. The current periodic table for natural elements would have to be supplemented by a second table of synthetic stable elements. In chapter 6 we presented the reader with a format of nucleus structure and its related Mp-shell configuration. We expect the future to build upon this theme and to evolve a science that can predict the properties in elements for any given nucleus design.

On the earth we would expect there to be a total control of weather conditions by the creation of 'low barometric pressure' zones. This could be implemented by using the principle of the uncovered proton shield at the required location. We do not expect there to be any arid or desert type areas on the earth since large volumes of water could be transported or pumped there with relative ease. With the proper directional use of E-Flow sources man could cause rivers to literally flow upstream and hence to any desired location. As such water channels need not have to rely entirely upon the earth's gravitation for flow direction. Since there would be no major emissions into the atmosphere - from factories or vehicles - the overall environment should be much cleaner than it is today. The carbon dioxide levels would be accurately monitored across the globe and controlled by man to a required level, as would all other aspects of the atmosphere, We would expect the ultraviolet shielding ozone layer to be either reinforced or replaced by an equally effective equivalent. In colonising other suitable planets, we would expect man to alter their atmosphere conditions to that like on earth. The planet Venus could be the first to undergo such a transformation. It may have to be repositioned into an earth type orbit but at some distance from the earth itself - perhaps on the opposite side of the sun. Its rotation and axis inclination could also be made earth-like so that eventually it became a home from home for the future colonists from earth. In time other suitable planets in distant star systems or even other galaxies could be so converted to suit mans' needs. Yet we consider that the earth will always remain as the focus for events in the realm, and as man spreads out into the vastness of the 'New-Universe' he would feel an irresistible urge to make a pilgrimage to the planet of his origin.

Mans' birthright is the earth and hence he will always safeguard its welfare jealously. When the sun's nuclear fires begin to diminish, man will probably cause the earth to be relocated in an orbit around a much younger star. As we approach 'the end of time' suitable realm EILs may only occur near to the many proton giants that would still exist. Obviously, the earths of that era would all have to be brought into suitable orbits near these high EIL zones. Our expectations from the human race of the future need know no limitations since man will always meet challenges with ingenuity and skill.

We would expect the proven 7 circle theory concepts to be taught in schools where children would acquire an early appreciation of its basic principles. As our technology developed in all spheres of learning we would expect the teenager of tomorrow to know more in his specialised field than the comparable university graduate of today. If our young can dream of jumping at the stars with the ambition of developing and colonising new worlds, then their potential for achievement will be enormous. We would of course expect scientific and engineering practices to keep pace with the new technological advancements. Realm navigation would be aided by extremely powerful computers able to generate additional memory capacity whenever required. Perhaps these computers would be programmed to consider all the alternatives in any situation - even if the eventuality had never been entered into their memory. These computers should be able to observe and virtually to think for themselves, record all this to memory and then discard all data deemed unnecessary. This discarded data could in fact be beamed at some interim data storage device based on holographic principles. In medicine we would expect similar powerful computers to aid in the diagnosis of ailments and formulation of the proper immunisation factors. We would expect artificially grown cell tissue to be the norm and which causes the healing processes to be speeded up considerably. Imagine someone with a fractured leg one day and a perfectly healed bone in 48 hours. Transplant surgery would then be a technique of the past and quite unnecessary. With most body organs also being renewable, we should be able to extend the life span of man more or less indefinitely.

Finally we expect the human brain to develop a consciousness that reflects the advances in technology. This development could be unique. Earlier in chapter 7 we discussed the aspect of forcephoidal emissions from current carrying conductors. Since the human brain and body neuron network function with the aid of tiny electric currents, then similar forcephoidal emissions should also occur from these. With natural development and recognition growth, it should be possible for a second human brain to pick up these emissions and accurately interpret them into mental images. Thus one person should in theory be able to 'see' what another person is thinking. Today we seem to come upon this talent of telepathy on rare occasions only. In the future we would expect this to develop into an art form that is easily learnt and which becomes the main form of communication between humans. Since this would accurately convey emotions and intentions in a complete and precise manner, then language barriers should no longer be an obstacle to effective communication between differing cultures. With the progressive development of telepathic skills it may be possible - in time - to tune into another's thoughts at increased distances. It is however considered that speech, reading and writing would remain an essential part of our communicating art.

In the next progressive step we would expect the human brain to develop even further to produce electric currents along defined intricate paths. These would then set up nodal fields at a distance to either attract or repel particular objects located there. We would expect a continued improvement in the art of telekinesis in the future. Perhaps this progress could be aided by means of a special learning program. As an example let us assume that John, aged five, is considered suitable for telekinesis tests and training. He would be taken to a testing centre where his brain emissions could be analysed. He is then positioned in front of a large number of material objects that are on floats (on a liquid surface) and is told to attempt to move them using his telekinetic ability. By means of delicate sensors it would be possible to ascertain the object or objects that are most affected. Once the material type most responsive to John's brain pattern had been ascertained, the next test would be for a range of object shapes and sizes in that material to be located on the floats in front of him. Once again the delicate sensors should show which object was most influenced to movement. This object then becomes John's constant companion for the next few years and he must continually practice his telekinetic influence upon it. Eventually John will be able to move this object at will and with considerable ease. Further testing may result in minor changes to the shape, size or material of the object. We would expect a much older John to make good use of the impulses induced in such an object. He could use it to push or pull something that was on wheels or even in the air. He could make it lift his own weight by placing it in a pouch that was harnessed to his body. There are perhaps many other uses for such an object. All of this is speculation on a grand scale and it is hoped that it can be considered by the reader as a possible feature in a far and distant future.

A great many unexplainable occurrences have been reported in our world of today and we do not intend to discuss them here. However as our knowledge of the Universe increases we expect to find some of the answers. As we venture further out into our realm we are bound to come across other occurrences that puzzle us even more. At our present technological level there is no means by which we can even remotely anticipate the end result of mans' advancements. There are some indications today that the brain's sensory perceptions can continue outside its physical boundary under certain exceptional conditions, and which has been referred to as an 'out of body experience'. If such a condition could be controlled at will and maintained over an indefinite period, then we may call into question the necessity for mankind to continue forever in his current mass-body-atomic structural configuration. Perhaps there are even other alternatives that may yet be considered. We shall leave all of that to the readers' imagination.

Chapter Summary.

1. The normal E-Flow velocity V_F is considerably greater than the velocity at which lightwaves propagate.

2. As the proton ages and reduces in mass, its net E-Flow emission rate will diminish. Subsequently, gravitational impulses induced in other protons will decrease with time.

3. The greater the total number of protons within a nucleus the higher is its EIL environment and the slower will be the ageing process of those protons.

4. Unilateral gravitational impulses can be generated by E-Flow sources which produce EIL gradients artificially. Mass A is attracted towards the E-Flow Source B, but B is not attracted towards A. By this means artificial gravity may be created inside a spacecraft.

5. It is possible to shield an object from another's E-Flow by means of a single layer of protons arranged in a sheet configuration. This sheet material is similar to that found in Neutron Stars.

6. Proton shields must be fitted to the upstream face of all spacecraft if high velocity realm travel is intended, in order to prevent absorption of excessive EA quantities by both spacecraft and its occupants.

7. The search must begin for synthetic materials that are able to produce unilateral magnetic type impulses.

8. Navigation through the realm should be possible using visual as well as grid gradient charting. EIL gradients result from E-Flows which travel in absolutely straight-line paths, and so are a more accurate means of indicating the true grid position of other realm masses.

9. Communications within the realm cannot rely upon conventional lightwave velocities as this would be too slow. Consideration must be given to the use of E-Flow blocks as a speedier means of sending messages across the realm.

10. We cannot rely upon red shift data in order to establish our realm's current circle 6 status or age. We must instead be able to evaluate the total energy medium content of single protons that are known to have existed from the start of decay under particular TN-line conditions in order to compute their age and subsequently the age of our realm. We must also consider that the earliest emissions are from the primary atoms that prevailed universally and that these emissions would be received at all locations in an isotropic manner. With time these emissions would arrive from progressively further afield, and measurement of their arrival status today should indicate a travelled distance and hence an elapsed period.

11. Mankind of the future should be able to harness the energy medium as a source of near infinite power, and will also be able to play with Time - the pace of event duration. Travel would be revolutionised and would be mainly above the earth and at phenomenal velocities. One of mans' most important materials would be that of the Neutron Star. When he is able to manufacture this then he will be able to create new synthetic elements designed to give only desired properties. Every technological field would in time benefit from this.

GLOSSARY OF TERMS.

ABCD Diagram: See Shell Diagram.

Absolute decay line. TN-line: Graphical representation of the optimum rate of proton decay. This can only occur when a single proton is in a void region and is under no influence from any realm EIL or E-Flows from other protons.

Absorption spheres: When a Mp or Mp-event occurs, it has a tendency to absorb all the energy medium within a defined spherical volume. This volume is termed the Absorption Sphere.

Additional Energy Medium Absorption. EA: During the imploding phase of the E-M/X sequence, any external E-Flows that enter the contracting mass medium are also absorbed into the Mp. This energy medium quantity is additional to that in the absorption sphere and is denoted by EA.

Age of a proton: This is defined as the quantity of energy medium that a proton has lost in decay since it was at its optimum mass TMAX.

Alternating current: A flow of electrons in a conductor loop which periodically reverses its direction.

Astrophysics: Subject dealing with the physical properties of astronomical objects.

Atomic clock: Duration measuring device making use of the oscillations in atomic nuclei i.e. caesium. These clocks have an extremely high level of accuracy, about one part in 100 billion.

Centralising impulse: The inward E-Flows that result from the repetitiveness of the Mp-shell of an atom cause a centralising impulse to be exerted upon its proton nucleus.

Contraction velocity: The rate at which an E-M/X absorption sphere volume implodes into a Mp.

Cosmic void: Inter-realm area of space that contains 'nothing'. It is devoid of all property or medium. Light cannot propagate through a void, and space or time can have no meaning in such an area.

Cosmology: Subject dealing with the large scale structure of the Universe, its origin and evolution.

Cosmos: This may be defined as that infinite volume of space within which we exist and which has no beginning and no end. It extends beyond human comprehension and is even larger than imagination allows. It is estimated that over 99.9% of the Cosmos is Void.

Critical energy medium levels: There are three critical EILs for the energy medium. These are,

C0. A minimum EIL below which the energy medium cannot reduce.

C1. A critical EIL at which energy medium sublimates into a mass medium and the level at which Mps form.

C2. A maximum EIL which the energy medium can attain. At this level all energy medium converts to a mass medium.

Dark Matter: See Energy Medium

Decay line. TA-Line: Graphical representation of the actual rate of proton decay in a particular realm environment.

Domino effect: When electrons from a stimulated nucleus propagate across to an adjacent atom's nucleus and cause it to be stimulated in turn to emit further electrons in an approximately similar direction, and so on and on, then the phenomenon results in a flow of electrons or electric current and is termed the Domino Effect.

Doppler effect: The apparent frequency of a moving object depends upon whether the observer has a relative velocity towards or away from the object. If towards, then the frequency appears higher; and if away the frequency appears lower. This is termed the Doppler Effect of observed

frequencies.

E-Flow: Energy medium outflow from the Mp during its M-E/X phase. A hypothetical layer by layer expulsion of the mass medium of the Mp results in a fairly uniform emission of energy medium at a constant relative velocity VF.

E-Flow Generator: Artificial device to produce an E-Flow from a non-mass source; the intention being to create an EIL gradient and hence artificial gravity.

EIL: Energy medium intensity level. An indication of the concentration of the energy medium. See also Critical Energy medium Levels.

EIL Gradient: The rate of change of EIL from point to point in free realm space. The same EIL can be represented by grid lines or curved planes within the realm EIL grid to show up a pattern of EILs and EIL gradients similar in format to the contour lines on a geographical map.

EIL Nodes: See Nodes.

EIL Oscillation: This occurs when the energy medium at a set plane undergoes a repetitive series of E-M/Xs and M-E/Xs. This is a result of the Mp-event and is indirectly responsible for the propagation of lightwaves through free realm space.

Electron: An Mp-Cluster emitted from the equatorial belt zone of an I-Proton, and consisting of a series of parallel wavefront sections spaced approximately one quantum step apart. Propagation is according to the principles laid down for wavefronts.

Electrostatics: When a body is made attractive by rubbing or friction it is said to be electrified or charged. The study of this type of phenomenon is called Electrostatics.

E-M/X: At or above C_1 EILs the energy medium converts/sublimates to a mass medium which then implodes into a Mp. This 'energy medium to mass exchange' is abbreviated E-M/X, and is the basic functional relationship between a mass entity and its convertible energy medium content.

Energy Medium: This is the 'energy' portion referred to in the law of interchangeability between mass and energy. Mass is made from concentrations of the subtle medium 'energy' and as such 'energy' must be considered as a medium having property and form. The energy medium is to be considered as a perfectly homogeneous entity that has no structure or individual particles. It is the subtle medium that the ancient philosophers speculated upon to explain the activity of Nature. It is indivisible in its volume and yet an infinitesimal part of it can be absorbed into a mass particle. It has varying concentration levels. It can also be referred to as **Dark Matter**

Energy Medium Phasing: The phenomenon in which two or more energy medium volumes pass through the same space at the same instant is termed Energy Medium Phasing. Each energy medium volume continues in its original EIL status without reaction to the other. When each leaves the shared space, it continues as it did prior to phasing. When referring to very large sections of energy medium such as a realm, then the above phasing phenomenon is termed Realm Phasing,

Energy Medium Realm: This is an expansive volume of energy medium moving at a set velocity thorough the Cosmic Void. Each energy medium realm travels in an absolute straight line path and is separated from other such entities by the vast distances of the Cosmic Void. Each has a defined volume, defined relative velocity and defined levels of energy medium intensity within its volumetric boundary. These entities are termed Energy Medium Realms or simply as Realms.

E-Proton: In the atom nucleus large numbers of protons are bonded to one another and form proton strings or chains. As such, protons at the end of a string are called End-Protons or simply E-Protons. Protons that are held between other protons are called Intermediary-Protons or I-Protons. E-Protons give rise to Mp-shell sections but I-Protons do not. However only I-Protons can be the producers of electrons.

Ether: A subtle inter-elemental medium introduced by man in his attempt to explain observed phenomena. In order to rationalise certain unknown factors within a problem or observed phenomenon, the physicists, chemists and theologians of the day found consolation in the presence of an unquantified, vaguely characterised and often obscure medium upon whose action the particular behaviour was wholly related. Comparisons between the varying mediums could not be performed easily since each applied itself to a different problem.

Euclidean Geometry: Geometry based upon Euclid's postulates, as opposed to Non- Euclidean Geometry.

Event Horizon: The distance from the surface of a proton giant at which a zone of C1 EIL or greater prevails. Lightwaves cannot propagate in such a zone and as such events taking place within it are not visible to an outside observer.

Event Time: Under varying conditions of EIL our concept of the duration of events will always appear to be the same. This concept of event duration is termed Event Time. A function of the TA-Line is to indicate the variation in Event time rates at differing EILs.

Forcephoids: The E-Flows from Mps on the proton surface do not occur in a random manner but rather in a set sequence from defined positions. This results in an overlapping of E-Flows causing EIL peaks to occur as outward traversing spheroids/waves. A series of these EIL peaks exist around every proton in decay, with peak EILs that diminish with distance traversed. These spherically shaped EIL peaks are termed Forcephoids. The outward velocity of these Forcephoids is the same as the E-Flow velocity VF.

Frequency: The number of occasions that a repetitive event occurs in a given time. The measurement of true frequency is a virtual impossibility because of the variance pattern in the proton time rate by which we gauge all event duration. Our concept of the velocity of wavefronts and their frequencies is therefore a relative one.

Generator Block: An isolated conductor block possessing a plane of stimulated atom nuclei that produce an emission of electrons in the same direction. This is also called a voltage generator block and has the capacity to send a flow of electrons or electric current through a closed conductor loop.

Gravitational Impulse: When a proton exists within an EIL gradient its centrally directed impulses are upset causing a resultant impulse in the direction of the higher EILs. This occurs in all the protons of the mass object and the net impulse is called gravity. Since all of the realm masses emit E-Flows, then EIL gradients exist in the realm space around them causing each to induce attractive impulses within one another's protons. This mutual attraction is termed Gravitational Attraction.

Gravity Wheel: This can be compared in principle to a simple waterwheel in which the weight of water pouring into the buckets results in one half of the wheel becoming heavier than the other thus causing the wheel to rotate. In a gravity wheel the same result can be achieved by making one side of the wheel permanently lighter than the other. This is done by positioning a proton shielding under one half side of the wheel. As such the gravity wheel could become a nearly inexhaustible rotating power source.

Hubble's Constant: The ratio of the apparent velocity of recession of a distant galaxy to its distance. Its present estimate is about 53 Km per second per Mega-parsec (obtained from C times the Red Shift).

Impulse: This is defined as the rate of change of the Linear Momentum of a mass body. The momentum being the product of its mass and velocity. Linear Momentum is a vector quantity, the momentum being in the same direction as the velocity.

Inertia: The property of mass that characterises its resistance to a change of state of rest or of uniform motion in a straight line.

I-Proton: Intermediary proton within a nucleus proton string. See E-Proton,

Isotropic: Having the same property in every direction.

Life of the proton: The total decay duration of the proton is termed as its 'Life'. The Life of one proton is not necessarily the same as that of another since the duration depends upon the rate of decay – which is variable with realm EIL.

Light Speed: Realm velocity of an object equal to the propagation velocity of lightwaves.

Light Year: Distance traversed by a lightwave in the period of one year and is approximately equal to 9 million, million kilometres.

Linear Momentum: The Linear Momentum of an object is the product of its mass and velocity and is in the same direction as the velocity. The Linear Momentum of an entity remains constant until Linear Momentum from another entity is added to it.

Mass Point. (Denoted Mp): This is the most basic mass entity within the realm that is formed when a section of energy medium attains C1 EIL or greater. The energy medium converts to a stretch-strained mass medium. This results in an implosive effect which causes the mass block to become sectionalised with each section imploding into a Mass Point with a life of 10^{-43} seconds in Planck time.

M~E/X: This is the 'mass to energy medium exchange' process that occurs when the Mp reverts to energy medium in the form of an E-Flow. The mass to energy medium conversion is reflected in the formula $E = MC^2$.

MTU: Mass point Time Unit. The duration of one Mp-cyclic event.

Mp-Cluster: See Electron,.

Mp-Congregation: Profusion of Mps formed in the Circle 4 and 5 phases which have formed a compact group held together by centrally acting impulses which can result in the formation of the Proton.

Mp-Shell: This is the sphere of Mps in perfect cyclic synchronism that forms repetitively around the proton at a set distance. The Mp-shell and its proton nucleus are together the basis of atomic structures within the realm. For multiple proton nuclei the overall Mp-shell occurs as one shell section per E-Proton, thus giving each atom element its unique shape and property.

Maximum Energy Radius: The distance at which the E-Flow from a mass particle reduces in EIL to the critical level Co.

Molecular Bounce: When two atoms are bonded together then the repetitive character of their Mp-shells in contact results in alternative attractive and repulsive impulses between them. This phenomenon is termed Molecular Bounce.

Mother Realm: A hypothetical gross energy medium entity within the Cosmos from which all other lesser realm entities (such as our own realm) emanate.

Neutron Star: Star composed of matter in the form of I-Protons or Neutrons. Such stars have very high densities similar to that of the inner zone of a nucleus consisting of I-Protons compacted together with only small interstitial spaces - like marbles in a bag. The overall mass density is marginally less than that of a proton.

Nodal Field: The zone of EIL Nodes arranged in a set pattern near a mass structure.

Nodal Rows: Successive rows of EIL Nodes within the Nodal Field. Nodal rows are numbered from the surface of the structure of origin and then outwards with increasing row number.

Nodes: High EIL peak points caused by multiple forcephoidal intersections. These recur at the same location at a frequency dependant upon the arrival rate of traversing forcephoids.

Non-Euclidean Geometry: A geometry based upon postulates differing from Euclid's in one or more respects.

Nucleus: The central mass of the atom consisting of one or more bonded protons. In the more complex atoms the nucleus may consist of an inner more compacted zone of bonded I-protons and an outer zone consisting mainly of radiating proton-strings.

Nucleus Orientation: The atoms of a mass object are bonded to one another at their Mp-shells in a preferential manner. The centralising impulses therefore maintain the nucleus (which has an eccentric configuration) in a set directional position. This directionality is defined as its orientation. External influences can be exerted upon the nucleus to alter or re-direct this orientation.

Nucleus Window: In general the string configuration in the outer zone of the nucleus is such that the proton-strings occur in an approximate radial format. This means that electron emissions are directed laterally towards other proton-strings. Thus, although there may be a profusion of electron emission within each nucleus, very few actually escape out towards the Mp-shell. The paths through which the electrons emerge from the nucleus zone are termed Nucleus Windows. The stimulation of the atom by external means can result in a considerable increase in the number of nucleus windows for extensive electron emission.

Parsec: A measure of realm distance slightly more than 3 Light-years. It is the distance at which a star would have a parallex of one second of arc (= 3600th part of a degree). A mega-parsec (Mpc) equals a million parsecs.

Pendulum Time: Imaginary measurement of event duration by association with an oscillating pendulum system.

Primary Atom: Atomic structure consisting of one proton surrounded by a single spherical Mp-shell. This is the most basic atom structure and is the Hydrogen element.

Primary Mass Body: Ferrite mass body with a permanent Nodal Field about it. A permanent magnetic body.

Propagation Diagram: The diagram representing the relative quantum step size of adjacent wavefront centres for the purpose of determining their propagation differences in an EIL gradient zone.

Propagation Plate: Small section of a wavefront used for explanatory purposes. Abbreviated p/p.

Proton: See Mp-congregation.

Proton Age: See Age of a Proton.

Proton Decay: Gradual diminishment of the total energy medium quantity within the mass of the proton.

Proton Giant: This is like a Neutron Star that has been compressed even further such that its density is the same as that of the proton. That is no more interstitial spaces. Proton Giants near the size of our entire solar system in diameter were formed extensively during the circle 4 phase of the realm energy medium cycle when EILs approached C2 levels. These then became the core around which stars accumulated during the circle 6 phase of our realm to form into galaxies.

Proton Shield: Neutron Star material repositioned into a theoretical sheet like structure with a width of one or two protons only. This is meant to block the passage of external realm E-Flows from causing a temperature rise in objects travelling at hyper light velocities.

Proton String: An open ended chain configuration of bonded protons in the outer zone of the Nucleus zone.

Proton Time: An automatic yardstick of duration measurement that is linked to the decaying aspect of the proton in the current Circle 6 phase of the realm energy medium cycle. The rate of decay reflects the rate of Time. See Time.

Quantum Step: This may be defined as the linear distance between the successive plate positions at which the EIL oscillations or wavefront is set up as it propagates through the realm. Can also be referred to as propagation steps. The quantum step is equated to the Planck distance of 10^{-35} metres.

Realm: See Energy Medium Realm.

Realm Datum: This is the EIL built up from all the E-Flows arriving and phasing at each and every realm grid location. Since all E-Flows arise from mass objects, we may consider the Realm Datum as the EIL status resulting from the combined phasing of all the local mass decay E-Flows.

Realm Grid: A 3-Dimension Map of the EILs within the realm. Realm grid planes or lines represent a particular EIL and successive planes would indicate the EIL gradient. The principle is similar to the 2-D geographical survey map with its contour lines representing heights and slopes.

Realm Gridology: The science and practical art of mapping the Realm Grid EILs and ensuring their accuracy with Time.

Realm Phasing: See Energy Medium Phasing.

Realm Velocity: The velocity of an object through free realm space relative to the local realm grid datum,

Red Shift: When an astronomer examines the spectrum of light coming from an extra-galactic object, he usually finds Emission or Absorption Lines - or both. However these lines do not always occur at the wavelength positions expected, but rather are found to have moved in a fixed ratio towards the longer wavelength side of the spectrum. The name Red Shift has been attached to this phenomenon because the colour red occurs at that end of the spectrum which has the longest wavelengths.

Secondary Mass Body: Ferrite mass body without any particular orientation pattern in its atoms. Hence this body has no initial nodal field around it. A non-magnetised body of soft iron.

Seven Circle Theory: A theory of the life cycle of the realm based upon its progressive energy medium intensity levels projected on a theoretical energy medium curve. This energy medium curve traces a graphical path through a Seven Circle configuration drawn between the three critical EILs of C0, Cl and C2. The principle of the theory is that realm EILs initially were caused to increase from C0 to C1 at which EIL mass formation takes place. The realm EILs continue to increase towards C2 when the more permanent mass structures are formed. Thereafter the EILs progressively reduce and the mass structures commence to decay. Ultimately the realm EILs reduce towards C0. Our current realm status is estimated as being in the early part of the Circle 6 phase.

Shell Distortion: This is the displacement that occurs to a portion of the atom's Mp-shell through external influences such as a collision with another atom.

Shell Diagram: Also termed the ABCD diagram which represents the positional repetitive phases of the Mp-shell.

Shell Section: See Mp-Shell.

Skin Effect: The internal reflection of the domino effect paths at any conductor surface results from an internally directed orientation of all its surface atom nuclei.

Temperature: This is simply a measure of the realm energy medium that is gained by the Mp-shell structure over and above its basic repetitive requirement. And which also represents the corresponding amount of energy medium that can be given up by it. Represents the rise in energy medium content of the Mp-shell.

Time: This is a non-entity function which specifies the duration of an event by comparison with a basic event sequence. Time is without property or form but in our definition we give it a basic unit of duration in the MTU, the mass point event unit.

Thought Experiment: This is not an actual experiment but an imaginary one in which the experimental setup is assumed without worrying about how this could be done in real life. The outcome and conclusions can then be determined from logical reasoning and known physical laws.

Unilateral Impulse: All mass objects are E-Flow sources and the impulse quantities generated in neighbouring masses are equal and opposite. If however a mass object A can be induced with an attractive impulse towards an E-Flow source B but without it being attracted in turn towards A, then a unilateral impulse may be said to have been induced in object A.

VF-Meter: Futuristic hypothetical instrumentation to detect and measure the velocity of E-Flows.

Voltage Generator: See Generator Block.

Wavefront: This is the plane of EIL oscillations originating at the Mp-shell and traversing outwards, and in which all the Mp-event sequences occur in synchronism. The wavefront is repetitively formed at quantum step intervals in the direction of propagation.

Wavelength: The distance between two successive propagation plates.

APPENDIX 1

Calculation of E-Flow Velocity

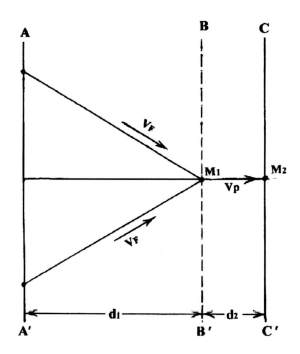

The E-Flow velocity is always assumed relative to the Mp from which it emanated. Any former velocity status of the Mp must also be directionally imparted to that E-Flow.

In a wavefront the synchronous E-Flows VF tend to intersect at a compounded forward point M1 at which a C1 EIL condition will eventually develop. This EIL is a compound phasing of all the E-Flows arriving at this position in addition to the prevailing realm EIL grid datum levels for that spatial location. Since the wavefront

E-Flows cannot on their own attain anywhere near C1 levels, then we consider that it is this realm grid that becomes the major factor for the achievement of C1 EIL. According to the 7-Circle principles we are currently relatively early-on in the circle 6 phase and so the realm datum is considered to be at a very high level – this being represented graphically to logarithmic scale. Subsequently, we would consider that at the position M1 in the above diagram when C1 EIL is achieved, that the realm datum is the main contributor of energy medium in the new wavefront that forms there. We consider this contribution to be in excess of 95% (and more nearly 98%) of the total wavefront energy medium content.

If the realm datum were to reduce (as it will in time), then wavefront E-Flows from a wider p/p zone would need to arrive at M1 to build up to C1 EIL. This would cause the quantum step to increase marginally to accommodate the phasing of all these E-Flows, which in turn would result in a reduction of propagation velocity. In the longer term (as we approach the circle 7 phase of our realm cycle) the realm datum would drop to a level that could not support wavefront propagation.

Let us assume for now that at our current realm EIL datum the quantum step is as dictated by the shortest Planck distance so that d1 = 10^{-35} metres. Let us also assume that the relevant E-Flows from the wavefront plane AA' are emitted from Mps within a conical zone that has its base angle at 60° to the propagation direction. The Mps that form at the M1 type position in the forward plane BB' will possess a linear momentum that is the resultant of all the absorbed energy medium quantity in its E~M/X process. For the sake of simplicity in our calculations we must assume that this resultant linear momentum /velocity is in the same direction as the wavefront propagation direction. Let this velocity be denoted as Vp.

Since the realm grid EIL is a make-up from isotropic E-Flows from the entire realm in general, it does not impart any linear momentum from its absorption into a Mp. And since this contributes approximately 98% to the energy medium in the new Mps, then we consider that any forward velocity Vp in small in comparison to VF.

The VF contribution to Vp is assumed at approximately 2%. Also, directionally this contribution is even less since it arrives at an average 60° inclination to the propagation direction. So let us assume the VF contribution to be even less, at say 1🗁 %.

This velocity component is therefore VF Sine 60, or 0.866 VF.

Since this contributes approximately 1.5% to Vp, then we may assume that the Vp component is,

$$\text{Vp} = 0.015 \times 0.866\ \text{VF}$$

or, $$\text{Vp} = 0.01\ \text{VF}$$

As a rough first approximate calculation let us assume that,

$$\text{VF} = {}^{C}/\text{Sine } 60 = {}^{299,799}/0.866\ \text{km/sec.}$$

$$\text{VF} = 3.46 \times 10^{8}\ \text{metres /sec.}$$

$$\text{Vp} = 0.01 \qquad \text{VF} = 3460\ \text{km/sec} = {}^{3460}/299,799 = 0.011\ \text{of C velocity.}$$

Therefore distance d2 is traversed at ✍ of C velocity.

Since we know that the overall propagation velocity is C, then d1 distance must traverse at nearly 100 times C velocity in order for the propagation velocity to average out at C.

We know that the distance d1 is 10^{-35} metres. And that the distance d2 is traversed in the life of an Mp which is = 10^{-43} secs at a velocity ⤡ of C. Therefore,

Distance d2   = $3.46 \times 10^6 \times 10^{-43}$
                = 3.46×10^{-37}

Thus the distance d1 is 100 times greater than distance d2.

Simplifying the equation we have,

$(Vd1 + Vp)/_2 = C$ (where Vd1 = 0.866 VF)
$Vd1 = 2C - Vp$
$Vd1 = 2 \times 2.997 \times 10^8 - 3.46 \times 10^6$
$Vd1 = (599.4 - 3.46) 10^6$
$0.866 \quad VF = 5.96 \times 10^8$
$VF = {}^{5.96}/0.866 \times 10^8$ m/sec
$VF = 6.88 \times 10^8$ m/sec = 2.295 times light velocity.

The accuracy of the above value is rough and it is recommended that the readers carry out their own calculations based on assumptions as appropriate but to the above mentioned basic principles of wavefront propagation.

EPILOGUE

Master, here I am again. I have done as you wished. I spent ten days and nights in the hills observing the beauty that lay all around me. I watched the sunrise at the dawn of each new day. I listened to the awakening of the morning sounds and to the hush that settled with the blanket of the night. This was an experience in itself for I sat for hours just looking up at the starry sky. The moon and stars seemed so near that I was sure I could easily reach up and touch them. By day the clouds stood proud and bright yet they became darkly menacing at night. My senses were so aware of all the sights, sounds and odours around me that I decided to return there again as often as I could. I would like to spend more time in places such as that. Was this the next lesson that you wished me to learn - a true appreciation of nature's beauty?

No my son, that was not the lesson. Beauty is everywhere. That which you observe daily you take for granted and do not appreciate. New things are a fascination and cause fresh awareness - for a while. There is mystery and beauty in your own personality yet it is ignored by you. But enough of beauty, tell me of your hilltop visit. Describe all that you saw and heard and felt. Detail every aspect that was around you. Do not omit the slightest detail, even that which you may consider as trivial. I would like to possess the same mental images that you have retained from this experience so that I too may appreciate all that you saw.

Master, where do I begin? I see a whole scene before me. The picture is complete and there is no beginning or end part to it. Every aspect within it has a linkup effect with the others and I could not do it full justice by describing them one at a time. Yet I am only capable of doing just that, describing them one at a time. I felt that I had become an integrated part of that scenic landscape. I sensed a complete mental awareness and comprehension of nature's laws which gave me a feeling of confidence in the future. Perhaps the inner mind has more subconscious knowledge and intuition than we realise. Words cannot adequately convey to you all that I observed, nor can they express my innermost feelings and emotions as they varied on each day. Were I to spend weeks describing all of this to you I could not hope to build in your mind anything close to the mental images that I possess. There is only one way in which this might be achieved and that is for you also to spend a few days in the hills. Why do you smile master?

I smile because I am pleased that you have understood the lesson I wished you to learn. To be truly aware of another's hilltop experience you also must climb the same hill.

BIBLIOGRAPHY

From Euclid To Eddington by Sir Edmond Whittaker. Cambridge University Press. 1949

Thirty Years That Shook The World by George Gamow. Double Day & Co. 1966

The Nature Of Matter by Otto Frisch. Thomas & Hudson, London. 1972

Conceptions Of Ether by G N Cantor and M J S Hodge. Cambridge University Press. 1981

The Structure Of The Universe by Jayant Narlikar. Oxford University Press. 1977

The Lighter Side Of Gravity by Jayant V Narlikar. W H Freeman & Co. 1982

The First Three Minutes by Steven Weinberg. Fontana/Collins. 1981

The Evolution Of Physics by Albert Einstein and Leopold Infeld.
Cambridge University Press 1st Edition 1938 re-issued 1971

Modern Cosmology by D W Sciama. Cambridge University Press. 1973

Beyond Einstein by Michio Kaku & JenniferThompson. Oxford University Press. 1999

Hyperspace by Michio Kaku. Oxford University Press. 1999

The Strange Story Of The Quantum by Banesh Hoffman. Dover Publications Inc. 1959

Quantum Mechanics by Alastair I M Rae. IOP Publishing Ltd. 1990

Was Einstein Right by Clifford W Will. Oxford University Press. 1988

Quarks by Harald Fritzsch. Penguin Books. 1983

Special Relativity by A P French. Thomas Nelson & Sons Ltd. 1982

Introduction To Atomic Physics by Enge, Wehr and Richards.
Addison-Wesley Publishing Company Inc. 1979

Superforce by Paul Davies. Unwin Hyman Ltd. 1989

The Quantum Universe by Tony Hey & Patrick Walters. Cambridge University Press. 2000

A Brief History Of Time by Stephen W Hawking. Bantam Press. 1989

The Birth Of Time by John Gribbin. Phoenix. 1999

In Search Of The Big Bang by John Gribbin. Penguin Books. 1998

In Search Of Shrodinger's Cat by John Gribbin. Black Swan Books. 1991

Shrodinger's Kittens by John Gribbin. Phoenix. 2001

Science A History by John Gribbin. Penguin Books. 2003

ABOUT THE AUTHOR

C J Harvey was born in 1940 in Peshawar in the North West Frontier Province of then British India. He went to English-run boarding schools till the age of sixteen and continued his education at university to graduate in Mechanical Engineering in 1962 from the Peshawar University in Pakistan.

His working life started as part of the maintenance team in a Hydro-Electric Power Station for seven years before he emigrated to Birmingham in England in 1970 with his wife and infant daughters. He then joined the British Steel Works at Bromford in Birmingham that same year as a Work Study Engineer and worked his way up to Safety Adviser and Training Officer. He took early retirement in 1993 when the works was shutdown.

He has always maintained an interest in physics and cosmology with a wide ranging study of Texts by the world's prominent physicists such as Einstein, George Gamow, Narlikar, Otto Frisch, Schrödinger, John Gibbin, Michio Kaku etc to name a few.

Out of all this he developed his own concept for the 'Life Cycle' of our universe in his 'Seven Circles Theory' and presented here with the title of Faster than Light.

The author is indebted to the scientific discoveries and theories presented over the years starting with the earlier Ether theorists and on to the present day. Without these as essential clues the above theory could never have been realised.

Also published is a work of science fiction titled Pink Knight but with factual science included where relevant; also with the current publisher.